Innovative Smart Materials Used in Wireless Communication Technology

Ram Krishan
Mata Sundri University Girls College, India

Manpreet Kaur
Yadavindra Department of Engineering, Guru Kashi College, Punjabi University, India

Shilpa Mehta
Auckland University of Technology, New Zealand

A volume in the Advances in Wireless Technologies and Telecommunication (AWTT) Book Series

Published in the United States of America by
IGI Global
Information Science Reference (an imprint of IGI Global)
701 E. Chocolate Avenue
Hershey PA, USA 17033
Tel: 717-533-8845
Fax: 717-533-8661
E-mail: cust@igi-global.com
Web site: http://www.igi-global.com

Library of Congress Cataloging-in-Publication Data

Names: Krishan, Ram, DATE-editor. | Kaur, Manpreet, DATE- editor. |
  Mehta, Shilpa, DATE- editor.
Title: Innovative smart materials used in wireless communication technology
  / edited by Ram Krishan, Manpreet Kaur, Shilpa Mehta.
Description: Hershey, PA : Information Science Reference, 2023. | Includes
  bibliographical references and index. | Summary: "The objective of the
  book is to: Learn - about the latest developments going on in the field
  of wireless networks and innovative technologies. Discover- how to apply
  design principles and technologies for creating smart designs.
  Understand- how to integrate the advanced technologies with wireless
  controls systems to make the devices smarter?"-- Provided by publisher.
Identifiers: LCCN 2022045272 (print) | LCCN 2022045273 (ebook) | ISBN
  9781668470008 (hardcover) | ISBN 9781668470046 (paperback) | ISBN
  9781668470015 (ebook)
Subjects: LCSH: Wireless communication systems--Equipment and supplies. |
  Wireless communication systems--Technological innovations. | Smart
  materials.
Classification: LCC TK5103 .I38 2023 (print) | LCC TK5103 (ebook) | DDC
  621.382--dc23/eng/20221128
LC record available at https://lccn.loc.gov/2022045272
LC ebook record available at https://lccn.loc.gov/2022045273

This book is published in the IGI Global book series Advances in Wireless Technologies and Telecommunication (AWTT) (ISSN: 2327-3305; eISSN: 2327-3313)

British Cataloguing in Publication Data
A Cataloguing in Publication record for this book is available from the British Library.

For electronic access to this publication, please contact: eresources@igi-global.com.

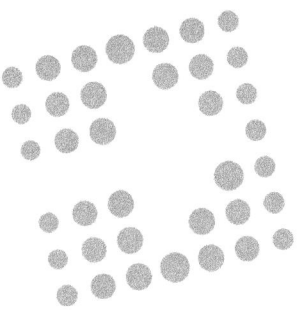

Advances in Wireless Technologies and Telecommunication (AWTT) Book Series

Xiaoge Xu
University of Nottingham Ningbo China, China

ISSN:2327-3305
EISSN:2327-3313

MISSION

The wireless computing industry is constantly evolving, redesigning the ways in which individuals share information. Wireless technology and telecommunication remain one of the most important technologies in business organizations. The utilization of these technologies has enhanced business efficiency by enabling dynamic resources in all aspects of society.

The **Advances in Wireless Technologies and Telecommunication Book Series** aims to provide researchers and academic communities with quality research on the concepts and developments in the wireless technology fields. Developers, engineers, students, research strategists, and IT managers will find this series useful to gain insight into next generation wireless technologies and telecommunication.

COVERAGE

- Cellular Networks
- Wireless Sensor Networks
- Network Management
- Telecommunications
- Digital Communication
- Mobile Technology
- Global Telecommunications
- Virtual Network Operations
- Mobile Communications
- Wireless Broadband

IGI Global is currently accepting manuscripts for publication within this series. To submit a proposal for a volume in this series, please contact our Acquisition Editors at Acquisitions@igi-global.com or visit: http://www.igi-global.com/publish/.

The Advances in Wireless Technologies and Telecommunication (AWTT) Book Series (ISSN 2327-3305) is published by IGI Global, 701 E. Chocolate Avenue, Hershey, PA 17033-1240, USA, www.igi-global.com. This series is composed of titles available for purchase individually; each title is edited to be contextually exclusive from any other title within the series. For pricing and ordering information please visit http://www.igi-global.com/book-series/advances-wireless-technologies-telecommunication/73684. Postmaster: Send all address changes to above address. Copyright © 2023 IGI Global. All rights, including translation in other languages reserved by the publisher. No part of this series may be reproduced or used in any form or by any means – graphics, electronic, or mechanical, including photocopying, recording, taping, or information and retrieval systems – without written permission from the publisher, except for non commercial, educational use, including classroom teaching purposes. The views expressed in this series are those of the authors, but not necessarily of IGI Global.

Titles in this Series

For a list of additional titles in this series, please visit: www.igi-global.com/book-series

Economic and Social Implications of Information and Communication Technologies
Yilmaz Bayar (Bandirma Onyedi Eylul University, Turkey) and Lina Karabetyan (Independent Researcher, Trkey)
Information Science Reference • © 2023 • 318pp • H/C (ISBN: 9781668466209) • US $235.00

Modelling and Simulation of Fast-Moving Ad-Hoc Networks (FANETs and VANETs)
T.S. Pradeep Kumar (Vellore Institute of Technology, India) and M. Alamelu (Kumaraguru College of Technology, ndia)
Information Science Reference • © 2023 • 251pp • H/C (ISBN: 9781668436103) • US $240.00

Challenges and Risks Involved in Deploying 6G and NextGen Networks
A.M. Viswa Bharathy (GITAM University, Bengaluru, India) and Basim Alhadidi (Al-Balqa Applied University, Jordan)
Information Science Reference • © 2022 • 258pp • H/C (ISBN: 9781668438046) • US $250.00

Achieving Full Realization and Mitigating the Challenges of the Internet of Things
Marcel Ohanga Odhiambo (Mangosuthu University of Technology, South Africa) and Weston Mwashita (Vaal University of Technology, South Africa)
Engineering Science Reference • © 2022 • 263pp • H/C (ISBN: 9781799893127) • US $240.00

Handbook of Research on Design, Deployment, Automation, and Testing Strategies for 6G Mobile Core Network
D. Satish Kumar (Nehru Institute of Engineering and Technology , India) G. Prabhakar (Thiagarajar College of Engineering, India) and R. Anand (Nehru Institute of Engineering and Technology, India)
Engineering Science Reference • © 2022 • 490pp • H/C (ISBN: 9781799896364) • US $360.00

Implementing Data Analytics and Architectures for Next Generation Wireless Communications
Chintan Bhatt (Charotar University of Science and Technology, India) Neeraj Kumar (Thapar University, India) Ali Kashif Bashir (Manchester Metropolitan University, UK) and Mamoun Alazab (Charles Darwin University, Australia)
Information Science Reference • © 2022 • 227pp • H/C (ISBN: 9781799869887) • US $240.00

Handbook of Research on Advances in Data Analytics and Complex Communication Networks
P. Venkata Krishna (Sri Padmavati Mahila University, India)
Engineering Science Reference • © 2022 • 275pp • H/C (ISBN: 9781799876854) • US $380.00

IGI Global
PUBLISHER of TIMELY KNOWLEDGE

701 East Chocolate Avenue, Hershey, PA 17033, USA
Tel: 717-533-8845 x100 • Fax: 717-533-8661
E-Mail: cust@igi-global.com • www.igi-global.com

Table of Contents

Detailed Table of Contents

    Rohit Anand, G.B. Pant DSEU Okhla-I Campus, India
    Shahanawaj Ahamad, University of Hail, Saudi Arabia
    Vivek Veeraiah, Adichunchanagiri University, India
    Sushil Kumar Janardan, Rungta College of Engineering and Technology, India
    Dharmesh Dhabliya, Vishwakarma Institute of Information Technology, India
    Nidhi Sindhwani, Amity University, Noida, India
    Ankur Gupta, Vaish College of Engineering, India

Optimizing a function helps solve challenging problems. Any optimization problem that can't be solved quickly utilizing deterministic approaches is hard. Metaheuristic optimization uses metaheuristics to discover optimum solutions. Meta and heuristic mean "higher level" and "solution;" using complicated approaches to address unsolvable problems.6G will replace 5G wireless networks. 6G networks' higher frequencies will boost their capacity and minimize latency. The 6G internet aspires for one-microsecond latency. This is 1,000 times faster at 1 millisecond.6G technology is expected to advance imaging, presence technologies, and location awareness. WSNs integrate hardware and software networks to monitor and record environmental variables and other observable quantities. Wireless sensor networks (WSN) are utilized commercially and domestically for effective communication.

    Pradeep Kumar, Vellore Institute of Technology, India
    Abhijit Bhowmick, Vellore Institute of Technology, India

In cellular networks, device-to-device (D2D) communication is a new paradigm. In this chapter, the end-to-end performance (outage, throughput, improvement in spectrum utilization, and energy efficiency) of a non-linear energy harvesting D2D network underlaying a cellular network will be studied under a channel quality constraint. A D2D node harvests energy from RF signal of cellular users (CU) and power transfer units (PTUs) and transmits its information in hybrid mode. Spectrum sensing, censoring of users, and energy harvesting are incorporated to make the study more realistic. Spectrum sensing provides information about the CU spectrum status, censoring technique eliminates the D2D users from the transmission process of those who have bad quality channels. Analytical expressions of an outage, throughput, improvement in spectrum utilization (ISU), and energy efficiency are provided. The impact of the interference threshold on CU, the censoring threshold, and the saturation level of the energy harvester on network performance parameters is addressed.

Chapter 3

Mehaboob Mujawar, Glocal University, India
Subuh Pramono, Universitas Sebelas Maret, Indonesia

In this chapter the authors will be discussing about what wireless body area networks(WBAN) are, where are they used, what are the advantages of using millimeter wave frequencies, and novel millimeter wave antenna design suitable for wearable applications using CST software. The specific frequency that is used in the millimeter wave frequency is 60GHz with bandwidth of 7 GHz. Results from simulation are satisfactory in terms of gain and effectiveness. The performance of millimetre wave antennas is less affected by the presence of a human body because of their miniature-like dimensions and shallow penetration depth into human body layers. This makes them perfect for body-centric wireless communication (BCWC) applications. The essential property data was thus used to virtually generate a human body model. The antenna is kept close to the created human body model as simulations are run again at the same frequencies as previously. The results were promising as the VSWR and return loss curves remained almost identical.

Chapter 4

Anindita Bhattacharjee, Tripura University, India
Anirban Karmakar, Tripura University, India
Anuradha Saha, Netaji Subhash Engineering College, India

The advancement of antennas that can access high data speeds, and have higher gain, is largely responsible for the advancements in wireless technology. One such antenna called the multiple input multiple output (MIMO). This antenna has attracted the utmost attention of the researchers owing to its features like higher capacity, increased data rate, low bit error rate, and stable radiation characteristics. Over the years, different strategies have been proposed by several researchers to enhance the performance of the MIMO antenna. This article's main goal is to provide a summary of the changes in MIMO antenna design during the last few decades, with an emphasis on the most significant contributions, which are mainly from IEEE publications. This review work intends to show antenna researchers a promising route for their development using MIMO antennas.

Chapter 5

Devendra Kumar Somwanshi, Poornima College of Engineering, India

In this chapter, SDN is utilized as a unified answer for measuring the packet loss, memory optimization, and network delay. Presently, there is no single arrangement available to measure all the issues at once, but numerous different strategies are used. By using different strategies, separate frameworks are required, and it increases the establishments and cost. While these strategies are not the ideal solution, it is just an arrangement to give the solution of a particular issue. Besides, those techniques are not proficient to meet every one of the necessities for a checking arrangement—some are not exact or sufficiently granular, and others are adding extra organization burden or absence of versatility. Techniques have been proposed to find out the packet loss, memory optimization, and network delay. The ongoing checking of SDN Regulator framework is utilized, and it gives data and key execution markers for network. It assists with making an execution model for Regulator as a component of basic organization boundaries.

Chapter 6

Jyoti Saharan, JCDM College of Engineering, India
Silki Baghla, JCDM College of Engineering, India
Dinesh Kumar Gupta, JCDM College of Engineering, India

In a 5G wireless communication system, demands of massive connectivity, ultra-low latency, higher data rates, higher reliability, and many more are fulfilled by NOMA (non-orthogonal multiple access) which is one of the potential candidates that can carry multiple users at the same time and frequency. As NOMA uses the power allocation method to transmit signals, a problem of mismatch between quality of service (QoS) and channel conditions of the users occurs. To resolve this issue, a scheme cooperative rate-splitting non-orthogonal multiple access (C-RS-NOMA) has been proposed in this work to enhance the performance of downlink power-domain NOMA in terms of outage probability, system capacity, spectrum efficiency, and energy efficiency. The performance of the proposed techniques has also been compared with the existing cooperative-NOMA and traditional NOMA techniques. The simulation results showed the improvement in different key parameters of C-RS-NOMA with respect to the existing NOMA and cooperative NOMA schemes.

Chapter 7

Manpreet Kaur, Yadavindra Department of Engineering, Guru Kashi College, Punjabi University, India

The aim of this work is to design and characterize a compact 5G antenna with an appreciable bandwidth. In the suggested antenna, two fractal curves are applied onto the conducting patch that helped to achieve the targeted operational characteristics. The whole patch is printed on the chosen FR4 material. The modified rectangular patch is placed on the top side, whereas a partial ground is placed on the bottom side. The antenna size observed is 45 mm2. The overall adjustments of the specific design parameters are carried out for better S11 characteristics at the fundamental resonance. The operational band stretches from 25.99 GHz to 27.72 GHz with a fractional bandwidth of 6.4%. The designed antenna showed resonance at 26.75 GHz with S11 value -21.00 dB. The proposed antenna possesses good S11 characteristics, appreciable gain, and significant bandwidth at the operational frequency band. Moreover, the influence of variations in the antenna geometry has also been demonstrated for deep understanding.

Chapter 8

Purva Joshi, Information Engineering, University of Pisa, Italy

Fifth generation (5G) technology is being incorporated into the infrastructure of many sectors to provide high-quality communication services, smart factories, vehicle-to-vehicle connection, and many other new services. Next-generation networking technology ultra-wideband (UWB) connects many devices and provides location-based services. This chapter reviews 5G enabling technologies and the cellular network as backhaul network interface and solution. Wireless backhauling, which replaces fiber cables for core network communication, is discussed in this chapter. Wireless backhaul sends communications between nodes using radio or microwave frequencies. Wireless backhaul can be installed faster and cheaper than fiber when fiber is too costly or impracticable (which involves trenching). 5G's enhanced connectivity, AI, massive IoT, and other disruptive technologies for essential sectors and public services will be crucial to these solutions. This chapter compares 5G and 6G.

Chapter 9

M. N. Suma, BMS College of Engineering, India
Sudhindra K. R., BMS College of Engineering, India

Communication link budget is a balance sheet of all gains and losses, where the calculation of link parameters such as received power, path losses, noise power, signal to noise ratio, receiver figure of merit and link margin. There are software available in the market, which can be utilized to design, but are not cost friendly and also have limitations in the form of some predefined data values, which the user cannot alter. In this paper the authors propose a new testbed which is designed to formulate the link budget calculations using Python GUI for space communications link, considering all theoretical parameters needed for computations.

Chapter 10

Utsab Banerjee, MVJ College of Engineering, India
Anirban Karmakar, Tripura University, India
Anuradha Saha, Netaji Subhash Engineering College, India

This article is intended to present a comprehensive, technical review of circularly polarized (CP) antennas for an array of applications in various fields of wireless communication, emphasizing on the recent developments in the projected research. The article should also present a comparative study of various approaches reported in the open literature, with an aim to highlight the contribution of CP antenna systems and their chronological development in the domain of wireless communication technology. The primary motive of this literature is to (a) highlight the methodologies used by different attempts to portray and analyze the different aspects in which circularly polarized antennas find their applications in modern day wireless communication, (b) provide a practical viewpoint of the future scope of study, based upon the past and present state-of-art research trends, and (c) provide a conceptual and technical support to present day antenna designers to help the process of enhancement of innovation and multiple system integration.

    Jay Kumar Pandey, Shri Ramswaroop Memorial University,India
    Shahanawaj Ahamad, University of Hail, Saudi Arabia
    Vivek Veeraiah, Adichunchanagiri University, India
    Nishchal Adil, Rungta College of Engineering and Technology, India
    Dharmesh Dhabliya, Vishwakarma Institute of Information Technology, India
    Ashok Koujalagi, Godavari Institute of Engineering and Technology (Autonomous), India
    Ankur Gupta, Vaish College of Engineering, India

The 5G network is the main topic of this investigation. 5G is expected to be far more advanced than 4G since it makes use of three distinct bands of the network spectrum. The acronym "5G" refers to the latest generation of wireless communications. With 5G, communications will improve all across the world. The purpose of this research is to examine the consequences of the increasing call drop ratio in 5G networks. In other words, if a user's current session is interrupted, they will need to make a new connection to continue using the service. One area where 5G excels over its predecessors, 4G (and LTE), is in reducing latency. Also, there would be fewer lost calls for individuals utilizing VoIP because network uptime will have risen significantly. Reduced call failure rates lead to happier customers.

    Vikas Goyal, Malout Institute of Management and Information Technology, India
    Geetanjali Goyal, Malout Institute of Management and Information Technology, India
    Hritik Ranjan Nanda, Malout Institute of Management and Information Technology, India

HealthCare is an application which connects the patients/users and health care professionals registered on the HealthCare application, acting as merely intermediary. The data, which is provided by the users/ patients or from any health care professional, belongs simply to them. Here a person is given access with some credential, non-exclusive license to use the HealthCare Services as described therein. The authors are using Flutter and Firebase as the technology to develop the App and WebSite. The authors are using Flutter technology, which is provided by Google for creating front-end/back-end, along with crafting beautiful designs (FutureLearn,); moreover, it is a compiled environment for mobile, web, and desktop by using a single codebase. Furthermore, the authors are using Google's Firebase technology that is provided by Google as a back-end service that helps in the development of software, which lets the developers to make iOS, Android, and web applications. It basically provides the tools which helps keep a tab on investigation, addressing, fixing application bugs, making marketing strategies, and product testing.

    V. V. Satyanarayana Tallapragada, Mohan Babu University, India
    M. Venkatanaresh, Mohan Babu University, India
    N. Gireesh, Mohan Babu University, India
    M. Naresh, Matrusri Engineering College, India

Media independent handover (IEEE 802.21 MIH) services are an area of particular focus for IEEE. The primary objective of IEEE 802.21 MIH is to streamline the handover procedure, resulting in a more reliable and less time-consuming handover service. There are two distinct kinds of handoffs, called horizontal and vertical, respectively. Among these, the performance of vertical handover (VH) can be improved through parameter optimization. In order to achieve the best possible outcome and reduce the rate of handover failure, careful parameter selection is required. Despite the fact that many options exist for VH management optimisation processes, many current works only consider one or two parameters for VH optimisation. Therefore, the ideal handover solution that emerges is inferior in terms of reliability, responsiveness, and precision. So, a technique called multi-objective emperor penguin handover optimisation (MOEPHO), which takes into account nearly all network characteristics for VH optimisation is discussed in this chapter.

Chapter 14

Rafael Vargas-Bernal, Instituto Tecnológico Superior de Irapuato, Mexico

A crucial technology to meet the growing demand for faster wireless communication is terahertz band (0.1–10 THz) communication. New generations of technology, such as 5G and 6G telecommunications, which have faster transmission speeds, larger network capacities, and shorter delays, will find extensive use soon. Due to fabrication and installation restrictions, particularly for smaller sizes, conventional telecommunications devices cannot support the new frequency. Two-dimensional materials have been proposed as the most suitable possibility to design and implement wireless telecommunications devices that meet these requirements. This chapter describes the advances made in the design of antennas, resonators, and electromagnetic interference shielding systems based on graphene and MXenes. Despite the advances achieved so far, future research directions required to commercialize the developed test-stage devices are also described. These materials must be investigated in this century to guarantee the success of 5G and 6G communications.

Chapter 15

Sobana Sikkanan, Karpagam College of Engineering, India
Seerangurayar T., Hindusthan College of Engineering and Technology, India
Krishna Prabha S., PSNA College of Engineering and Technology, India
Kasthuri M., PSNA College of Engineering and Technology, India

The 6G wireless communication network has shown its tremendous advantages in digital transformation of societies. 6G enables reliable, pervasive, and near instant wireless connectivity by integrating aerial, maritime, and terrestrial communications. The development of cutting-edge technologies like machine learning, blockchain, millimeter wave communication, non-orthogonal multiple access, tera-Hertz communication, quantum communication/quantum machine learning, fog/edge computing, and tactile Internet encourages the demand of beyond 5G and 6G communication. An effective wireless communication system must be capable of satisfying the user requirements such as increase in capacity, efficiency flexibility, coverage, and quality of experience. As the number of users and coverage area increases, the complexity of designing a 6G network is also getting increased. In recent years several research works are applying ML to wireless communication. This chapter discusses about the application of ML in the area of massive MIMO, NOMA, OWC, polar codes, and security of wireless communication.

The outbreak of COVID-19 pandemic is an unprecedented shock to the entire economy, majorly the hospital sector. In the existing situation, embedding sustainability in the hospital sector is the most crucial aspect for reducing harm to the environment. Thus, the question is aroused in front of researchers to study more about the health care sector and to find out certain practices that emphasizes on the sustainability of hospitals sector. During research it has been identified that one of the most important factors known as information technology is playing a very vital role in the success of heath care sector and specially in the era of pandemic. Various technological advancements like digital health care, internet of medical things, smart health monitoring, telemedicine, chatbot systems, emotive sensory web and robotics can play a huge role towards strengthening the healthcare sector leading to sustainability of the entire world economy, paving a way towards better well-being of health care professionals.

Foreword

I am immensely delighted at introducing this great literary work. I have personally and professionally known the editors of this book and I can proudly say they have done a great job in bringing this out. This book presents new ideas and technologies in the field of wireless communication and would surely be a great eye-opener to the readers. When I first heard about a collection of academic articles in the form of this book getting published by a publisher of such repute, I made up my mind to read it and contribute to it in any possible way. The variety of subjects covered in this book opens a doorway for a novice as well as an experienced researcher to the world of innovation and start of the art technology. The readers will fully rejoice in these peer-reviewed articles.

The pace at which technology is changing has not been seen in human history before. The research opportunity is huge. The right experimentation and analysis paves a way for greater industrial adoption and thus impact the lives of millions of people. As a reader, I get access to a lot of peer reviewed literature and reference material in some of the key research areas in this field. I personally liked the articles on 5G, 6G, and software-defined networking (SDN). I appreciated the presentation of the material in an easy-flowing manner that is easy to understand for beginners.

Like the previous books, the editors have given utmost importance to the quality of the content and its relevance to the current problems. I hope the readers will be inspired like I am, after reading this. This book deserves a place at the book shelf of a keen scholar looking for challenging opportunities ahead.

Rohit Goyal
Zscaler, Mohali, India
(The views of the writer are personal only.)

Preface

One of the earliest research domains to gain attention was wireless communications in the early 1860s when electromagnetic waves were first demonstrated by Maxwell. A few years later, Hertz verified this experimentally. Later in 1897, Marconi performed the first wireless transmission. This demonstrated the feasibility of wireless communication. It led to the emergence of a variety of radio systems that can operate in a wide range of frequencies. Earlier radio systems operated in analog mode. In other words, modulation and filtering processes are handled primarily through analog circuits and components. These systems were not suitable for mass production. Future technological advancements resolved this problem, enabling mass production and cost-effective systems to be developed. As a result, signal processing operations were made easier than in conventional radio systems (Mehta, 2022).

1980 marked the launch of the first generation (1G) of mobile networks. Despite the slow speed of only a few kilobits per second, it was a breakthrough at the time because it allowed people to make telephone calls and send texts almost anywhere. This was followed by the launch of second-generation (2G) digital mobile networks in the early 1990s. Besides enhancing data speeds to a few hundred kbps, 2G also brought new features like text messaging.

Technology advancements attributed to the internet, including wireless fidelity (Wi-Fi), have led to several technological advances. An IEEE 802.11 standard for wireless transmission was introduced in 1997. Wireless internet connections use radio waves to communicate wirelessly with devices. Therefore, WiFi became popular as soon as it appeared, as it eliminated the need for cables and made internet access more convenient.

A century after the first wireless communication using Morse code, wireless technology is still evolving. Sixth-generation (6G) technology is being researched by researchers for future wireless networks. 6G technology is predicted to be fully implemented by 2030.

As Wi-Fi networks become more sophisticated, data rates will reach 10 gigabits per second and latency will go down to single-digit milliseconds. It has also become possible to accommodate billions of devices via wireless networks without sacrificing the quality of service or security because wireless networks have evolved technologically to accommodate increasing demands. Technology advancements and innovative approaches will further enhance current networks to meet the needs of an ever-increasing population (Rajiv, 2022).

The wireless industry needs smart materials to remain competitive. Materials of this type can both sense and respond to environmental stimuli in real time as well as control their responses(Karana). While focusing on smart materials in the current wireless industry it is critical to understand the useful semiconductor materials, specifications, and requirements of advanced communication systems, supporting wireless standards, expected performance and how it can be attained, how to secure advanced networks, wearable and flexible wireless applications for the smart world. To meet the current demands of industry, it is imperative to understand these concepts and how they can be useful to society in various contexts. Currently, the primary focus is on 5G and 6G communication and its relevant concepts and technologies. Internet of Things (IoT), Artificial Intelligence (AI) and Machine Learning (ML) are the most important which pave the way for these future advancements. Smart IoT applications have paved the way for wireless networks to evolve. Through emerging technologies, the IoT combines things, data, people, and processes to offer a wide range of smart services, including autonomous connected vehicles, brain-computer interfaces, augmented reality, haptics, and flying vehicles, among others. Unmanned mobility management, ultra-high reliability, and long-distance communications are some of the components of these services. IoE-based smart services will be enabled by 5G wireless networks. 5G targeted tactile networks will be accessible via a variety of approaches, including simultaneous use of unlicensed and licensed bands, intelligent spectrum management, and 5G new radios, which will enable different smart applications to be developed. It has been difficult for 5G to achieve its target goals due to several inherent limitations. Thus, 6G wireless systems will be required to overcome 5G's limitations. AI will be incorporated into 6G to enable the optimization of a number of wireless network issues. It is common for wireless network optimization problems to be solved mathematically; methods used here include convex optimization schemes, matching theory, game theory, heuristics, and brute force algorithms. It is possible that these solutions approaches will be complex, resulting in a degradation of the system's performance. Moreover, ML can optimize all sorts of complex mathematical problems, even those that can't be modeled by mathematical equations. A 6G network will have much higher capacity and substantially lower latency compared to a 5G network and be able to use higher frequencies than a 5G network (C.T., 2022). Thus, once one is aware of the advanced networks and how their performance can be optimized, it is equally important to understand how to secure the network. Moreover, how the advanced communication systems should behave and how they can be developed. Communication between earlier systems was limited to two or three signals operating at different frequencies. The incompatibility between radio systems between different groups of people prevented communication between them. Software-defined radios (SDR) can solve this problem (Sinha, 2016). For an SDR system to be capable of simultaneously operating at different frequencies, it requires reconfigurable or concurrent front ends (mixers, low noise amplifiers (LNA), antennas, and filters. In the last few years, a great deal of research has been directed toward developing both individual front ends or an SDR system that is suitable for a range of applications. Research is also underway to develop smart, flexible, and integrated devices, components, and systems with the best semiconductor materials.

The purpose of this book is to provide an overview of advancements in smart material utilization in wireless communication technologies. Specifically, it examines the design, use, and construction of smart materials that are designed for wireless applications. Various experts have contributed their research and review works to this book. By providing the latest insights and thoughts of leading researchers on major topics in this emerging discipline, this book will serve as a valuable resource that provides information about the current state of the field as well as inspires ideas for future challenges. Various experts have contributed their research and review works to this book. By providing the latest insights and thoughts of leading researchers on major topics in this emerging discipline, this book will serve as a valuable resource that provides information about the current state of the field as well as inspires ideas for future challenges.

This book provides an invaluable overview of a wide range of smart materials that have been widely underrepresented in existing publications. This book, which offers comprehensive and dedicated coverage of smart materials, attempts to fill the gap in available titles by providing a comprehensive discussion of some of the most essential and important issues in the field, including topics of both theoretical and practical interest. Furthermore, the book provides a balanced and quality discussion of theoretical ideas and practical research that will be of interest to readers. The book is divided into 16 chapters. The chapters do not have to be read in the order in which they appear, so readers may focus directly on those that are of interest to them. Below are brief descriptions of each chapter of this book.

Chapter 1

This chapter focuses on optimizing the security of 6G wireless networks. A number of optimization algorithms have been discussed by the authors and they have identified that integrating PSO with MVO is one of the most effective hybrid approaches for optimizing the security of 6G networks.

Chapter 2

D2D (Device-to-Device) communication has become more prevalent in cellular networks. The focus of this chapter is on D2D communication, wherein authors have investigated the performance of a non-linear energy harvesting network underneath a cellular network under a channel quality constraint. The study incorporates spectrum sensing, censoring of users, and energy harvesting to provide a more realistic representation. Additionally, some of the significant parameters are expressed analytically, including outage, throughput, improvement in spectrum utilization, and energy efficiency.

Chapter 3

This chapter talks about millimeter wave antennas for wearable applications. In this chapter, authors explore the benefits of millimeter waves in Wireless Body Area Networks (WBANs). Their study led them to propose a novel millimeter-wave antenna design suitable for wearable applications. The simulated results of the proposed antenna were promising when compared to similar works in the literature.

Chapter 4

In recent years, researchers have been attracted to the multiple input multiple output (MIMO) antenna because of its features such as higher capacity, increased data rate, low bit error rate, and stable radiation characteristics. An overview of the major contributions to MIMO antenna design during the past few decades is presented in this chapter.

Chapter 5

In this chapter, the focus is on implementing Software Defined Network (SDN) network infrastructure, which enables simultaneous measurement of multiple parameters, including packet loss, memory optimization, and network delay, which had previously been impossible. A number of different strategies have been used to measure these parameters, which require separate frameworks, which not only increase establishments but also increase costs.

Chapter 6

One of the potential candidates for 5G communication systems is a NOMA (Non-Orthogonal Multiple Access) system, which can carry multiple users at the same time and frequency and can meet the current requirements of 5G communication systems, including massive connectivity, ultra-low latency, high data rates, and high reliability. A major problem with NOMA is that the QoS (Quality of Service) does not match the channel conditions of the users. In this chapter, a scheme called C-RS-NOMA (Cooperative Rate-Splitting Non-Orthogonal Multiple Access) is proposed to improve the performance of downlink power-domain NOMA systems. In comparison to the existing NOMA and Cooperative NOMA schemes, simulation results of C-RS-NOMA were promising.

Chapter 7

In this chapter, a novel microstrip patch antenna with dual fractal curves is proposed for 5G applications. In addition to good return loss characteristics, this antenna attains significant gain and significant bandwidth at its operating frequency range.

Chapter 8

The purpose of this chapter is to provide an overview of 5G enabling technologies as well as the cellular network as a solution for backhaul networks. Wireless backhaul is able to send information between nodes via radio or microwave frequencies and is easier to install, cheaper, and more convenient than fiber.

Chapter 9

This chapter proposes a novel software testbed for performing link budget calculations. Python GUI was used to develop the testbed, which is primarily relevant to space communications links. The proposed testbed is not only cost-effective but also has the capability of altering data values, which is not possible with existing testbeds.

Chapter 10

Circularly polarized antennas have the potential to be one of the most promising antennas for 5G. It mainly focuses on presenting a comprehensive overview of these antennas. As part of the current study, the authors provide an overview of different methodologies that have been presented in the literature. This can help us identify future research directions and application areas.

Chapter 11

Several research works are being conducted on 5G networks due to their improved communication capabilities and ability to utilize three distinct spectrum bands as opposed to previous networks. A major focus of this chapter is to analyze the consequences of increasing the call drop ratio in 5G networks.

Chapter 12

This chapter discusses how Agile methodologies can be used to improve the performance of health-tech applications. In this chapter, a Healthcare application is presented, which can assist users and healthcare professionals with accessing information by entering some credentials when connected through this app. The application and website were developed using Flutter and Firebase technologies.

Chapter 13

IEEE 802.11 media independent handover requires a less time-consuming and more reliable service. The main focus of this chapter is a multi-objective emperor penguin optimization approach for handling virtual handovers, which is considered one of the best solutions since it can optimize all network characteristics.

Chapter 14

This chapter examines two-dimensional materials that can be useful for designing advanced wireless devices suitable for 5G and beyond applications. In this chapter, we describe the latest developments in the design and manufacturing of antennas, resonators, and electromagnetic interference shielding systems using graphene and MXene. A discussion of future research directions is also included in this chapter.

Chapter 15

This chapter discusses the fundamentals of a number of emerging wireless technologies, such as massive multiple input multiple output (MIMO), optical wireless communication, non-orthogonal multiple access (NOMA), power line communication, and reconfigurable intelligent surfaces. In addition, the chapter discusses how machine learning frameworks can be used for handling the challenges associated with implementing 6G wireless communication systems and dealing with the design and security issues that 6G systems present.

Chapter 16

The focus of this chapter is on improving the sustainability of the healthcare sector after the COVID-19 pandemic. The growth of information technology is one of the most significant factors in the healthcare sector and especially in the era of pandemics. With advancements in technology such as the internet of medical things, smart health monitoring, telemedicine, chatbots, emotive sensory webs, and robotics, the healthcare sector can be strengthened for the entire world economy, paving the way for health professionals to be more well-off.

Ram Krishan
Mata Sundri University Girls College, Mansa, India

Manpreet Kaur
Yadavindra Department of Engineering, Guru Kashi College, Punjabi University, Talwandi Sabo, India

Shilpa Mehta
Auckland University of Technology, New Zealand

REFERENCES

Karana, E., & Kandachar, P. (n.d.). *Smart Surroundings': A New Era For Communication And Information Technologies.* Delft University of Technology. https://www.irbnet.de/daten/iconda/06059007007.pdf

Mehta, S. (2022). *Reconfigurable Mixer Design for Software-Defined Radios* [Doctoral dissertation]. Auckland University of Technology.

Rajiv. (2022). History of Wireless communication – Morse Code to 5G Technology. *RF Page.* https://www.rfpage.com/history-of-wireless-communication-morse-code-to-5g-technology/

Ranju, C. T., & Sreekanth, C. (2022). Comparison of 5G and 6G Wireless Systems and Proposing of 7G, A New Era. *International Journal of Engineering Research & Technology*, *11*(6), 285–291.

Sinha, D., Verma, A., & Kumar, S. (2016). S*oftware defined radio: Operation, challenges and possible solutions*. Semantic Scholar. doi:10.1109/ISCO.2016.7727079

Chapter 1
Optimizing 6G Wireless Network Security for Effective Communication

Rohit Anand
G.B. Pant DSEU Okhla-I Campus, India

Shahanawaj Ahamad
University of Hail, Saudi Arabia

Vivek Veeraiah
Adichunchanagiri University, India

Sushil Kumar Janardan
Rungta College of Engineering and Technology, India

Dharmesh Dhabliya
Vishwakarma Institute of Information Technology, India

Nidhi Sindhwani
Amity University, Noida, India

Ankur Gupta
(iD) https://orcid.org/0000-0002-4651-5830
Vaish College of Engineering, India

ABSTRACT

Optimizing a function helps solve challenging problems. Any optimization problem that can't be solved quickly utilizing deterministic approaches is hard. Metaheuristic optimization uses metaheuristics to discover optimum solutions. Meta and heuristic mean "higher level" and "solution;" using complicated approaches to address unsolvable problems.6G will replace 5G wireless networks. 6G networks' higher frequencies will boost their capacity and minimize latency. The 6G internet aspires for one-microsecond latency. This is 1,000 times faster at 1 millisecond.6G technology is expected to advance imaging, presence technologies, and location awareness. WSNs integrate hardware and software networks to monitor and record environmental variables and other observable quantities. Wireless sensor networks (WSN) are utilized commercially and domestically for effective communication.

DOI: 10.4018/978-1-6684-7000-8.ch001

1 INTRODUCTION

Optimizing a function helps solve challenging problems. Any optimization problem that can't be solved quickly utilizing deterministic approaches is hard. Metaheuristic optimization uses metaheuristics to discover optimum solutions. Meta and heuristic mean "higher level" and "solution" Using complicated approaches to address unsolvable problems. Single- or multi-objective difficulties may arise. Strong objective functions produce the finest outcomes. With one goal, where the particles converge is optimum. 6G internet wants one-microsecond latency. This is 1 millisecond quicker. 6G will improve imaging, presence, and location awareness. WSNs monitor and record environmental variables and other observables. Commercial and home communication using wireless sensor networks (WSN) (Ahmed et al., 2017).

The present work is related to the optimization of 6G wireless network security. Section 1 is introducing the optimization mechanism along with the Meta heuristic approach and nature-inspired approach. PSO, ACO, ABC, BFO, Cuckoo search and firefly algorithm are considered nature-inspired optimization approach. Then 6G wireless network has been illustrated considering its benefits and working. The requirement of a 6G network has been considered for edge computing and IoT. Finally, wireless sensor networks are presented as a hybrid system. Existing research on WSN, 6G networks, and network security is presented in Section 2 (Arora et al., 2019).

Many studies have been conducted on 6G wireless networks and wireless sensors (Kirubasri et al., 2021). Further, other optimization mechanisms have been shown in the literature. This section details the author's contributions to the study of wireless networks and optimization. Some of the studies examined network safety, while others focused on throughput and routing. After considering the contribution of existing research third section is presenting the problem statement. Here are issues related to research work such as the compatibility of optimization algorithms. Moreover, there is the issue of accuracy as well as error. Section 4 considers the process flow of the proposed work. Here the study of the existing optimization mechanism is considered and issues related to previous work have been presented. Then hybrid approach is built considering the objective function and integration of PSO-MVO. Finally, an evaluation of accuracy and performance is made (Anand et al., 2022). Section 5 is focused on the simulation of time consumption and accuracy in the case of the hybrid model. Section 6 is presenting the conclusion of the research work while section 7 presents the future scope of the research work (Behnaam Aazhang et al., 2019).

1.1 Optimization

To solve difficult issues, optimization is used to maximize or minimize an objective function. Any optimization issue that cannot be solved in a certain amount of time using deterministic methods is said to be hard or complex. Metaheuristic optimization is another name for optimization because it employs metaheuristic techniques to find optimal solutions (Anand & Chawla, 2022). Together, meta and heuristic denote "upper level" and "solution or technique," respectively. That involves using more complex methods to issues that cannot be solved with absolute certainty (Duan et al., 2020).

These issues might be single- or multi-objective in nature. To get the best possible results, the objective function should be as strong as possible. With a single objective, the ideal solution is wherever the particles converge. Using a search space, the metaheuristic algorithm attempts to locate a workable answer. For their searches, some algorithms choose to stay close to home, while others choose to look far and wide. Several relevant parameters or metrics are taken into account when formulating the goal function. There are two main categories of metaheuristics: those based on a single solution and those based on a population (Dushyant et al., 2022). When a problem has just one solution, the focus is on making money, which limits the search area (to refine the solution). The population-based investigation is limited to a worldwide search (to find a new good solution). Either evolutionary algorithms or swarm intelligence are examples of population-based methods. The genetic algorithm and differential evolution are two examples of evolutionary algorithms.

Figure 1. Optimization techniques

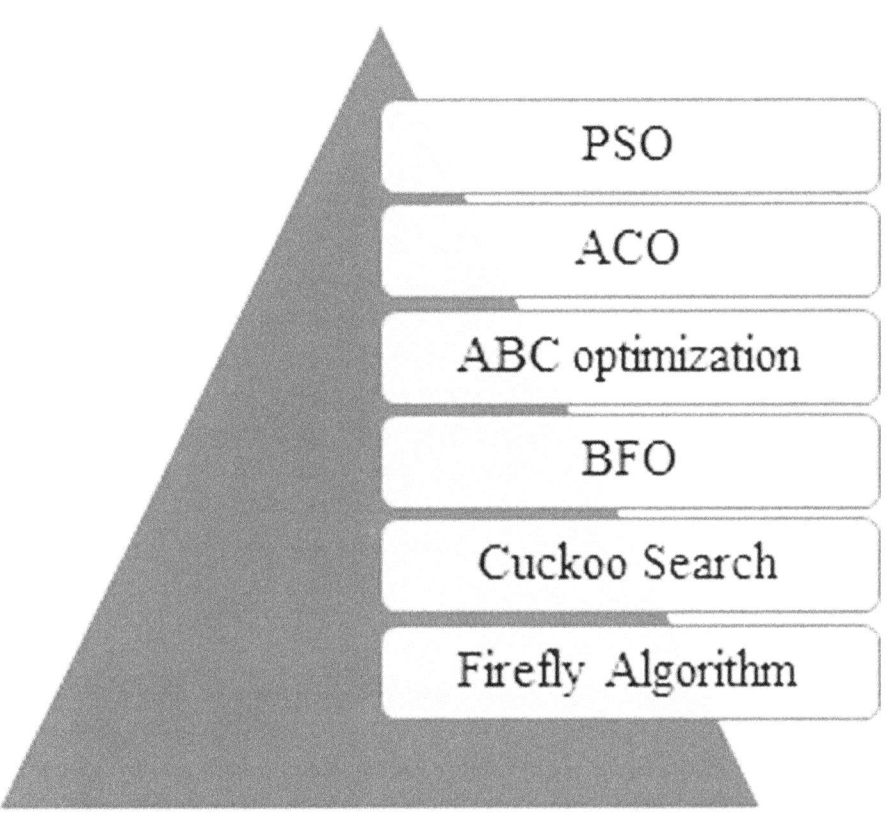

1.1.1 Particle Swarm Optimization

PSO is an idea suggested by Kennedy and Eberhart in 1995. (PSO). Researchers in the field of sociobiology have theorized that animals and birds that travel in large groups "may profit from the experience of all other individuals." A flock of birds, for instance, can share information about where members have found food, allowing them to select the best available option. We may assume that the best solution found by all the birds in a flock is the best answer since each bird is contributing to the search for the best solution in a high-dimensional solution space. Because there is no hard data to suggest otherwise, we have to use some kind of heuristic. In practice, however, we find that PSO's solution is often very close to the global optimum (Jiaheng Wang et al., 2017). This area is dedicated to helping you maximize your earnings or minimize your losses. A function may have several local maxima and minima. In any case, there is only one maximum and one minimum on the planet (Babu et al., 2022). If you have a function that is very involved, it may be challenging to find its global maximum. PSO attempts to find the global maximum or minimum. Despite its inaccuracy, it gets close to representing the global maximum and lowest. For this reason, PSO is classified as a heuristic model. When trying to describe a mathematical function that is based on a real-world situation, we sometimes need to include extra variables or higher-dimensional vector space. Conditions such as temperature, humidity, the container, the solvent, etc. may influence bacterial growth in a jar. It's trickier to get the precise global maximum and minimum of this kind of function (Liang et al., 2019).

1.1.2 Ant Colony Optimization

Pheromones are a natural phenomenon that ants employ for communication, and they serve as the inspiration for several algorithms, including Ant colony optimization (ACO). By giving and receiving outputs in response to inputs, the ants hoped to establish long-distance communication. Pheromone (a chemical-like substance) is dispersed while they are on the quest for food. When additional ants from the colony follow this track, they might save time and effort while foraging for food (Long et al., 2019). A network of connected curves can be constructed to stand in for the edges of objects, markers, and surface orientation breaks when an ideal edge detector is used. Edge detection is a strong tool for preserving important structural features of an image while reducing the amount of data that must be processed. Among the most popular approaches to detecting edges, search-based and zero-crossing-based algorithms stand out. Finding the edges of moving objects is a common task for many image analysis and machine vision programs. Because each set of procedures is run for each pixel, standard ways to edge detection are computationally intensive. When employing nonstandard approaches, computation time scales very poorly with image size. For solving the TSP, Ant Colony algorithms are utilized (Nawaz et al., 2020).

1.1.3 Artificial Bee Colony (ABC) Optimization

To generate the best possible list of Cluster Heads, the ABC algorithm is employed. The first packet sent by the sensor node is a "hello" message. Each sensor node stores the id of the sending sensor and the RSSI value in a nearby database whenever it receives the broadcast message. The nodes will now update the BS on their current energy status, table data from their immediate vicinity, and unique identifier. The BS now uses the formula to compile the CH list. Nodes with energies over a particular threshold are considered CHs. Once a list of CHs has been obtained, they are clustered according to their RSSI values. Following that, one's healthiness is evaluated. The best possible CH candidate is then selected from the pool of candidates using the ABC algorithm. This means the optimum numbers of CHs are chosen. The information is then sent from the nodes to the CHs via TDMA slots. CDMA MAC protocol is used to transmit data to the BS. Here, we assume a CH energy level larger than the threshold as part of the fitness function. The number of CHs should be reduced as much as possible to save energy use (Nguyen et al., 2020).

1.1.4 Bacterial Foraging Optimization (BFO)

Bacterial initialization is the first phase, and it is performed arbitrarily. Some fraction of the bacterium is composed of CHs. Using node ids and updated 2D coordinates, chemotaxis may accurately track the location of a CH. The equation is then used to guide further swarming. The bacteria are then ranked from most fit to least fit. To reproduce, bacteria from the upper half of the population must be transported to the bottom half. It is possible to remove the weakest bacteria and start a new bacterium with the newly added nodes. In this case, the fitness function is composed of the CH's residual energy, the intra-cluster distance, and the CH's separation from the BS. Each bacterium's dimension for routing is the same as the number of CHs, plus one more for the BS. The next step in locating the BS is to use the mapping function. The routing fitness function is composed of the CH-to-next-hop and next-hop-to-BS Euclidean distances as well as the next-residual hop's energy (Sharma et al., 2017).

1.1.5 Cuckoo Search

Recently, the meta-heuristic optimization method Cuckoo Search has been created for use in addressing optimization issues. Natural occurrences such as the brood parasitism of certain cuckoo species and Levy flights random walks provide motivation for this metaheuristic method. Long-term use of the same cuckoo search settings might degrade the performance of the algorithm. A suitable technique for tweaking the cuckoo search parameters has to be established to address this problem.

1.1.6 Firefly Algorithm (FA)

Each cluster's CH-to-BS path may be calculated with the help of the firefly method. The firefly is first put through its paces during the initialization phase. In this case, each flamingo symbolizes a different path that might be taken to reach the BS from the CH. Each node here stands in for the CH's next jump on the way to the BS. The firefly with the lower brightness is updated in position relative to the firefly with the higher brightness at each iteration. Maximum iterations are performed to ensure optimal results.

1.1.7 Genetic Algorithm (GA)

An optimal number of CHs is determined with the help of a Genetic Algorithm. The method has two phases: initialization and operation in a steady state. During setup, BS determines how many CHs will be present and assigns nodes accordingly. Second, the nodes transmit data to the CHs, which relay it to the base station (BS). Chromosomes are generated by the BS, and then the genetic algorithm is utilized to choose the best candidate for CH. The Fitness/Objective function may be used to evaluate a person's physical health and fitness (Wu et al., 2017). The objective function in this study is the aggregate of hop counts between each node and the CH and between each CH and the BS. Here, we want to maximize the goal function to achieve the shortest possible distance for data transfer. In this case, a third agent, CH, is utilized to increase that distance to a maximum. When choosing an agent CH, the one with the most residual energy is prioritized.

1.2 6G

Wireless networks built on the 5G standard will be phased out in favor of the 6G standard. Because 6G networks will run at higher frequencies than 5G networks, their capacity will be substantially boosted and their latency will be greatly decreased. As a goal, the 6G internet should provide connections with a latency of one microsecond or less. This is a thousand times faster, with a throughput of one millisecond. Industry experts believe that the advent of 6G networks will pave the way for revolutionary improvements in areas such as imaging, presence technology, and location awareness. The 6G computational infrastructure, in tandem with AI, will be able to determine the optimal location for all aspects of computing, from data storage and processing to the dissemination of findings. Remember that 6G is not a fully operational technology as yet. Although some companies are making preparations for the next generation of wireless technology (Anand et al., 2022), industry standards for 6G-enabled network equipment are still some years away.

1.2.1 Advantages of 6G over 5G

6G networks rely on sending and receiving signals at very high frequencies in the radio spectrum to operate. Though it is too soon to approximate 6G data rates, a senior professor at the University of Sydney has postulated that a theoretical peak data rate of 1 terabyte per second for wireless communications may be achievable. For relatively brief data transmissions across short distances that estimate is accurate. In the same year, LG, a company located in South Korea, introduced a method based on adaptive beam shaping. This combination of capacity and latency will greatly improve the performance of 5G apps. In addition, it will broaden the capability to support state-of-the-art programs in areas including wireless communication, cognition, sensing, and imaging. Using orthogonal frequency-division multiple access, 6G will allow access points to handle connections from several users at once (Wu et al., 2018).

1.2.2 Expected Working of 6G

It is anticipated that 6G wireless sensing systems would utilize a variety of frequencies for absorption measurements and frequency adaptation. This technique works because all substances have a unique frequency range at which their atoms and molecules generate and absorb electromagnetic radiation. Numerous public and private sector approach to ensuring public safety and safeguarding vital assets will be significantly altered by 6G. Emerging technologies like smart cities, driverless cars, virtual reality, and augmented reality will all benefit from advancements in these areas, in addition to smartphones and other mobile network technology. The limitations of 5G for supporting new problems need the creation of a sixth-generation (6G) wireless technology with a novel, alluring characteristics. 6G will be driven by the convergence of many trends that have been developing in recent years: increased network density; higher throughput; greater reliability; lower energy consumption; and larger connections. New services with the inclusion of new technology would also be a hallmark of the 6G system, which would be a continuation of the pattern established by the previous generations. Artificial intelligence, smart wearable, implant, autonomous car, computing reality gadgets, sensors, and 3D maps are all examples of the new services.

1.2.3 Need of 6G

Several factors need the development of 6G networks. The following are some of them:

- **Technology Convergence.** Deep learning and big data analytics, for example, will be combined with other technologies in the sixth generation of cellular networks (Jain et al., 2022). Many of these synergies are now possible because of the advent of 5G.

- **Edge Computing.** The need for ultra-reliable, low-latency communications solutions necessitates the use of edge computing, which has been a driving force behind the development of 6G. This allows for both higher data rates and lower latency (Gupta et al., 2023).

Figure 2. Need of 6G

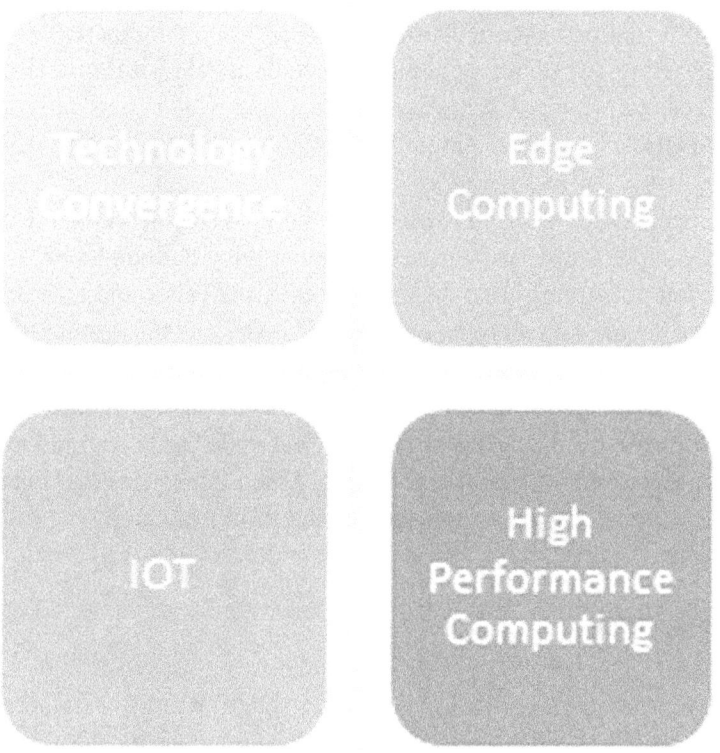

- **Internet of things (IoT).** The requirement to enable machine-to-machine communication in IoT is another factor contributing to its development (Kaur et al., 2023; Tripathi et al., 2023).
- **High-performance computing (HPC).** Sixth-generation wireless connectivity (6G) and high-performance computing (6G HPC) have been found to have a robust correlation (Meivel et al., 2023). Although some of the data generated by IoT and mobile devices may be processed by edge computing resources, the majority will require the use of more traditional centralized high-performance computing (HPC) capabilities.

1.2.4 Availability of 6G

In 2030, we should witness the first commercial deployment of 6G internet. The technology optimizes the utilization of the distributed RAN and THz spectrum to improve capacity, lower latency, and increase spectrum sharing. Although there had been some early discussions to identify the technology, sixth-generation R&D activities did not get off the ground until 2020. 6G will need the development of cutting-edge mobile communications technologies including cognitive and highly secure data networks. Furthermore, a far faster pace of spectral bandwidth expansion than 5G will be required.

1.3 WSN

Wireless sensor networks (WSNs) are hybrid systems that combine hardware and software networks to collect data from a network of sensors to monitor and record environmental conditions and other measurable quantities. Wireless sensor networks (WSN) are becoming increasingly used in both commercial and domestic settings. Since privacy is so important in WSN, it is crucial to protect any and all data that is processed, sent or stored. Users are vulnerable to a wide variety of security risks (Anand et al., 2018) while using wireless communication technologies like Bluetooth, wi-fi, etc.WSNs typically employ one of two wireless network topologies known as "relay nodes" or "leaf nodes." Both the source and the target are linked to other nodes in the relay network. As the network's transmission range falls short of the gap between the source and the destination, a relay node must be used. With the help of relay nodes, information may travel great distances even if the caller and the recipient are physically separated. Determining the most direct route between the nodes and the destination computer expedites the transfer of data. An internet-connected laptop is an instance of a relay node. Information sent from a laptop through a wireless network can be relayed from one network to another until it reaches its final destination. A data structure with no cycles is called a "leaf node" or "tree," and it is made up of nodes/vertices and edges. A null or empty tree has no nodes. A non-empty tree has many tiers of leaf nodes and extra nodes that can connect them to one another and form a hierarchy.

2 LITERATURE REVIEW

There are several researches in the area of wireless sensors and 6G wireless networks. Moreover, different optimization mechanisms have been illustrated in previous research work. This section is presenting the contribution of the author in the area of research related to wireless networks and optimization. Some of the research works focused on routing and throughput while some authors considered network security.

Ahmed et al. (2017) presented information that has to be quickly sent from the field to first responders and neighboring emergency centers, wireless sensor networks (WSN) was seen as a viable option for providing such services. The nodes, however, were vulnerable to capture, damage, and misbehavior in such application contexts. The infected nodes may also contribute to traffic congestion by spreading misinformation. In times of crisis, information mustn't be lost, but that's exactly what may happen when there's a lot of traffic and certain malicious actors try to exploit vulnerabilities in the system. Due to their high price tag and inability to defend against nodes' misbehavior assaults, conventional cryptography and authentication-based security systems cannot be implemented. Current trust-based routing protocols also have considerable control overheads in trust estimation and broadcast, in addition to a high degree of route instability due to the route-finding approach used (Pandey et al., 2022). ESRT is a technique introduced in this research that helps maintain network security by isolating malicious nodes. The routing choices in ESRT take into account trust, energy, and hop counts. This multi-faceted approach to routing allows for more efficient use of energy resources across reliable nodes while still making use of optimal channel lengths for the transmission of data. The simulation findings show that the ESRT system outperforms prior efforts (Gupta et al., 2019).

Arora et al. (2019) introduced the lifespan of wireless networks, energy-efficient routing algorithms must intelligently deal with the power-limitation problem of the sensor nodes. As a result, it is critical to gather and efficiently share sensor data. This algorithm finds a productive path for intra-cluster communication. Cluster formation, multi-path generation, and data transmission are the three main steps in the AOSTEB scheme's operation. To build a cluster, a subset of the sensor nodes are selected to act as cluster heads (CHs), and their neighbors form a cluster around them. Additionally, the Ant Colony Optimization technique is used to unearth the many connections between the CH and the member nodes. When initiating data exchange inside a cluster, a dynamic energy-efficient optimal path is chosen based on the shortest distance and lowest energy usage (Gupta et al., 2020). Extensive simulation results verify the efficacy of the proposed approach by showing improved network lifespan, stability period, and lower energy consumption compared to previously published studies in wireless sensor networks.

Behnaam Aazhang et al. (2019) focused on Limitless wireless communication which will be essential for many highly anticipated future applications, such as eHealth and driverless cars. It is anticipated that future advancements in mobile communication technology will allow for wirelessly enabled apps that vastly improve the efficacy of enterprises and the quality of people's daily lives. The research community has to prioritize the creation of beyond-5G solutions and the 2030 age, i.e. 6G, while fifth-generation (5G) research advances toward a worldwide standard. What exactly 6G will entail is still unclear. It will include technologies that are either not yet developed enough for 5G or fall outside of the boundaries set by 5G. To be more explicit, 6G will be largely motivated by the evolution of data collection, processing, transmission, and consumption via wireless networks. Key drivers, research needs, obstacles, and vital research topics about 6G were initially identified during the inaugural 6G Wireless Summit in March 2019. The overarching theme of this white paper, which will be updated yearly, may be summed up in the term "ubiquitous wireless intelligence" taken from the inaugural 6G Wireless Summit. We expect that in addition to the present 5G technical KPIs, we will require additional KPI drivers. Solutions aimed towards the year 2030 are driven in large part by social megatrends, UN sustainability targets, decreased carbon dioxide emissions, development of new technology enablers, and ever-increasing productivity needs. Lightweight glasses that provide dynamic range are poised to replace current smartphones as the primary device for XR experiences. This is made feasible through smart spectrum use in the THz range. Communications and emerging technologies, such as 3D imaging and sensing, may be combined thanks to the THz spectrum extension. Although potential exists for semiconductors, and optics, among others, new paradigms for transceiver design and computation are required to realize them (Gupta et al., 2021).

Duan et al. (2020) focused on the development of 5G networks and terminal equipment over the last several years, and this technology is now entering the commercial phase. Academic and commercial interests in the 6G mobile communication technology have grown rapidly in recent years. When compared to 5G, 6G will be significantly more energy and spectrum efficient while also providing a substantially greater access rate (10-100 times higher), smaller access latency, and broader and deeper communication coverage. It paves the way for constant development in antenna materials, processes, technologies, and designs, as well as RF system configurations. Several noteworthy trends may be seen in the evolution of antennas and RF systems.

Jiaheng Wang et al. (2017) introduced blockchain-enabled WSNs. The distributed ledger technology (blockchain) that has emerged in the last decade holds great promise as a viable answer. Blockchain's decentralized, transparent, anonymous, immutable, traceable, and resilient nature allows it to establish cooperative trust among different entities within a network, paving the way for advancements in areas like secure access control, data privacy, and more in wireless networks as we move towards 6G. In this study, we focus on wireless communication technologies that can make use of blockchain protocols.

Liang et al. (2019) reviewed LEACH routing protocol will lead to unstable network functioning since it balances the network's energy usage by picking cluster heads at random in a loop. Therefore, to resolve this issue, it was important to extend the lifespan of networks by decreasing the energy required for data transmission in the routing protocol. However, there are still unresolved issues with the LEACH's cluster heads counting with a broad range and the cluster head forwarding data using substantial power. Here, we propose a method for bettering the routing protocol. First, to minimize the possibility of uneven cluster head distribution, the ideal number of cluster heads is computed by taking into account the total energy usage every round.

Long et al. (2019) looked software defined 5g and 6G networks: a survey. The current state of mobile communications was inadequate to meet the skyrocketing demand for data from its users.

Nawaz et al. (2020) provided a review of the vision and challenges of 6G technology. Wireless information traffic has significantly increased due to the fast development of smart terminals and increasing new applications, and current cellular networks (including 5G) were unable to fully cope with the constantly evolving technological requirements. Between 2027 and 2030, a new wireless communication framework, the sixth-generation (6G) framework, is expected to be equipped with the floating assistance of artificial intelligence. Several major technological issues, as well as some potential answers, of 6G are outlined in this study, along with physical layer transmission protocols, network architecture, and security measures.

Nguyen et al. (2020) presented privacy-aware block-chain innovation for 6G: challenges and opportunities. 6G wireless networks are superior to 5G in several ways, including increased dependability, network speed, and capacity. These incremental advancements, in tandem with others like high-precision 3D localization, ultra-high dependability, and extreme mobility, usher in a brand-new age of 6G-native software. Examples of technologies that might support this kind of application include ultra-reliable and distributed pervasive AI. Privacy and security of networks and apps must be guaranteed with improved connectivity and cutting-edge uses. While distributed ledger technologies like blockchain provide a potential remedy for app security and privacy concerns, they also pose new threats. In this paper, we examine the advantages and disadvantages of using block-chain technology in 6G networks and provide a road map for overcoming the disadvantages.

Sharma et al. (2017) introduced a rendezvous-based routing protocol. Each sensor node in a WSN was responsible for determining the most efficient route for sending data to the network's central hub, known as the sink. Data is sent from the source node to the sink node, either directly or through the intermediate nodes. Because of the sensor node's limited battery life, an efficient routing method is required to prolong the network's operation. This study proposes a rendezvous-based routing system that builds a tree inside a centrally located rendezvous zone in the network. In the proposed protocol, there are two channels for exchanging information.

Wu et al. (2017) explained IoT refers to the concept that everything on the global network is also linked and accessible. Among the many major applications of this idea, wireless sensor networks (WSN) have found a home in practically every field. Improved authentication for WSNs was provided by Hsieh et al. in 2014. However, it is vulnerable to a variety of attacks, such as insider attacks, off-line guessing attacks, user forging attacks, and sensor capture attacks since there is no session key and mutual authentication is not used. They provide a novel authentication mechanism that was likewise suitable for WSNs and which avoids the flaws. The formal proof is then shown using the random oracle model, and the protocol analysis tool Proverif is used to provide a detailed inventory of the formal verification steps. By focusing on the security features of the Internet of Things, the proposed method is superior to other current schemes for WSNs in terms of addressing these concerns.

Wu et al. (2018) presented the safety of cyberspace, both difficulties and possibilities exist. As the ocean, land, air, and space had all been explored, cyberspace had become the new "fifth frontier." Cyberspace has never before been up against so many obstacles, As such, the shift from classic to modern internet was unavoidable. Cyberspace in the future will be open, heterogeneous, mobile, dynamic, and secure, among other crucial qualities.

Table 1. Literature review

Author/Year	Title	Methodology	Limitation
Ahmed/2017	Trustworthy energy-conscious routing for WSN in emergency response.	wireless sensor networks	Lack of efficiency
Arora /2019	Studies aimed at better understanding how the LEACH protocol might be used to optimize routing in WSNs.	ACO, wireless sensor network	Lack of technical work
Behnaam Aazhang/2019	development of 6G pervasive wireless intelligence, and overcome in the field of research	white paper, 7G	There is less technical work
Duan/2020	Innovations in blockchain technology that protect user privacy at 6G speeds	5G, 6G, antenna systems	Need to enhance the scope of work
Jiaheng Wang /2017	As we go toward 6G, a new paradigm emerges in wireless communications supported by blockchain technology.	6G, wireless communications	This work is not long-lasting
Liang /2019	Constraints on 6G Ubiquitous Wireless Networks	wireless sensor network	There is a lack of performance
Long /2019	The AOSTEB algorithm is an ACO-optimized self-organized tree-based energy balance method for WSNs.	5G, 6G	The performance of this research is very low
Nawaz Oliveira/2020	An Overview of Software-Defined 5G and 6G Networks Communications and Applications in Mobile Networks.	6G	Did not consider Real-life solution
Nguyen /2019	Discusses the potential and difficulties of 6G networking.	Block-chain, 6G	Need to consider optimization technique
Sharma /2017	Protocol for mobile sinks in WSN based on rendezvous	wireless sensor networks	There is not performed in future
Wu/2017	WSN leverages IoT security that protects user privacy while allowing for verifiable logins	wireless sensor networks	Need to improve the performance and accuracy
Wu /2018	Cybersecurity threats and opportunities.	Security	Research is limited to traffic flow

3 PROBLEM STATEMENT

A research study discusses the several optimization methods that have been used for 6G. Existing optimization techniques have been shown to have several drawbacks. ACO-based strategy can offer dependability in 6G but does not support Application Complexity. The shortest path's length can only be calculated using an ACO-based module. However, PSO with MVO offers the finest answer quickly. Therefore, it has become crucial to provide an improved 6G setup module that can address the current problems in this area. In addition, MVO-based PSO has been found to perform quickly in comparison to other optimization methods. To do a comparative study of various optimization strategies, it is necessary to take into account the amount of time and correctness of the result. Issues with the previous optimization mechanism are accuracy and performance. It has been observed that there is a need to integrate conventional nature-inspired algorithms with meta-heuristic optimization algorithms. The proposed work is providing a solution for accuracy and performance by providing a hybrid PSO-MVO approach. The proposed research is providing an optimized network security model for effective communication.

4 PROPOSED WORK

Providing a better 6G setup module that can fix the existing issues in this field is now an absolute need (Arora et al., 2023). When compared to other optimization techniques, MVO-based PSO has also been demonstrated to be quite efficient. When evaluating different optimization approaches, it's important to think about how much time they take and how reliable their results are. Accuracy and efficiency have been problems with earlier optimization methods. It has been realized that the standard nature-inspired method and the meta-heuristic optimization algorithm should be combined. The proposed study offers a hybrid PSO-MVO strategy as a means of improving both accuracy and performance. The proposed study offers a model of network security that may be tuned for maximum efficiency in conversation. The proposed study takes into account recent studies in the field of 6G and optimization. PSO, ACO, and MVO are three optimization methods that are the subject of current research. The proposed work has some issues due to the limitations of the optimization process as well as its precision and effectiveness. Think about the hybrid technology that is being proposed; it should be able to incorporate the PSO-MVO, objective function, and several parameters, such as iteration, upper bond, and lower bond. Last but not least, research involves measuring the time required and error rate compared to standard procedures (Veeraiah et al., 2022).

Fig 3 is presenting the research methodology used for the implementation of the proposed work. It lays down the steps that will be taken to complete the planned job. Here, we take into account research into preexisting optimization mechanisms and describe problems associated with prior efforts in this area. After the goal function and PSO-MVO integration are taken into account, a hybrid strategy is developed. The process concludes with an assessment of precision and efficiency. Finally, a comparison of performance has been made for PSO, MVO, and hybrid PSO-MVO

4.1 Deriving Hybrid Multiverse Optimization (PSOMVO)

Deriving Equation for base MVO

Phase 1: The mechanism used in the Particle Swarm Optimization (PSO) phase one equation was influenced by the social behavior of fish or birds. PSO is made up of P_{best} and G_{best}. These mathematical equations allow for the updating of not just locations but also speeds and direction changes.

$$v_{ij}^{t+1} = wv_{ij}^t + C_1R_1\left(Pbest^t - X^t\right) + C_2R_2\left(Gbest^t - X^t\right) \tag{1}$$

Figure 3. Research Methodologies

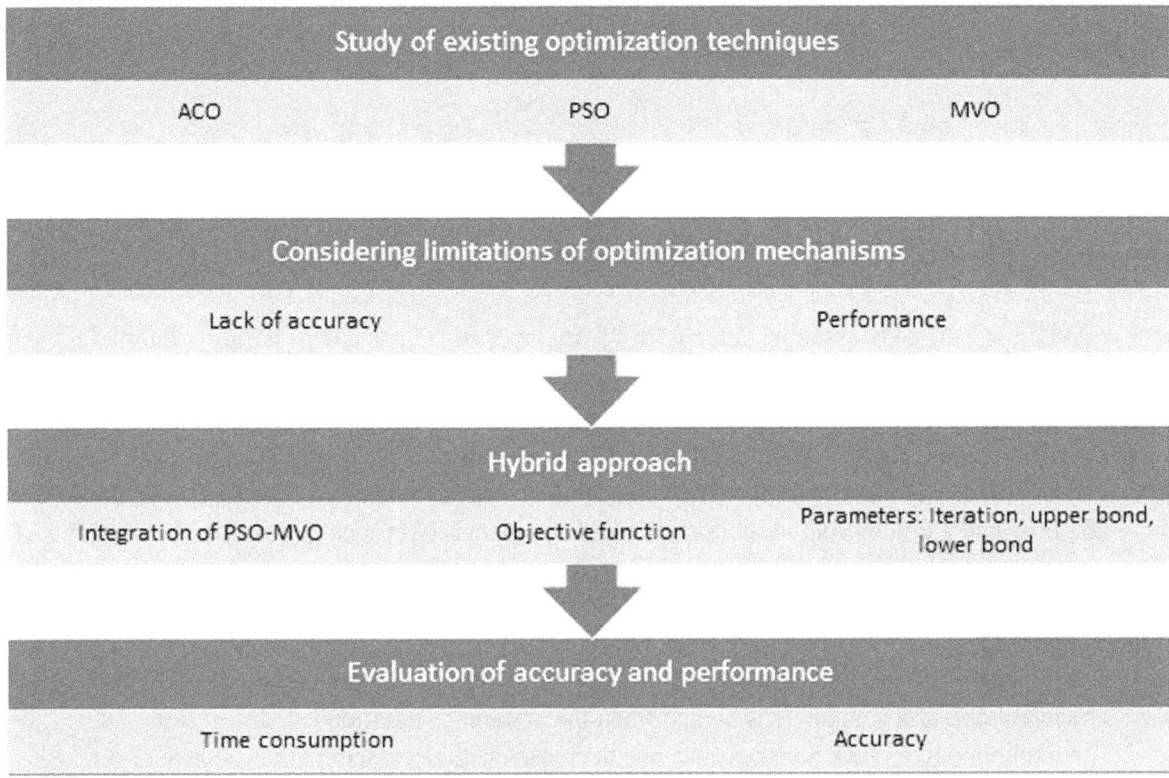

Here

v_{ij}^{t+1} = velocity of particle I at time t+1

w = inertia factor rage 0.4 to 1.4

wv_{ij}^{t} = current motion

X^{t} = particle state at time t

Pbest^{t}= personal best at time t

C1 = self-confidence range 1.5 to 2

R1 = random number 1

$R_{1}(Pbest^{t} - X^{t})$= particle memory influence

C2= swarm confidence 2 to 2.5 range

R2 = random number 2

Gbest^{t}= global best at time t

$R_{2}(Gbest^{t} - X^{t})$= swarm influence

$$X^{t+1} = x^{t} + v^{2t+1} \left(i = 1,2..NP \right) And \left(J = 1,2..NG \right) \tag{2}$$

Here

X^{t+1} = particle state at time t+1

Where

$$W = w^{max} - \frac{\left(w^{max} - w^{min}\right) * iteration}{\max iteration} \tag{3}$$

$w^{max} = 0.4$

$w^{min} = 0.9$. v_{ij}^{t}, v_{ij}^{t+1} Already assumed velocity of "j" member of "i" particle in iteration number (t) and (t + 1). (Normally C_1 is equal to C_2 is equal to 2), r_1 and r_2 Random number (0, 1).

Phase 2: The black hole, white hole, and wormhole letters serve as the inspiration for the multiverse optimizer equation (MVO) method. Each of these alphabets is presented in an arithmetical order. In that sequence, it examines operations, research studies, and domestic discoveries. It is believed that the white hole had a crucial role in the formation of the planet. Everyone is drawn in by black holes' stunning gravitational effects. Wormholes serve as passageways for time and space travel. By using this method, it is possible to ship items quickly throughout the world. The following stages are used in the creation of MVO:

- The likelihood of a white hole is higher when the amount of inflation is high.
- The likelihood of a black hole's occurrence decreases with decreasing inflation levels.
- Populations with high inflation rates distribute chemicals via white holes.
- Populations that do not sustain high inflation levels absorb more substances via black holes.

Every element in this universe wants to advance in favor of a population that is healthy. Wormholes are to blame for doing it. The amount of inflation is not taken into consideration for this. From a universe with a high rate of spread, every element moves towards the direction of the population where the rate of spread is low (Verma et al., 2021). It takes on responsibility for the increase in the overall cosmoses' average spreading pace in the presence of change. As soon as a change takes place, the population is categorized based on how quickly it spreads. Then it picks one out of them. The usage of a white hole on a roulette wheel is made for this reason. The use of subsequent phases supports this methodology.

4.2 Derived hybrid PSO-MVO equation

Hybrid PSO-MVO sets combine the best features of each of these algorithms. Hybrid PSO-MVO combines the advantages of both PSO and MVO to find an optimal solution that meets certain criteria. The PSO Pbest value is being swapped out for the MVO Universe value.

$$v_{ij}^{t+1} = wv_{ij}^{t} + C_1 R_1 \left(Universes^{t} - X^{t}\right) + C_2 R_2 \left(Gbest^{t} - X^{t}\right)$$

Performance Algorithm

1. Get the transaction initialization time t1 for PSO
2. Apply PSO optimization mechanism
3. Get optimization time t2
4. Get the transaction initialization time t1 for MVO
5. Apply the MVO optimization mechanism
6. Get optimization time t3
7. Get the transaction initialization time t1 for Hybrid approach
8. Apply hybrid optimization mechanism
9. Get optimization time t4
10. Get performance for PSO = t2-t1
11. Get performance for MVO = t3-t1
12. Get performance for Hybrid PSO-MVO = t4-t1
13. Compare the performance of PSO, MVO, and hybrid PSO-MVO.

5 RESULTS AND DISCUSSION

To find the optimal solution, simulation work took into account optimization methods PSO, ACO, and MVO. PSO, ACO, and MVO mechanisms have been compared to see how long each takes to operate. The time consumption of hybrid PSO-MVO is smaller than that of traditional optimization processes, according to the results. Comparisons of time consumption and accuracy are shown in Tables 1.2 and 1.3, respectively.

Table 2. Time Consumption

Configuration	PSO	ACO	MVO	HYBRID PSO-MVO
1	1.95548295	1.89085	1.55944	1.45332
2	2.57886463	3.05439	1.44933	1.24213
3	4.36254442	3.93237	4.05107	3.12123
4	5.3199872	5.05179	4.70883	4.00851
5	6.13927672	5.82743	5.26284	4.57169
6	7.06001566	6.56305	6.37277	5.33787

6 CONCLUSION

According to simulation, hybrid techniques are more effective than traditional ones. When making decisions about 6G, it is giving higher accuracy. Additionally, less time is used. As a result, the hybrid strategy has improved 6G's accuracy and performance.

Figure 4. Comparison of Performance in the case of PSO, ACO, MVO, Hybrid PSO-MVO

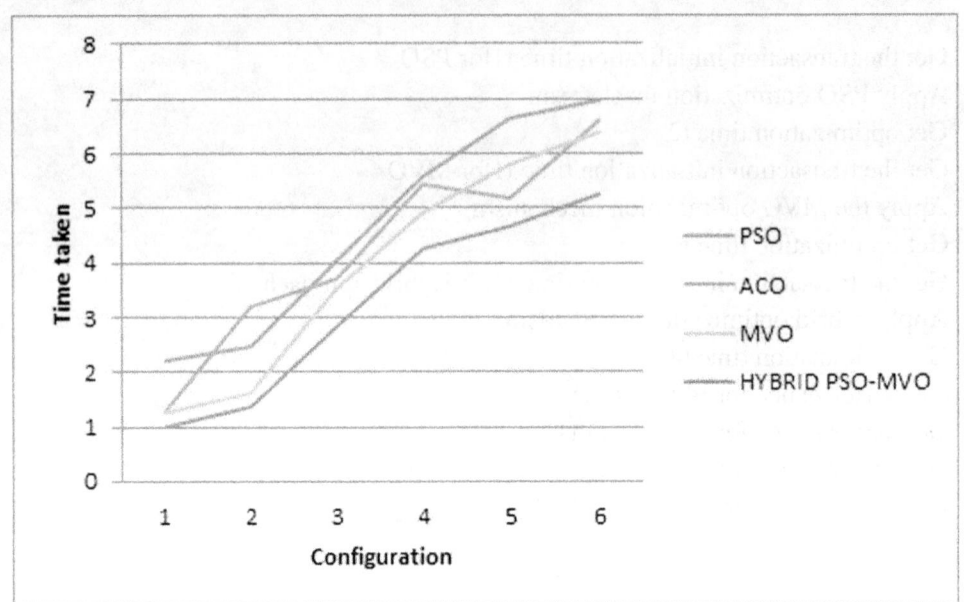

Table 3. Accuracy comparison

Configuration	PSO	ACO	MVO	HYBRID PSO-MVO
1	85.3071127	85.6916	87.0203	87.9847
2	85.3636379	86.4019	86.8533	87.3833
3	86.1815196	86.3162	87.3187	87.8634
4	85.801944	86.6398	87.5865	88.6169
5	85.9907827	86.1881	86.7097	87.1698
6	85.7280618	86.4012	87.075	87.5548

Figure 5. Comparison of accuracy

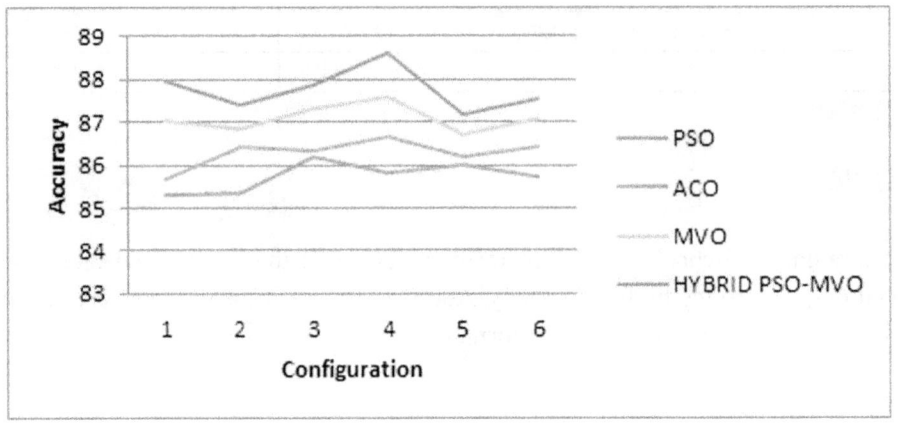

7 FUTURE SCOPE

This novel 6G are distinct from its larger predecessors in that they communicate using radio waves rather than electromagnetic fields. The approach used to specify inputs such as resource demands, available resources, and planned utilization is the scope of resource scheduling optimization. The optimum times are also provided. IT companies in the telecoms business are looking at new applications for 6G in the next generation of consumer electronics despite 5G's outdated image. To reduce size, weight, and strength, 6G is now being developed via 3D printing or enhanced manufacturing procedures. To usher in a new era of smaller cell towers, telecom companies are aiming to introduce 5G technology. Although these breakthroughs are thrilling, traditional mobile phone towers will continue to have a significant impact on the future of human communication.

REFERENCES

Aazhang, B. (2019). Key drivers and research challenges for 6G ubiquitous wireless intelligence (white paper). *Human-Centric Computing and Information Sciences, 9*(1), 1–10. doi:10.1186/s13673-019-0198-1

Ahmed, A., Bakar, K. A., Channa, M. I., Khan, A. W., & Haseeb, K. (2017). Energy-aware and secure routing with trust for disaster response wireless sensor network. *Peer-to-Peer Networking and Applications, 10*(1), 216–237. doi:10.100712083-015-0421-4

Anand, R., & Chawla, P. (2022). Bandwidth Optimization of a Novel Slotted Fractal Antenna Using Modified Lightning Attachment Procedure Optimization. In *Smart Antennas* (pp. 379–392). Springer. doi:10.1007/978-3-030-76636-8_28

Anand, R., Shrivastava, G., Gupta, S., Peng, S. L., & Sindhwani, N. (2018). Audio watermarking with reduced number of random samples. In *Handbook of Research on Network Forensics and Analysis Techniques* (pp. 372–394). IGI Global. doi:10.4018/978-1-5225-4100-4.ch020

Anand, R., Singh, J., Pandey, D., Pandey, B. K., Nassa, V. K., & Pramanik, S. (2022). Modern Technique for Interactive Communication in LEACH-Based Ad Hoc Wireless Sensor Network. In *Software Defined Networking for Ad Hoc Networks* (pp. 55–73). Springer. doi:10.1007/978-3-030-91149-2_3

Arora, S., Sharma, S., & Anand, R. (2023). A Survey on UWB Textile Antenna for Wireless Body Area Network (WBAN) Applications. In *Artificial Intelligence on Medical Data* (pp. 173–183). Springer. doi:10.1007/978-981-19-0151-5_14

Arora, V. K., Sharma, V., & Sachdeva, M. (2019). ACO optimized self-organized tree-based energy balance algorithm for wireless sensor network: AOSTEB. *Journal of Ambient Intelligence and Humanized Computing, 10*(12), 4963–4975. doi:10.100712652-019-01186-5

Babu, S. Z. D. (2022). Analysation of Big Data in Smart Healthcare. In M. Gupta, S. Ghatak, A. Gupta, & A. L. Mukherjee (Eds.), *Artificial Intelligence on Medical Data. Lecture Notes in Computational Vision and Biomechanics* (Vol. 37). Springer. doi:10.1007/978-981-19-0151-5_21

Bansal, R., Gupta, A., Singh, R., & Nassa, V. K. (2021). Role and Impact of Digital Technologies in E-Learning amidst COVID-19 Pandemic. *2021 Fourth International Conference on Computational Intelligence and Communication Technologies (CCICT)*, (pp. 194-202). IEEE. 10.1109/CCICT53244.2021.00046

Duan, B. (2020). Evolution and innovation of antenna systems beyond 5G and 6G. *Frontiers of Information Technology and Electronic Engineering*, *21*(1), 1–3. doi:10.1631/FITEE.2010000

Dushyant, K., Muskan, G., Gupta, A., & Pramanik, S. (2022). Utilizing Machine Learning and Deep Learning in Cyber security: An Innovative Approach. In M. M. Ghonge, S. Pramanik, R. Mangrulkar, & D. N. Le (Eds.), *Cyber security and Digital Forensics*. Wiley. doi:10.1002/9781119795667.ch12

Gupta, A. (2019). Script classification at word level for a Multilingual Document. *International Journal of Advanced Science and Technology*, *28*(20), 1247–1252. http://sersc.org/journals/index.php/IJAST/article/view/3835

Gupta, A. (2020). An Analysis of Digital Image Compression Technique in Image Processing. *International Journal of Advanced Science and Technology*, *28*(20), 1261–1265. http://sersc.org/journals/index.php/IJAST/article/view/3837

Gupta, A., Goyal, B., Dogra, A., & Anand, R. (2023). Proximity Coupled Antenna with Stable Performance and High Body Antenna Isolation for IoT-Based Devices. In *Communication, Software and Networks* (pp. 591–600). Springer. doi:10.1007/978-981-19-4990-6_55

Gupta, A., Kaushik, D., Garg, M., & Verma, A. (2020).Machine Learning model for Breast Cancer Prediction. 2020 Fourth International Conference on I-SMAC (IoT in Social, Mobile, Analytics and Cloud) (I-SMAC), (pp. 472-477). IEEE. 10.1109/I-SMAC49090.2020.9243323

Gupta, A., Singh, R., Nassa, V. K., Bansal, R., Sharma, P., & Koti, K. (2021) Investigating Application and Challenges of Big Data Analytics with Clustering. *2021 International Conference on Advancements in Electrical, Electronics, Communication, Computing and Automation (ICAECA)*, (pp. 1-6). IEEE.10.1109/ICAECA52838.2021.9675483

Gupta, N., Khosravy, M., Patel, N., Dey, N., Gupta, S., Darbari, H., & Crespo, R. G. (2020). Economic data analytic AI technique on IoT edge devices for health monitoring of agriculture machines. *Applied Intelligence*, *50*(11), 3990–4016. doi:10.100710489-020-01744-x

Jain, S., Sindhwani, N., Anand, R., & Kannan, R. (2022). COVID Detection Using Chest X-Ray and Transfer Learning. In *International Conference on Intelligent Systems Design and Applications* (pp. 933-943). Springer. 10.1007/978-3-030-96308-8_87

Kaur, J., Sindhwani, N., Anand, R., & Pandey, D. (2023). Implementation of IoT in Various Domains. In *IoT Based Smart Applications* (pp. 165–178). Springer. doi:10.1007/978-3-031-04524-0_10

Kaushik, D., & Gupta, A. (2021). Ultra-secure transmissions for 5G-V2X communications. *Materials Today: Proceedings*. doi:10.1016/j.matpr.2020.12.130

Kaushik, K., & Garg, M., Gupta, A., & Pramanik, S. (2021). Application of Machine Learning and Deep Learning. In M. Ghonge, S. Pramanik, R. Mangrulkar, & D. N. Le, (eds.), Cyber security: An Innovative Approach, in Cybersecurity and Digital Forensics: Challenges and Future Trends. Wiley.

Kirubasri, G., Sankar, S., Pandey, D., Pandey, B. K., Singh, H., & Anand, R. (2021, September). A Recent Survey on 6G Vehicular Technology, Applications and Challenges. In *2021 9th International Conference on Reliability, Infocom Technologies and Optimization (Trends and Future Directions) (ICRITO)* (pp. 1-5). IEEE.

Liang, H., Yang, S., Li, L., &Gao, J. (2019). Research on routing optimization of WSNs based on improved LEACH protocol. *Eurasip Journal on Wireless Communications and Networking, 2019*(1). doi:10.1186/s13638-019-1509-y

Long, Q., Chen, Y., Zhang, H., & Lei, X. (2019). Software Defined 5G and 6G Networks: A Survey. *Mobile Networks and Applications*. doi:10.100711036-019-01397-2

Meivel, S., Sindhwani, N., Valarmathi, S., Dhivya, G., Atchaya, M., Anand, R., & Maurya, S. (2023). Design and Method of 16.24 GHz Microstrip Network Antenna Using Underwater Wireless Communication Algorithm. In *Cyber Technologies and Emerging Sciences* (pp. 363–371). Springer. doi:10.1007/978-981-19-2538-2_36

Nawaz, F., Ibrahim, J., Junaid, M., Kousar, S., Parveen, T., & Ali, M. A. (2020). A review of vision and challenges of 6G technology. *International Journal of Advanced Computer Science and Applications, 11*(2), 643–649. doi:10.14569/IJACSA.2020.0110281

Nguyen, T., Tran, N., Loven, L., Partala, J., Kechadi, M. T., & Pirttikangas, S. (2020). Privacy-aware blockchain innovation for 6G: Challenges and opportunities. *2nd 6G Wireless Summit 2020: Gain Edge for the 6G Era, 6G SUMMIT 2020*, (pp. 1–5). IEEE. doi:10.1109/6GSUMMIT49458.2020.9083832

Pandey, B. K. (2022). Effective and Secure Transmission of Health Information Using Advanced Morphological Component Analysis and Image Hiding. In M. Gupta, S. Ghatak, A. Gupta, & A. L. Mukherjee (Eds.), *Artificial Intelligence on Medical Data. Lecture Notes in Computational Vision and Biomechanics* (Vol. 37). Springer. doi:10.1007/978-981-19-0151-5_19

Pathania, V. (2022). A Database Application of Monitoring COVID-19 in India. In M. Gupta, S. Ghatak, A. Gupta, & A. L. Mukherjee (Eds.), *Artificial Intelligence on Medical Data. Lecture Notes in Computational Vision and Biomechanics* (Vol. 37). Springer. doi:10.1007/978-981-19-0151-5_23

Sharma, S., Puthal, D., Jena, S. K., Zomaya, A. Y., & Ranjan, R. (2017). Rendezvous-based routing protocol for wireless sensor networks with mobile sink. *The Journal of Supercomputing, 73*(3), 168–1188. doi:10.100711227-016-1801-0

Shukla, A., Ahamad, S., Rao, G. N., Al-Asadi, A. J., Gupta, A., & Kumbhkar, M. (2021). Artificial Intelligence Assisted IoT Data Intrusion Detection. *2021 4th International Conference on Computing and Communications Technologies (ICCCT)*, (pp. 330-335). IEEE. 10.1109/ICCCT53315.2021.9711795

Sreekanth, N., Rama Devi, J., Shukla, A., Mohanty, D. K., Srinivas, A., Rao, G. N., Alam, A., & Gupta, A. (2022). (2022).Evaluation of estimation in software development using deep learning-modified neural network. *Applied Nanoscience*. doi:10.100713204-021-02204-9

Tripathi, A., Sindhwani, N., Anand, R., & Dahiya, A. (2023). Role of IoT in Smart Homes and Smart Cities: Challenges, Benefits, and Applications. In *IoT Based Smart Applications* (pp. 199–217). Springer. doi:10.1007/978-3-031-04524-0_12

Veeraiah, V., Ahamad, G. P. S., Talukdar, S. B., Gupta, A., & Talukdar, V. (2022) Enhancement of Meta Verse Capabilities by IoT Integration. *2022 2nd International Conference on Advance Computing and Innovative Technologies in Engineering (ICACITE)*, (pp. 1493-1498). IEEE. 10.1109/ICACITE53722.2022.9823766

Veeraiah, V., Khan, H., Kumar, A., Ahamad, S., Mahajan, A., & Gupta, A. (2022). Integration of PSO and Deep Learning for Trend Analysis of Meta-Verse. *2022 2nd International Conference on Advance Computing and Innovative Technologies in Engineering (ICACITE)*, (pp. 713-718). IEEE. 10.1109/ICACITE53722.2022.9823883

Veeraiah, V., Kumar, K. R., Lalitha, K. P., Ahamad, S., Bansal, R., & Gupta, A. (2022). Application of Biometric System to Enhance the Security in Virtual World. *2022 2nd International Conference on Advance Computing and Innovative Technologies in Engineering (ICACITE)*, (pp. 719-723). IEEE. 10.1109/ICACITE53722.2022.9823850

Veeraiah, V., Rajaboina, N. B., Rao, G. N., Ahamad, S., Gupta, A., & Suri, C. S. (2022).Securing Online Web Application for IoT Management. *2022 2nd International Conference on Advance Computing and Innovative Technologies in Engineering (ICACITE)*, (pp. 1499-1504). IEEE. 10.1109/ICACITE53722.2022.9823733

Verma, A., Gupta, A., Kaushik, D., & Garg, M. (2021). Performance enhancement of IOT based accident detection system by integration of edge detection. *Materials Today: Proceedings*. doi:10.1016/j.matpr.2021.01.468

Wang, J. A. D. O. (2017). Blockchain-enabled wireless communications: a new paradigm towards 6G. *American Journal of Roentgenology, 186*(2), 227–236. https://pubmed.ncbi.nlm.nih.gov/28459981

Wu, F., Xu, L., Kumari, S., & Li, X. (2017). A privacy-preserving and provable user authentication scheme for wireless sensor networks based on Internet of Things security. *Journal of Ambient Intelligence and Humanized Computing, 8*(1), 101–116. doi:10.100712652-016-0345-8

Wu, J., Li, J., & Ji, X. (2018). Security for cyberspace: Challenges and opportunities. *Frontiers of Information Technology and Electronic Engineering, 19*(12), 1459–1461. doi:10.1631/FITEE.1840000

Chapter 2
Device–to–Device (D2D) Communication for Advanced Wireless Communication:
CR–Enabled D2D Communication to Enhance and Extend the Wireless Network Performance

Pradeep Kumar

Vellore Institute of Technology, India

Abhijit Bhowmick

Vellore Institute of Technology, India

ABSTRACT

In cellular networks, device-to-device (D2D) communication is a new paradigm. In this chapter, the end-to-end performance (outage, throughput, improvement in spectrum utilization, and energy efficiency) of a non-linear energy harvesting D2D network underlaying a cellular network will be studied under a channel quality constraint. A D2D node harvests energy from RF signal of cellular users (CU) and power transfer units (PTUs) and transmits its information in hybrid mode. Spectrum sensing, censoring of users, and energy harvesting are incorporated to make the study more realistic. Spectrum sensing provides information about the CU spectrum status, censoring technique eliminates the D2D users from the transmission process of those who have bad quality channels. Analytical expressions of an outage, throughput, improvement in spectrum utilization (ISU), and energy efficiency are provided. The impact of the interference threshold on CU, the censoring threshold, and the saturation level of the energy harvester on network performance parameters is addressed.

DOI: 10.4018/978-1-6684-7000-8.ch002

1.1 INTRODUCTION

Due to the rapid growth of wireless operations in recent years, the, base station (BS) of CN suffers from over traffic burden and results in congestion. The D2D communication permits two users to establish direct communication between them and bypassing a BS. This is indifferent from conventional CU network where all transmissions take place through a BS even though the CUs are in close proximity. Thus, the traditional communication method supports such services only that require low-data-rate like phone calls and text. In such case, the D2D communication (DDC) technology can be a better alternative to reduce latency as well as traffic burden from the BS. It can boost network performance and enhance throughput, energy efficiency, spectrum efficiency, and system fairness. D2D communication has been accepted for use in out-of-band wireless sensor networks, but it has not been studied in the in-band transmission during the early cellular communication eras. The 3rd Generation Partnership Project (3GPP) has now legitimized in-band DDC.

1.2 RELATED WORKS

Wireless networks have been experiencing congestion in recent years. The demands for several high data rate services are increasing day by day. As the congestion rises exponentially, now, it becomes very hard to support the new services. On the other side, the available resources, such as unlicensed spectrum bands, are very restricted (Hua et al., 2011). The D2D communication technique utilizes the cellular network resources for their operations and it could be a potential solution to properly reuse resources and increase the coverage range (Prasad et al., 2014). In DDC, local user equipments (UEs) make pairs of D2D users (DDUs). However, it is very challenging to coordinate DDC with existing cellular communication on the same channel. A DDU can use a CU link for its transmission in one of the modes between underlay and overlay.

The CR technology, helps secondary users (SUs) to use underused primary users (PUs) band to transmit information when it suffers from congestion (Ghasemi and Sousa, 2005; Chatterjee et al., 2015). The combination of CR technology and D2D communication technique could provide an additional stability to the communication system. D2D transmission was introduced in (Lin and Hsu, 2000), where authors examined a multi-hop relay network that uses a CN in underlay mode. In DDC, two UEs near to each other, establishes a direct channel between them. It reduce the communication latency and enhances the throughput (Ali et al., 2018). A comprehensive overview of D2D communication is discussed in (Lien et al., 2016) for public safety networks (3GPP Release 12). It is regarded as one of the most promising strategies in 5G communication (Zhang et al., 2017). It provides stable communication and reduces in energy consumption (Prasad et al., 2014; Höyhtyä et al., 2018). In (Fodor et al., 2014; Chu et al., 2017), the authors explored D2D communication to expand the network coverage in areas where disasters or emergencies have occurred. They proposed a dynamic clustering approach that combines both BS and mobile devices.

D2D devices are powered by batteries. These batteries can be recharged or replaced for uninterrupted operation. Replacement or recharging of sensor batteries is challenging in a few applications, such as biodiversity measurement, when a large number of sensors are scattered across a huge area. Thus, the technique like energy harvesting makes a communication node free from energy constraint. The EH technique has evolved as a prominent energy solution for wireless networks (Liu et al., 2017; Yang &

Ulukus, 2011). In (Bhowmick et al., 2015; Prasad et al., 2016), performance of a CR network has been studied where CR user EH is based on sensing decision. If PU is detected, a CR user harvests energy (HsE) from the radio frequency (RF) signal of PU, otherwise it harvests from non-RF signals received from ambient (Bhowmick et al., 2015).

1.3 PROBLEMS IN D2D NETWORK AND NEED OF CR TECHNIQUE, ENERGY HARVESTING, AND CENSORING

- The concept of D2D network (DDN) is evolved to reduce the traffic load on BS and end-to-end latency in a CN. In DDC, a D2D device uses or shares a CU band. Thus, it is much needed to sense a CU channel and adapts transmission policy accordingly. Otherwise, it could cause problems for CU and degrades its quality of service (QoS).
- In a practical cooperative D2D network, some channels are in good conditions and some are heavily faded. The transmissions between the users those who are linked with the bad channels (heavily faded channels) are not useful. Thus, it is better to exclude the users linked with the heavily faded channels from the transmission process to use the resources effectively.
- All the DDUs/sensors are battery powered. In present scenario, it is impractical to power them periodically. To improve the network life time, and to reduce the probability of energy outage, it is needed to make the communication nodes energy independent.

To resolve the problems stated above, some techniques can be adopted namely: (i) **Spectrum sensing** to sense the CU channels (ii) **Censoring** to eliminate bad channels (iii) **Energy harvesting** to make nodes energy independent. In the following sub-sections, analytical modeling of (a) conventional D2D network with spectrum sensing, and (b) Improved D2D network are discussed. Following that, a detailed analysis is made on the basis of the obtained results, followed by a conclusion and recommendations for further work.

1.4 D2D COMMUNICATION FOR NETWORK EXTENSION IN THE DISASTER/EMERGENCY ZONE

In recent years, disasters and calamities have become very common due to global warming and climate change. Despite being one of the most commonly used communication infrastructures, the cellular communication system is fragile as it relies heavily on the base station (BS), mobile switching centre (MSC), etc., for communication. During disasters, the BS and MSC have a higher chance of failing, disrupting the entire region's transmission. Moreover, a quick and effective way to re-establish the communication network is required in such situations. In this era where every individual possesses many wireless devices and is surrounded by IoT-enabled gadgets, D2D communication can be a potential alternative during disasters or in emergency situation. D2D communication doesn't require infrastructure and can quickly re-establish communication networks during natural disasters or in emergency. Many researchers have studied the D2D communication in different network scenarios and proposed their work to improve the system performance.

In (Raza et al., 2020), a D2D-based system which uses an ad hoc clustering strategy to establish communication during natural disasters is studied. The proposed ad hoc clustering method can effectively improve communications in affected areas. An optimization strategy was proposed to improve end users' throughput and localization performance. The paper (Moghaddam et al., 2018) proposes a cognitive radio-enabled D2D communications-based disaster response network to activate necessary communication in the disaster zone. A multi-hop methodology based on clustering topology was used to connect the two devices in a relaying network. The authors showed that the D2D devices can communicate without any infrastructure support using relay selection and cooperative beamforming. In (Ali et al., 2018), the authors studied the wireless signal and power transfer technique to extend the network's lifetime. They proposed that D2D communications are carried out through harvested energy from the RF signal of a BS. The cellular network is combined with clustering-based D2D communications to provide service even when cellular infrastructure break downs. The observations showed that the proposed EH D2D clustering-based system can efficiently extend the coverage in the area of interest.

In the works mentioned above, D2D communications is used to develop an emergency communication network by adding several coverage enhancements. Researchers have also examined D2D communications with other networking techniques to enhance communication in the affected area. In (Firozjae et al., 2022), a CR enabled D2D communication is investigated as a disaster response network where Unmanned Aerial Vehicle (UAV)-supports in information forwarding. In (Nishiyama et al., 2014), a testbed has been proposed considering natural disasters like those in Great East Japan, for a multi-hop DDC. The prototype has been designed successfully using smartphones. It has been demonstrated that the developed model enable users to communicate messages using smartphones, computers, and tablet PCs without any cellular infrastructure support. From the previous discussion, it's clear that DDC can play an important role in extension or deployment of network in a zone that is under disaster or emergency (Al-Hourani et al., 2016).

1.5 CR ENABLED D2D NETWORK

1.5.1.1 Conventional D2D Network With Spectrum Sensing (SS)

The system model of a conventional D2D network with SS is shown in Fig.1 and there are N numbers of D2D pairs, and a CU. Here, $D_{i,Tx}$ represents i-th the D2D transmitter (DDT) and $D_{i,Rx}$ represents i-th D2D receiver (DDR), where $i = 1,2,3...N$. The channels between the communication nodes are considered to be Rayleigh faded with independent and identical distributions. Here, $h_{i,cd}$ denotes the channel or link between CU and the i-th DDT, $h_{i,dd}$ represents the link between i-th DDT and i-th DDR. The link gains $g_{i,cd}$ and $g_{i,dd}$, respectively, from the channels $h_{i,cd}$ and $h_{i,dd}$, where $g_{i,cd}=|h_{i,cd}|^2$ and $g_{i,dd}=|h_{i,dd}|^2$.

The DDUs are battery powered and exploit TDMA technique for their transmission through a CU channel. The channel's QoS cannot be permitted to fall below a specific threshold because it belongs to the CU. Thus, the DDUs are therefore needs to be enabled with CR technology. DDUs equipped with CR technology first detects the status of CU in the channel and then choose the mode of transmission based on the CU's presence in the given channel. As a result, DDUs select to communicate in hybrid mode (either underlay mode or overlay mode). In underlay condition, channel is already occupied by the CU and DDU has to share it while in overlay condition CU is not using the channel and DDU can use it. All DDT detects the CU link, makes an individual decision, and communicate it to FC. As a central unit, FC receives the individual decisions from all DDTs via feedback channels (channel among DDTs and FC) and takes the overall final decision by combining all DDT's individual decisions using a fusion rule (OR rule). The feedback channel is then used to communicate the overall decision to the DDTs. The communication time frame structure for the considered network consists of sensing time ($\tau s_{)}$, and transmission time (T$r_{)}$, T=τs+$_{)}$r. The communication frame structure is given below (B. Prasad et al., 2016; Liang et al., 2008).

Figure 1. Conventional D2D network with spectrum sensing.

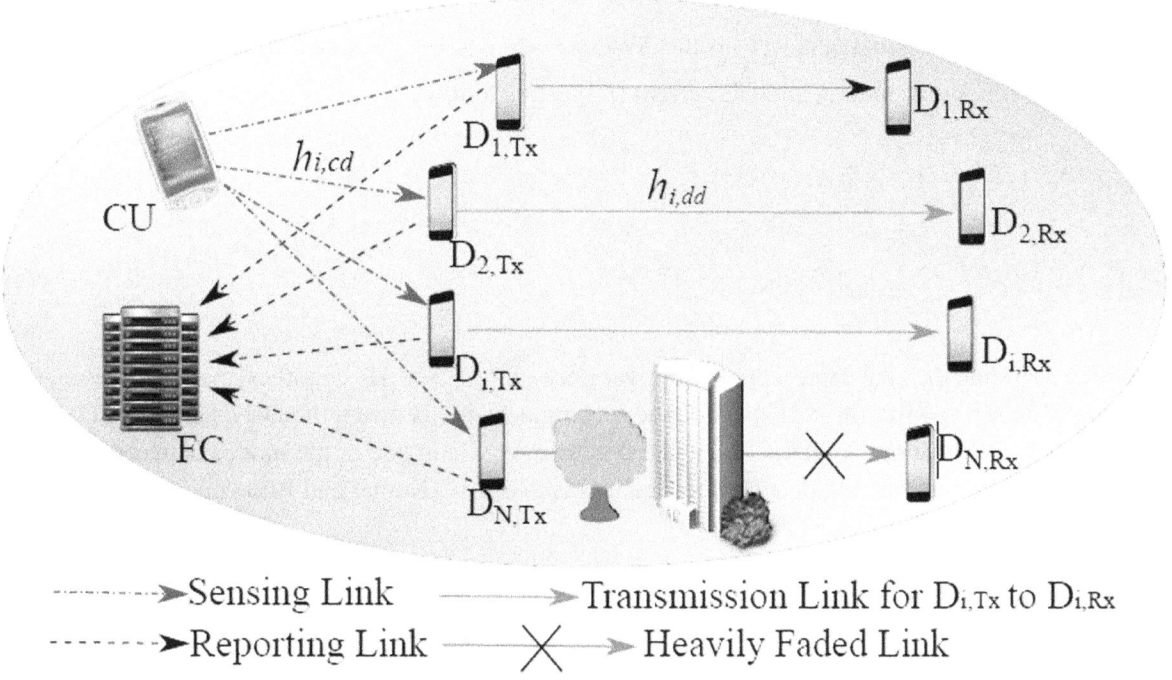

Table 1. Communication time frame structure

sensing time (τs$_{)}$	transmission time ($T_{)}$

1.5.1.2 Spectrum Sensing (SS) Process

The activity of CU affects the overall sensing decision. The CU is assumed to have a random activity model according to that the busy state (presence of CU) as well as idle state follows an exponential distribution with mean values of $\alpha 0$ for H0 hypothesis and $\alpha 1$ $_{f}$or H1 $_{h}$ypothesis respectively. Here, the H0 $_{i}$ndicates the absence of CU, i.e., idle state, while H1 $_{i}$ndicates the busy state, i.e., the CU is active in the given channel. The PDFs for busy (Pb($_{)}$)) and idle (Pi($_{)}$)) states are as follows (Bhowmick et al., 2020):

$$\begin{cases} P_i(t) = \alpha_0^{-1} e^{-t/\alpha_0} \\ P_b(t) = \alpha_1^{-1} e^{-t/\alpha_1}. \end{cases} \tag{1}$$

During sensing, the signal received at a DDU can be expressed as (Bhowmick et al., 2020),

$$y(\tau_s) = \begin{cases} w(\tau_s); & H_0 \\ x(\tau_s) h_{i,cd} + w(\tau_s); & H_1 \end{cases} \tag{2}$$

where $x(\tau_s)$ is the CU signal, and $w(\tau_s)$ is the AWGN with $N\left(0, \sigma_w^2\right)$. The sensing decisions in terms of probability of detection (P_d), and false alarm (P_{fa}) are as follows:

$$P_d(\tau_s) = P(H_1 \mid H_1) = \left\{ Q[(\lambda - \mu_1)/\sigma_1] \right\}^N, \tag{3}$$

$$P_{fa}(\tau_s) = P(H_1 \mid H_0) = \left\{ Q[(\lambda - \mu_0)/\sigma_0] \right\}^N, \tag{4}$$

where (μ_0, σ_0) and (μ_1, σ_1) represents (mean, variance) for H_0 and H_1 hypothesis respectively and λ represents the sensing threshold [17]. Once the sensing decision is made, the DDUs exploit TDMA to transmit their information. A DDUs utilize the spectrum of CU in one of the modes between underlay and overlay. Therefore, the total network throughput is given by (Kumar and Bhowmick, 2021),

$$T_{total} = P(H_1) R_{un} + P(H_0) R_{ov} \tag{5}$$

where R_{un} and R_{ov} denote the throughput for underlay and overlay modes,

$$R_{ov} = \frac{T_r}{T} \cdot \left(1 - P_{fa}(\tau_s)\right) \ \log_2\left(1 + \frac{P_D g_{i,dd}}{\sigma_w^2}\right), \tag{6}$$

$$R_{un} = \frac{T_r}{T} P_d(\tau_s) \log_2\left(1 + \frac{P_D^*}{\sigma_w^2 + P_{CU} g_{i,cd}}\right), \tag{7}$$

where, P_D denotes the transmission power (TxP) of DDU, σ_w^2 denotes the noise variance, P_{CU} denotes the TxP of CU, the received power at a DDR is $P_D g_{i,dd}$, $P_{CU} g_{i,cd}$ is the interfering CU power at DDR, and P_D^* denotes the controlled transmission power in underlay mode.

$$P_D^* = I_{th} \cdot \frac{\beta}{\widehat{g_{i,cd}}}, \tag{8}$$

where, β denotes the scaling factor such that $0 < \beta < 1$, I_{th} denotes the interference threshold, and $\widehat{g_{i,cd}}$ denotes estimated coefficient of CU-D2D channel ($\widehat{g_{i,cd}} = |\widehat{h_{i,cd}}|^2$) where $|\widehat{h_{i,cd}}|^2 = \rho h_{i,cd} + \xi \sqrt{1-\rho^2}$ and $0 \le \rho, \xi \le 1$ (Ali et al., 2018).

1.5.1.3 Outage Analysis

Whenever, the network total throughput drops down to a targeted rate R_{th}, resulting in outage. Thus, the outage of the network can be given by,

$$P_{out} = P(T_{total} \le R_{th}).$$ (9)

On substituting the expressions related to throughput from (5), (6), and (7) in (9), and using some algebra, the expression of outage can be rearranged as

$$P_{out} = \left[1 - e^{-\frac{K}{\gamma \Delta \Delta}}\right],$$ (10)

where

$$K = \frac{(2^m - 1)}{P_D}, \quad m = \frac{\frac{TR_{th}}{T_r} - P_d(\tau_s)P(H_1)C_b}{(1 - P_{fa}(\tau_s))P(H_0)},$$

and $\overline{\gamma}$ is the average channel SNR.

1.5.1.4 Optimal Sensing Time (τ_s^*)

The expression of outage obtained in (10) can be reformulated as

$$P_{out} = 1 - C_1 e^{-2^{\frac{C_2}{(1-P_{fa}(\tau_s))}}},$$ (11)

where

$$C_1 = e^{-\frac{1}{A}}, \quad A = P_D \overline{\gamma}, \quad C_2 = \frac{\frac{TR_{th}}{T_r} - P_d(\tau_s)P(H_1)C_b}{P(H_0)}.$$

After double differentiation of (11) with respect to τ_s, the second derivative of outage can be written as,

$$P_{out}'' = P_{out}' \left(\frac{1}{(1-P_{fa}(\tau_s))^2} \right) \left[C_2 \ln(2) 2^{\frac{C_2}{1-P_{fa}(\tau_s)}} P_{fa}'(\tau_s) - C_2 \ln(2) \right. $$

$$\left. + \left(P_{fa}''(\tau_s)(1-P_{fa}(\tau_s))^2 + 2(1-P_{fa}(\tau_s))(P_{fa}'(\tau_s))^2 \right) / P_{fa}'(\tau_s) \right] \tag{12}$$

In practical view $P_{fa}(\tau_s) \le 0.5$ therefore,

$$\tau_s \ge \frac{1}{f_s} \left(\frac{(2\gamma_s+1)Q^{-1}(P_d(\tau_s))}{\gamma_s} \right)^2 .$$

Here, P_{out}'' is positive for

$$\frac{1}{f_s} \left(\frac{(2\gamma_s+1)Q^{-1}(P_d(\tau_s))}{\gamma_s} \right)^2 \le \tau_s \le \tau_c$$

where f_s denotes sampling frequency, and γ_s denotes the received SNR. Thus, P_{out} is a convex function in a range of τ_s. Therefore, there is an optimal value of τ_s, i.e. τ_s^* for which the outage is minimum.

1.5.1.5 Results and Analysis

Now, on the basis of previous discussion, results and related analysis are presented in. In order to study network performance, the following parameters are set to their default values: $f_s = 6\text{MHz}$, $\sigma_1^2 = \sigma_2^2 = 1$, $\alpha_1 = 0.7$, $\alpha_2 = 0.3$, T = 100ms, $P_{td} = 0.9$, $SNR_c = -16$ dB unless specified. The throughput performance obtained for faded links is compared with without-faded links which is shown in Fig. 2. The network performance is studied for $P(H_0) = 0.8$, $P(H_1) = 0.2$, $SNR_s = 20\text{dB}$, $SNR_c = -15\text{dB}$, and $I_{th} = 16\text{dB}$. It is noticed that initially throughput increases as the sensing time increases, attains a maximum value, and then reduces with the increase in sensing time. This indicates that throughput is highest at an optimal sensing time. In addition, throughput performance is significantly affected by channel fading. When there is fading in the channel, the total throughput reduces at a fixed sensing time. The results obtained in simulation matches well with the theoretical results and validates the simulation model. The result obtained for throughput of the present study is compared with the throughput obtained in (Liang et al., 2008).

The outage of the network is influenced by the interference threshold (I_{th}) (Fig. 3). It is noticed that the outage decreases with increases in I_{th}. Moreover, for given values of I_{th}, the outage is observed to be minimum for a particular value of τ_s. The acceptable transmit power for DDUs reduces as I_{th} decreases at a fixed value of sensing time, which results in an increase in outages.

In the previous section, the conventional DDN is not equipped with the techniques such as SS, censoring, and energy harvesting. These all three techniques can be considered to improve network performance. Thus, in the next section, performance of an improved D2D network with harvesting and censoring is studied.

Figure 2. Throughput of D2D network for faded and without faded channels.

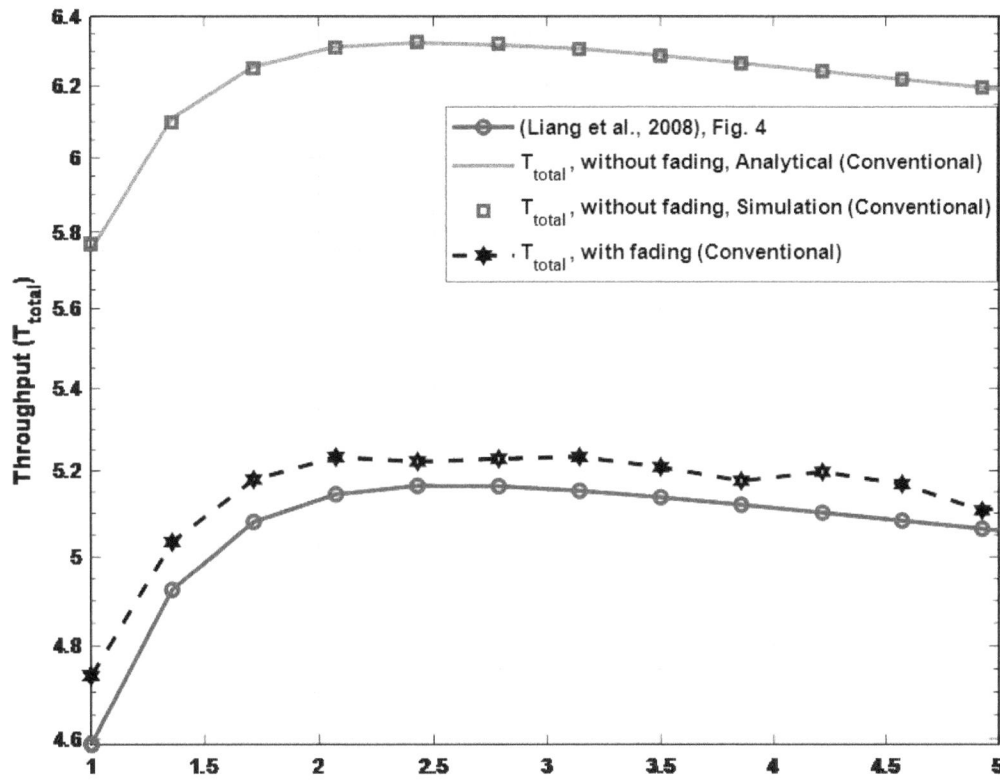

Figure 3. Outage with change in interference threshold (I_{th}).

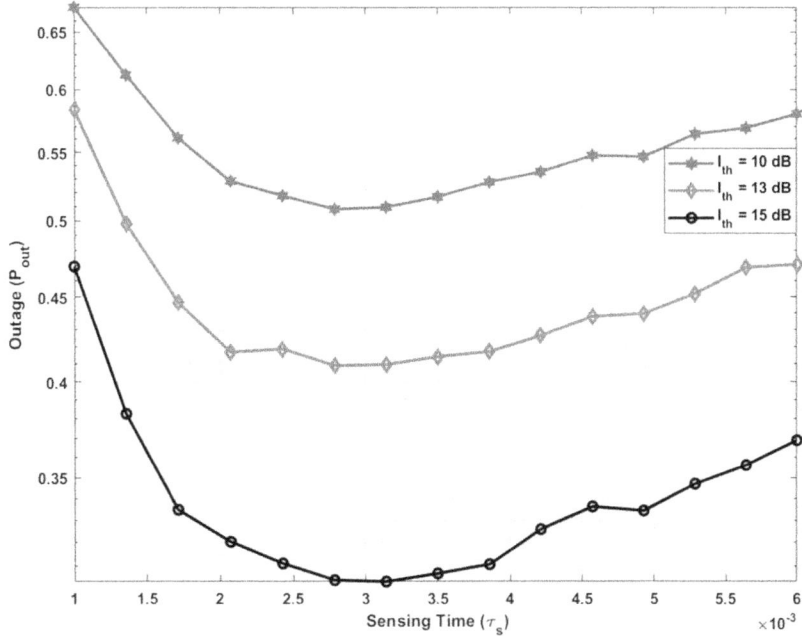

1.5.2 Improved D2D Network With Energy Harvesting and Censoring

In this section, energy harvesting and censoring schemes are included in this present study. The network model is shown in Fig. 4 and 5. Along with the DDUs, and a CU, There are Z numbers of power transfer units (PTUs) distributed over a geographic region to power the DDUs.

Figure 4. Sensing of CU channel in cooperative D2D network.

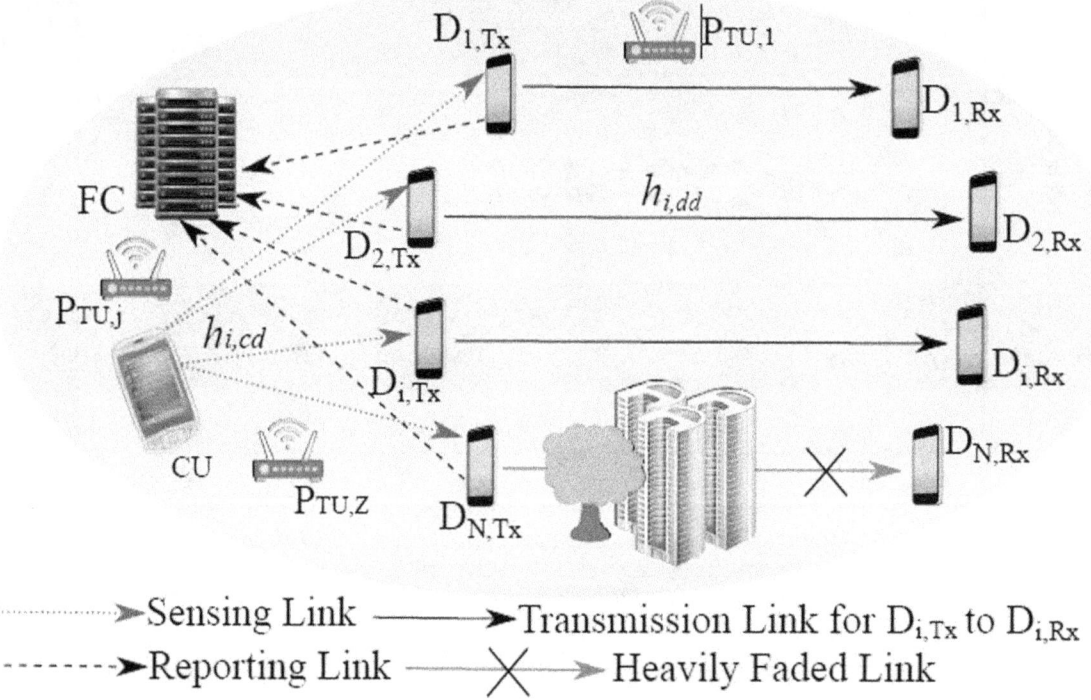

As mentioned in the above sections, DDUs sense the CU channel before using it. After sensing, they harvest energy either from the CU signal or a PTU. The communication time frame structure for the studied network consists of sensing time (τ_s), energy harvesting time (τ_h), and transmission time (T_r), T= $\tau_s+\tau_h+T_r$. The communication time frame structure is given below (Nishiyama et al., 2014).

In practical scenario, the connection between a DDT and a DDR may be faded. If the link is substantially faded, the received signal at the receiver may vary from transmitted signal. If the channel condition is very bad, in that circumstance, the DDT should stop transmission and allow other DDTs to transmit. Thus, it is much needed to censor the bad channels and eliminate the users those who are associated with the bad channels. During censoring, DDTs with poor transmission channel quality are removed from the transmission process for that particular time frame. In the next subsections, the sensing process is discussed followed by censoring and energy harvesting.

Figure 5. Harvesting from RF signal of CU and PTU.

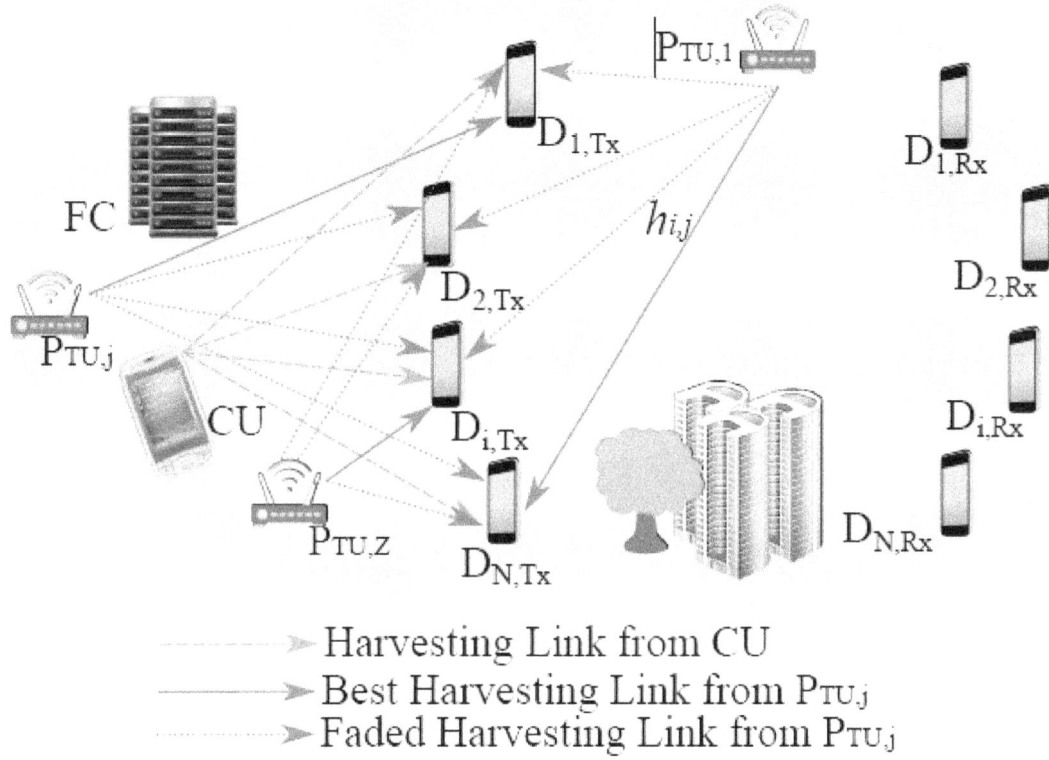

Table 2. Communication time frame structure

sensing time (τ_s)	harvesting time (τ_h)	transmission time (T_r)

1.5.2.1 Sensing with Censoring

During censoring time, every individual DDT transmits a parity bit (W_i) to its paired DDR, where $W_i = \sqrt{E_b}$, E_b represents a training symbol bit. The receiver estimates the CSI using minimum mean squared error and then sends back the received CSI to its paired DDT. It is considered that the channel remains stable during the estimation time. Thus, the estimated channel coefficient ($\widehat{h_i}$) can be expressed as (Nallagonda et al., 2013),

$$\widehat{h_i} = E\left[h_{i,dd} \mid y_i\right] = \frac{\sqrt{E_b}}{E_b + \sigma_n^2} y_i , \tag{13}$$

where $i = 1, 2,..N$, $y_i = h_{i,dd}W_i + n_i$ where n_i is the AWGN noise with $\left(0, \sigma_n^2\right)$. Channel estimation error ($h_{i,e}$) is defined as the difference of the actual coefficient of link (h_i) and the coefficient of estimated link

($\hat{h}_i$), $h_{i,e} = h_i - \hat{h}_i$. It considered that the error of estimation is a complex Gaussian (CG) with $\left(0, \sigma_{\tilde{h}}^2\right)$ (Nallagonda et al., 2013),

$$\sigma_{\tilde{h}}^2 = \left[\frac{E_b}{\sigma_n^2} + 1\right]^{-1}.$$

(14)

Here, a DDUis allowed to transmit in the scheduled time frame only if the amplitude of the estimated channel coefficient ($\hat{h}_i$) exceeds a specific threshold (C_{th}). Here, C_{th} denotes the lowest acceptable linkl quality. As a result, the probability of allowing a channel for transmission can be expressed as (Nallagonda et al., 2013):

$$P_{CS} = P\left(\left|\hat{h}_{i,dd}^{E}\right| > C_{th}\right) = exp\left(-\frac{C_{th}^2}{2\sigma_n^2}\right).$$

(15)

As there are N DDU pairs, only those having good channels are permitted to schedule their transmission. Let be the number of good channels is u out of N channel. Hence, selection of u channels probability can be defined as (Nallagonda et al., 2013),

$$P(u) = \frac{N!}{(N-u)!u!}[P_{CS}]^u [1 - P_{CS}]^{(N-u)}.$$

(16)

The average f cooperative detection probability and false alarm for a given $P(u)$ can be expressed as (Kumar and Bhowmick, 2022),

$$\overline{Q_d} = \sum_{u=1}^{N} P(u) Q_d(u),$$

(17)

$$\overline{Q_f} = \sum_{u=1}^{N} P(u) Q_f(u).,$$

(18)

where $Q_d(u)$ and $Q_f(u)$ are given below:

$$Q_d(u) = 1 - \left[1 - P_{d,i}(\tau_s)\right]^u,$$

(19)

$$Q_f(u) = 1 - \left[1 - P_{f,i}(\tau_s)\right]^u,$$

(20)

where $P_{d,i}(\tau_s)$ and $P_{f,i}(\tau_s)$ represent local detection and false alarm probabilities respectively (Kumar & Bhowmick, 2022).

1.5.2.2 Energy Harvesting

All DDTs HsE from the CU signal and PTUs, as shown in Fig. 5, during the harvesting time (τ_h). In this network, PTUs of Z numbers are spread over a specific region. Each DDU is assumed to be connected to all Z PTUs. The channels between PTUs and DDUs are assumed to be faded. Here, $h_{(i,j)}$ describes the link between i-th DDU and j-th PTU and $g_{(i,j)}$ denotes the channel gain of the corresponding channel such that $g_{(i,j)}=|h_{(i,j)}|^2$ where $i = 1, 2, ...N$ and $j = 1, 2, ...Z$. The DDU harvests energy from a particular PTU while the channel gain between them is maximum as compared to the other channel gains.

The DDTs harvests from CU signal when it is available in the given channel else it harvests from signal of a PTU. Therefore, the energy harvested at the i-th DDT can be given by,

$$E_{h,i} = \eta \left[P(H_0)\left(1-\overline{Q_f}\right)P_{ts}g_{max} + P(H_1)\overline{Q_d}P_{CU}g_{i,cd} \right]t_h, \tag{21}$$

where P_{ts} denotes PTU's transmit power and $g_{max}=max(g_{i,j})$; η is the parameter for energy conversion, $0 \leq \eta \leq 1$. In reality, harvesting circuit is nonlinear since the energy harvester's circuits are designed with semiconductor components and there is a saturation energy level (P_{max}). Let all the DDUs are equipped with same type of harvesting devices. Therefore, the expression for effective harvested energy is given by (Kumar and Bhowmick, 2022),

$$E_{h,i} = \begin{cases} \eta M t_h, & if \ M \leq P_{max} \\ \eta P_{max}t_h, & if \ M \geq P_{max} \end{cases} \tag{22}$$

where,

$$M = \left[P(H_0)\left(1-\overline{Q_f}\right)P_{ts}g_{max} + P(H_1)\overline{Q_d}P_{CU}g_{i,cd} \right]$$

and at a DDT, it the highest possible harvested power.

1.5.2.3 Throughput

During transmission time, DDTs transmit messages to their peers by exploiting TDMA scheme after the sensing, censoring, and harvesting periods. DDTs communicate in or in overlay. DDTs are restricted to transmit with limited power in underlay mode since they share the channel with CU. Thus, their transmit power should not exceed the certain interference limit (I_{Th}) at CU in underlay mode. On contrary, they are allowed to transmit with maximum power in overlay mode. Therefore, the total throughput (TT) of the DDN is the combination of throughput obtained in underlay and overlay transmission. Hence, the TT for the DDN can be written as (Kumar and Bhowmick, 2022),

$$T_{tot} = \begin{cases} \dfrac{(N-L)T_r}{NT}\Big[P\big(H_1\big)R_{un}+P\big(H_0\big)R_{ov}\Big], & \text{Conventional} \\[4mm] \dfrac{T_r}{T}\Big[P\big(H_1\big)R_{un}+P\big(H_0\big)R_{ov}\Big], & \text{Proposed} \end{cases} \tag{23}$$

where R_{un} and R_{ov} are the throughput obtained in underlay and overlay modes. According to the conventional scenario, the transmission time T_r is evenly distributed among all N DDTs. If L DDTs have severely faded channels where $L \leq N$, then in reality, only $(N-L)$ channels are contributing to the effective throughput (useful throughput) which is equivalent to the throughput obtained for $[(N-L)T_r]/N$ unit time. In the improved scheme, all L users are eliminated from the transmission process and T_r is distributed evenly among $(N-L)$ DDTs. The expression of R_{ov} and R_{un} can be expressed as follows:

$$R_{ov} = \big(1-\overline{Q_f}\big)\left(log_2\left(1+\frac{P_{TDO}g_{i,dd}}{\sigma_d^2}\right)\right), \tag{24}$$

$$R_{un} = \overline{Q_d}\left(log_2\left(1+\frac{P_{TDU}g_{i,dd}}{P_{CU}g_{i,cd}+\sigma_d^2}\right)\right), \tag{25}$$

where $P_{TDO} = \dfrac{\widehat{E_{h,i}}}{T_r}$ is the TxP in overlay mode and σ_d^2 is the variance of noise, and P_{TDU} is the TxP in underlay mode.

$$P_{TDU} = \min\big(P_{TDO}, P_{allow}\big) = min\left(\frac{\widehat{E_{h,i}}}{T_r}, \frac{\eta_d I_{th}}{\widehat{g_{i,cd}}}\right), \tag{26}$$

where, P_{allow} is the permitted transmission power for the DDT under I_{Th} (B. Prasad, Roy, et al., 2014); and ηd is the scaling factor such that $(0 \leq \eta_d \leq 1)$, and $\widehat{g_{i,cd}}$ is given by

$$\widehat{g_{i,cd}} = \left|\widehat{h_{i,cd}}\right|^2, \tag{27}$$

where $\widehat{h_{i,cd}}$ denotes the approximated channel coefficient between CU and i-th DDT (B. Prasad, Roy, et al., 2014).

$$\widehat{h_{i,cd}} = \rho h_{i,cd} + \varepsilon\sqrt{1-\rho^2}, \tag{28}$$

where, ρ is the coefficient of correlation and ε is the CG random variable with $\left(0, \sigma_\varepsilon^2\right)$ such that $(0 \leq \rho, \varepsilon \leq 1)$.

1.5.2.4 Improvement in Spectrum Utilization (ISU)

In EH and censoring-based (improved network) network, spectrum is utilized better as compared to the conventional techniques. The expression for ISU can be expressed as (Kumar and Bhowmick, 2022),

$$
\begin{aligned}
I_{SU} &= \frac{T_{total}\big|_{Proposed} - T_{total}\big|_{conventional}}{T_{total}\big|_{Proposed}} \\
&= \frac{\dfrac{T_r}{T}\Big[P\big(H_1\big)R_{un} + P\big(H_0\big)R_{ov}\Big] - \dfrac{(N-L)T_r}{NT}\Big[P\big(H_1\big)R_{un} + P\big(H_0\big)R_{ov}\Big]}{\dfrac{T_r}{T}\Big[P\big(H_1\big)R_{un} + P\big(H_0\big)R_{ov}\Big]}.
\end{aligned}
\tag{29}
$$

1.5.2.5 Energy Efficiency

Energy-efficiency (EE) can be described as the ratio of the throughput and consumed energy. There are four types of energy consumptions: (i) E_C for circuit usage, (ii) E_S for spectrum sensing, (iii) E_h for running energy harvester, and (iv) E_{tr} for data transmission. Here, $(E_S + E_C)\tau$ denotes the energy consumption during τ_s time, $(\xi E_{tr} + E_C)(T_f - \tau)$ denotes the amount of required energy for data transmission, $\tau = \tau_s + \tau_h$. Here, ξ is the inverse of the efficiency of a amplifier. Therefore, the total energy consumption during a communication frame time can be expressed as,

$$
E_{tot} = \sum_{u=1}^{N-L} E_{h,i} = \big(E_S + E_C\big)\tau + E_h \tau_h + \big(\xi E_{tr} + E_C +\big)\big(T - \tau_S - \tau_h\big).
\tag{30}
$$

The EE (η_{eff}) of the DDN can be written as,

$$
\eta_{eff} = \frac{T_{tot}}{\sum_{i=1}^{(N-L)} E_{h,i}}.
\tag{31}
$$

1.5.2.6 Algorithm Maximization of Energy Efficiency

Algorithm: Determination of optimal sensing time (τ_s^*) to maximize the Energy Efficiency

Require: Length of detection cycle (T), censoring threshold (C_{th}), targeted rate R_{th}, simulation times, parameters related to sensing
    1.   Sensing time (τ_s) = 1 ms: 6 ms (n values)

2. **for** $i = 1 : n$
3. count = 0;
4. DDUs sense the CU link and take their local sensing decisions.
5. Local decisions are fused at FC to take overall final sensing decision D (H_0 or H_1).
6. Repeat the steps 1 to 2 until the sensing time τs is completed.
7. During censoring time, all DDT estimates the channel coefficient ($\hat{h}_i$)
8. **If** $\hat{h}_i > C_{th}$
9. DDU associated with that channel is selected for transmission
10. **else**
11. DDU associated with that channel is eliminated from the transmission process.
12. Repeat the steps 8 or 10 until the censoring time τc is completed. After censoring, the selected DDUs harvest energy
13. **end if**
14. **If** $D = H_0$
15. Nodes harvest energy from PTU
16. **else**
17. Nodes harvest energy from CU signal.

Total harvested energy $E_{tot} \leftarrow \sum_{u=1}^{N-L} E_{h,i}$ is calculated using (22) and (29).

18. **end if**
19. **If** $D = H_0$
20. DDUs transmit information in overlay mode, i.e., *Throughput* $\leftarrow P(H_0)T_{over}$.
(DDUs transmit with maximum power)
21. **else**
22. DDUs transmit information in underlay mode, i.e., *Throughput* $\leftarrow P(H_1)T_{under}$.

(DDUs transmit with controlled power $P_D^* = I_{th} \cdot \dfrac{\beta}{\widehat{g_{i,cd}}}$)

23. **end if**
24. Step 18 or 22 continues for transmission time T_r.
25. Total network throughput is given by, $T_{total} = P(H_1)T_{under} + P(H_0)T_{over}$.

26. $\eta_{eff} \leftarrow \dfrac{T_{tot}}{\sum_{i=1}^{(N-L)} E_{h,i}}$

27. **end**
28. Find τ_s^* for $max\{\eta_{eff}\}$

1.5.2.7 Results and Analysis

In this section, results obtained for harvesting and censoring scenario are discussed below. Here, Table-3 shows the simulation parameters.

Table 3.

Parameter	Value
f_s	6 MHz
$\sigma^2_0 = \sigma^1_0$	1
α_0	0.64
α_1	0.36
P_{CU} W	1 W
T	100 ms
P_{LS} W	1 W
Z	5
τ_h	1 ms
SNR	-16 dB
Target detection probability (P^*_d)	0.9

In Fig.6, TT is studied for C_{th} = 0.7, 1.4, N =10, γ_s = -16 dB, ε = 0.3, ρ = 0.99, η = 0.9, I_{th}= 5 dBW. It is observed that the TT follows a concave pattern with sensing time. Thus, there is an optimal τ_s. As C_{th} increases, the throughput of conventional system decreases whereas for the improved network model it remains likely to same. As the value of C_{th} increases, the required acceptable quality of link increases. As C_{th} increases, in conventional model, more number of channels falls below the acceptable quality and useful throughput reduces (due to bad impact of fading). Whereas, in the improved network scenario, since the DDUs associated with the heavily faded channels are removed from the communication process, now complete transmission time is allotted to those DDUs having good channels resulting in higher throughput. With the increases in C_{th}, a greater number of DDUs are eliminated from transmission. Since the entire transmission time is utilized among the allowed DDUs only, the throughput remains same.

In Fig. 7, TT is studied for both conventional and improved network for a range of censoring threshold (C_{th}). It can be noticed from Fig. 7 that at a fixed value of C_{th}, TT increases with increases in I_{th}. Increase in I_{th}, allow users to transmit with greater power. In general, high power transmission improves the received SNR as well as channel capacity. Thus, TT improves with increases in I_{th}. Impact of γ_s (lower region) on TT is shown in Fig. 8 and it is observed that increase in γ_s results in a significant improvement in TT.

In Fig. 9, effect of scaling factor (η_d) on the variation of throughput is studied. It is found that for a fixed value of τ_s, the TT improves as η_d increases. As η_d increases, allowable interference threshold to CU increases. Thus, DDUs can transmit with high power which results in higher throughput.

ISU is studied in Fig. 10. It is noticed that for any particular number of cooperative DDUs (N), ISU improves as C_{th} increases. It is also observed that for a particular C_{th}, I_{SU} improves as the N increases. EE, i.e., η_{eff} of the considered DDN is studied in Fig. 11. The ηeff improves as η_d increases. There is no significant impact of Cth on η_{eff} and there is an optimal τ_s for *which* the η_{eff} is maximum.

Figure 6. Impact of C_{th} on total throughput (TT).

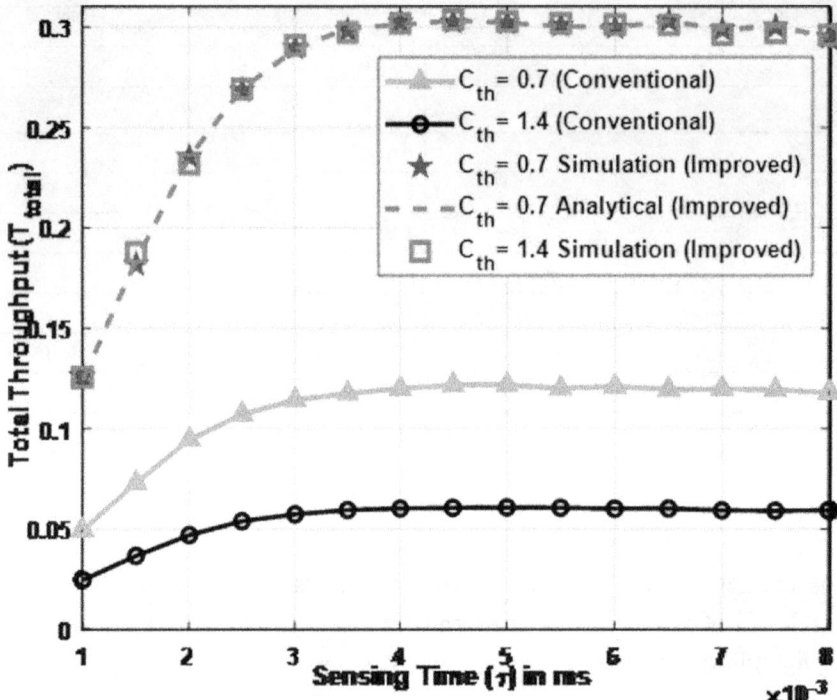

Figure 7. Impact of C_{th} on total throughput (TT).

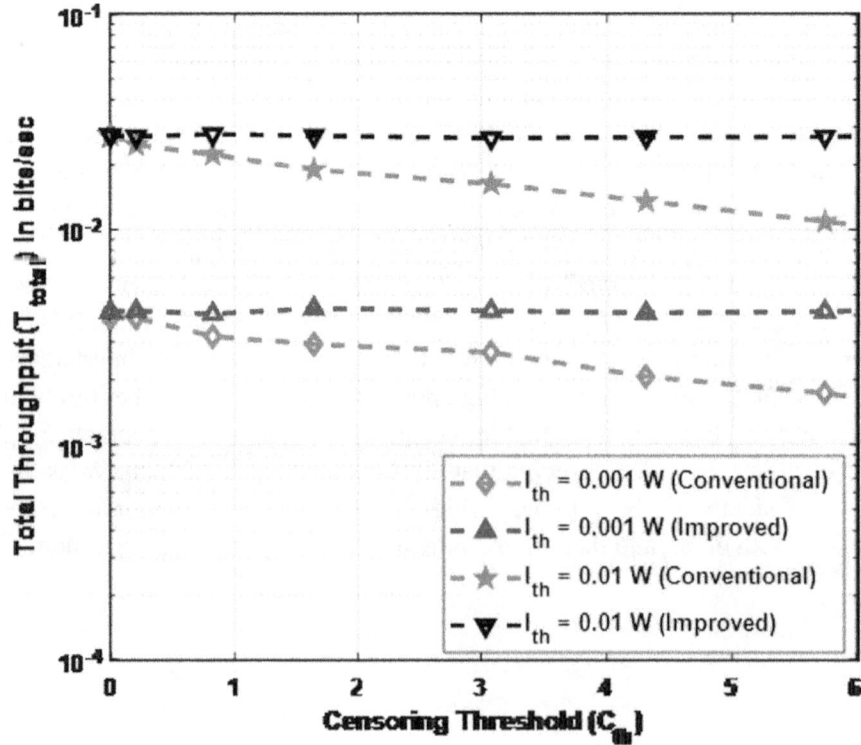

Figure 8. Impact of C_{th} on total throughput.

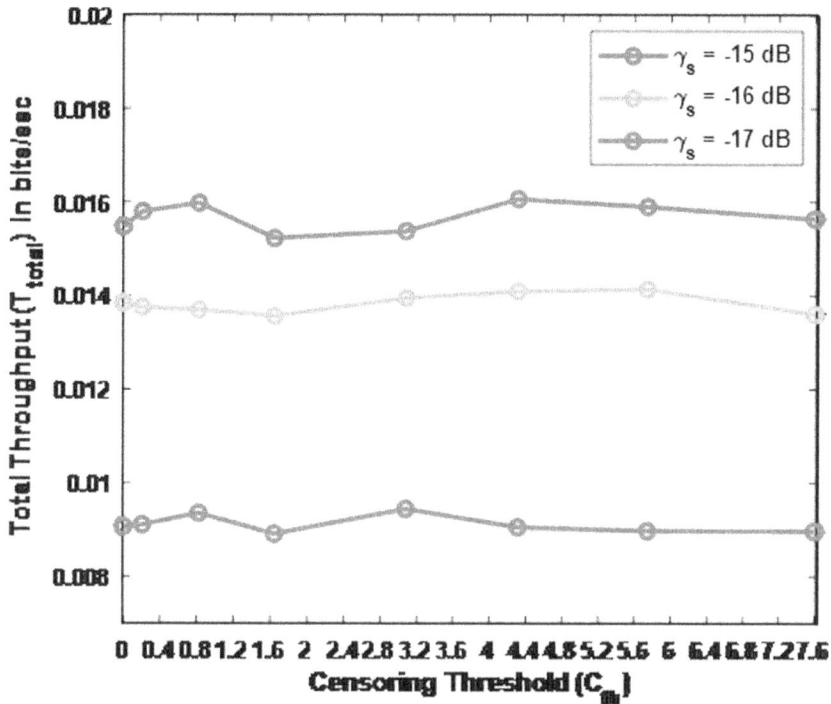

Figure 9. Impact of scaling factor on total throughput (TT).

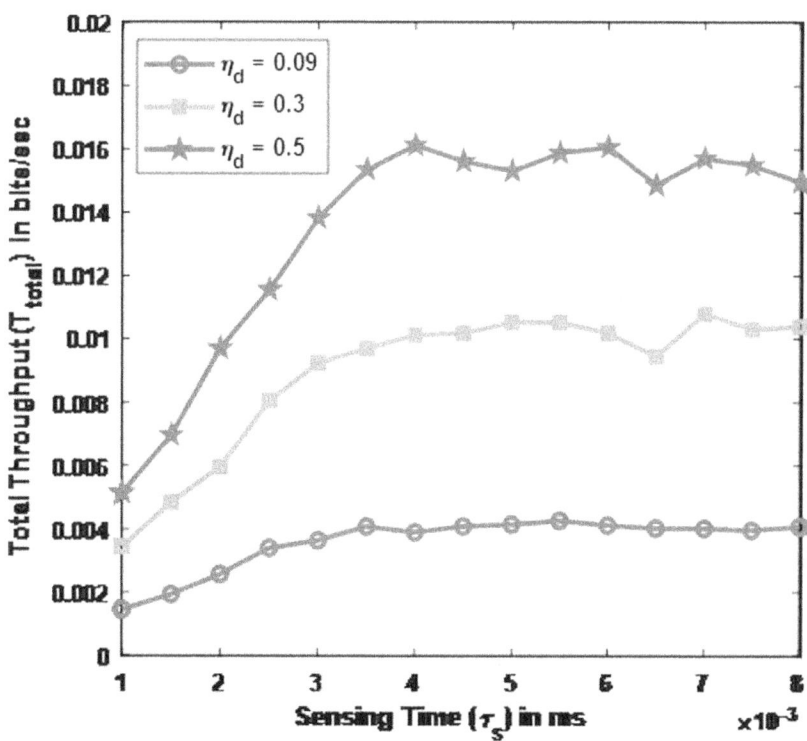

Figure 10. ISU vs. censoring threshold.

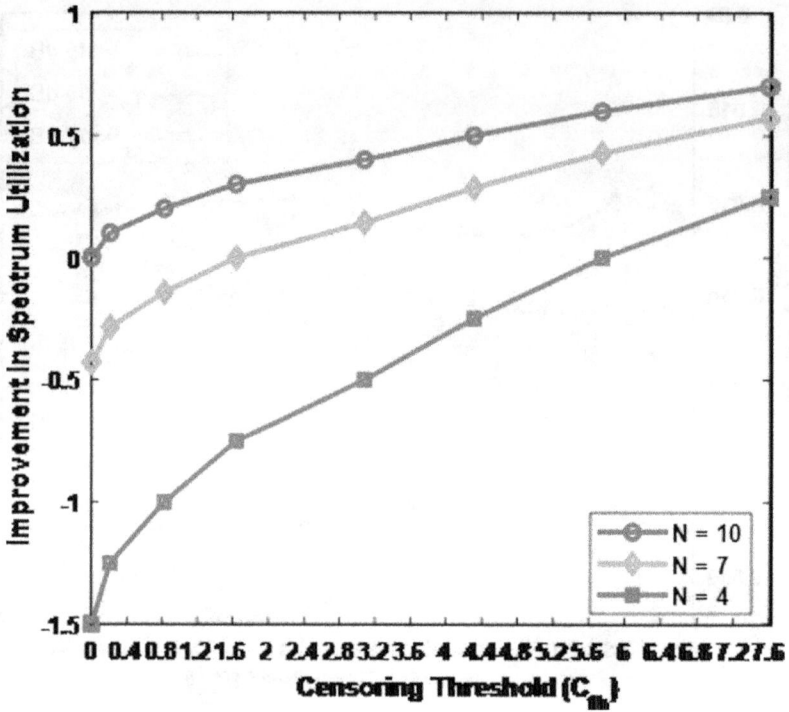

Figure 11. Energy efficiency vs. sensing time.

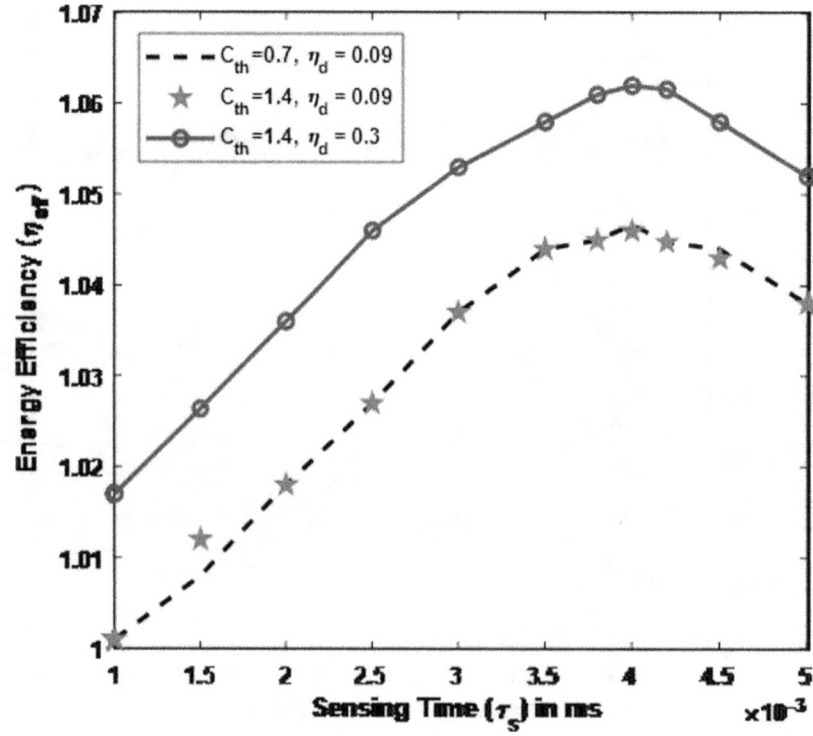

1.6 CONCLUSION AND FUTURE WORK

The present chapter study both conventional and improved CR-enabled D2D networks in which CU spectrum is shared among the DDUs in any of the modes between underlay and overlay, depending on the CU status. In conventional network, communication between DDUs may be interrupted due to the energy constraints of Devices. Also the transmission time allotted to those users which are having heavily faded channels is wasted. Therefore, energy-harvesting of the D2D devices and censoring of heavily faded channels are added in the improved network. A DDU can harvest energy wirelessly from of either CU signal or a PTU depending upon CU status, and later the harvested energy is used for sensing and transmission. While a censoring based transmission scheme is adopted to eliminate the DDUs associated with heavily faded channels from the transmission process. Therefore, DDUs with better channel conditions are only allowed to transmit during the transmission time. For the considered network scenarios, analytical expressions are developed for outage, throughput, ISU, and energy efficiency. An algorithm is proposed to maximize the energy efficiency. The impact of threshold of censoring, energy harvester's saturation level, interference threshold to CU, SNR, scaling factor, and number of cooperative DDUs on outage, throughput, energy efficiency, and ISU is presented. It is noticed that there is a particular span of sensing time for which the network shows minimum outage and maximum throughput. The network performance improves if allowable interference threshold and scaling factor increases. ISU improves as the number of DDUs in cooperation is increased. There also exists a particular sensing time for which the energy efficiency is maximum. In future, the present study can be studied with Non-orthogonal Multiple Access scheme and Reflecting Intelligent Surface-assisted communication.

REFERENCES

Al-Hourani, A., Kandeepan, S., & Jamalipour, A. (2016). Stochastic Geometry Study on Device-to-Device Communication as a Disaster Relief Solution. IEE*E Transactions on Vehicular Technology, 65(5)*, 3005–3017. doi:10.1109/TVT.2015.2450223

Ali, K., Nguyen, H. X., Vien, Q.-T., Shah, P., & Chu, Z. (2018). Disaster management using D2D communication with power transfer and clustering techniques. *IEEE Access: Practical Innovations, Open Solutions, 6*, 14643–14654. doi:10.1109/ACCESS.2018.2793532

Bhowmick, A., Das, G. C., Roy, S. D., Kundu, S., & Maity, S. P. (2020). Allocation of optimal energy in an energy-harvesting cooperative multi-band cognitive radio network. *Wireless Networks, 26(2)*, 1033–1043. doi:10.100711276-018-1849-2

Bhowmick, A., Roy, S. D., & Kundu, S. (2015). Performance of secondary user with combined RF and non-RF based energy-harvesting in cognitive radio network. In *2015 IEEE International Conference on Advanced Networks and Telecommuncations Systems (ANTS)*, (pp. 1–3). 10.1109/ANTS.2015.7413665

Chatterjee, S., Maity, S. P., & Acharya, T. (2015). On optimal sensing time and power allocation for energy efficient cooperative cognitive radio networks. In *2015 IEEE International Conference on Advanced Networks and Telecommuncations Systems (ANTS)*, (pp. 1–6). IEEE. 10.1109/ANTS.2015.7413620

Chu, Z. (2017). D2D cooperative communications for disaster management. In *2017 24th International Conference on Telecommunications (ICT)*, (pp. 1–5). IEEE. 10.1109/ICT.2017.7998227

Firozjae, H. M., Moghaddam, J. Z., & Ardebilipour, M. (2022). Performance Analysis of an UAV-assisted cognitive D2D communication-based Disaster Response Network. *In 2022 30th International Conference on Electrical Engineering (ICEE)*, (pp. 665–669). IEEE. 10.1109/ICEE55646.2022.9827018

Fodor, G., Parkvall, S., Sorrentino, S., Wallentin, P., Lu, Q., & Brahmi, N. (2014). Device-to-device communications for national security and public safety. *IEEE Access: Practical Innovations, Open Solutions, 2*, 1510–1520. doi:10.1109/ACCESS.2014.2379938

Ghasemi, A., & Sousa, E. S. (2005). Collaborative spectrum sensing for opportunistic access in fading environments. In *First IEEE International Symposium on New Frontiers in Dynamic Spectrum Access Networks*, (pp. 131–136). IEEE. 10.1109/DYSPAN.2005.1542627

Höyhtyä, M., Apilo, O., & Lasanen, M. (2018). Review of latest advances in 3GPP standardization: D2D communication in 5G systems and its energy consumption models. *Futur. Internet, 10*(1), 3. doi:10.3390/fi10010003

Hua, S., Guo, Y., Liu, Y., Liu, H., & Panwar, S. S. (2011). Scalable video multicast in hybrid 3G/Ad-Hoc networks. *IEEE Transactions on Multimedia, 13*(2), 402–413. doi:10.1109/TMM.2010.2103929

Kumar, P., & Bhowmick, A. (2021). Hybrid Spectrum Access in a CR Enabled Cooperative D2D Network. *IJCS, Wiley, 34*(Issue. 11), 1–14.

Kumar, P., & Bhowmick, A. (2022). Throughput performance of a non-linear energy-harvesting cognitive radio-enabled device-to-device network. *International Journal of Communication Systems, e5124*, doi:10.1002/dac.5124

Liang, Y.-C., Peh, E., Hoang, A., & Zeng, Y. (2007). Sensing-throughput tradeoff for cognitive radio networks. In *IEEE International Conference on Communications*. (pp. 5330-5335). IEEE.

Lien, S.-Y., Chien, C.-C., Tseng, F.-M., & Ho, T.-C. (2016). 3GPP device-to-device communications for beyond 4G cellular networks. *IEEE Communications Magazine, 54*(3), 29–35. doi:10.1109/MCOM.2016.7432168

Lin, Y.-D., & Hsu, Y.-C. (2000). Multihop cellular: A new architecture for wireless communications. In *Proceedings IEEE INFOCOM 2000. Conference on Computer Communications. Nineteenth Annual Joint Conference of the IEEE Computer and Communications Societies (Cat. No. 00CH37064)*, (vol. 3, pp. 1273–1282).

Liu, X., Li, F., & Na, Z. (2017). Optimal resource allocation in simultaneous cooperative spectrum sensing and energy harvesting for multichannel cognitive radio. *IEEE Access: Practical Innovations, Open Solutions, 5*, 3801–3812. doi:10.1109/ACCESS.2017.2677976

Moghaddam, J. Z., Usman, M., & Granelli, F. (2018). A Device-to-Device Communication-Based Disaster Response Network. *IEEE Trans. Cogn. Commun. Netw., 4*(2), 288–298. doi:10.1109/TCCN.2018.2801339

Nallagonda, S., Roy, S. D., Kundu, S., Ferrari, G., & Raheli, R. (2013). Cooperative spectrum sensing with censoring of cognitive radios in Rayleigh fading under majority logic fusion. In *Proc. of IEEE Nineteenth National conference on Communications (NCC),* (pp. 1-5). IEEE.

Nishiyama, H., Ito, M., & Kato, N. (2014). Relay-by-smartphone: Realizing multihop device-to-device communications. *IEEE Communications Magazine, 52*(4), 56–65. doi:10.1109/MCOM.2014.6807947

Prasad, A., Kunz, A., Velev, G., Samdanis, K., & Song, J. (2014). Energy-efficient D2D discovery for proximity services in 3GPP LTE-advanced networks: ProSe discovery mechanisms. *IEEE Vehicular Technology Magazine, 9*(4), 40–50. doi:10.1109/MVT.2014.2360652

Prasad, B., Bhowmick, A., Roy, S. D., & Kundu, S. (2016). Performance of cognitive relay network with novel hybrid spectrum access schemes with imperfect CSI. *International Journal of Communication Systems, 25*(11), 1761–1776. doi:10.1002/dac.3142

Prasad, B., Roy, S. D., & Kundu, S. (2014). Outage and SEP of secondary user with imperfect channel estimation and primary user interference. In *Proc. IEEE CONECCT,* (pp. 1-6). IEEE. 10.1109/CONECCT.2014.6740333

Raza, M., Awais, M., Ali, K., Aslam, N., Paranthaman, V. V., Imran, M., & Ali, F. (2020). Establishing effective communications in disaster affected areas and artificial intelligence based detection using social media platform. *Future Generation Computer Systems, 112,* 1057–1069. doi:10.1016/j.future.2020.06.040

Yang, J., & Ulukus, S. (2011). Optimal packet scheduling in an energy harvesting communication system. *IEEE Transactions on Communications, 60*(1), 220–230. doi:10.1109/TCOMM.2011.112811.100349

Zhang, H., Liao, Y., & Song, L. (2017). D2D-U: Device-to-device communications in unlicensed bands for 5G system. *IEEE Transactions on Wireless Communications, 16*(6), 3507–3519. doi:10.1109/TWC.2017.2683479

Chapter 3
Millimeter Wave Antennas for Wearable Applications

Mehaboob Mujawar
Glocal University, India

Subuh Pramono
Universitas Sebelas Maret, Indonesia

ABSTRACT

In this chapter the authors will be discussing about what wireless body area networks(WBAN) are, where are they used, what are the advantages of using millimeter wave frequencies, and novel millimeter wave antenna design suitable for wearable applications using CST software. The specific frequency that is used in the millimeter wave frequency is 60GHz with bandwidth of 7 GHz. Results from simulation are satisfactory in terms of gain and effectiveness. The performance of millimetre wave antennas is less affected by the presence of a human body because of their miniature-like dimensions and shallow penetration depth into human body layers. This makes them perfect for body-centric wireless communication (BCWC) applications. The essential property data was thus used to virtually generate a human body model. The antenna is kept close to the created human body model as simulations are run again at the same frequencies as previously. The results were promising as the VSWR and return loss curves remained almost identical.

INTRODUCTION

WBAN is a multi-tier network. WBAN can be used for medical purposes. Small biosensors can be worn around or implanted in the human body. Biosensors can be put around the wrist, chest or legs. The implementation of biosensors in human body is still an area of research. These biosensors monitor and record human health parameters. In this case sensors will be put around the human body in the form of antennas, which will record various health parameters. The focus in this chapter is not on the actual measurement of parameters but it could be more on antennas that are required for such an application. Sensor nodes are connected with sink node which are connected to the medical server through PDA or

DOI: 10.4018/978-1-6684-7000-8.ch003

Laptop. Medical server at remote location receives data which contains patient's daily report and his vitals. The sensors will be placed on the human body, body will communicate with sensors through PDA or laptop, it could be a mobile handset as well. Then the mobile handset will transmit the signals to medical servers, which will be accessible to doctors. Body area networks apart from medical applications can also be implemented for military and sports training, interactive gaming, and information sharing. The antenna design is one of the critical part in wearable antenna technology because antenna works as a sensor which will take certain information from body and then communicate it in a useful manner. In order to overcome the problems associated with wearable antenna technology some special antennas and technologies have been developed, which will be discussed in this chapter. There are two basic scenarios for body centric communication. One is On body to Off body propagation, the dots shown in the figure.1 are biosensors which are basically antennas. The biosensors on the body communicate with the personal device, then the signal via internet passes on to the medical server. In this case the propagation is usually perpendicular to the body and broadside pattern will be required. The antenna that would radiate minimum exposure towards the body will be required. Typically, the antenna radiates in both the directions i.e., forward and backward directions. The antenna that is communicating to personal device will also send some energy in the body. the body will also be exposed to electromagnetic energy. Therefore, we should have antenna that would give minimum exposure towards the body. The second scenario is On body propagation, in this case biosensors communicate with each other. For example, the antenna at the heart shown in figure 1 will communicate with the other antenna on the head or hand and so on to perform different functions. For on body propagation the pattern should be tangential to the body or it should be end fire pattern.

Figure 1. Body centric communication

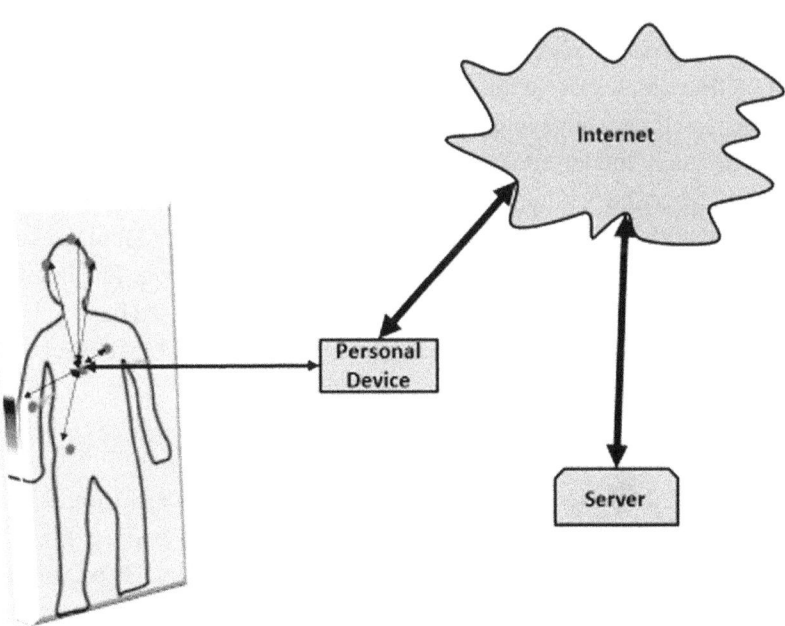

60 GHz Propagation

First wireless communication using 60 GHz carrier waves was studied by Sir J. C. Bose in the year 1895. Sir J. C. Bose is called the father of millimeter wave communication. This band is ISM band at 60 GHz with unlicensed bandwidth of 7 GHz. With this bandwidth these antennas can support high data rates. India generally follows US or European bands. Table.1 shows the unlicensed 60GHz band in different countries. The Indian researchers working in the field of millimeter wave generally consider the frequency band of 57 – 65 GHz for not only antenna design but also for other RF circuits in the millimeter wave range.

Table 1. The unlicensed 60GHz band in different countries

Country	Bandwidth (GHz)
USA	57-64
Canada	57-64
Japan	59-66
EU	57-66
Australia	59.4-62.9
South Korea	57-64

In normal cases the size of the antenna is approximately half wavelength than the operating frequency, when we go higher up in frequency the wavelength decreases as a result the size of the antenna also decreases. The spectrum below 10 GHz is already crowded because of many applications in the sub 6 GHz region, applications include Wi-Fi, Mobile, Wi-Max band, RFID, WLAN. The antenna size is very much smaller at 60 GHz which saves a lot of space. For example, if a person has to wear an antenna at 2.4 GHz on the arm, it will be a huge antenna. But the size of the antenna operating at 60 GHz will be very small and can be easily hidden on the body. Hence antennas operating at 60 GHz are suitable for wearable applications. since the bandwidth is very large that is 7 GHz, the data rates are high (Gao et al., 2016). When compared with 60 GHz at 2.4 GHz the interference is more. At 60 GHz the communication distance is not very large they are suitable for short range communication. The interference between the co-located BAN users will be much smaller. 60 GHz band is an unlicensed band. The major drawback of 60 GHz band is short range communication (10m to 100m), it is not suitable for higher range applications.

Phantoms

Phantoms are basically used for experimental purposes. We cannot do experiments on the human body, animals directly. Therefore, phantoms are used to mimic the human body parameters using artificial substances which are then shaped into the form of human body, so that experimentation of the wearable devices on the human body can be carried out. Instead of performing the experimentation on the real human body, it has been performed in the artificial environment so that the effect of the device on the human body can be estimated. The phantoms are classified as solid, liquid or semi-solid, which are available in market. The semi-solid phantom composition to mimic skin properties at 60 GHz are shown in Table 2. These phantoms take into account the effect of skin, tissues, bone, blood and so on. Liquid phantoms are used up to 6 GHz for handheld mobile devices. Semisolid phantoms and solid phantoms do not require container and can be used at millimeter waves. For low frequencies generally liquid phantoms are used and for higher frequencies either solid or semi-solid phantoms are used.

Table 2. Semi-solid phantom composition to mimic skin properties at 60 GHz (Mujawar, 2022)- (Mujawar et al., 2022)

Component	Quantity(%)	Use
Deionized water	81	Determines the phantom's dispersion
Polyethylene powder	16	Utilized to lower phantom permittivity
Agar	1	Keeping its shape
TX-151	1	Increase the viscosity
Sodium Azide	1	Preservative

The Debye's equation can be used to describe the skin permittivity. (Mujawar & Thangavel, 2022):

$$\varepsilon = \varepsilon_\infty + \frac{\varepsilon_s - \varepsilon_\infty}{1 + j\omega\tau}$$

Where, static permittivity, ε_s =37.2 permittivity, when optical permittivity, ε_∞ =3.88 and relaxation time τ= 7.1×10-12$^{s.}$

State of Antennas Reported at 60GHz

4 element printed array on RT-Duroid substrate with gain about 11.8 dBi and bandwidth of 6 GHz (Mujawar & Thangavel, 2022). In this antenna design, 4 elements are fed by power divider, which equally feeds power to all 4 antenna elements, supported by microstrip line. Yagi Uda antenna array using 4 way power divider on RT-Duroid substrate of thickness 0.127mm, Gain of about 15 dBi and 5 GHz band (Wu et al., 2011). This antenna is printed on very thin substrate of 0.127mm. The feedline power is divided into two parts, which is further divided into two each and hence there are 4 elements. H plane Substrate Integrated Waveguide (dielectric loaded waveguide) horn on RT-Duroid 5880 of thickness 0.787 mm, Gain of 6.6 dBi and bandwidth of 6.1 GHz (Razafimahatratra, 2015). Textile antenna: Yagi Uda antenna on Cotton fabric of thickness 0.2mm, Gain of about 9.2dBi and 7GHz of bandwidth is achieved (Mujawar, 2021a). Textile antenna: 4 element array on 0.2mm of fabric, gain of about 8 dBi and bandwidth of 7 GHz is achieved (Mujawar, 2021b). The antenna is made of FR4 with a 0.127 mm thickness, 8 Yagi-Uda antennas at the vertices of the octagonal plates, and center fed octagonal plates. The gain is 2.86 dBi (Lee & Choi, 2018), and the antenna is printed on flexible material with a gain of 10.6 dBi and a bandwidth of 9.8 GHz (Young et al., 1973).

Antenna Requirements for BAN

The main antenna requirements for body area network are, it should be low profile and small in size. Antenna should be light in weight, specifically when it is used in wearable applications. since it is not comfortable to carry heavy antennas on the arms, shoulders or any other parts of human body. Antennas should be conformal to the body, which basically means they should be in the shape of the body wherever it is mounted on the human body. It is should be a fairly high gain antenna because at 60 GHz range is very small (Pellegrini et al., 2013). In order to gain more range, the only way is to increase the power from the antenna, i.e. increase the gain of the antenna.

Antenna Design

The antenna design was created and tested using Computer Simulation Technology's (CST) Microwave Studio Suite. The three adjacent planner components in the model were the radiating element, substrate, and ground plane. Figure 2 depicts the antenna's overall length and width, which are 7.6 mm and 3.8 mm, respectively. The physical and electrical dimensions of the antenna at 60 GHz frequency are summarized in Table.3 below. Flame-resistant fiberglass reinforced epoxy (FR-4) with a dielectric constant of 3 serves as the substrate. The return loss and VSWR is also being analyzed with the help of phantom model in CST software as shown in figure.3.

$$W = \frac{c}{2f} \sqrt{\frac{2}{\varepsilon_r + 1}} \tag{1}$$

$$L = L_{eff} - 2 \cup L \tag{2}$$

$$L_{eff} = \frac{c}{2f \sqrt{\varepsilon_{reff}}} \tag{3}$$

$$\Delta L = 0.412h \frac{\left(\varepsilon_{reff} + 0.3\right)\left(\frac{w}{h} + 0.264\right)}{\left(\varepsilon_{reff} - 0.258\right)\left(\frac{w}{h} + 0.8\right)} \tag{4}$$

$$\varepsilon_{reff} = \frac{\varepsilon_r + 1}{2} + \frac{\varepsilon_r - 1}{2} \left[1 + 12\frac{h}{w}\right]^{-\frac{1}{2}} \tag{5}$$

$$L_g = L + 6h \tag{6}$$

$$W_g = W + 6h \tag{7}$$

$$h = \frac{0.0606\lambda}{\sqrt{\varepsilon_r}} \tag{8}$$

$$Feed\ length\left(L_f\right) = \frac{\lambda_g}{4} \tag{9}$$

$$\lambda_g = \frac{\lambda}{\sqrt{\varepsilon_{reff}}} \tag{10}$$

$$\eta = \frac{Gain}{Directivity} \times 100\% \tag{11}$$

$$Axis\ position = \frac{-\lambda_g}{6} + \frac{-\lambda_g}{6} + \frac{-\lambda_g}{6} \tag{12}$$

$$Length = \frac{\lambda_g}{6} + \frac{\lambda_g}{6} + L_g \tag{13}$$

$$Width = \frac{\lambda_g}{6} + \frac{\lambda_g}{6} + W_g \tag{14}$$

$$Height = \frac{\lambda_g}{6} + \frac{\lambda_g}{6} + h \tag{15}$$

Table 3. Physical size and Electrical size at 60 GHz frequency

Dimensions	Physical size (mm)	Electrical size at 60 GHz (λ)
Length (substrate)	7.6	0.529
Height (substrate)	1.56	3.09
Width (substrate)	3.8	0.78
Length (radiating element)	7.4	1.56
Height (radiating element)	0.035	142.85
Width (radiating element)	3.6	3.02

Results

From the simulated results obtained for the proposed antenna, the return loss as shown in fig.5 is obtained to be -24 dB for free space and -29dB for on body. For better performance of antenna return loss should be less than -10 dB. The simulated results from fig.6 indicates that the VSWR of the antenna is 1.2321 for free space and 1.0491 for on body.

Figure 2. The proposed antenna design

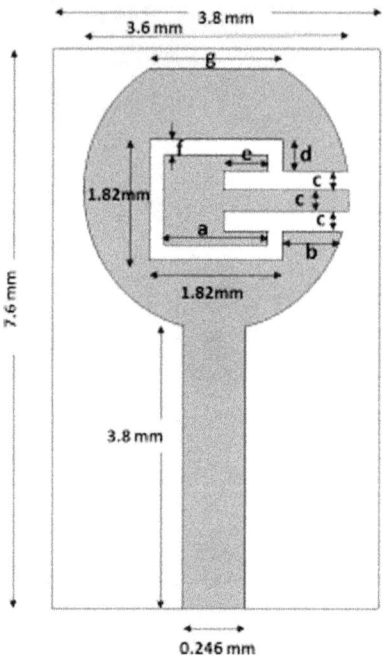

Table 4. Dimensions of the proposed antenna

Parameters	Dimension (mm)
a	1.692
b	0.89
c	0.082
d	0.787
e	0.83
f	0.064
g	0.738

Figure 3. Simulation of the proposed antenna with phantom model in CST 3D View

Figure 4. Simulation of the proposed antenna with phantom model in CST side view

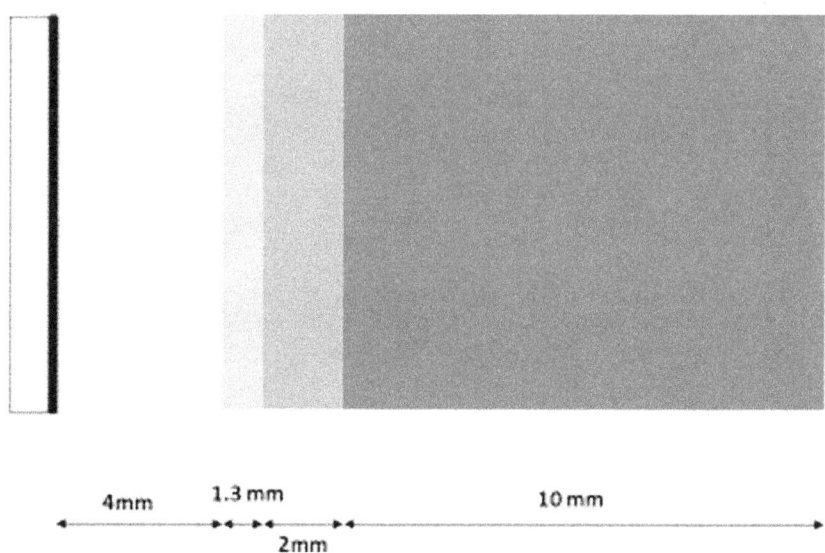

Table 5. Thickness of human tissues in phantom model (Gao et al., 2016)

Human tissue	Thickness in mm
Fat	2
Antenna Substrate	1.5
Skin	1.3
Muscle	10

Figure 5. Comparison of free space and on body return loss

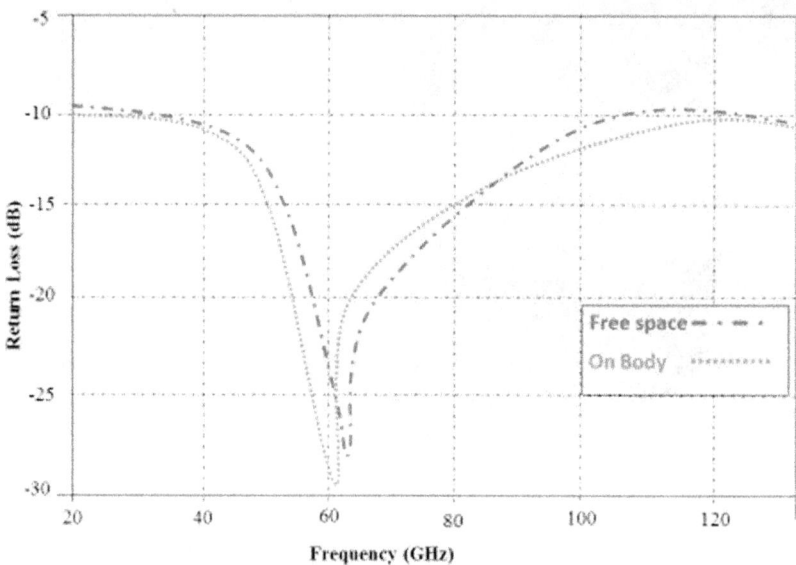

Figure 6. Comparison of free space and on body VSWR

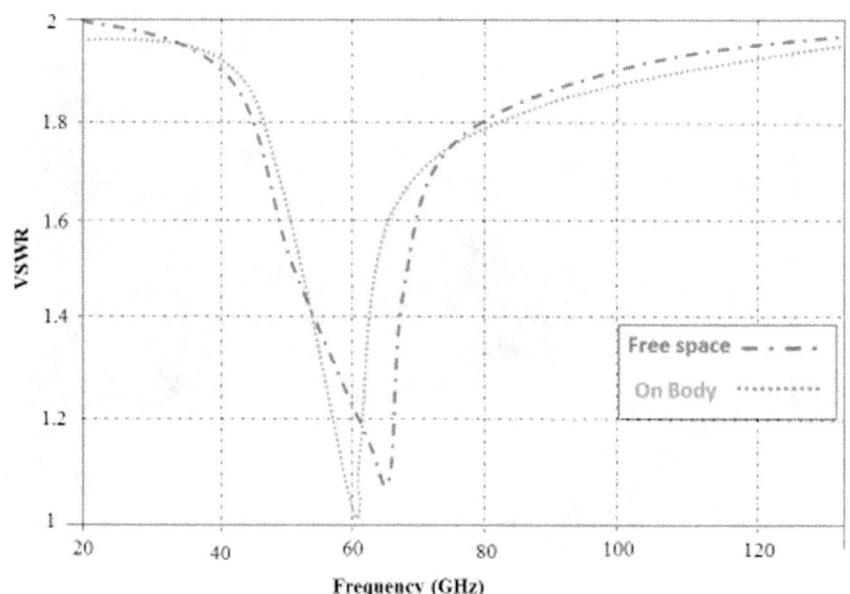

CONCLUSION

A new millimeter wave antenna has been designed that is appropriate for wearable applications. The simulated findings show that the proposed antenna is suitable for biomedical, telemetry applications, which is designed to operate at 60 GHz. The return loss is obtained to be -24 dB for free space and -29dB for on body. The return loss should be less than -10 dB, which is satisfied in the proposed antenna design. VSWR of the antenna is 1.2321 for free space and 1.0491 for on body. The analysis of the designed antenna using phantom model in CST gives clearer picture of the practical usage of the antenna for wearable applications.

REFERENCES

Gao, Y., Ma, R., Wang, Y., Zhang, Q., & Parini, C. (2016). Stacked Patch Antenna with Dual-Polarization and Low Mutual Coupling for Massive MIMO. *IEEE Transactions on Antennas and Propagation, 64*(10), 1–1. doi:10.1109/TAP.2016.2593869

Lee, S., & Choi, J. (2018). A 60-GHz Yagi-Uda circular array antenna with omni-direcitional pattern for millimeter-wave WBAN applications. In *2018 IEEE International Symposium on Antennas and Propagation USNC/URSI National Radio Science Meeting*, (pp. 1699–1700). IEEE. 10.1109/APUSNCURSINRSM.2018.8609346

Mujawar, M. (2021a). *Antenna Array Design for Massive MIMO System in 5G Application.* Taylor & Francis Group. . doi:10.1201/9781003175155-18

Mujawar, M. (2021b). *Compact Microstrip Patch Antenna Design with Three I-, Two L.* One E- and One F-Shaped Patch for Wireless Applications. doi:10.1201/9781003093558-7

Mujawar, M. (2022). *Arrow Shaped Dual-Band Wearable Antenna for ISM Applications.* Wiley. . doi:10.1002/9781119792581.ch8

Mujawar, M., Dommeti, V., Naz, M., & Muduli, A. (2022). Design and performance comparison of arrays of circular, square and hexagonal meta-material structures for wearable applications. *Journal of Magnetism and Magnetic Materials, 553*, 169235. doi:10.1016/j.jmmm.2022.169235

Mujawar, M., & Thangavel, G. (2022). *Multiband Slot Microstrip Antenna for Wireless Applications.* Springer. . doi:10.1007/978-3-030-76636-8_3

Pellegrini, A., Brizzi, A., Zhang, L., Ali, K., Hao, Y., Wu, X., Constantinou, C. C., Nechayev, Y., Hall, P. S., Chahat, N., Zhadobov, M., & Sauleau, R. (2013, August). Antennas and Propagation for Body-Centric Wireless Communications at Millimeter-Wave Frequencies: A Review [Wireless Corner]. *IEEE Antennas & Propagation Magazine, 55*(4), 262–287. doi:10.1109/MAP.2013.6645205

Razafimahatratra, S., (2015). On-body propagation characterization with an H-plane Substrate Integrated Waveguide (SIW) horn antenna at 60 GHz. In *2015 European Microwave Conference (EuMC)*, (pp. 211–214). IEEE. 10.1109/EuMC.2015.7345737

Wu, X. Y., Akhoondzadeh-Asl, L., & Hall, P. S. (2011). Printed Yagi–Uda array for on-body communication channels at 60 GHz. *Microwave and Optical Technology Letters*, *53*(12), 2728–2730. doi:10.1002/mop.26443

Young, L., Robinson, L., & Hacking, C. (1973). Meander-Line Polarizer. Antennas and Propagation. *IEEE Transactions on.*, *21*(3), 376–378. doi:10.1109/TAP.1973.1140503

Chapter 4
Multiple Input Multiple Output (MIMO) Antennas:
A Concise Review and Current State of the Art

Anindita Bhattacharjee
Tripura University, India

Anirban Karmakar
Tripura University, India

Anuradha Saha
Netaji Subhash Engineering College, India

ABSTRACT

The advancement of antennas that can access high data speeds, and have higher gain, is largely responsible for the advancements in wireless technology. One such antenna called the multiple input multiple output (MIMO). This antenna has attracted the utmost attention of the researchers owing to its features like higher capacity, increased data rate, low bit error rate, and stable radiation characteristics. Over the years, different strategies have been proposed by several researchers to enhance the performance of the MIMO antenna. This article's main goal is to provide a summary of the changes in MIMO antenna design during the last few decades, with an emphasis on the most significant contributions, which are mainly from IEEE publications. This review work intends to show antenna researchers a promising route for their development using MIMO antennas.

DOI: 10.4018/978-1-6684-7000-8.ch004

I. INTRODUCTION

Now-a-days in communication field, Wireless communication is a rapidly growing segment. The main goals of wireless system are to enhance data rate and increase transmission dependability. The advent of "green communication" significantly increases the wireless network's resource and energy efficiency without sacrificing the quality of services provided to consumers. By expanding the channel capacity without raising the broadcast power of the antenna, Multiple Input Multiple Output (MIMO) technology makes significant advancement in the field of wireless communication, resulting in green wireless network (Zheng, 2003).

Due to its powerful performance-enhancing capabilities, such as its ability to increase data rate within a constrained bandwidth, the use of MIMO technology—has speedily increased acceptance during the last few decades (Rajagopalan,2007) . The concept of utilizing several antenna configurations rather than a single one has been beneficial in improving radio networks' data transmission rate, coverage, and overall performance, among other things. In a perfect world, the channel capacity attained in a multipath environment is linearly analogous to the number of transmitting and receiving antennas (Sharawi, 2013). Equation (1) shows the relation between channel capacity and channel bandwidth,

$$C = B log_2(1 + M \times N \times SNR) \tag{1}$$

Signal-to-noise ratio is abbreviated as SNR, where M and N stand for the number of antennas on the transmitting and receiving sides, respectively, C stands for the channel capacity (measured in bits per second), and B stands for the channel bandwidth (measured in Hz) (absolute value).Size plays a big part in attempts to implant numerous antennas at the user terminal. In order for the antennas to be effective, they must also cover a variety of operating bands. This creates yet additional challenge for the antenna designer.Therefore, one of the key problems for wireless terminal designers is to identify compact MIMO antennas with favourable performance metrics.

To fully comprehend the growth of MIMO communication, we must examine the history of antenna diversity. Surprisingly, the concept of antenna diversity was born out of research done by RCA engineers Harold H. Beverage and Harold O. Peterson, who observed that the signal levels of radio broadcast signals at two stations about a half mile apart varied greatly (Brittain, 2008). The researchers created a diversity system to combat the fading, which they later realised was caused by multipath propagation. To improve audio quality, this system merged the audio outputs of two independent receivers, and experimental findings revealed that the two antennas required being at least a wavelength apart in order for the system to work correctly (Peterson, 1931). In 1941, a technique that utilised a single radio receiver with a switch that alternated among two antennas at a frequency between 300 and 1000 Hz was suggested. This method gave the receiver an average signal that helped it resist fading (Bartlet, 1941). It also laid the foundation for a vast amount of study on the characteristics of multipath propagation and the best antenna designs for effective diversity performance.

In 1996, Foschini at Bell Laboratories published a ground breaking paper (Foschini, 1996) in which he showed that the communication capacity made possible by the (now-classic) formula can be achieved even if the transmit radio is unaware of the multi-antenna transfer function H from the transmitter to the receiver (Jensen, 2016)

$$C_U = log_2[I + \frac{\rho}{N_T}HH \quad]$$

(2)

Where I represent the identity matrix, ρ represent the signal to- noise ratio of single-antenna, and $\{\}†$ denotes a conjugate transpose. His work was supported in 1998 not only by his own study on the subject (Foschini, 1998) but also concurrently by a paper from Raleigh and Cioffi that showed even greater potential capacity if the transmitter does know the channel (Raleigh, 1998). The article also offered a space-time coding scheme designed to once more take advantage of part of the multipath, multi-antenna channel capacity. Last but not least, a study by (Alamouti, 1998) demonstrated a transmit diversity scheme that offered the best transmit proceeding to take advantage of the multipath channel features.

Beam forming, spatial multiplexing, and space-time coding are the three core techniques on which the MIMO system is based. In order to achieve efficiency, which is assessed in terms of lower transmit power, greater range, increased noise immunity, or larger throughputs, these strategies are utilised individually or in some combination. Beam formation has the benefit of minimizing multipath fading by increasing the received signal strength at the receiver by causing signals from several antennas to constructively add up. Multiple copies of data can be transmitted via space-time coding through a number of antennas.

Additionally, it boosts the reliability of data transport with gains comparable to those that may be obtained with multiple antenna diversity receptions. MIMO utilizes several sets of antennas in order to achieve spatial multiplexing. With this technique, a high-rate signal is divided into several streams of lower-rate signals, each of which is broadcast employing the identical central frequency but from a distinct antenna. To correctly decode each transmitted stream, the number of receiver antennas must be higher than or equal to that of the transmit antennas.

In the literature, there is still a dearth of comprehensive discussions based on various wide band and multi band MIMO techniques, their instances, and comparative studies. The various MIMO antenna design strategies and all of their methods for achieving multiband and wide band features through diverse structures and mechanisms are thus discussed in this work with several examples and a comparison of their characteristics.

This review study provides an in-depth analysis of several MIMO antennas and helps improve the comprehension of both novice researchers and seasoned antenna designers regarding their respective topologies. This review chapter consist of different Performance parameters of MIMO antenna, MIMO antenna sections, Conclusion and Future scope.

II. PERFORMANCE PARAMETERS OF MIMO

To accurately assess a MIMO antenna's performance, not only some standard parameters like as bandwidth, resonance frequency, radiation patterns, gain, and efficiency must be reviewed, but also some critical parameters should be evaluated.

A. The Correlation Coefficient

The correlation coefficient (ρ_e) is a metric that narrates how isolated or correlated communication channels are from one another (Blanch, 2013). This parameter primarily considers the radiation pattern of the antenna system.In case of MIMO antenna, if the antennas are simultaneously operated this parameter measure how the patterns are affected by one another. The correlation coefficient squared is used to get the envelope correlation coefficient. The following formula is used to calculate this parameter (Blanch, 2013)

$$\rho_e = \frac{\left| \iint_{4\pi} \left[\vec{F_1}(\theta,\varphi) * \vec{F_2}(\theta,\varphi) \right] d\copyright \right|^2}{\iint_{4\pi} \left| \vec{F_1}(\theta,\varphi) \right|^2 d\copyright \iint_{4\pi} \left| \vec{F_2}(\theta,\varphi) \right|^2 d\copyright} \tag{3}$$

Here, $F_i(\theta,\phi)$denotes the antenna's three-dimensional field radiation pattern when the ith port is energised, and Ω denotes the solid angle. The Hermitian product operator is represented by the asterisk.

B. Total Active Reflection Coefficient (TARC)

Any antenna system's bandwidth is clarified using the S-parameter. However, with MIMO, the s-parameter does not adequately characterize the system's bandwidth and efficiency (HoChae, 2007). The TARC parameter is utilized for this. TARC stands for the square root of the reflected power divided by the square root of the total incident power for multiport systems (Manteghi, 2005). For the N-element system, TARC is given by,

$$t_a = \frac{\sqrt{\sum_{i=1}^{N} |b_i|^2}}{\sqrt{\sum_{i=1}^{N} |a_i|^2}} \tag{4}$$

In the above equation, a_i and b_i are represent incident and reflected signal, respectively. S-parameter data can be used to calculate these variables. For a multiport network, the following equation describes the relationship between the incident and reflected waves (Pozar,1998),

$$b = Sa \tag{5}$$

Here, **S** representsthe S-parameter matrix. In the above equation each variable represents vector quantity with magnitude and direction. The TARC has a value that ranges from 0 to 1. When the value is zero, radiated all power; when the matter is one, all incident power was reflected but no power was radiated. For two-port MIMO antenna system, TARC may be determined by using the following formula (Su, 2012),

$$t_a = \frac{\sqrt{\left(\left| S_{11} + S_{12}e^{j\theta} \right|^2 \right) + \left(\left| S_{21} + S_{22}e^{j\theta} \right|^2 \right)}}{\sqrt{2}} \tag{6}$$

In the above equation, where S_{xy} is the coupling between the two ports related with the antenna architecture, S_{xx} is the port's reflection coefficient, and θ is the input feeding phase. To investigate the impact of phase variation between the two ports on the performance of the antenna's resonance, the random phase of the two-port network is wiped between 0° and 180° after the S parameters are established. This leads to the generation of the TARC curves for effective bandwidth evaluation. If TARC is required to calculate for more than two port antennas, equations (4) and (5) are combined.

C. Mean Effective Gain (MEG)

Because antenna in these applications is not employed in an anechoic chamber, the solo antenna gain is not a fair indicator of antenna performance. In order to assess the antenna's genuine performance, it is crucial to investigate how the environment affects its radiation characteristics. Fabricating an antenna, using it in conjunction with a conventional antenna with known properties, and testing it under the required conditions are one method of achieving this (Sharawi, 2013).

D. Diversity Gain

When the transmitter gets numerous category of the input stream over distinct channel pathways, diversity is frequently obtained. The diversity gain (DG) is a metric for assessing how well diversity strategies perform. The error probability curve's slope, expressed in terms of the received SNR on a log-log scale, is known as the DG. Mathematically, it can be shown as (Foschini,1998),

$$DG = \frac{(SNR)_c}{(SNR)_r} \tag{7}$$

Where indices "c" and "r" are used to denote the combined and the reference. The following equation approximately describes the relationship between DG and the correlation coefficient (ρ_c),

$$DG = 10\sqrt{1 - |\rho_c|} \tag{8}$$

From the above equation it is clear that the higher the diversity gain the lower the correlation coefficient.

E. Branch Power Ratio

The MIMO antenna system performance is also impacted by the relative power levels coming from various branches of the antenna system. These power levels lie very close to each other. Branch Power Ratio parameter is introduced to denote the differences between power level. It is represented by (Sharawi, 2013),

$$k = \frac{P_{min}}{P_{max}} \tag{9}$$

In the above equation, P_{min} is denoting the lower power level of antenna and P_{max} is denoting the power obtained from high power antenna. The mean effective gain values of a MIMO antenna system can be used to estimate the branch power ratio (k) using the following formula (Sharawi, 2013),

$$k = \min\left(\frac{MEG_2}{MEG_1}, \frac{MEG_1}{MEG_2} \right) \tag{10}$$

Here, MEG_1 and MEG_2 are the mean effective gains of both the antenna respectively. This above formula is valid for two-element MIMO antenna system.

F. System Capacity

The use of the MIMO antenna system improves the channel capacity in a multipath environment as that of the SISO system. The radiation properties of the antenna and the environment of channel determine this property.This variable is influenced by the channel matrix.The channel capacity measured in bits/sec/Hz. It is represented as,

$$C = log_2\left[det\left(I_N + \frac{\rho}{N} HH^T \right) \right] \tag{11}$$

ρ is the average SNR, H represent the normalized channel covariance matrix, I_N is an identity matrix of order $N \times N$, N is the quantity of antenna at the transmitting as well as receiving sites.

An identity matrix for T HH is produced in the situation of uncorrelated transmitting waves by antenna elements with zero correlation coefficients at the transmitter and receiver, identical powers, and normalised mean effective gain values.In contrast to a SISO system, the channel capacity gradually increases as the number of antenna elements rises (Sharawi, 2013).

$$C = N \times log_2\left(1 + \frac{\rho}{N} \right) \tag{12}$$

It represents the MIMO system's optimal channel capacity limit. But there is always some correlation; this limit cannot be reached. The correlation coefficient among the antenna elements is never zero. The antenna elements' correlation increases with decreasing the level of mutual coupling. Due to this the performance of the channel degrades even more.

III. CHALLENGES OF MIMO

The main concern when designing a MIMO antenna is minimising mutual coupling in modern miniature printed and other antennas.Higher mutual coupling in digital infrastructure MIMO systems has a negative impact on channel capacity and error rate. To reduce mutual coupling, a variety of MIMO pre-coding and decoding techniques, including partial swam optimization, evolutionary algorithms, and galaxy-based search algorithms, are offered. Decoupling networks (Zhu, 2017), neutralisation lines (Ou, 2017), parasitic elements (Bilalet, 2017), complementary split rings resonators (Khan, 2017), electromagnetic band gap structures (Abdalla,2017), and defected ground structures (Chen, 2016) are some of the straightforward methods for reducing mutual coupling in MIMO systems.These methods can adjust the coupling by reducing, blocking, or attenuating the flow of surface current.To counteract the negative effects of mutual coupling, prominent antenna topologies include printable, reconfigurable, metamaterials and dielectric resonator antennas may be used.

IV. MIMO ANTENNAS

Different multi band and wide band MIMO antenna designs are discussed in this section. A comparison table and numerous current examples are used to explain each strategy. Every MIMO antenna's performance is compared in terms of gain, envelop correlation coefficient (ECC), diversity gain (DG), isolation level, bandwidth, substrate materials, and efficiency.

A. Multiband MIMO Antennas

In the current scenario, a wireless device operates in more than one frequency band. An antenna that covers multiple wireless communication bands is considered a multiband antenna.

In (Manteghi, 2006), authors proposed a triple band two and four-element MIMO antenna based on a Printed Inverted "F" Antenna (PIFA) design. This design introduced a J-shape and quarter wavelength slot to achieve a tri-band in the operating range. In (Bhatti, 2009), a quad-band MIMO antenna array for portable communication was presented. This design uses a C-shape slot introduced in the PIFA to achieve dual-band. Then, a T-shape slot is engraving on the antenna structure to attain a tri-band. At last, to achieve the quad-band, a resonator was placed to the central radiator near the feed point at a height of 0.8 mm above the lower antenna part. A matching stub is also connected to the resonator to maximize impedance matching in the 5-GHz range. A multi-band antenna was proposed in (Nezhab, 2010). E-shape monopoles with a narrow slot were responsible for the tri-band characteristics. The antenna generates resonances at 2.4, 5.4, and 5.8 GHz. A MIMO array for portable wireless communication was proposed in (Ling, 2011). The proposed antenna array consists of two similar multi-branch monopoles connected in a back-to-back combination on the upper side of the substrate and on the back side ground plane. In this design, to lessen the mutual coupling between antenna elements, a strip line connected the two monopoles and a stub attached to the ground plane with the help of a shorting pin. The proposed antenna exhibits a dual-band nature. In (Zhou,2012), a dual-band MIMO system for Global System for Mobile Communication (GSM)/Universal Mobile telecommunication System (UMTS)/Long Term Evolution (LTE) and Wireless Local Area Network (WLAN)Handset application was proposed. The configuration of the dual-band antenna comprised two open loops. The inner loop is printed on the substrate's top side and acts as a half wave dipole for a higher frequency lower band. The ground plane on the substrate's reverse face is used to generate the outer loop, which functions as a half-wave dipole for the lower band of the lower frequency. The proposed antenna covers the GSM1800/1900, UMTS2000, LTE2300/2600, and WLAN2.4/5-GHz bands. In (Sharawi, 2012), a dual element dual-band MIMO antenna for LTE wireless handheld and portable terminals applications was presented. This design achieved two bands at 803–823 MHz and 2440–2900MHz, respectively. A multi-band meander loop antenna for wireless router applications was presented in (Fernandez, 2013). This low profile loop antenna covers different bands, namely, 4G LTE (699–798 MHz) band, UMTS (DCS/PCS) bands (1.7–2.0 GHz), Worldwide Interoperability for Microwave Access (Wi-MAX) bands (2.3 and 3.5 GHz), and WLAN bands (2.4 and 5 GHz). Again in (Karimian, 2013), a four-element MIMO antenna was proposed for quad-band applications. The radiating elements were orthogonal to one another in order to achieve proper isolation. Again a cross shape slot was engraved in the ground plane to boost isolation further.

In (Shoaib, 2014), a linearly polarized MIMO antenna system supported different frequency bands namely; GSM 850/900, DCS, PCS, UMTS, WLAN, and Wi-MAX are presented. Each monopole antenna comprises a capacitive feed attached to a 50-Ω microstrip feed line, two twisted lines, a parasitic loop, and a shorting strip. Five resonant modes can be initiated by the printed monopole antenna element at frequencies of 900, 1800, 2100, 3500, and 5400 MHz. For dual-band application, a MIMO antenna with slitted ground structure is demonstrated (Wu, 2014). This study employed the printed slot antenna technique to enable dual-band functionality without growing the antenna size. The antenna is comprised of two identical slots of C-shaped, and as shown in Figure. 1, each C-shaped slot was fed by a 50Ω microstrip line with one via-grounded end.

The T-shaped and C-shaped slot antennae are coupled, yielding a dual-band response. The antenna design introduced two similar T-shaped slots to achieve the proper impedance matching. To reduce the upper band's mutual coupling, six pairs of slits were carved. The two operational bands provided by the antenna are centred at 2.70 and 3.95 GHz.

A compact MIMO array with dual-band features suitable for LTE application is presented (Li, 2014). Two different categories of planar inverse-F antennas (PIFAs) are among the eight components of this MIMO system.One kind of PIFA inserts a U-shape slit to get dual-band characteristics, while another kind of PIFA inserts an L-slit. The suggested antenna has a frequency range of 3.5 GHz for Wi-MAX and 2.7 GHz for LTE.To provide polarisation variety for coupling mitigation, the two types of PIFAs are orthogonal to one another.

Figure 1. MIMO Antennas with slitted ground for Dual-band (All the parameters in mm) (Wu, 2014)

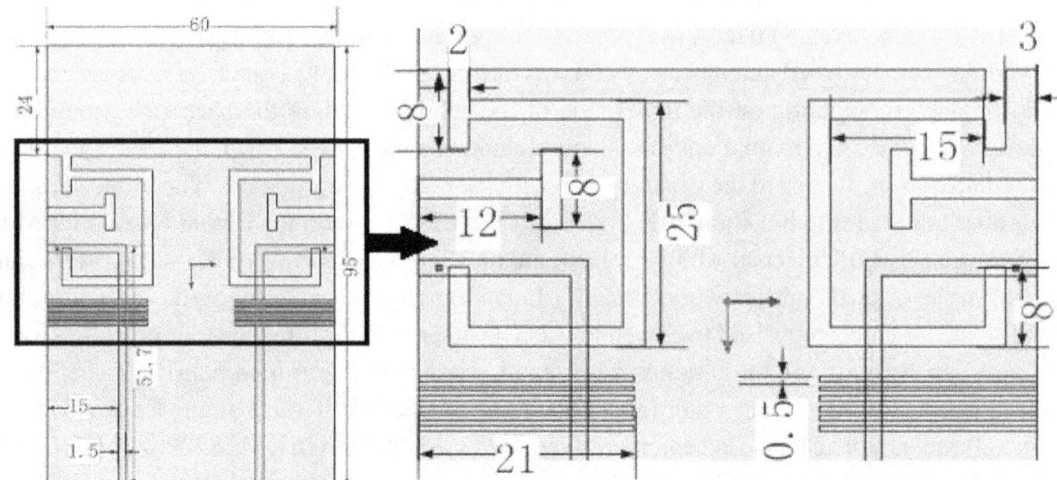

In (Sarkar,2015), a MIMO antenna system consisting of a uniplanar quad-element is presented for quad-band characteristics. A CPW-fed monopole is loaded with SRR and IDC using meta-material concepts, as shown in Figure. 2, to achieve simultaneous miniaturization and quad-band functioning.

Figure 2. Four element MIMO antennas for quad band application with $L_a=110$ mm, #1, #2,#3,#4 represents antenna (Sarkar,2015)

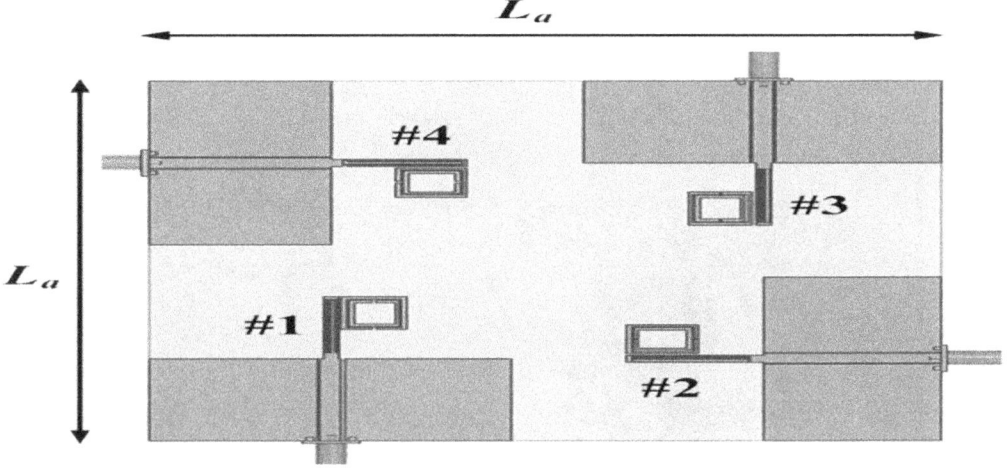

Four quad-band monopole elements arranged orthogonally allow for polarisation variety in MIMO operation. The distance between elements was optimized to boost isolation between antenna elements.

A triple-band MIMO antenna for mobile wireless communication is presented in (Sun, 2016). Three inverted -L shape meander-line was used as radiators. To achieve 2.6 GHz and 0.9 GHz frequency bands, a T-shape slot is carved onto the ground. A MIMO antenna suitable for portable device applications was proposed (Peristerianos, 2016). The antenna operates at ISM (2.4GHz - 2.489GHz) and between 5GHz and 6GHz.

For Wi-max and WLAN applications, a tri-band MIMO antenna is presented (Nandi, 2017). In this design CRLH unit cell associated with a square loop antenna is used to achieve triple band characteristics. The CRLH unit cell is loaded into the rectangle loop to generate the first two bands at 2.38-2.58, 3.37-3.64 GHz; the third band from 5.13-6.38 GHz is the regular frequency of the loop itself. I-shape slot with one open end is engraved on the ground plane to minimize mutual coupling. In (Nandi, 2017), MIMO antenna consists of two-element suitable for dual-band characteristics are proposed. Two-unit cell with a microstrip line stepped feeding structure is responsible for better impedance matching as well as dual-band feature. The meander line shape slot creates resonance at 2.5 GHz, and the other slot creates resonance at 5.5 GHz. At 2.35-2.69 GHz and 5.4-5.68 GHz, two operating bands (|S11|£-10 dB) were obtained. In (Liu,2018), a dual-band platform free antenna for 5G mobile device applications was introduced. The antenna is mimicked by placing a second planar inverted-F antenna (PIFA) to feed it through a coupling process and adding a metallic patch in vertical position to the antenna.The vertical position patch's existence can lessen the amount of ground-plane fringe currents, adjusting the antenna's electromagnetic compatibility with various working platforms. A dual-band MIMO antenna system based on a dielectric resonator is proposed in (Das,2018). Figure.3. represents the radiating elements that make up the MIMO antenna. The Dual-band feature was produced using a hybrid antenna that incorporates a CDRA and an altered annular ring printed line. Annular ring-based printed lines and a CDRA printed on the substrate's upper side.

Figure 3. Two element MIMO antennas based on dielectric resonator for dual band application with L_S=60.0, T_1= 2.5, W_S= 60.0, L_4= 23, H_s= 1.6, L_5= 33, L_1= 19, L_6= 15, W_1= 2, W_3= 1.5, L_2= 8, D= 20, W_2= 2, H= 10, D_1= 20(All parameters in mm). Feeding Structure (Das, 2018)

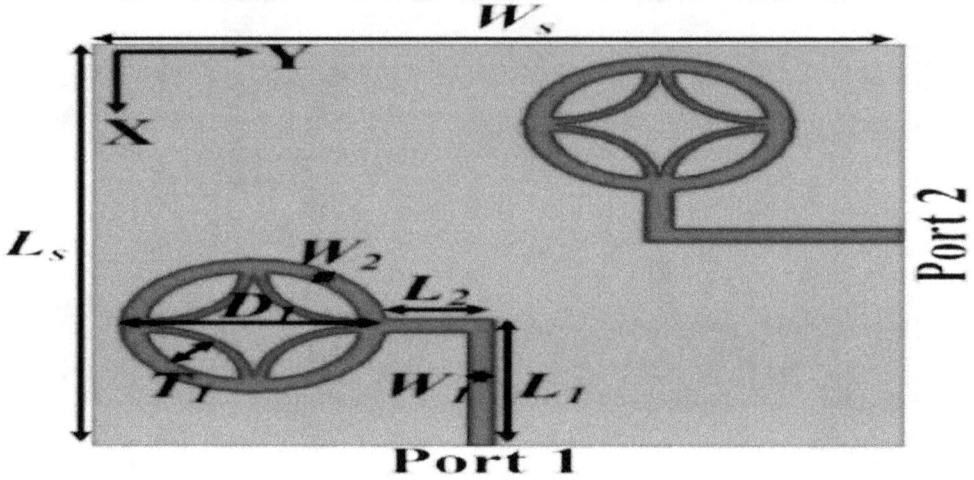

Figure 4. Two element MIMO antennas based on dielectric resonator for dual band application with L_S=60.0, T_1= 2.5, W_S= 60.0, L_4= 23, H_s= 1.6, L_5= 33, L_1= 19, L_6= 15, W_1= 2, W_3= 1.5, L_2= 8, D= 20, W_2= 2, H= 10, D_1= 20(All parameters in mm). Ground Plane (Das, 2018)

Figure 5. Two element MIMO antennas based on dielectric resonator for dual band application with L_s=60.0, T_1= 2.5, W_s= 60.0, L_4= 23, H_s= 1.6, L_5= 33, L_1= 19, L_6= 15, W_1= 2, W_3= 1.5, L_2= 8, D= 20, W_2= 2, H= 10, D_1= 20(All parameters in mm). MIMO with dielectric (Das, 2018)

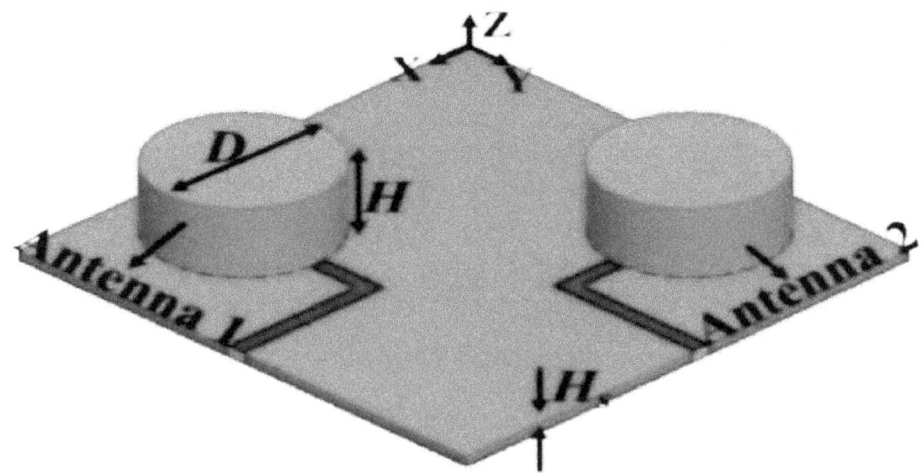

Modified rings create the HEM_{11} and TE_{01} modal fields in the CDRAs. The orthogonally placed radiators reduced the mutual coupling. To boost the isolation in the lower frequency region, tiny rectangular shape slot is carved into the ground plane. The restriction of surface current flow improved the isolation; the isolation further improved when the slit was etched. The two frequency bands that the proposed MIMO radiator generates are 1.75-2.4 GHz and 3.5-5.5 GHz.

An eight-element MIMO antenna for dual-band application is presented (Cui, 2019). The proposed antenna operates in two bands, one in 5G New Radio (NR) band n77 (3300-4200 MHz) and another in the 5 GHz band (4800-5000 MHz) in mobile handsets. Four sets of dual-antenna arrays (DAA) that are consistently printed across the Smartphone's two long side-edge frames can be used to create each of the 8-element MIMO antennas. In (Hussain,2019), Multiband application is achieved by four annular slot-line-based MIMO elements. A rectangular slot inside each annular slot extends its electrical length and lowers the resonance frequency even more, as represented in Figure 4.

Additionally, T shape defective ground structure can reduce isolation among the antenna elements as shown in Figure.4(b). A distinct split ring resonator (SRR) is inserted into each element, resulting in multi-band functioning. In (Ameen,2020), a miniature and triple-band two-element MIMO antenna is presented combining meander line and numerous straight line slots on the ground plane, as shown in Figure.5.

The top view of the antenna consit of two feed line as depicted in Figure. 5(a). Multiple slots were etched to create a DGS in the backside ground plane to provide triple-band antenna performance and antenna miniaturization as represented in Figure. 5(b).The ground plane is etched with a slot of meander line-shaped to obtain the first resonance at 2.42 GHz and a rectangular slot to obtain the second resonance at 3.56 GHz. The third resonance produce by adding a second rectangular slot close to the previous one at 5.43 GHz. In this design, the MS reflector and two-element MIMO antenna improve the antenna's impedance bandwidth (IBW), gain, and isolation properties.

Figure 6. Quad element MIMO antenna using SRR slot. Top View (Hussain,2019)

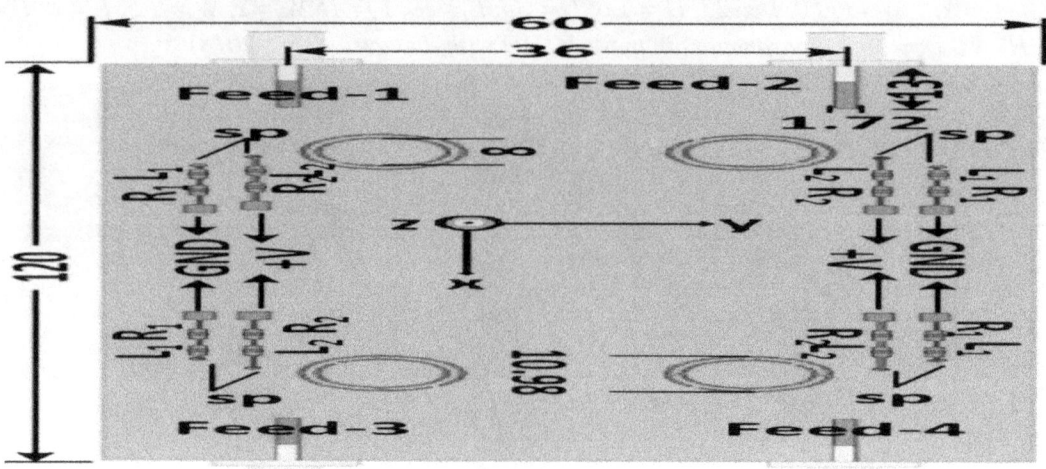

Figure 7. Quad element MIMO antenna using SRR slot. Back View(All the parameters in mm) (Hussain,2019)

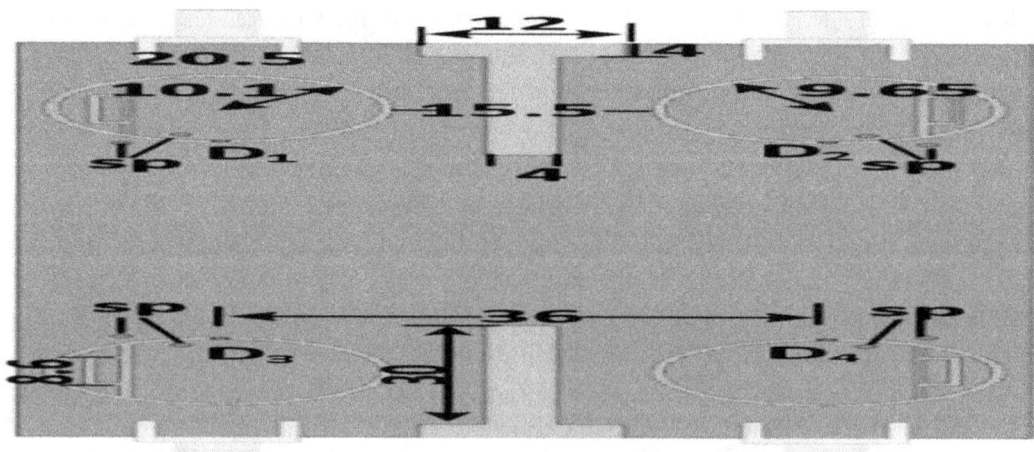

B. Wide Band MIMO Antennas

The demand for Wireless communication systems has significantly developed in recent times, and the ultra-wideband (UWB) technology has become popular. The Federal Communications Commission (FCC) designated 3.1 GHz from 10.6 GHz to the UWB spectrum for public use in February 2002 (Zhang, 2009). In order to modernise high data rate transmission, UWB technology has been acknowledged as one of the most essential and important options since 2002. Although UWB technologies have also faced some issues like the problems of reliability and multipath fading. MIMO technology has been proposed to resolve all these difficulties as a technical solution. In attempt to do this, UWB and MIMO technology integration is required, which can both increase the data transmission capacity and lessen the impact of the multipath issue.

Figure 8. Two-element MIMO antenna for triple band application (a) Top View with W₁=44, L₁=22

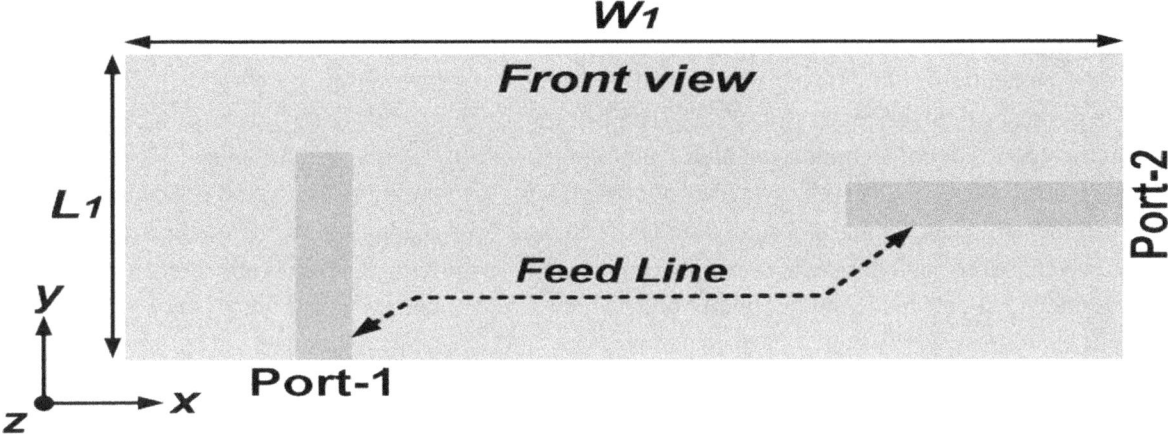

Figure 9. Two-element MIMO antenna for triple band application (b) Back View with Wₘ=10, Lₘ= 0.3 (All the parameters in mm) (Ameen,2020)

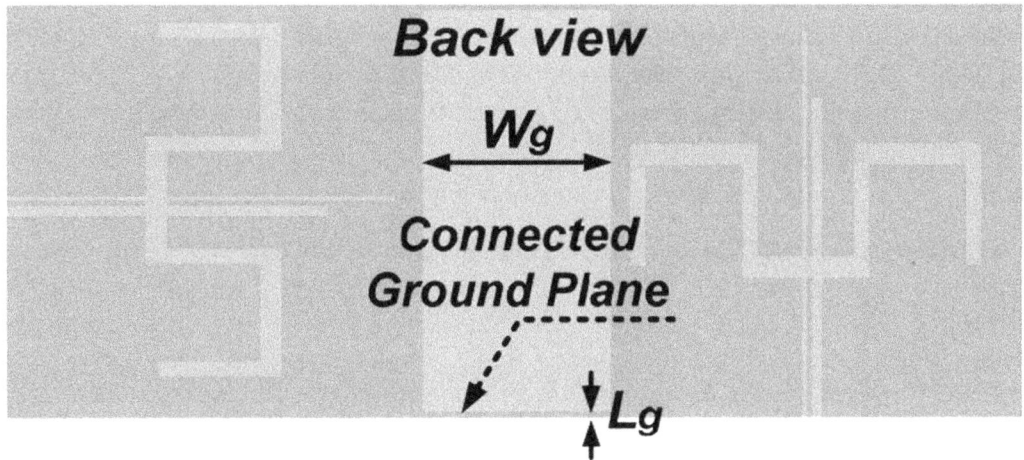

The antenna based on the tree structure used in the UWB application is presented (Zhang, 2009). In this design, a tree-like structure was used to achieve ultra-wide bandwidth. On each radiator, an L-shaped slot was engraved to broaden the bandwidth and generate more resonance. Each slot's length controls its resonance. This design enhances the antenna's bandwidth after introducing each branch.

Depending on bent slit geometry, a MIMO antenna with wideband features is presented (Li, 2012). In this design, two monopole antennas consisting of a dual branch structure were used, keeping some distance between them. To improve the wide impedance bandwidth and isolation, two bent slits are included to the ground plane. To further limit the effect of mutual coupling caused by near-field, a metal strip was inserted between the two monopoles. A MIMO antenna for wireless device application is presented (Lee, 2012). In this design, two symmetrical antennas were placed on the mobile device's substrate, and a meander-shaped parasitic line was introduced between the monopole to achieve better isolation. The suggested antenna can cover many bands and has an UWB of 3.1 to 10.6 GHz, namely WCDMA (1.92–2.17 GHz), WiMAX (2.3, 2.5 GHz) and WLAN (2.4 GHz). The interference of the WLAN band was reduced by introducing an open stub in each monopole design.

Based on neutralization line concept, A MIMO antenna system appropriate towards wireless USB-dongle purposes is suggested in (Su, 2012). The two monopoles in this design were positioned on the PCB's two opposite corners and separated from one another by a small ground piece. As an optional strategy, this ground area serves as an area of coverage for the feeding network and connectors for the usage of independent antennas. It originates that by etching 1.5 mm long inwards from the upper portion in the ground section and linking the two antennas inside with the help of a thin printed line, the isolation of recommended antenna could be greatly boosted.A MIMO antenna eligible for portable application is proposed in (Liu, 2013). A pair of planar-monopole (PM) antenna components fed by microstrip line was position orthogonal to one another. This orthogonal arrangement minimizes the mutual coupling among the antenna elements.To increase isolation and bandwidth, two extended ground stubs that operate as parasitic monopoles and a short ground strip are utilised.

In (Lee, 2013), a compact UWB MIMO antenna to fulfil the demand of wireless application is presented. In this design, two protruded ground planes are added together with antenna elements to enhance impedance matching and isolation. It provides UWB bandwidth from 3.0 to 11.0 GHz. To avoid interference of the WLAN band as well to achieve notching at 5.15 -5.85 GHz band, a loop path of 1.0 λ length is made of 1/3 λ rectangular metal strip is used with radiating elements. Again to achieve another notching from 3.30-3.70 GHz, open slots also etched on the each radiator.

Depending on parasitic element theory, a UWB diversity antenna with floating digitated decoupling structure is represented in Figure.6 (Khan, 2014).

Figure 10. MIMO antenna with floating parasitic element with ws = 45.5, ls = 33, wp$_1$ = 10, wp$_2$ = 4, lp$_1$ = 12, lp$_2$ = 4, sp = 5.5, lm = 2.1, wf = 1.7, lf = 9.9, wg = 10.5, lg = 6, sg = 13, wbb = 19.5, lbb = 3, wb = 1, lb1 = 3, lb2 = 7, lb3 = 8, lr = 3, wr = 7.5, lb = 15 and sb =6 (All parameters in mm). Top view (Khan, 2014)

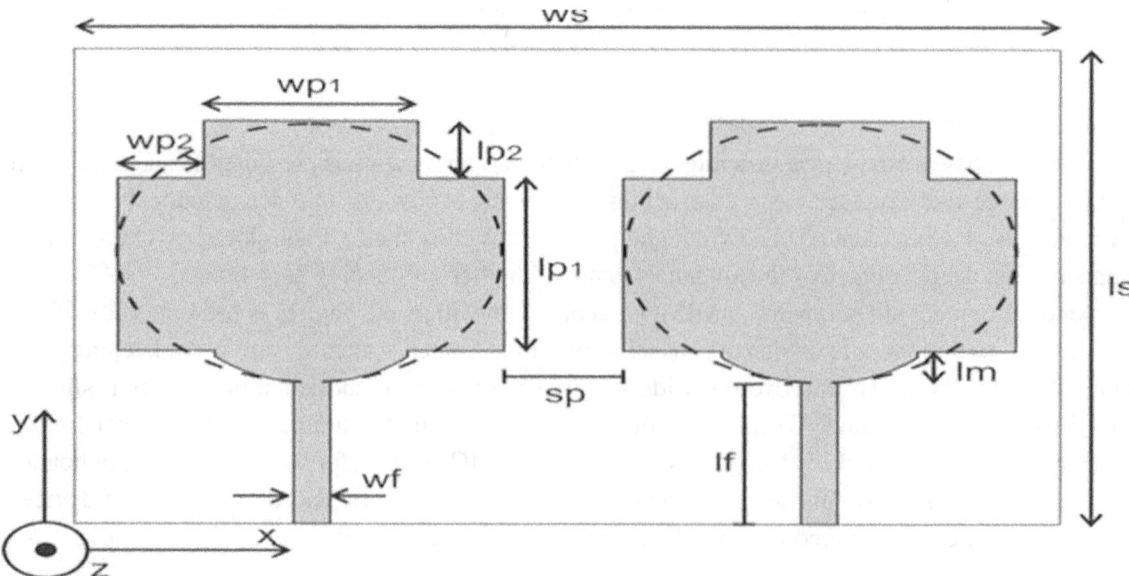

Table 1. Performance comparison of multiband MIMO antenna

Reference	Dimensions (mm³)/ Material	Bandwidth (GHz)	Isolation Level (dB)	Gain / Radiation Efficiency	ECC/DG
Manteghi, 2006	50×13×8	2.4-2.5, 5.15-5.35, 5.7-5.85	Not available	Not available	Not available
Bhatti, 2009	11.5× 13.8 ×4 FR-4	2.4–2.5, 3.4–3.6,5.15–5.35 and 5.75–5.875.	<-15	86%	9.6 dB
Nezhad, 2010	35×38×0.8 FR-4	2.4, 5.4, and 5.8	<-15	3.6 dB /90%	Lower than 0.002
Ling, 2011	15×50×0.8 FR-4	2.45, 5.8	<-20	4.12-4.86 dBi	Lower than 0.01
Zhou, 2012	50×63×0.8 FR-4	1.5–2.8,4.7–8.5	Not available	3 ~ 7dBi / 70%	Lower than 0.01
Sharawi, 2012	50×100×1.56 FR-4	803 - 823 MHz 2440-2900 MHz	<-17	(-4) ~ (-2.4) dBi	Lower than 0.21
Fernandez,2013	150×150×0.8 FR-4	699–798 MHz, 1.7–2.0, 2.3 and 3.5, 2.4 and 5.		5 dBi / 90%	Lower than 0.02
Karimian, 2013	33×36×1.524 Rogers RT Duroid 4003	2.28-2.66,3.35-3.65, 5.07-5.3, 5.75-5.85	< -14	5 – 3 dBi / 98%	Lower than 0.05
Shoaib, 2014	125×85×0.8 FR-4	826-1005 MHz 1527-2480 MHz 3439-3690 MHz 5340-5749 MHz	<-15	97.9%	Lower than 0.025
Wu, 2014	95×60×0.8 FR-4	2.50 to 3.25, 3.75 to 4.20	< -18	Not available	Lower than 0.2/9.8 dB
Li, 2014	140×70×9.55	2.6-2.8 and 3.4-3.6 GHz	<-20	3.7 dBi / 70%	Lower than 0.5
Sarkar, 2015	110×110×0.8 FR-4	1.95, 2.39, 2.64 and 3.27	< -20	93.5%	Lower than 0.01
Sun, 2016	100×65×1.6 FR-4	0.88 - 0.92, 1.78 - 1.81, 2.45 - 2.62	<-15	3.28dBi/74.99%	Not available
Peristerianos, 2016	121.8×61.45×0.812 Rogers RO3003	2.4 -2.489, 5 - 6	<-19	Not available	Lower than 0.1
Nandi,2017	45×25×1.57 Rogers RT/ Duroid 5880	2.37-2.64, 3.39-3.58 and 4.86-6.98	<-15	3.2 dBi / 95%	Lower than 0.012
Nandi, 2017	24×25×1.5 FR-4	2.35-2.69 and 5.4-5.68	<-20	Not available	Lower than 0.5
Das, 2018	60×60×1.6 FR-4	1.75–2.4 and 3.5–5.5	<-20	3 dB /82%	Lower than 0.16
Cui, 2019	18.6 × 7.0×0.8 FR-4	3300-4200 MHz 4800-5000 MHz	<-15	70%	Lower than 0.1
Hussain, 2019	60 × 120 × 0.76 RO3450	1.7–2.28, 2.35–2.85 and 2.9–3.1	<−11.5	2.95 dBi/ 82%	Lower than 0.2
Ameen,2020	22 ×22 ×1.6 FR-4	(2.44–2.47), (3.44– 3.50) and (5.15–5.48)	<-15	2 dBi/ 68.9%	Lower than 0.03/ >9.97 dB

Figure 11. MIMO antenna with floating parasitic element with ws = 45.5, ls = 33, wp$_1$ = 10, wp$_2$ = 4, lp$_1$ = 12, lp$_2$ = 4, sp = 5.5, lm = 2.1, wf = 1.7, lf = 9.9, wg = 10.5, lg = 6, sg = 13, wbb = 19.5, lbb = 3, wb = 1, lb1 = 3, lb2 = 7, lb3 = 8, lr = 3, wr = 7.5, lb = 15 and sb =6 (All parameters in mm). Bottom side with parasitic (Khan, 2014)

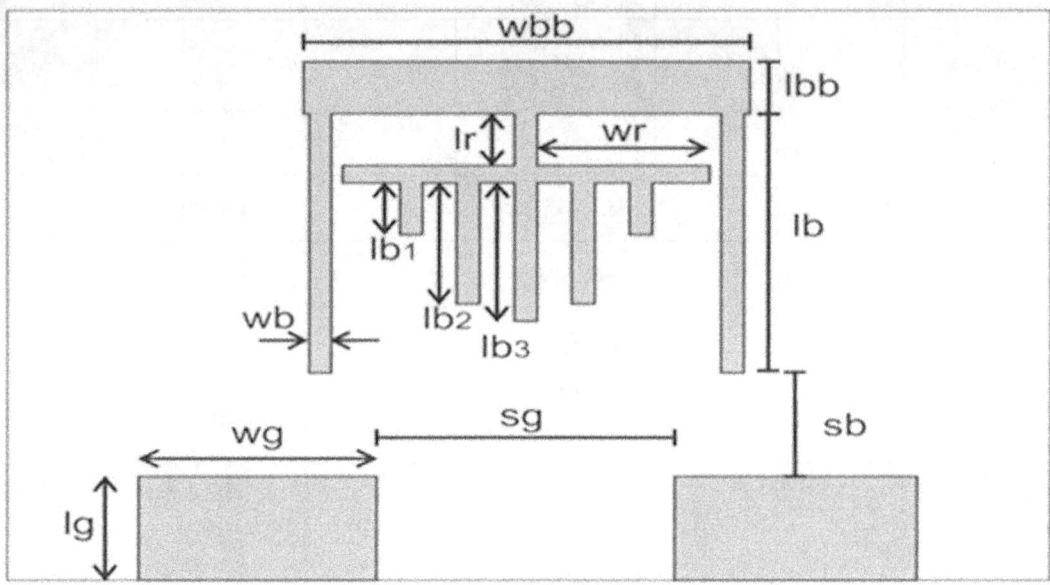

Figure 12. UWB MIMO antenna based on QSCA structure withL=21, L$_G$=7, l$_{f1}$=3, l$_{f2}$=4, l$_s$=2, ds=4.5, r=6, W=38, W$_G$=18, w$_{f1}$=3.6, w$_{f2}$=1, g=1, dr=1.5 (all are in mm) (Liu, 2014)

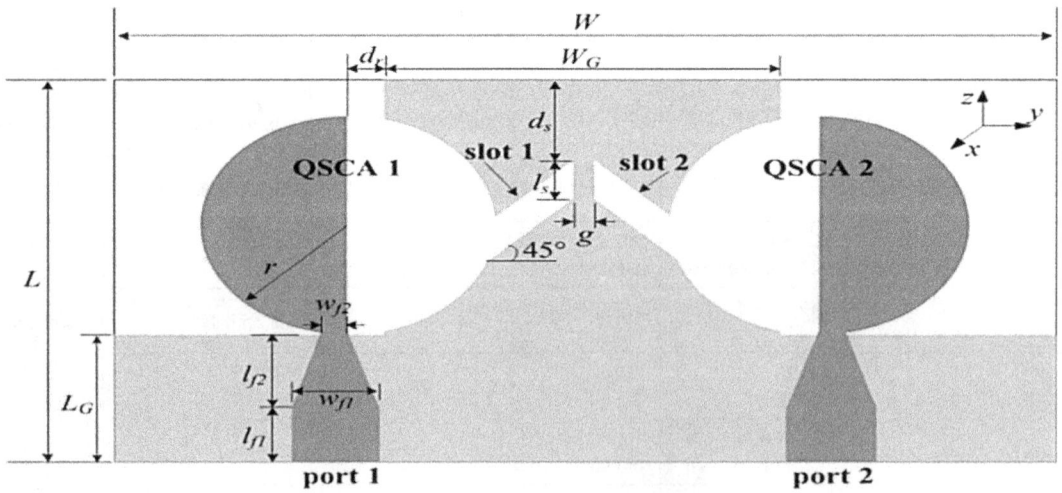

In this design, circular shape modified patch with straight edges on the three sides and arched feeding section was used to boost the impedance bandwidth as well as to obtain wide operating bandwidth ranging from 3.1 to 10.6 GHz. In order to create the floating decoupling structure on the substrate back side, vertical stubs of different lengths were joined to a horizontal conducting strip. These strips on the parasitic element operate as resonant components for various frequencies when they are placed in the shape of digits, and they improve isolation by introducing additional resonances in the radiating band. Better than -20.0 dB of isolation is provided among antennas elements owing to the decoupling structure.

Depending on the orthogonal arrangement, a slot MIMO antenna for diversity application purposes is proposed in (Gao, 2014). Here, two staircase-shaped monopole antennae have been arranged in an orthogonal pattern, and a coplanar waveguide has been employed to feed them. A rectangular stub slanted at 45 degrees was placed between the monopoles to achieve higher isolation. In this design, two split-ring resonators (SRR) slots were also introduced on each radiator to minimize the effect of interference at the WLAN band.

In (Liu, 2014), a quasi-self-complementary antenna (QSCA) element with a half-circular radiating patch on substrate's upper side and a slot with a complement of a similar shape on the other side is presented as shown in Figure. 7. Wide bandwidth can be achieved with a small footprint using QSCA.

Figure 13. Geometric configuration of the antenna. Front View with W_s= 40, L_S=37.5, L=20, W=23, L_T=1.6, L_{TE}=3, L_{TI}=2, W_T=1.5, W_{TE}=6.5, W_{TI}=4.9, F_L=17, F_W=3 (All parameters are in mm) (Jafri, 2016)

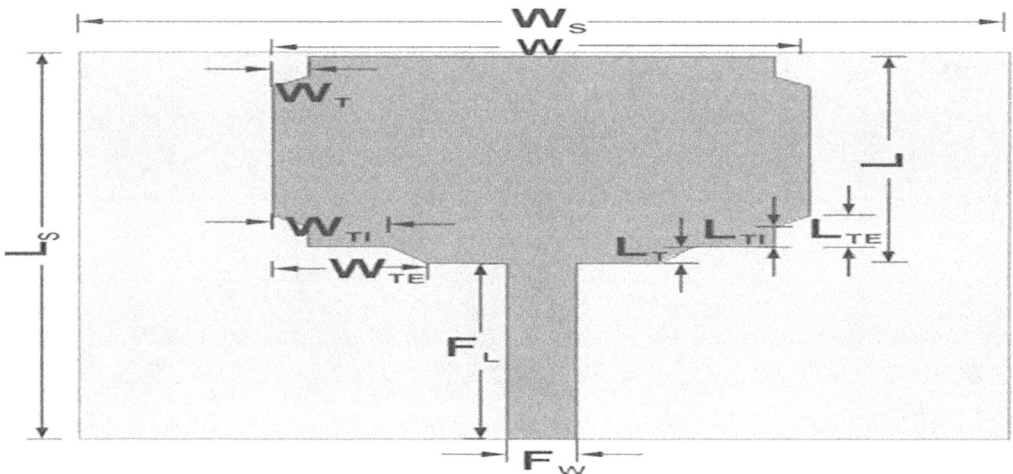

The rectangular shape slit present on the ground plane minimizes the mutual coupling effect and provides isolation of less than -15dB.

In (Tang, 2014), a MIMO antenna with dual-band notch characteristics and ultra-wide bandwidth is presented. The two rectangular shape radiating patches were used to make the antenna suitable for UWB application, but the isolation worsens owing to the close emplacement of the antennas. A decoupling approach is created on the substrate's bottom side to overcome mutual coupling. The decoupling structure helps to deteriorate the mutual coupling by forming another coupling path, resulting in solid isolation during UWB operation.

A wideband multiple-input–multiple-output (MIMO) antenna appropriate for Wi-Fi/2.4 GHz and Long-Term Evolution (LTE)/2.6 GHz wireless access point (WAP) applications in presented (Mora-diKordaliv, 2014). In this design, four wideband microstrip feedline printed monopole antennas with a standard rectangular shape radiating element and a ring-shaped ground plane constitute the proposed MIMO antenna system. To expand the impedance bandwidth, the radiator of the MIMO antenna system was designed as a modified rectangle with four-stepped lines at every end. Four slots were introduced to each corner of the ground plane to lessen mutual coupling.

For portable UWB applications, a MIMO band-notched antenna with a compact dimension is presented (Liu, 2015). Two square monopole antenna components, a T-shaped ground stub, a vertical slot to prevent mutual coupling, and two ground plane strips that create a notched frequency band constitute the antenna.Better impedance matching is provided by the T-shape ground stub, which also lowers the low-cut off frequency to about 2.8 GHz and creates a resonance at about 5 GHz.

Figure 14. Geometric configuration of the antenna. Back View with G_L=40, G_W=16.6, L_G=1.1, W_{GL}=5, W_{GU}=7 (All parameters are in mm) (Jafri, 2016)

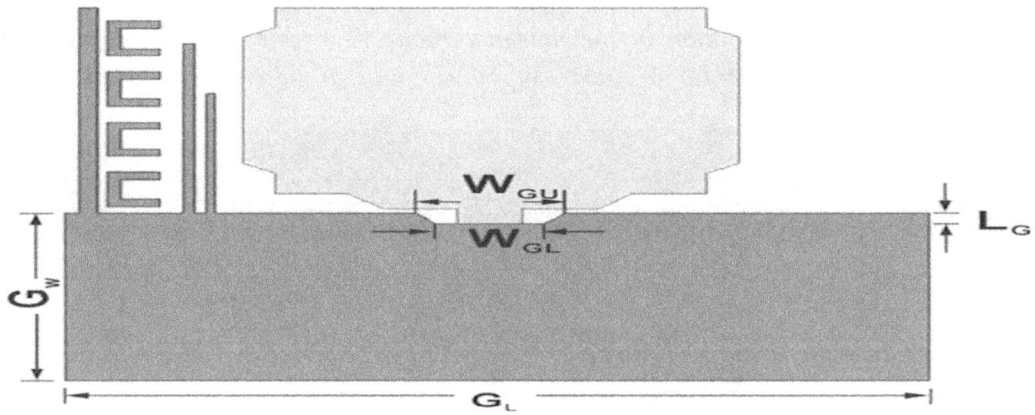

Figure 15. Geometric configuration of the antenna. (c) Decoupling Structure with L_{S1}=20.4, W_{S1}=0.9, L_{S2}=16.9, W_{S2}=0.5, L_{S3}=11.9, W_{S3}=0.5, L_C= 3.4, W_C= 2.5, T_C= 0.7, D_C= 1.6, S_{C1}= 0.7, S_{C2}= 1.1, S_2=S_3= 0.5(All parameters are in mm) (Jafri, 2016)

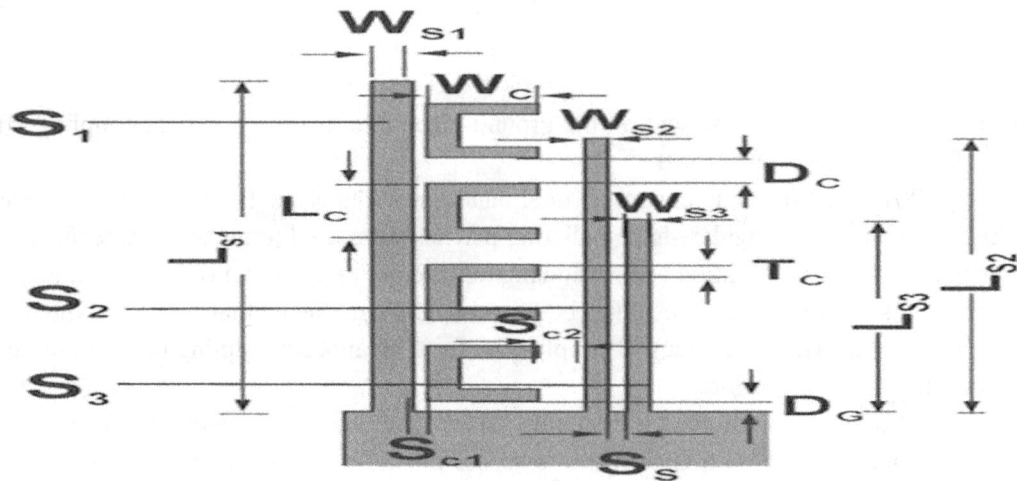

For ultra-wideband (UWB) applications, a band-notched MIMO antenna is described in (Kang, 2015). In this design, two offset microstrip-fed antenna elements are placed orthogonal and a rhombic shape slot on the ground plane. The offset microstrip-fed slot antenna can create a wider impedance bandwidth than the traditional centre-fed slot antenna. The orthogonally fed MIMO antenna delivers sound isolation as well as polarisation diversity. In order to further limit interference, a parasitic T-shaped strip is inserted between the antenna elements, and a notched band is made by cutting two L-shaped slits into the ground.

Based on fractal geometry, a Koch fractal shape octagonal MIMO antenna is presented (Tripathi, 2015). Koch geometry and an octagonal monopole were combined to prolong the electrical path length. For maximum isolation, these fractal monopoles are positioned orthogonally to one another. Additionally, grounded stubs were introduced to improve the isolation. The band rejection phenomena were produced by etching a C-shaped groove onto the antenna's monopole.

An extremely compact MIMO antenna with strong isolation is proposed in (Luo, 2015) for UWB systems. A step slot antenna was used as an antenna element in this design. An open-ended slit was engraved on the back of the feed to improve the impedance matching characteristic at a lower frequency. The T-shaped slot etched into the ground had two purposes: it extends the current channel and suppresses surface currents, improving isolation in the 4–10.6 GHz frequency range. A line slot is a neutralisation line to cancel the old coupling and lessen coupling at the 3–4 GHz band.

A wideband printed dipole with V-shaped ground branches for MIMO antennas presented (Wang, 2015). This design introduced a dipole with an integrated balun and V-shaped ground branches to achieve better impedance matching. In this design, the mutual coupling between adjacent elements was tightly managed by the orthogonal arrangement of elements. The V-shaped ground branches also minimize the spatial coupling between adjacent and opposite elements.

A compact UWB MIMO antenna with a neutralization line as a decoupling structure is presented (Zhang, 2016). In this design, two circular shape monopoles are connected by a wideband neutralizing line between them. Two metal strips and a metal circular disc comprise the neutralizing line. A number of decoupling current pathways of different lengths can be used on the circular disc to neutralize the ground plane coupling current. Wideband isolation is therefore feasible.

A tiny ultra-wideband MIMO antenna with strong isolation is proposed in this research (Deng, 2016)—meandering monopoles developed into the antenna. An ultra-wide bandwidth was created by combining two inverted L-shape parasitic slices with two smaller L-shape stubs. By carving a slot in the ground planes, it is feasible to achieve a higher level of isolation between two antenna elements. Band-notched features were presented for a dual-polarized UWB QSC MIMO/diversity antenna (Zhu, 2016). Dual polarisation was achieved by positioning antenna elements orthogonally and feeding them perpendicularly. This design can help to lessen mutual coupling and improve the MIMO system's reliability. Etching bent slits can create a notched band in the radiating elements.

A quad-element MIMO antenna array useful for portable wireless communication is achieved (Srivastava, 2016). As a UWB antenna element, the microstrip-fed stepped slot antenna is functional. The antenna provides a bandwidth ranging from 3.1 GHz to 12 GHz. Due to the intrinsic directional radiation qualities of slot antennas (SA) and their asymmetrical placements, high isolation (-22dB) between the UWB antenna elements was obtained in this design without the usage of a decoupling network.

A compact planar wideband MIMO antenna with high element isolation across a broad frequency range is presented (Anitha, 2016). A modified ground plane and an SRR-loaded printed monopole antenna are employed as an element. Using orthogonal parts placement, a quad-element antenna with low waveform correlation is formed.

Based on the reconfigurable antenna concept, a two-element MIMO antenna for UWB application is presented in (Jafri, 2016). In this work, modified square patches with chamfered corner profiles were used to increase the impedance bandwidth.

Antenna elements can be placed orthogonally for corner placement or back-to-back for compact three-dimensional (3D) modules to reconfigure the array. To lessen mutual coupling between antenna elements and provide mutual coupling larger than -20 dB, a decoupling device made up of four C-shaped strips sandwiched in the middle of three vertical stubs was attached to the substrates back as shown in Figure. 8(c).

Figure 16. MIMO antenna with rectangular DR for wide band application. Top View of the antenna with L=29, $W_2=W_3=1.1$, $L_1=3.5$, $W_4=0.75$, $L_2=11.5$, $L_3=10$, $L_4=3.3$, a=11, b=7 (All parameters in mm) (Abedian, 2017)

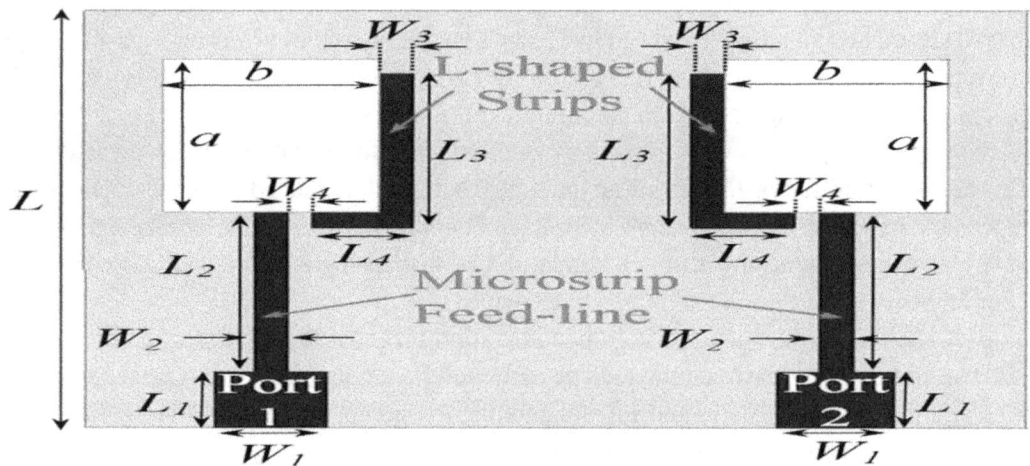

Figure 17. MIMO antenna with rectangular DR for wide band application. Bottom view of the antenna with stub with $L_g=9$, $W = W_g=29$, $W_5=3.8$, $L_s=14.5$, $W_s=3.4$ (All parameters in mm) (Abedian, 2017)

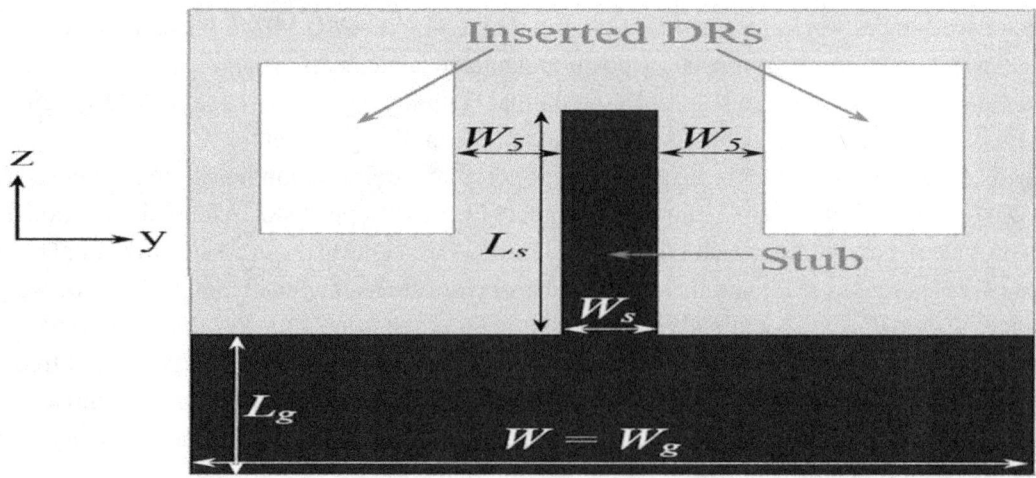

Figure 18. UWB MIMO antenna with mushroom shape DRA L = 52, W = 26, s = 13 (All the parameters in mm) (Sharma, 2017)

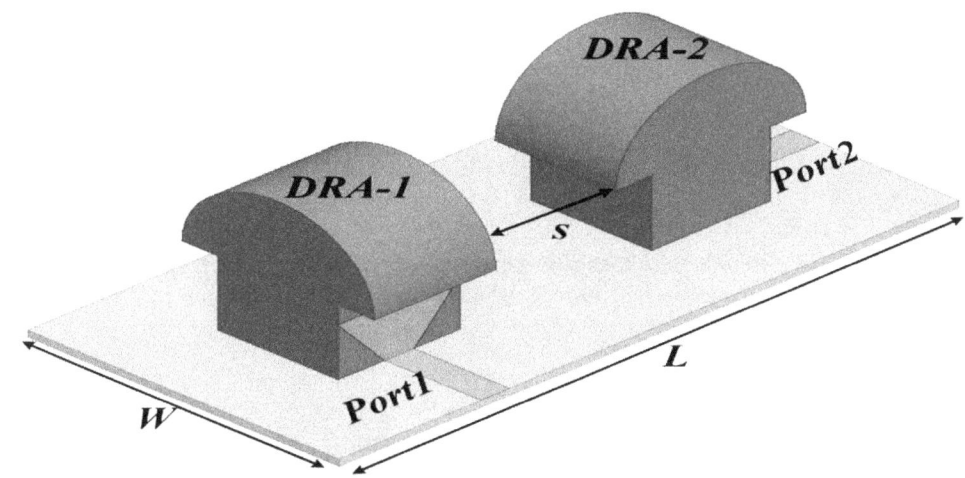

Figure 19. MIMO antenna with G-shaped Radiator (All the parameters in mm) (Toktas, 2017)

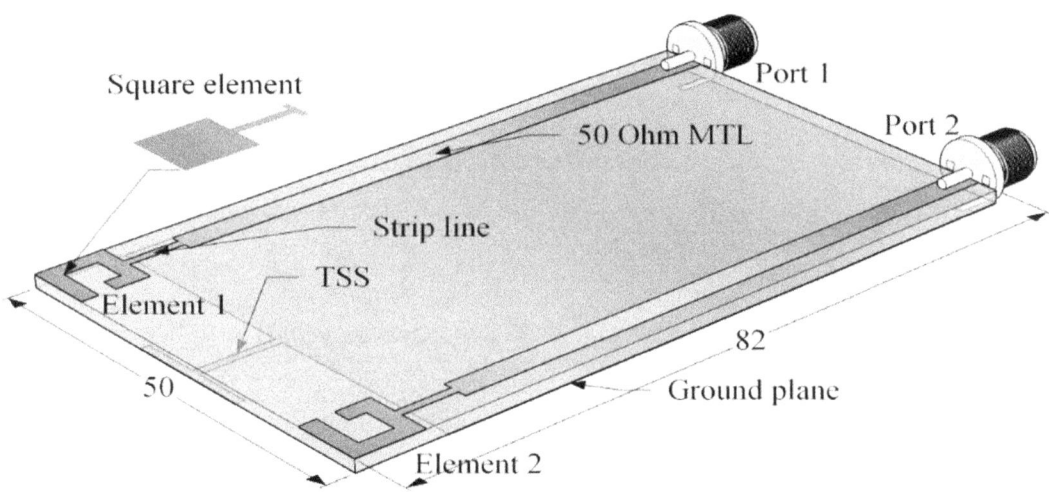

In (Abedian, 2017), Dielectric resonator (DRAs)-based UWBMIMO antennas with WLAN band rejection had been proposed. Two identical rectangular DRs were used in the antenna, which was fired using microstrip feeds, as represented in Figure. 9.

A ground-connected stub was added to the bottom plane to improve impedance bandwidth and isolation. In order to lessen interference with WLAN networks two L-shaped parasitic strips attached to the dielectric resonators.

In (Lin, 2017), a UWB MIMO antenna based on carbon black film technology is presented for isolation improvement. The carbon black film absorbs electromagnetic signals between the two antenna elements. According to Schelkunoff's hypothesis, there are three categories that make up the shielding effect of electromagnetic materials: reflection loss, absorption loss, and multi-reflection loss. The latter category includes the most significant effect factor. MIMO antenna with carbon black film varnished is expected to minimize total interference signal and increase isolation.

In (Tao, 2017), the half-slot structure with CPW fed is used for UWB application owing to its benefits of small size, wide bandwidth, and solid directional characteristics. The protruding ground plane, which contributes to the half-slot structure, serves as both a radiating element and an impedance-matching circuit. A Y-shaped slit was strategically carved into the bottom centre of the common ground plane to effectively stop current from flowing among two ports at the lower UWB frequency band.

Figure 20. UWB MIMO antenna with band notch structure Dimensions are L=15, L_{p1}=4.5, L_{p2}=1.5, L_{p3}=6, L_{g1}=6, L_{g2}=7,L_{g3}=2, ds=4.5, r=6, W=38, W_G=18, w_{f1}=3.6, w_{f2}=1, g=1, dr=1.5 (all are in mm) (Gautam, 2018)

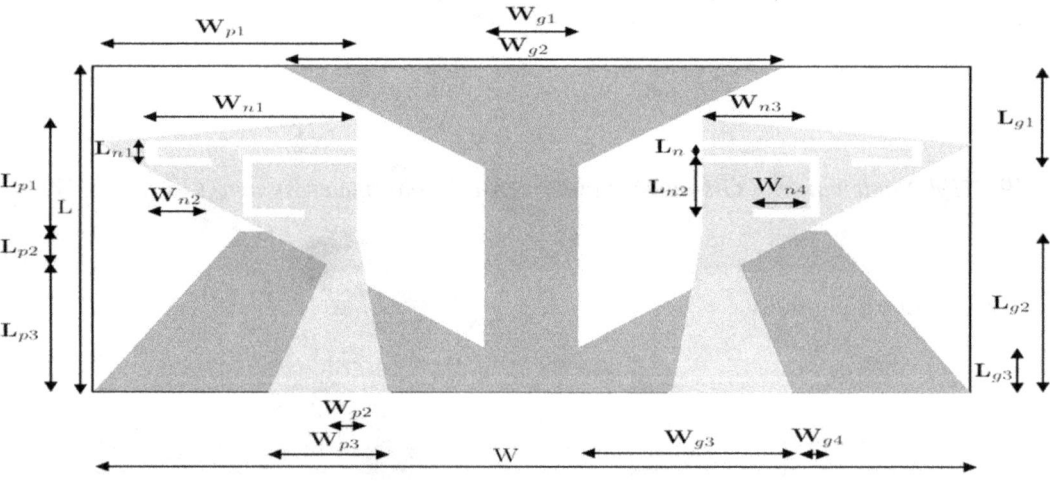

Figure 21. Tapered slot antenna for broadband MIMO applications with W_1= 9, W_2= 5, W_3=6.2, W_4= 10.6, W_5=7.5, W_6=5, W_7=11.8, W_8=2.36, L_1=21, L_2=17.5, L_3=12.5, L_4=5.5, d=1.8,d_1=2, p=2.4, L=55, W=46.8 (All parameters in mm) (Wu, 2018)

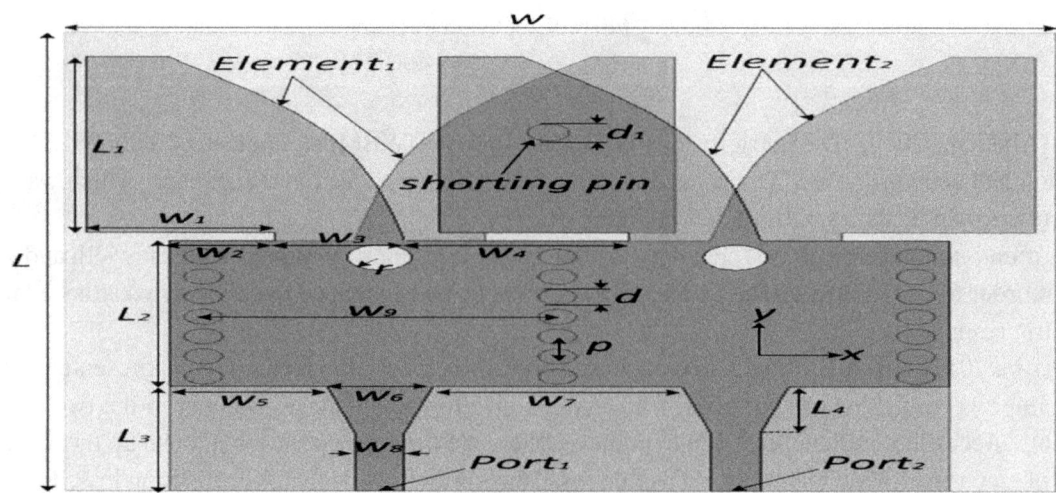

In (Sharma, 2017), two-element MIMO based on a mushroom-shaped dielectric resonator for wideband characteristics is presented. The mushroom-shaped dielectric resonator is activated by patch antenna of conformal trapezoidal shape used to achieve wideband functioning, as given in Figure 10.

Orthogonal arrangements of two wideband radiators provide polarization diversity and diminish mutual coupling among radiating elements. The mutual coupling between the adjacent antenna segments was controlled by the inter-element spacing 's'. It has been noted that the isolation grows as s increases. The proposed antenna provides wide bandwidth from 5.08 -9.80 GHz.

A triangular shape radiator for UWB application is reported in (Khanet, 2017). The first decrease in the space between the two parts was made possible using triangular forms, which prevent the monopoles from facing (and radiating) directly towards one another. In the ground plane, two inverted L-shaped stubs of length λ/4 were used to boost isolation, which serves as wave traps and increases resonance. The segmented L-shaped stubs initiated two resonances, at 3.8GHz and 6.8 GHz.

The band-notched MIMO antenna applicable for mobile terminals has been presented (Toktas, 2017). This system comprises G-shaped radiating elements, as shown in Figure.11.

Firstly, two symmetrical square elements form the MIMO antenna structure, which covers a wide frequency range of 2.2-13.3 GHz. To avoid the disrupting from critical WLAN band, the square components were changed utilising the slot–loading method. Split ring configurations were predicted to produce band-rejection operation. Therefore, by adding a G-shaped element to the square element, good results were obtained. The isolation level between the two parts was adequate for MIMO functioning, but the isolation for low frequency was further improved by using a T-shaped strip on the ground.

A UWB-MIMO spline antenna system of quad-element is reported (Bilal, 2017). The quad elements encircled a cuboid polystyrene block to form the MIMO design. A new inverted-L structure provided the isolation between the antenna ports with an array of slotted Y-shaped FSS. For bandwidth enhancement, chamfering and defective ground techniques were used.

Two triangular shape patches and a ground plane of funnel-shaped are used to achieve the UWB performance of an antenna, as shown in Figure.12 (Gautam, 2018). Funnel shape ground plane on the substrate's bottom significantly lessen the mutual coupling between the radiators and provide isolation of less than -24 dB.

Figure 22. MIMO antenna with hybrid ring dielectric resonator. Top View

Figure 23. MIMO antenna with hybrid ring dielectric. Back Viewwith L_2=3, L_3=0.5, L_4=2, L_5=0.5, L_6=1 (All parameters in mm) (Das, 2018)

Figure 24. UWB MIMO antenna with neutralization line. Top View with Ws=55, Ls= 35, D_1= 5, R_1= 10.5, R_2= 5, L_f= 9.6, W_f=3

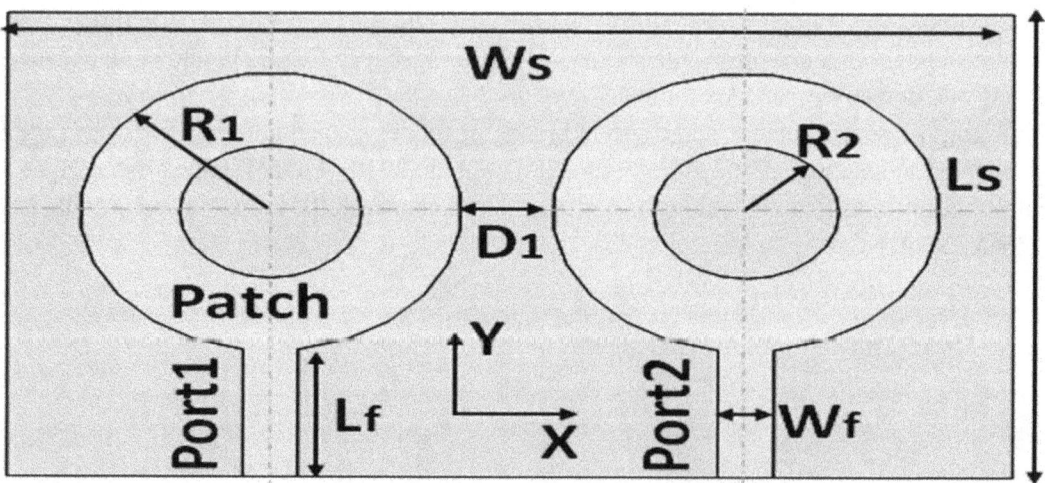

Two inverted J-shaped slits in the radiating patch were used to accomplish dual band rejection in the WLAN band of 5.1 to 5.8 GHz and the IEEE INSAT/Super-Extended C-band of 6.7 to 7.1 GHz. The density of current can be changed to provide band-notch features due to a J-shaped slit that created a quarter-guided wavelength resonator.

For broadband applications, a substrate-integrated waveguide (SIW) tapered slot antenna is presented in (Wu, 2018). Two broadband tapered slot antenna components, two SIW feeding configurations, and two microstrip-to-SIW transitions made up the MIMO antenna. Figure.13 shows the geometrical constructions.

Figure 25. UWB MIMO antenna with neutralization line. Back View with '8 shape' stubG_1=8.6, G_2=5.9, G_3=5.5,G_4=21,G_5=4.5,G_6=18,G_7=7,G_8=3 (All parameters in mm) (Biswas, 2019)

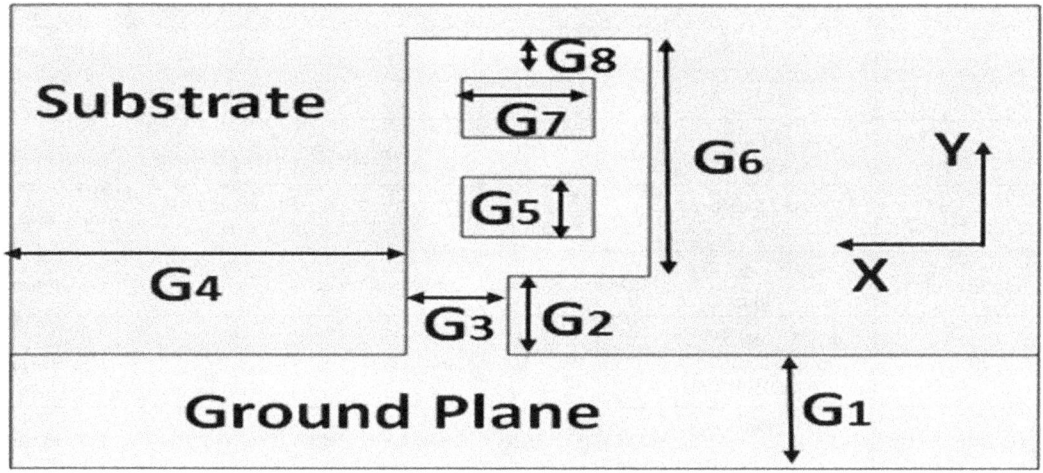

Two microstrip-to-SIW transitions were employed in the design for easy manufacturing and effective impedance matching. On the top and bottom grounds, symmetrically, four round apertures and four slits were engraved to provide better impedance matching and isolation. The antenna performs well from 8.0 to 15.0 GHz, with |S_{11}| £ −10 dB andS_{12} lower than −17 dB.

In (Das, 2018), a wideband MIMO antenna along with dielectric resonator of ring shape is presented. In order to activate the slots and the ring dielectric resonator, a 50 Ω microstrip line was used. Figure.14 illustrates the construction details.

At the upper layer of the substrate, two microstrip lines and two ring shape dielectric resonators were placed. The metallic ground plane of the antenna is comprised of stepped slots. Since the principal radiating directions of the suggested antenna have opposed one another, it is called self-complementary. The stepped slot and the ring dielectric resonator were both excited by the radiator to produce a wide impedance bandwidth.

A compact MIMO antenna suitable for handheld communication in the UWB region is presented (Chandel, 2018). In this design, two polygon shape radiators with tapered microstrip lines are used as feeding structures. The antenna's ground plane is made up of two inverted L-shaped stubs and a rectangular shape on the opposite side of the substrate. In order to create band-notched features in the WLAN band (5.09-5.8 GHz) and the IEEE INSAT/Super-Extended C-band, respectively, two L-shape slots were etched on each radiator. The recommended antenna has a 2.93 to 20 GHz bandwidth.

In (Biswas, 2019), wearable MIMO antenna applicable for UWB applications is presented. The patch components of the MIMO were printed on readily accessible, reasonably priced Jeans material. Figure.15 illustrates the construction of the antenna.

In this design, two antenna elements of ring-shaped were used. The antenna's ground plane comprised of one '8'-shaped stub attached to the substrate's back side's ground plane. The presence of the stub also enhanced the isolation. The antenna offers >26 dB port isolation throughout the whole UWB frequency range and a bandwidth span of 2.74 to 12.33 GHz (or around 127.27%).

Table 2. Performance comparison of wideband MIMO antenna

References	Dimensions (mm³)/ Materials	Band width(GHz)	Isolation Level (dB)	Gain/ Radiation efficiency	ECC/ DG
Zhang, 2009	35× 40×0.8 Taconic ORCER RF-35	3.1 to 10.6 GHz	Better than -16 dB	3.1 dBi	Lower than 0.01/ Larger than 9.95 dB
Li, 2012	78 ×40 ×1.6 FR4	2.4 to 6.55 GHz	Better than -18 dB	Not available	Not available
Lee, 2012	55×13.5×0.8 FR-4	1.85 to 11.9 GHz	Lower than -17 dB	4.96 dBi/ 91.36%,	lower than 0.18
Su, 2012	30×65×0.8 FR4	2.38 to 2.49 GHz	Lower than -19 dB	2.1 dBi/ 70%	Lower than 0.5
Liu, 2013	40×26×0.8 Rogers substrate, RO4350B	3.1 to 10.6 GHz	Lower than -15 dB	6.5 dBi/ 80%	Lower than 0.2
Li, 2013	27×30×0.8 FR4	3.0 to 11.0 GHz	Lower than -20 dB	70%	Lower than 0.002
Khan, 2014	45.5×33×1.524 Roger RO4003	3 to 10.6 GHz	Lower than -15 dB	5.3dBi/ 85%	Lower than 0.09
Gao, 2014	48×48×0.8 FR4	2.5 to 12 GHz	Lower than -15 dB	Not available	Lower than 0.005
Liu, 2014	21×38×1.6 Rogers substrate, RO4350B	2.7 to 12 GHz	Lower than -15 dB	4.2 dBi/ 70%	Lower than 0.15
Tang, 2014	48×48×0.8 FR4	2.5 to 12 GHz	-20 dB	3 dB	Lower than 0.005
MoradiKordaliv, 2014	21×30×1 FR4	2.14 to 11.4 GHz	< -20 dB	6.5 dBi/ 55%-88%	Lower than 0.1
Liu, 2015	32×32×0.8 FR4	3.1 to 10.6 GHz	Lower than -15 dB	4.2 dB/ 60%	Lower than 0.04
Kang, 2015	21 × 38×1.6 Rogers RO4350B	3.1 to 12 GHz	<−15	4.2 dBi/ 70%	lower than 0.15
Tripathi, 2015	45×45×1.6 FR4	2 to 10.6 GHz	<-17 dB	4 dBi	Lower than 0.003
Luo, 2015	22×26×0.8 FR4	3 to 10.6 GHz	<-18	3.8 dBi/ 90%	Lower than 0.004
Wang, 2015	40×40×1.6 FR4	2.30 to 4.40 GHz	Not available	2.1 dBi/ 68%	Lower than 0.045
Zhang, 2016	35×16×0.8 FR4	3.1 to 5 GHz	<-22 dB	7 dBi/ 50%	lower than 0.1
Deng, 2016	30×40×0.8 FR4	3 to 10.6 GHz	<-15	75%	Lower than 0.05
Zhu,2016	35×35×1 FR4	3 to12 GHz	<-22 dB	Not available	Lower than 0.5
Srivastava, 2016	42×25×1.6 FR4	3.1 to 12 GHz	<-22 dB	6 dBi/ 90%	Lower than 0.5
Anitha,2016	45×45×1.6 FR4	2.2 to 6.28 GHz	<-14	4 dBi/ 91%	Lower than 0.25

continues on following page

Table 2. Continued

References	Dimensions (mm³)/ Materials	Band width(GHz)	Isolation Level (dB)	Gain/ Radiation efficiency	ECC/ DG
Jafri, 2016	40 × 37.5 × 1.5 FR4	3.1 to 11 GHz	-20 dB	1.7 – 4.2 dB	Lower than 0.001
Abedian, 2017	29×29×1.524 Taconic RF-35	3.29 to 10.74 GHz	Lower than −15 dB	5.3 dBi/ 90%	Lower than 0.16
Lin, 2017	50×40×1.6 FR4	2.5 to 11 GHz	Lower than −15 dB	2.11 dBi/ 69.2%	Lower than 0.02
Tao, 2017	18×23×0.8 F4b-2	3 to 12.4 GHz	Lower than −15 dB	4 dB/ 70%	Lower than 0.015
Sharma, 2017	52×26	4.97 to 9.85 GHz	Lower than −15 dB	Not available	lower than 0.01/ 10 dB
Khanet, 2017	23 × 29×1.524 Rogers TMM4	3 to 12 GHz	Lower than −15 dB	5.9 dBi/ 82%	Lower than 0.15
Toktas, 2017	82×50×1.6 FR4	2.2 to 13.3 GHz	Lower than −15 dB	1.2 dBi/ 60%	lower than 0.04/ 9.8 dB
Bilal, 2017	36 × 32×1.5 FR4	3 to 10 GHz	Lower than -17 dB	70%	Lower than 0.0025
Gautam, 2018	26×15×1.6 FR4	3.1 to 35 GHz	Lower than -24 dB	8 dB/ 50%	lower than 0.2/ 9.9 dB
Wu, 2018	55 × 46.8 × 0.79 ArlonIsoClad 917	8.0 to 15.0 GHz	Lower than -17 dB	2.2 -8.7 dBi	Lower than 0.006
Das, 2018	50×31×1.6 FR4	3.3 to 7.65 GHz	Lower than -20 dB	4.5 dB/ 90%	lower than 0.01/ 9.9 4dB
Chandel, 2018	18×34×1.6 FR4	2.9 to 20 GHz	-22 dB	7 dB/ 75% -85%	Lower than 0.01/ > 9.95 dB
Biswas, 2019	55×35×1.5 Jeans material	1.51 to 12 GHz	Lower than -26 dB	6.9 dB	lower than 0.025/ 9.9 dB

A slot-based MIMO antenna for cognitive radio and UWB multi-function operation is presented in (Hussain, 2019). The biasing circuit, feed network, and SRR comprise the antenna's top layer, and UWB and FR antennas make up the bottom ground (GND) plane. To produce wideband frequency sweeps over the tuned bands, only one varactor diode is utilized for each antenna element. The suggested UWB architecture functions between 1.48 and 4.56 GHz. The proposed work's distinctive feature is the FR and UWB sensing antenna combination on the substrate board's top and bottom edges.

V. CONCLUSION AND FUTURE WORK

MIMO antennas are among the promising candidates for different wireless applications due to their higher data carrying capacity, reduced signal distortion due to the multipath phenomenon and time diversity and frequency diversity techniques. This review article describes the classification of MIMO antennas in terms of operating bandwidth. Each category is briefly discussed with the help of recent research to prove a particular technique's impact on antenna parameters. This review comes to the conclusion that various methods, such as various shape slots, metamaterial, an annular ring printed line with a CDRA, etc., enable multi-band and wideband MIMO antenna performance.Each category presents a comparison based on dimensions, the material used in substrate, bandwidth, isolation, gain, efficiency, ECC and DG. In keeping with this study, the new researchers will be enlightened with Multiband and wideband MIMO. They will serve as a basis for decision-making and a point of reference when choosing an appropriate approach to antenna design. The author regrets to the researcher society for any unintentional omission of innovative contributions in this subject throughout this investigation. As a result, this study offers a thorough overview of the beginner and specialised antenna designers.

In the upcoming years, there won't be a decrease in the need for faster data rates and video-on-demand. The 5G protocol is anticipated to offer data rates that are at least 1,000 times quicker than 4G. A handful of the enabling technologies developed, most of them are MIMO-based from the antenna design standpoint. According to a recent study, the communication sector is currently going forward with the goal of developing smart antennas, specialised, enhanced classes of antennas, and antennas with higher data rate capabilities and MIMO antenna fulfil all the facilities.

REFERENCES

Abdalla & Ibrahim. (2017). Design and performance evaluation of metamaterial inspired MIMO antennas for wireless applications. *Wireless Personal Communication, 95*(2), 1001–1017.

Abedian., M. (2017). Compact ultra wideband MIMO dielectric resonator antennas with WLAN band rejection. *IET Microwaves, Antennas & Propagation, 11*(11), 1524–1529.

Alamouti, S. M. (1998). A simple transmit diversity technique for wireless communications. *IEEE J. Selected Areas Communication, 16*, 1451-1458.

Ameen., M. (2020). Bandwidth and gain enhancement of triple band MIMO antenna incorporating metasurface-based reflector for WLAN/WiMAX applications. *IET Microwaves, Antennas & Propagation, 14*(13), 1493–1503.

Anitha., R. (2016). A Compact Quad Element Slotted Ground Wideband Antenna for MIMO Applications. *IEEE Transactions on Antennas and Propagation, 64*(10), 4550–4553.

BARTLETT. (1941).A dual diversity preselector. *QST*. XXV. 37–39.

Bhatti, R. A., Choi, J., & Park, S. (2009). Quad-Band MIMO Antenna Array for Portable Wireless Communications Terminals. *IEEE Antennas and Wireless Propagation Letters*. 8. 129-132.

Bilal., M. (2017). An FSS-Based Non-planar Quad-Element UWB-MIMO Antenna System. *IEEE Antennas and Wireless Propagation Letters, 16*, 987–990.

Bilal, M., Saleem, R., Abbasi, H. H., Shafique, M. F., & Brown, A. K.Bilalet al. (2017). An FSS-Based Nonplanar Quad-Element UWB-MIMO Antenna System. *IEEE Antennas and Wireless Propagation Letters, 16*, 987–990. doi:10.1109/LAWP.2016.2615884

Blanch, J. (2003). Exact Representation of Antenna System Diversity Performance from Input Parameter Description. *IET Electronic Letters, 39*(9), 705-707.

Chae, S. H., Oh, S., & Park, S.-O. (2007). Analysis of Mutual Coupling, Correlations, and TARC in WiBro MIMO Array Antenna. *IEEE Antennas and Wireless Propagation Letters*, 6, 122–125. doi:10.1109/LAWP.2007.893109

Chandel, & ... Gautam, A., & Rambabu, K. (2018). Tapered Fed Compact UWB MIMO-Diversity Antenna with Dual Band-Notched Characteristics. *IEEE Transactions on Antennas and Propagation*, 66(4), 1677–1684.

Vaughan, R. & Andersen, J. (2003). Channels Propagation and Antennas for Mobile Communications. IET.

Chen, Y. S. & Chang, C. (2016). Design of a four-element multiple-input– multiple-output antenna for compact long-term evolution small-cell base stations. *IET Microwave Antennas Propagation*, 10(4), 385–392.

Biswas, A. & Chakraborty, U. (2019). Compact wearable MIMO antenna with improved port isolation for ultra-wideband applications. *IET Microwaves, Antennas & Propagation*, 13(4), 498–504.

Nandi, S. & Mohan, A. (2017). CRLH Unit Cell Loaded Triband Compact MIMO Antenna for WLAN/WiMAX Applications. *IEEE Antennas and Wireless Propagation Letters*, 16, 1816–1819.

Cui., L. (2019). An 8-Element Dual-Band MIMO Antenna with Decoupling Stub for 5G Smartphone Applications. *IEEE Antennas and Wireless Propagation Letters*, 18(10), 2095–2099.

Das., G., Sharma, A., & Gangwar, R. (2018). Wideband self-complementary hybrid ringdielectric resonator antenna for MIMO applications. *IET Microwaves, Antennas & Propagation*, 12(1), 108–114.

Das, G., Sharma, A., & Gangwar, R. (2018). Dielectric resonator-based two-element MIMO antenna system with dual band characteristics. *IET Microwaves, Antennas & Propagation*, 12(5), 734-741.

Deng, G. (2016). An Ultra-wideband MIMO Antenna with a High Isolation. *IEEE Antennas and Wireless Propagation Letters*, 15, 182-185.

Brittain, J. (2008). Electrical Engineering Hall of Fame: Harold H. Beverage. *Proceedings of the IEEE*, 96, 9.

Fernandez, S. C., & Sharma, S. K. (2013). Multiband Printed Meandered Loop Antennas with MIMO Implementations for Wireless Routers. *IEEE Antennas and Wireless Propagation Letters*, 12, 96–99. doi:10.1109/LAWP.2013.2243104

Foschini, G. J., & Gans, M. J.Foschini and Gans. (1998). On limits of wireless communications in a fading environment when using multiple antennas. *Wireless Personal Communications*, 6(3), 311–335. doi:10.1023/A:1008889222784

Foschini, G. J. (1996).Layered space-time architecture for wireless communication in a fading environment when using multi-element antennas. *Bell Labs Technical Journal*, 41–59.

Gao., P. (2014). Compact Printed UWB Diversity Slot Antenna with 5.5-GHz Band-Notched Characteristics. *IEEE Antennas and Wireless Propagation Letters*, 13, 376–379.

Gautam, A. K. (2018). Design of ultra-compact UWB antenna with band-notched characteristics for MIMO applications. *IET Microwave Antennas Propagation*, 12(12).1895-1900.

Hussain., R. (2019). Split-ring-resonator-loaded multiband frequency agile slot-based MIMO antenna system. *IET Microwaves, Antennas & Propagation*, *13*(14), 2449–2456.

Hussain, R. & Sharawi, M. (2019). An Integrated Slot-Based Frequency-Agile and UWB Multifunction MIMO Antenna System. *IEEE Antennas and Wireless Propagation Letters, 18*(10).2150-2154.

Luo, H. & Zhong, L. (2015). Isolation Enhancement of a Very Compact UWB-MIMO Slot Antenna with Two Defected Ground Structures. *IEEE Antennas and Wireless Propagation Letters*, 14.

Jafri., S. (2016). Compact reconfigurable multiple-input multiple-output antenna for ultra-wideband applications. *IET Microwaves, Antennas & Propagation*, *10*(4), 413–419.

Jensen, M. A. (2016).A history of MIMO wireless communications. *IEEE International Symposium on Antennas and Propagation (APSURSI)* (pp. 681-682). IEEE.

Kang., L. (2015). Compact Offset Microstrip-Fed MIMO Antenna for Band-Notched UWB Applications. *IEEE Antennas and Wireless Propagation Letters*, *14*, 1754–1757.

Karimian, Oraizi, H., Fakhte, S., & Farahani, M. (2013). Novel F-Shaped Quad-Band Printed Slot Antenna for WLAN and WiMAX MIMO Systems. *IEEE Antennas and Wireless Propagation Letters*, *12*, 405–408. doi:10.1109/LAWP.2013.2252140

Khan., M. (2014). Compact ultra-wideband diversity antenna with a floating parasitic digitated decoupling structure. *IET Microwaves, Antennas & Propagation*, *8*(10), 747–753.

Khan, Capobianco, A.-D., Asif, S. M., Anagnostou, D. E., Shubair, R. M., & Braaten, B. D. (2017). A Compact CSRR-Enabled UWB Diversity Antenna. *IEEE Antennas and Wireless Propagation Letters*, *16*, 808–812. doi:10.1109/LAWP.2016.2604843

Khanet, L. (2017). A Compact CSRR-Enabled UWB Diversity Antenna. *IEEE Antennas and Wireless Propagation Letters*, *16*, 808–812.

Kordaliv, M. (2014). Common Elements Wideband MIMO Antenna System for WiFi/LTE Access-Point Applications. *IEEE Antennas and Wireless Propagation Letters*, *13*, 1601–1604.

Lee., J. M. (2012). A Compact Ultra wideband MIMO Antenna With WLAN Band-Rejected Operation for Mobile Devices. *IEEE Antennas and Wireless Propagation Letters*, *11*, 990–993.

Li., J. F. (2013). Compact Dual Band-Notched UWB MIMO Antenna with High Isolation. *IEEE Transactions on Antennas and Propagation*, *61*(9), 4759–4766.

Li., G., Zhai, H., Ma, Z., Liang, C., Yu, R., & Liu, S. (2014). Isolation-Improved Dual-Band MIMO Antenna Array for LTE/WiMAX Mobile Terminals. *IEEE Antennas and Wireless Propagation Letters*, *13*, 1128–1131. doi:10.1109/LAWP.2014.2330065

Li, J., Chu, Q., & Huang, T. (2012).A Compact Wideband MIMO Antenna with Two Novel Bent Slits. *IEEE Transactions on Antennas and Propagation*, *60*(2), 482-489.

Lin., G. S. (2017). Isolation Improvement in UWB MIMO Antenna System Using Carbon Black Film. *IEEE Antennas and Wireless Propagation Letters*, *16*, 222–225.

Ling, X. M., & Li, R. L. (2011). A Novel Dual-Band MIMO Antenna Array with Low Mutual Coupling for Portable Wireless Devices. *IEEE Antennas and Wireless Propagation Letters*, *10*, 1039–1042. doi:10.1109/LAWP.2011.2169035

Liu., D. Q. (2018). Dual-Band Platform-Free PIFA for 5G MIMO Application of Mobile Devices. *IEEE Transactions on Antennas and Propagation*, *66*(11), 6328–6333.

Liu, L. Cheung, A. & Yuk, T. (2013). Compact MIMO Antenna for Portable Devices in UWB Applications. *IEEE Transactions on Antennas and Propagation*, *61*(8), 4257-4264.

Liu, L., Cheung, A., & Yuk, T. (2014). Compact multiple-input–multiple-output antenna using quasi-self-complementary antenna structures for ultra-wideband applications. *IET Microwave Antennas Propagation*, *8*(13), 1021-1029.

Liu, L., Cheung, A., & Yuk, T. (2015). Compact MIMO Antenna for Portable UWB Applications with Band-Notched Characteristic. *IEEE Transactions on Antennas and Propagation*, *63*(5), 1917-1924.

Manteghi, M. & Rahamat, Y.. (2005). Multiport Characteristics of a Wideband Cavity Backed Annular Patch Antenna for Multi polarization Operations. *IEEE Transactions on Antennas and Propagation*, *1*, 466–474.

Manteghi, M. & Tahmat-Samii, Y. (2006). Novel Compact Tri-Band Two-Element and Four-Element MIMO Antenna Designs. *IEEE Antennas and Propagation Society International Symposium*, (pp. 4443-4446). IEEE.

Nandi., S. & Mohan, A. (2017). A Compact Dual-Band MIMO Slot Antenna for WLAN Applications. *IEEE Antennas and Wireless Propagation Letters*, *16*, 2457–2460.

Nezhad, MHassani, H. R. (2010). A Novel Tri-band E-Shaped Printed Monopole Antenna for MIMO Application. *IEEE Antennas and Wireless Propagation Letters*, *9*, 576–579. doi:10.1109/LAWP.2010.2051131

Ou, C., Cai, X., & Qian, K. (2017). Two-Element Compact Antennas Decoupled with a Simple Neutralization Line. *Progress In Electromagnetics Research*, *65*, 63–68. doi:10.2528/PIERL16111801

Peristerianos, Theopoulos, A., Koutinos, A. G., Kaifas, T., & Siakavara, K. (2016). Dual-Band Fractal Semi-Printed Element Antenna Arrays for MIMO Applications. *IEEE Antennas and Wireless Propagation Letters*, *15*, 730–733. doi:10.1109/LAWP.2015.2470681

Pozar, D. (1998). *Microwave Engineering* (2nd ed.). John Wiley.

Rajagopalan, A. & Gupta, G. (2007). Increasing Channel Capacity of an Ultra-wideband MIMO System Using Vector Antennas. *IEEE Transactions on Antennas and Propagation*, *55*(10), 2880-2887.

Raleigh, G. G., & Cioffi, J. M. (1998). Spatio-temporal coding for wireless communication. *IEEE Transactions on Communications*, *46*(3), 357–366. doi:10.1109/26.662641

RCA Review. (1931).Diversity telephone receiving system of R.C.A. Communications, Inc. *Proc. IRE*, *19*, 562–584.

Sarkar, Singh, A., Saurav, K., & Srivastava, K. V. (2015). Four-element quad-band multiple-input–multiple-output antenna employing split-ring resonator and inter-digital capacitor. *IET Microwaves, Antennas & Propagation*, *9*(13), 1453–1460. doi:10.1049/iet-map.2015.0189

Sharawi, M. S., Numan, A. B., Khan, M. U., & Aloi, D. N. (2012). A Dual-Element Dual-Band MIMO Antenna System with Enhanced Isolation for Mobile Terminals. *IEEE Antennas and Wireless Propagation Letters*, *11*, 1006–1009. doi:10.1109/LAWP.2012.2214433

Sharawi, M. S. (2013). Printed Multi-Band MIMO Antenna Systems and Their Performance Metrics. *IEEE Antennas and Propagation Magazine, 55*(5), 218-232.

Sharma, A. & Biswas, A. (2017). Wideband multiple-input–multiple-output dielectric resonator antenna. *IET Microwaves, Antennas & Propagation*, *11*(4), 496–502.

Shoaib, Shoaib, I., Shoaib, N., Xiaodong Chen, & Parini, C. G. (2014). Design and Performance Study of a Dual-Element Multiband Printed Monopole Antenna Array for MIMO Terminals. *IEEE Antennas and Wireless Propagation Letters*, *13*, 329–332. doi:10.1109/LAWP.2014.2305798

Srivastava, GMohan, A. (2016). Compact MIMO Slot Antenna for UWB Applications. *IEEE Antennas and Wireless Propagation Letters*, *15*, 1057–1060.

Su, S., Lee, C., & Chang, F.-S. (2012). Printed MIMO-Antenna System Using Neutralization-Line Technique for Wireless USB-Dongle Applications. *IEEE Transactions on Antennas and Propagation, 60*(2),456-463.

Su, S., Lee, S., & Chang, F. (2012). Printed MIMO-Antenna System Using Neutralization-Line Technique for Wireless USB-Dongle Applications. *IEEE Transactions on Antennas and Propagation,* 60(2), 456-463.

Sun, Fang, H.-S., Lin, P.-Y., & Chuang, C.-S. (2016). Triple-Band MIMO Antenna for Mobile Wireless Applications. *IEEE Antennas and Wireless Propagation Letters*, *15*, 500–503. doi:10.1109/LAWP.2015.2454536

Tang, TLin, K. (2014). An Ultra wideband MIMO Antenna with Dual Band-Notched Function. *IEEE Antennas and Wireless Propagation Letters*, *13*, 1076–1079.

Tao, J. & Feng, Q. (2017). Compact Ultra wideband MIMO Antenna with Half-Slot Structure. *IEEE Antennas and Wireless Propagation Letters*, *16*, 792-795.

Toktas, A. (2017). G-shaped band-notched ultra-wideband MIMO antenna system for mobile terminals. *IET Microwave Antennas Propagation, 11*(5), 718-725.

Tripathi, S., Mohan, A., & Yadav, A. (2015). A Compact Koch Fractal UWB MIMO Antenna with WLAN Band-Rejection. *IEEE Antennas and Wireless Propagation Letters*, *14*.

Wang., H. (2015). A Wideband Compact WLAN/WiMAX MIMO Antenna Based on Dipole With V-shaped Ground Branch. *IEEE Transactions on Antennas and Propagation*, *63*(5), 2290–2295.

Wu., Y., Zhang, B., & Ding, K. (2018). SIW-tapered slot antenna for broadband MIMO applications. *IET Microwaves, Antennas & Propagation*, *12*(4), 612–616.

Wu, Y.-T., & Chu, Q.-X. Wu and Chu. (2014). Dual-band multiple input multiple output antenna with slitted ground. *IET Microwaves, Antennas & Propagation*, *8*(13), 1007–1013. doi:10.1049/iet-map.2013.0340

Zhang, S. (2009). Ultra-wideband MIMO/Diversity Antennas with a Tree-Like Structure to Enhance Wideband Isolation. *IEEE Antennas and Wireless Propagation Letters*, *8*, 1279–1282.

Zhang, SPedersen, G. (2016). Mutual Coupling Reduction for UWB MIMO Antennas with a Wideband Neutralization Line. *IEEE Antennas and Wireless Propagation Letters*, *15*, 166–169.

Zheng, L. & Tse, D. (2003). Diversity and multiplexing: A fundamental trade-off in multiple-antenna channels. *IEEE Transaction Information Theory*. *49*(5), 1073–1096.

Zhou, X., Quan, X., & Li, R. (2012). A Dual-Broadband MIMO Antenna System for GSM/UMTS/LTE and WLAN Handsets. *IEEE Antennas and Wireless Propagation Letters*, *11*, 551-554.

Zhu., J. (2016). Compact Dual-Polarized UWB Quasi-Self-Complementary MIMO/Diversity Antenna with Band-Rejection Capability. *IEEE Antennas and Wireless Propagation Letters*, *15*, 905–908.

Zhu, Yang, X., Song, Q., & Lui, B. (2017). Compact UWB-MIMO antenna with metamaterial FSS decoupling structure. *J Wireless Communication Network*, *2017*(1), 1. doi:10.118613638-017-0894-3

Chapter 5
Design of Software–Defined Network (SDN)–Enabled Network Infrastructure

Devendra Kumar Somwanshi

 https://orcid.org/0000-0003-4331-0917

Poornima College of Engineering, India

ABSTRACT

In this chapter, SDN is utilized as a unified answer for measuring the packet loss, memory optimization, and network delay. Presently, there is no single arrangement available to measure all the issues at once, but numerous different strategies are used. By using different strategies, separate frameworks are required, and it increases the establishments and cost. While these strategies are not the ideal solution, it is just an arrangement to give the solution of a particular issue. Besides, those techniques are not proficient to meet every one of the necessities for a checking arrangement—some are not exact or sufficiently granular, and others are adding extra organization burden or absence of versatility. Techniques have been proposed to find out the packet loss, memory optimization, and network delay. The ongoing checking of SDN Regulator framework is utilized, and it gives data and key execution markers for network. It assists with making an execution model for Regulator as a component of basic organization boundaries.

1. INTRODUCTION

In recent times, data centers have gained prominence enough to be considered an essential framework for their caliber to stock massive amounts of information and facilitate governance practices. Now a days organizations use server farms to calculate the huge reach of their and IT organizations. Server virtualization and distributed computing have an impact on the approach to using server farms. To use IT assets more competently with finesse and in a more controlled manner, virtualization is used. Distributed computing completes the organization's requirements to make agile models which works on demand and organizations does not need any dedicated resources. Together, these advances make it easier for associations to meet authority needs and provide server farms with significantly more ingenuity.

DOI: 10.4018/978-1-6684-7000-8.ch005

1.1 Software-Defined Network

Software-Defined Network brings new capabilities and helps support many of the challenges facing legacy organizations. This approach involves isolating the organization's vision from the package exchange utility and placing it in a legally incorporated regulatory body. The throttle is responsible for sending selects that are fed into the switches using standard conventions, such as OpenFlow. SDN's inspiration was to use an organizational framework, where enterprise organizations could do it without adding additional programming to each exchange component and considering creating control applications. (Rao et al., 2016).

OpenFlow Protocol is the consequence of a six-year research cooperation between Stanford College and the College of California at Berkeley. It is a programmable network Protocol which works as a set of connection and coordinate the traffic between switches and remote controllers. It provides more controls on switches and routers.

There are three major parts in SDN, 1. Open Flow Protocol, 2. Steam Table 3. Controller. OpenFlow Protocol works on the basis of flow tables. The flow tables are used to install on the switches. The controllers control the switches on the basis of the Protocol and decide the flow through switches. The controller works on the specific characteristic to optimize the network and control the path on the basis of reduced latency, or speed and number of hops.

1.2 Traditional Network Architectures

In Traditional Networks to address the separate issues of the businesses isolated network are worked in which two planes are used i.e., control plane and the information sending plane on a similar device.

- Control Plane is the piece of an organization that conveys flagging traffic and is liable for directing. Control parcels begin from or are bound for a switch.
- Sending Plane is where the activity happens. It incorporates things like the sending tables, steering tables, ARP tables, line's labeling and yet again labeling, and so forth. The information plane does the orders of the control plane.

Although such sort of organization design has turned out great previously, however with today virtualized world it will be testing in the event that certainly feasible for customary organization to meet the new virtual necessities. With the present restricted or level spending plans.

1.2.1 Traditional Networks Limitations

While the current conventional organization designs were not underlying a method for meeting the present necessities for end-clients, specialist co-ops and ventures; a few restrictions of customary organization engineering are (Caesar et al., 2005)

- Past PC network innovations had forever been based on a bunch of steering conventions that are designed to interface has in a dependable way over significant distances with high velocities and different organization plans. To meet the business necessities, for example, high accessibility, security and broadened network, throughout the past many years, conventions have been planned in a ton of ways that lead to detachment, where every convention is to take care of a particular sort of issues, without remembering to profit from deliberations. Such methodology of configuration has prompted one of the principal issues that network overseers are confronting these days, to be specific organization the board intricacy. One model is that including or eliminating a gadget an organization has turned into a weight for network chairmen, where a few pieces of the organization must be reconfigured, for example, Access records, VLANs nature of administration strategies, directing conventions and organization geographies.

- Adjacent to the referenced above, gear merchant and programming adaptations similarity must be considered prior to making any adjustment to the organization. Subsequently, network managers keep their organization somewhat static, to stay away from or limit the help margin time.

- Traditionally virtualization administration was presented, a solitary server interfaces with chose clients. What's more, by and large VMs need to move to get adjusted jobs; such usefulness of virtualized stages put a ton of difficulties on the customary systems administration.

- Trouble to apply arrangements of policies as in now a days it is quite challenging for Network administrator to complete the continuous demands of the network with optimization.

- Versatility issues: Normally server farms have a popularity to develop quickly, simultaneously the organization needs to develop at a similar speed. In any case, truly, networks have become very mind boggling because of the expansion of hundreds on the off chance that not a huge number of switches.

- Gear produces trustworthiness: Internet specialist co-ops and Datacenters generally anticipate executing new elements and administrations to fulfill the changing business necessities or end client's requests. Typically, the capacity to reaction is limited by the gear seller's life cycles for the created administration and hardware, which now and again can be around three years or considerably more.

1.3 The SDN Architecture

The basic idea of SDN is pretty basic. Regular switches, switches or some other organizational gadgets have two planes, in which data plane gives up sending the information; It was then known as an information plane or a means of transport. While the Control Plane is capable of providing all the organizational and dynamic knowledge about the direction of traffic.

The SDN system comprises of three layers. The first layer is known as Infrastructure Layer, it also known as information plane or data plane. This layer manages the transmitter plane or forward plane with the programming. It is the middle layer therefore it collects the information provided the primary layer and on the basis of information, network operation is to be decided and the information is transmitted through scheduling controllers to transmitter network. The third and last layer is Application layer which add the new features in the network such as security, manageability. This layer guides the control layer to configure the network as per the global view of the network.

1.4 Benefits of SDN

Programming defined networking will meaningfully have an impact on the way that organization specialists and fashioners fabricate and work their organizations to accomplish business prerequisites. The benefits of SDN are as follows:(Feamster et al., 2013).

- Management of Network: In SDN the whole network can be managed in a simple manner as it can be controlled viewed and controlled through a single node.
- Service Deployment time reduction: new highlights can be sent in a quick way inside the space of hours rather than numerous days thus it reduces the times in deployment.
- Computerized Configuration: Manually design errands like allotting VLAN and arranging QoS can be automatically configured.
- Network Virtualization: Due to virtualization of networks, the networks can be deployed easily with the requirement of organizations.
- Operational Expense Reduction: Due to automation is applied in network configuration and networks are deployed virtualized, a change on the organization has never been more straightforward, subsequently lessening the expense of the organization activity.

Table 1. Categorical Review Process Outcome

Solution Approaches	Input Parameters		Performance Parameters		Results
	Switch	**Host**	**Topology**	**SDN Controller**	
Open Flow Security Engine (OFSE) (Giustiniano et al., 2013)	Two switches HP E3800 and one HP 5406zl	NA	Tree	NOX controller	Works perfectly without any infrastructure
Advanced Message Queuing Protocol (AMQP) (Chen-Xiao & Ya-Bin, 2016)	Enable Switch	2 Host	Tree	Open Day light Controller	Switch limit increased by double as compared to Traditional hardware switching
NOSIX, ForCES (Feamster et al., 2013)	8 Switch	1 Host	Tree	POX Controller	Throughput improved upto 25% and Delay improved upto 20%
Intrusion detection systems (IDS) (Zander & Forchheimer, 1988)	OpenFlow Switch enable	2 Host	Grid	NOX Controller	Efficiency was high
ECMP (Caesar et al., 2005)	Enable Switch	NA	NA	POX Controller	Processing was increased due to programmable switches.
Web Framework with Beacon the Inversion of Control (IoC) container (OpenFlow Switch Specification, n.d)	Enable Switch	NA	NA	OpenDaylight Controller	Beacon high performance with processing 12.8 million Packet In messages per second
NOS (Network Operating System) (Open Networking Foundation, n.d)	6 Switches,	2 Host	Tree	POX Controller	Easier topology with scalability issues.
Protocol Oblivious Forwarding (POF) (Sherwood et al., 2009)	Switch Enable	1 Host	Tree	NOX, POX, Floodlight	Simpliðed switches based flow forwarding
MPLS-based VPNs, Bidirectional Forwarding Detection(BFD) (Tennenhouse & Wetherall, 1996)	6 Switches	1 Host	Tree	NOX controller	Network management and operation were improved
Dijkstra's algorithm (, J. & Jehn-Ruey, 2011)	6, 12	4 Host	Tree	POX controller	Gives better output based on end-to-end latency, response time, throughput, and standard deviation

2. COMPARATIVE LITERATURE ANALYSIS

After an exhaustive literature review, more than 50 papers were reviewed belonging to the area in SDN. The table 1 presents comparison of research work in SDN.

2.1 Strengths of Research Works Reviewed

- The fundamental benefits of SDN with respect to conventional methodology are that SDN permits rapidly test and send new applications in genuine organization, limit capital and working costs and permits concentrated administration of every switch (Open Networking Foundation, n.d).
- The upsides of the heap balancer calculation to work out the incorporated burden for each way and pick one least stacked way as the outcome to get once again to SDN regulator (Chen-Xiao & Ya-Bin, 2016).
- The unique burden adjusting calculation has a superior asset booking capacity to keep away from uneven burden among servers, subsequently further developing by and large asset use of the framework (Bosshart et al., 2014)
- The sFlow-based technique, benefit of the bundle testing capacity of sFlow. The sFlow strategy decoupling the stream sending rationale with measurements, since factual data is not generally bound with stream section, the bundle examining gives the vital data of stream.

2.2 Weakness of Research Works Reviewed

- The fundamental burden of this approach is that switch is absolutely subject to regulator choice. So when a switch loses the association with the regulator, it can't deal with that bundle (Open Networking Foundation, n.d).
- Dijkstra's calculation could not accomplish a similar outcome by simply adding hub loads into edge loads. This is on the grounds that the hub weight ought to be viewed as just at the active edge of a moderate hub on the way (J. & Jehn-Ruey, 2011).
- ACO technique gathers the data to ascertain the connection use and screens the postpone in joins. Be that as it may, this technique applies a straightforward strategy to gather network way data with a solitary assessment rule (Chen-Xiao & Ya-Bin, 2016).

2.4 Gaps in Published Research

The researcher used many algorithms in SDN to monitor the network performance.

- Most of the work in the literature is based on implementation of SDN Controller Architecture.
- Most of the researchers worked on Controller design, load-balancing, and performance of network with different topologies under SDN environment.
- Being an emerging area of research there is lot of scope of research in all domains of computer networks and internet.

2.5 Objectives

For the research following objectives were framed:

- To create simulation environment for Software Defined Network using Mininet emulator, Floodlight controller and Pox Controller.
- To Design and Implement experimental scenario for network monitoring using SDN environment.
- To Analyze the performance of SDN Controller for variable loads in Real Time.

3 PROCESS FLOW DIAGRAM

Figure 1 shows the process flow design of the experimentation carried to achieve the objects. First of all Simulation Environment is setup. After this step Network Topologies is created on which the simulation is worked. To calculate the network performance traffic is generated. Once traffic is generated the network performance parameters from SDN Controller are collected and on the basis of the values of network parameters the Network performance is calculated.

Figure 2 shows the flow chart of the working of the open flow switch

Step 1: Start
Step 2: Packet received by a Switch
Step 3 Matching process for flow entries in flow table
Step 3: If match is exists in flow tables then step 4
Else forwarded to the controller step 5
Step 4: Execute actions set on Packet
Step 5: Controller calculate the applicable logic to the packet and execute
Step 6: End

3.1 Simulation Environment Setup

For the simulation following environment setup are used

1. Intel Core i3 Computer
2. 2.53 GHz Processor with 4 GB RAM
3. Mininet 2 which is used to create the realistic virtual topology.

3.2 Resource Monitoring (CPU, Memory, I/O)

Ganglia is used for resource monitoring. Ganglia is a versatile, circulated checking instrument for elite execution processing frameworks. By using this resource monitoring can be feasible live or recorded as well.

Figure 1. Process flow design of experiment

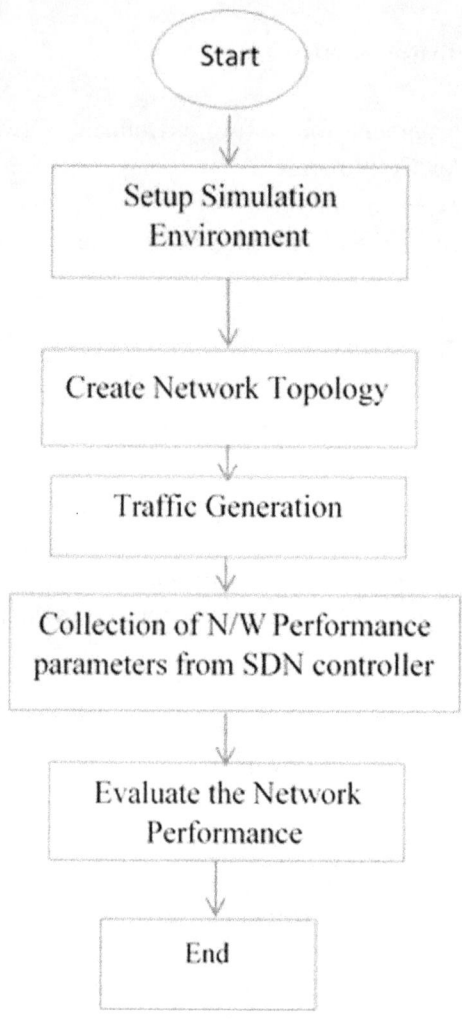

3.3 Network Topology Created

Now a days normally 2 or 3 layers topology for switches and routers are used. Normal designs today comprise of one or the other a few level trees of switches or switches. The main difference between 2 and 3 layers is in three layers topology aggregation tier is used in the middle layer. For this experimentation purpose tree layer topology is used.

3.4 Network Parameters Varied

In the experimentation performed, the number of switches, hosts, number of links, and link bandwidth has been varied. Simulation has been run for different network size and link bandwidth to analyze the performance of SDN enabled network infrastructure.

Figure 2. Flow chart of a Open Flow Switch

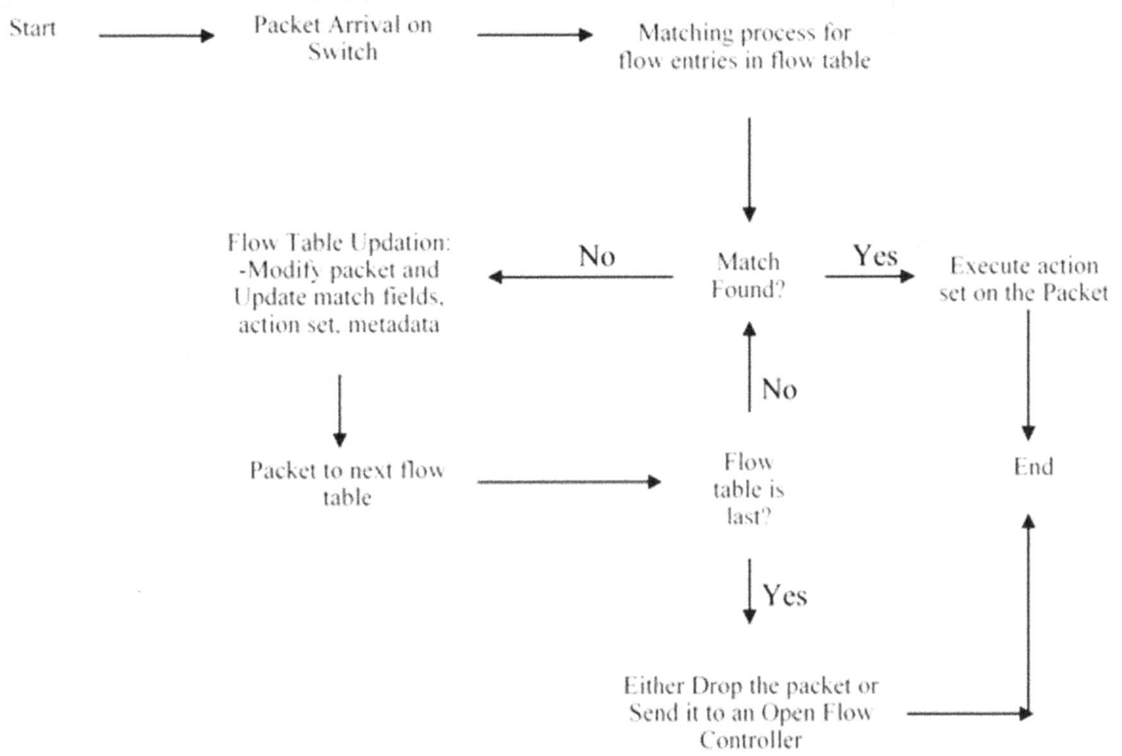

3.4.1 Network Delay

Network Delay is most important parameter to check the quality of the network. The total network delay is the sum up of all the delays which the network consist i.e. Propagation Delay, Queuing Delay, Processing Delay and Transmission Delay. The total network delay is calculated by the following formula which calculate the time of packets received from switch to controller to and fro.

$$t_{delay} = (t_{arrival} - t_{sent} - 1/2 (RTT_{s1} + RTT_{s2}))$$

3.4.2 CPU Utilization

CPU utilization indicates to a PC's utilization of handling assets, or how much work dealt with by a computer chip. Genuine computer processor usage shifts relying upon the sum and kind of overseen registering errands.

3.4.3 Memory Utilization

Memory utilization refers to a computer's usage of memory resources.

4 EXPERIMENTAL RESULT AND ANALYSIS

The tree network topology is used for performance analysis of SDN enabled network. The number of network nodes (switch and host) has been varied to simulate a network ranging from a small to large scale deployment. A tree based topology emulates campus network as well as data center network. The topology consist of 3 layers: core, distribution and access in Campus network scenario and core and aggregation data center scenario by emulated network topology also exhibit the property of CLOS architecture of network topology.

The switches are OpenFlow 1.0, 1.3 enabled and corresponds to Open Virtual Switch (OVS) Type. Mac based learning at layer 2, (Data link layer) of TCP-IP protocol stack has been used for forwarding decision.

4.1 Experimental Variations

Delay Analysis has been carried out for different links and for varying bandwidths from 200-50-500. Resource utilization has been studied for CPU utilization and Memory usage as discussed in following subsections.

Scenario 1: In this scenario, simulation is performed with 90 numbers of links, 8 numbers of switches and 60 numbers of hosts.

Topology	No. of Switches	No. of Host	No. of Links
Tree	8	60	90

Scenario 2: In this scenario, simulation is performed with 190 numbers of links, 18 numbers of switches and 110 numbers. of hosts.

Topology	No. of Switches	No. of Host	No. of Links
Tree	18	110	190

Scenario 3: In this scenario, simulation is performed with 294 numbers of links, 32 numbers of switches and 280 numbers of hosts.

Topology	No. of Switches	No. of Host	No. of Links
Tree	32	280	294

Scenario 4: In this scenario, simulation is performed with 392 numbers of links, 40 numbers of switches and 390 numbers of hosts.

Topology	No. of Switches	No. of Host	No. of Links
Tree	40	390	39

Scenario 5: In this scenario, simulation is performed with 489 numbers of links, 35 numbers of switches and 440 numbers of hosts.

Topology	No. of Switches	No. of Host	No. of Links
Tree	35	440	489

4.2 Delay Analysis

Table 1 shows Average delay and Average sum for each of the 100 links while simulating SDN concept in tree topology.

Table 1. Average Delay

Links	Average Delay (Sec)	Average Sum (Sec)
1-100	0.4438210	2.133146
100-200	0.4411373	3.176005
200-300	0.4632562	5.324561
300-400	0.4786378	6.527532
400-500	0.5987468	7.5382973

Following Figure 3 shows time versus delay for varying bandwidths.

In the network topology emulated analysis has been done for correlation between Bandwidth and delay as the bandwidth cross delay product significant parameters to determine of the capacity of the network. From the simulation result it has been evident that with the varying of the links bandwidth of the links there is no significant change in the delay. It clearly shows that there is less congestion in the SDN enables network.

It can be further optimized by tuning the TCP related parameters at the end host. Reno, Cubic, congestion control mechanism has been used on end host as a congestion control package. The minimum delay clearly shows that there is very less packet loss and packet error ratio in the simulated network.

Simulation has been performed with varying no. of links in the network. The no. of links has been varied from 200 to 500. With the varying no. of links and the delay on individual links as well as the total delay of the network is almost in similar range as defected from the above graphs.

Figure 3. Time vs Delay for varying Bandwidths

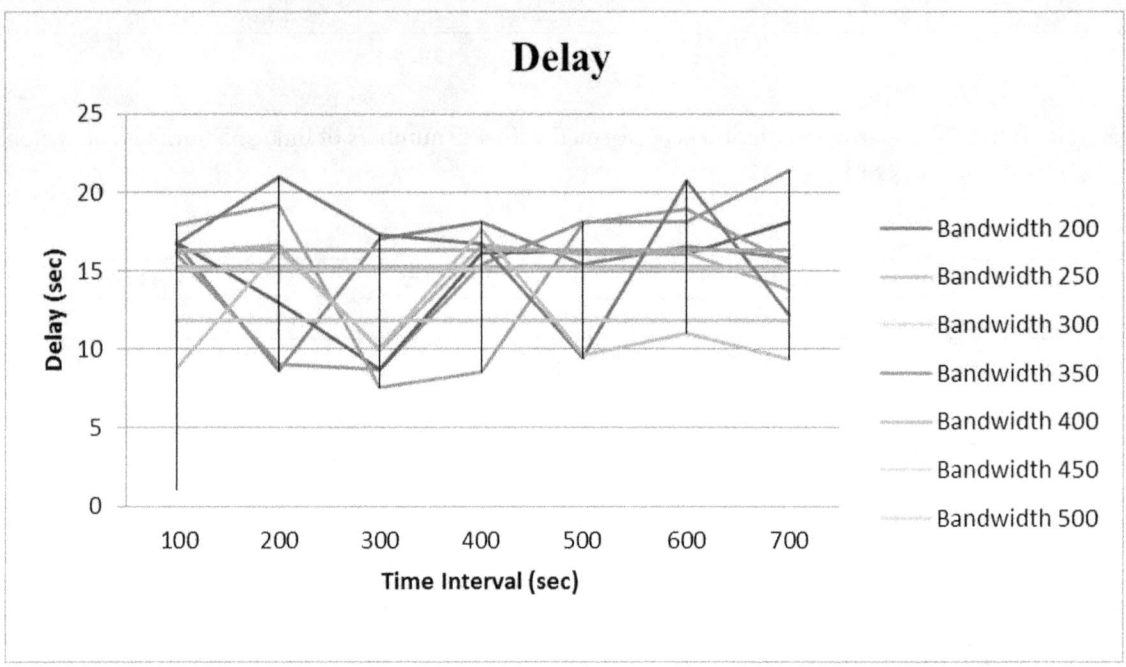

Figure 4. Delays for Network Link for bandwidth 200

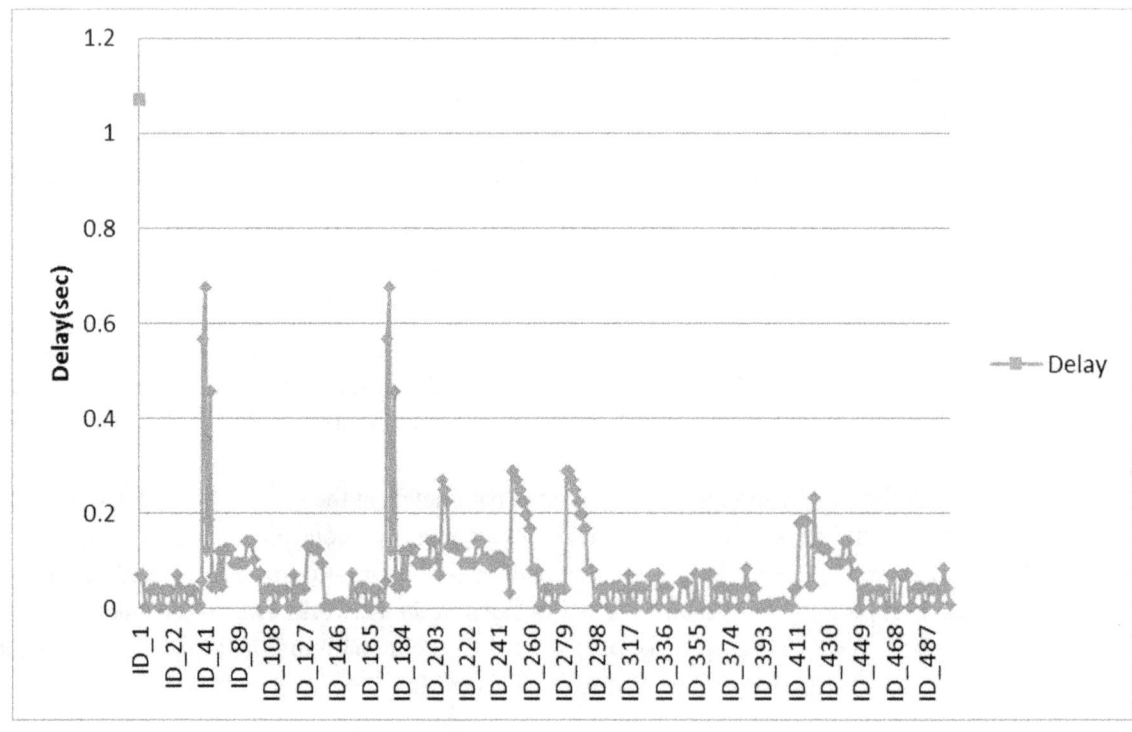

Figure 5. Delays for Network Links for bandwidth 300

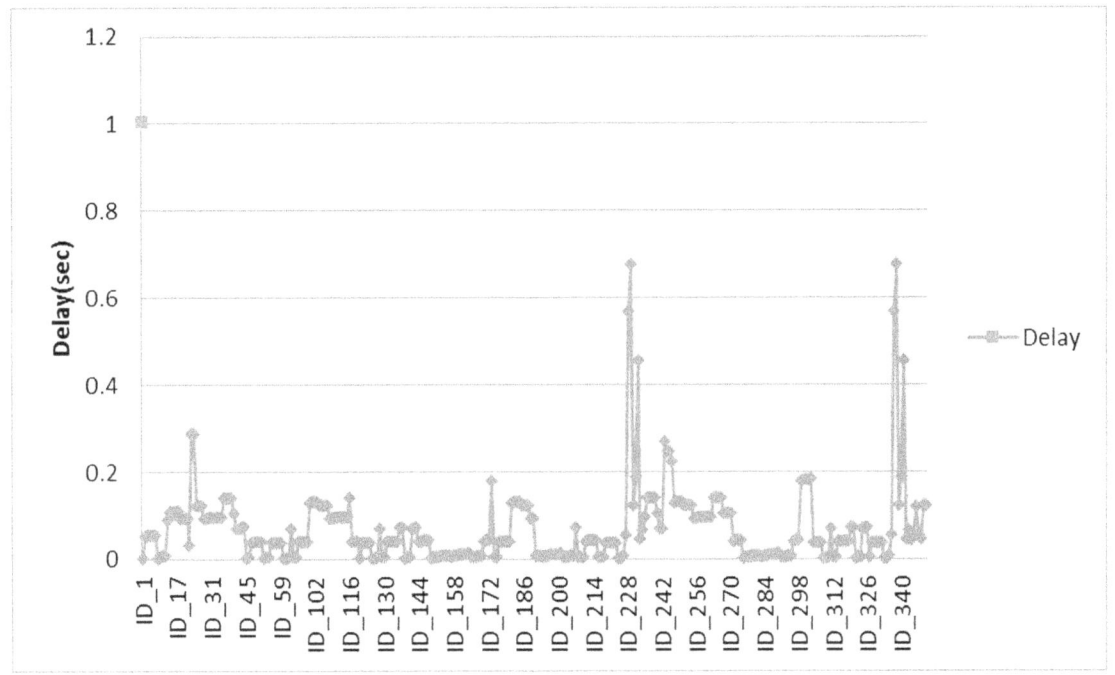

Figure 6. Delays for Network Link for bandwidth 350

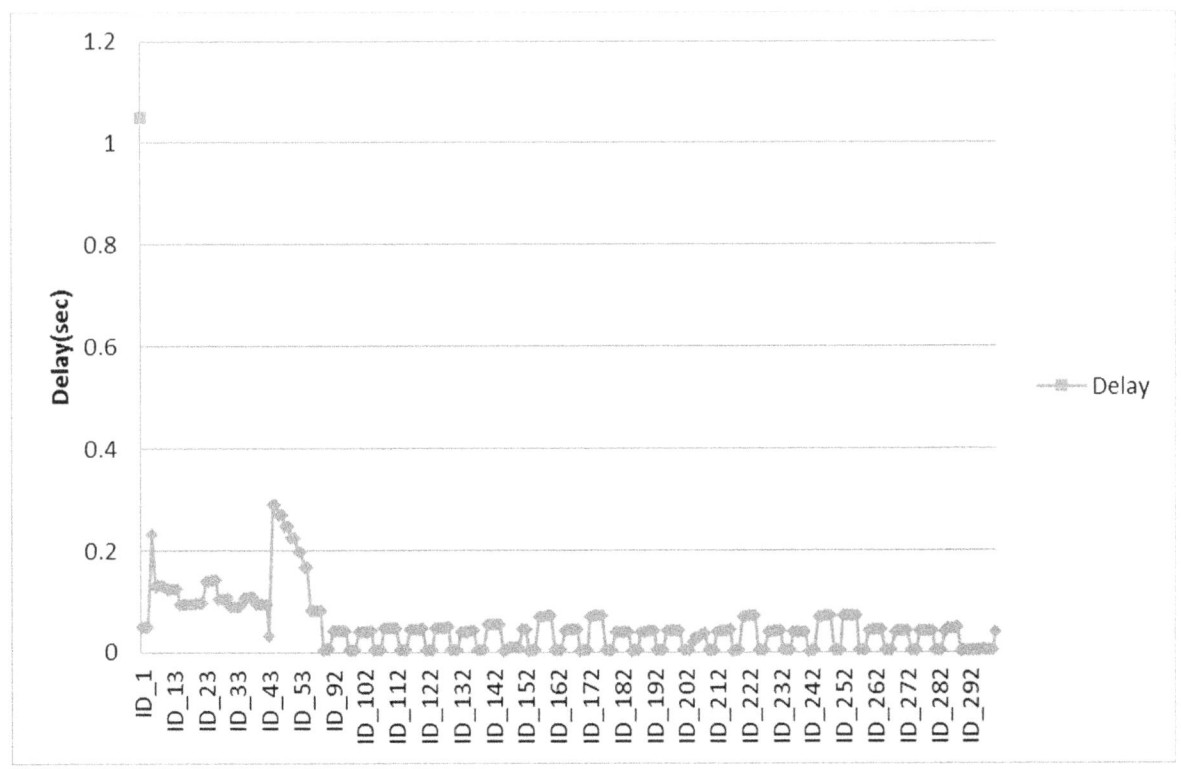

Figure 7. Delays for Network Link for bandwidth 400

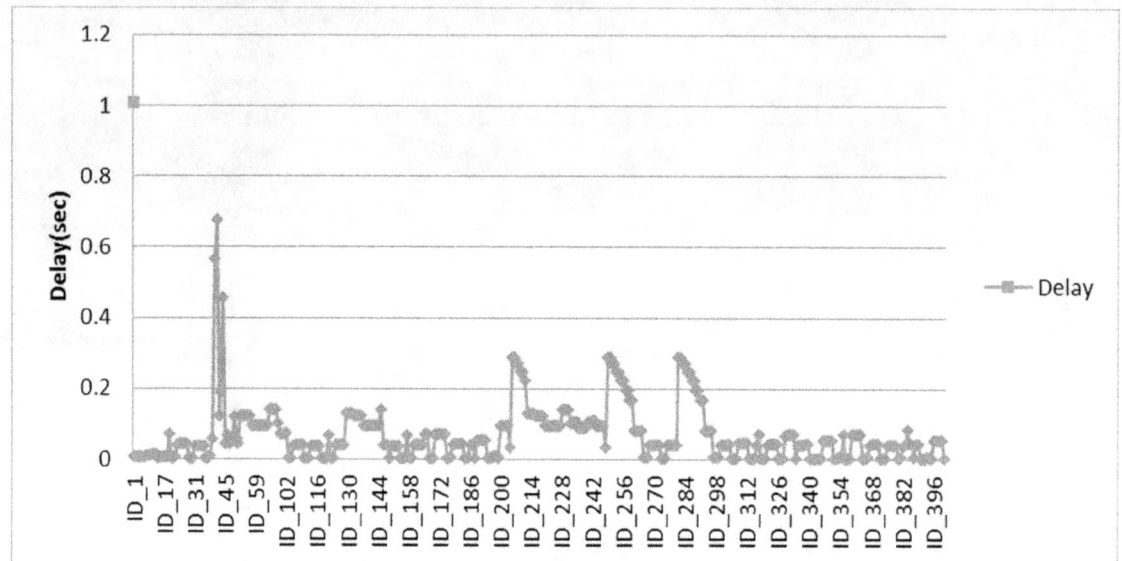

Figure 8. Delays for Network Link for bandwidth 450

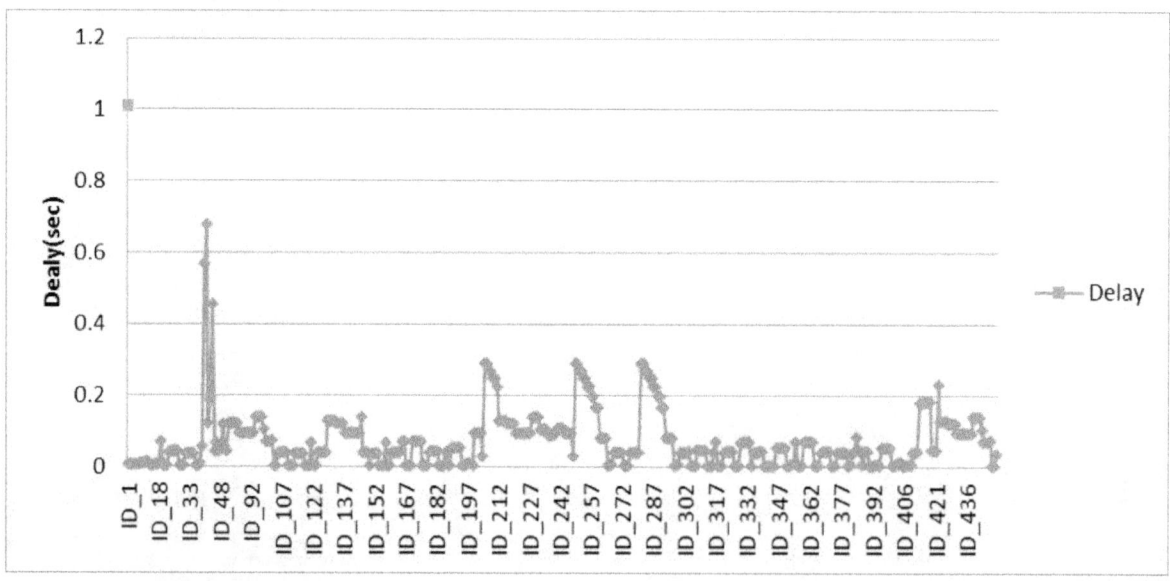

4.3 Resource Utilization Analysis

Monitoring of SDN controller infrastructure is a task of paramount importance of both network performances as well as for predicting an analysis of its own work load. Real time monitoring is a key tool for controlling and managing hardware and software infrastructure as it provides information and key performance indicator for both controller as well as the infrastructure it helps to quantify capacity and resources (e.g. CPU, Memory, Storage, Load etc.)

Figure 9. Delays for Network Link for bandwidth 500

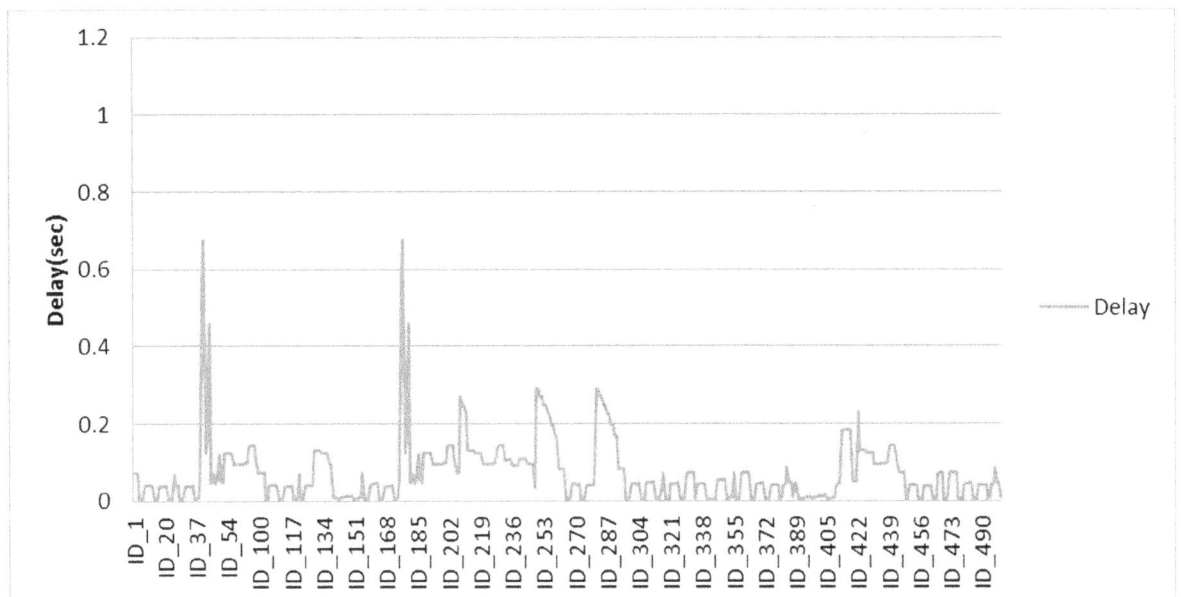

Figure 10. Memory last Hour

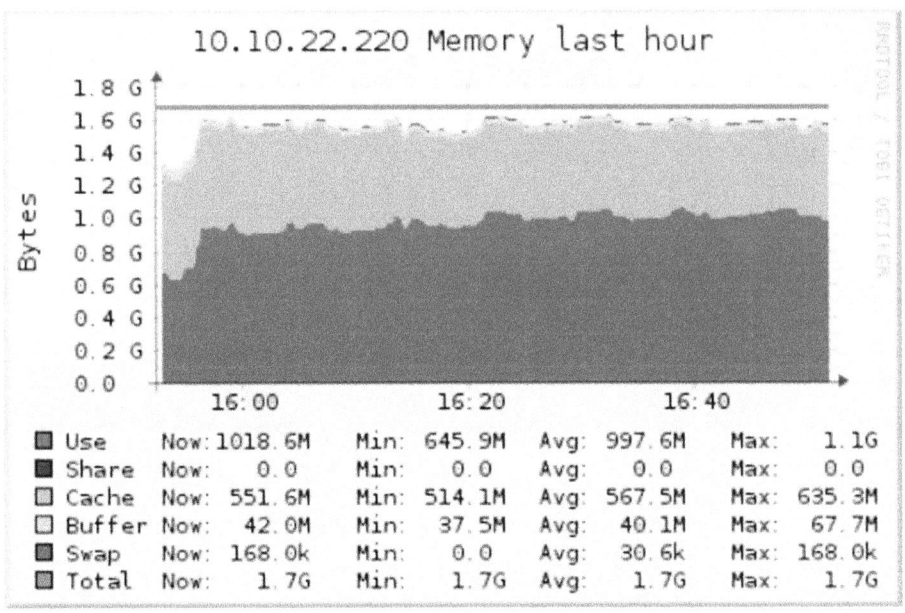

The Different performance parameters have been collected while SDN controller was running during the Simulation create. From the above graph it has been clearly shown that memory requirement is more or less constant and SDN controller application is using cache memory to optimize the system performance. There is no spike visible in the graph is clearly defects that the memory requirement is uniform and bursty traffic has been handle properly with any impact performance of overall system.

Figure 11. CPU last Hour

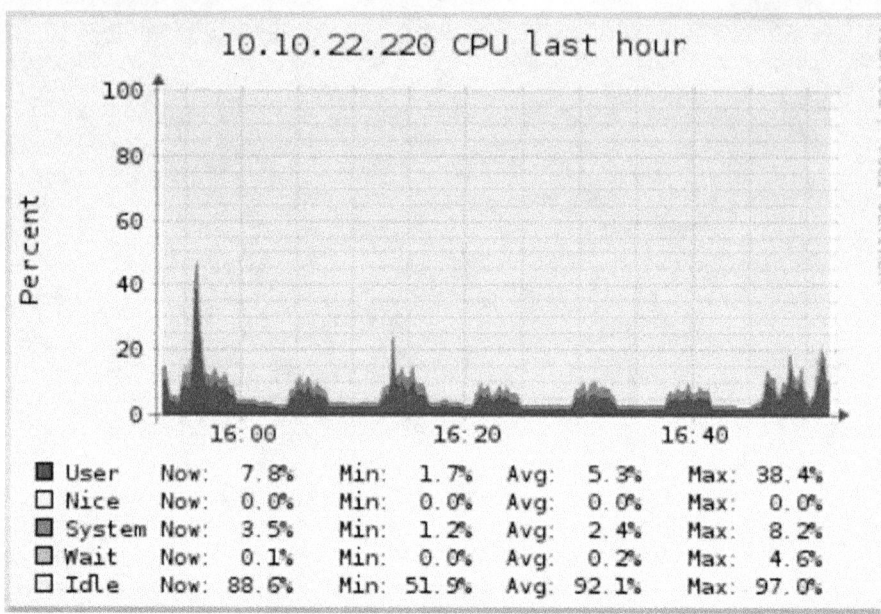

Figure 12. Cached Memory

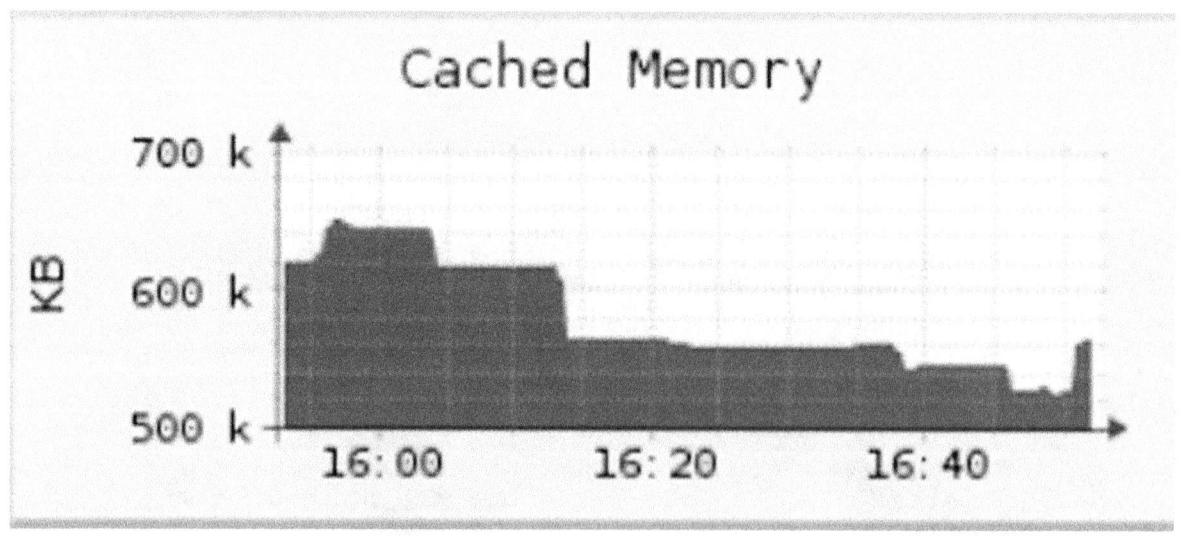

The average load on the SDN controller is uniforms; those networks of different size and capacity has been simulated.

It can be further optimized with the deployment of SDN controller in a cluster architecture where the load will be distributed and the bursty nature of internet traffic can be easily handled.

A conclusion can be drawn from the above graph that in SDN enabled network the impact of network diameter is less significant in the performance of the network. Some spikes in the above graphs are visible and that is due to the heavily loaded SDN controller.

Figure 13. Fifteen Minute Load Average

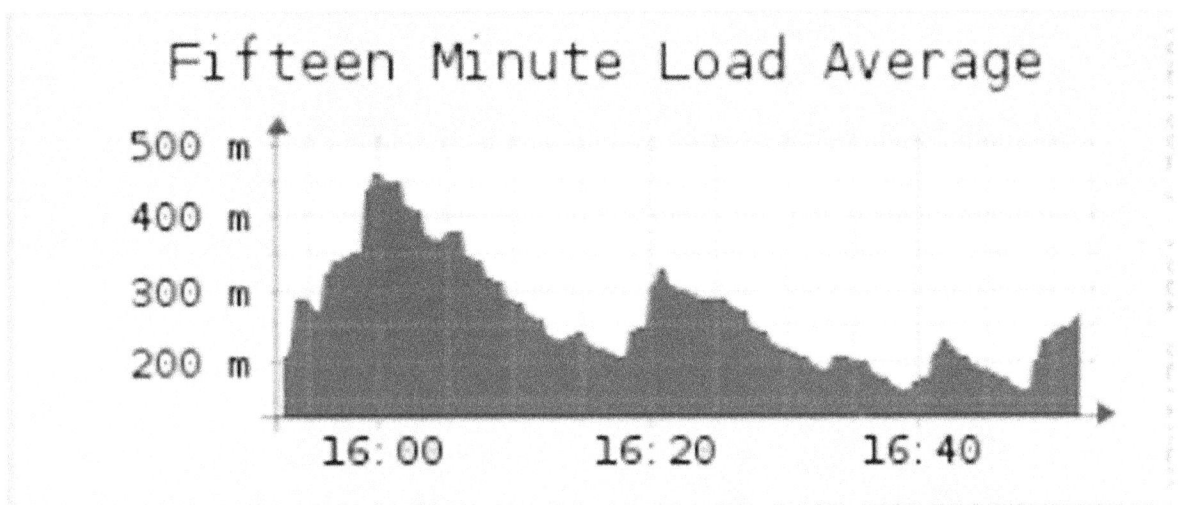

Figure 14. CPU Idle Stages

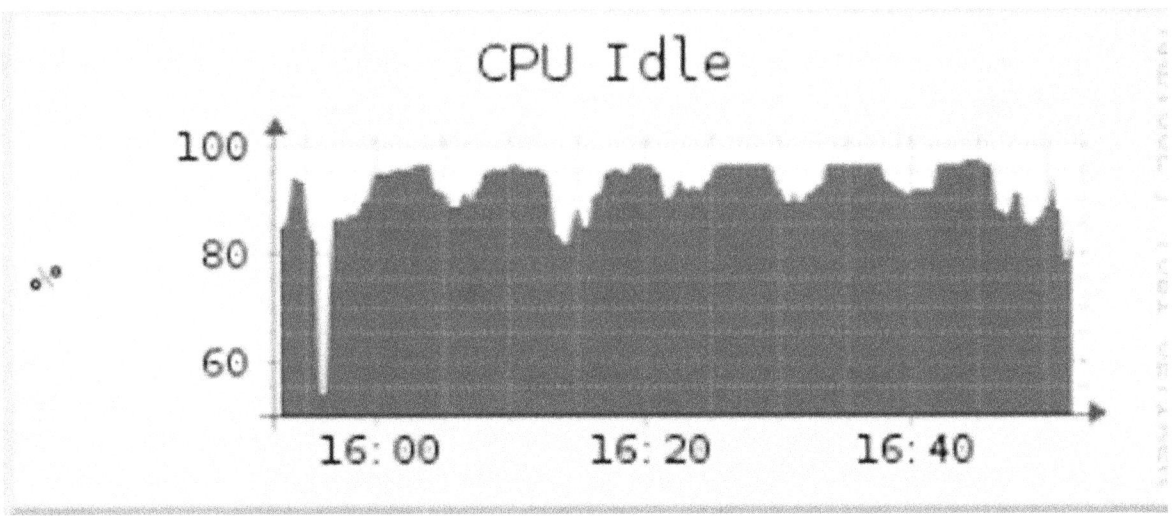

In SDN enable network the centralized controller determines the route for flow and install corresponding action for a flow in the flow table at switches at regular interval new type of flow has been generated in the network from different host connected as different switch to different destination .,which result into large no. of route calculation request at the controller. That is by spikes has been observed as THE SDN controller take some time to process the request and install rules at the switch for the forwarding decision.

4.4 Limitations

The presented work is limited to analysis of various links and the controller resource utilization for varying traffic conditions by considering tree topology.

Figure 15. Load Last Hour

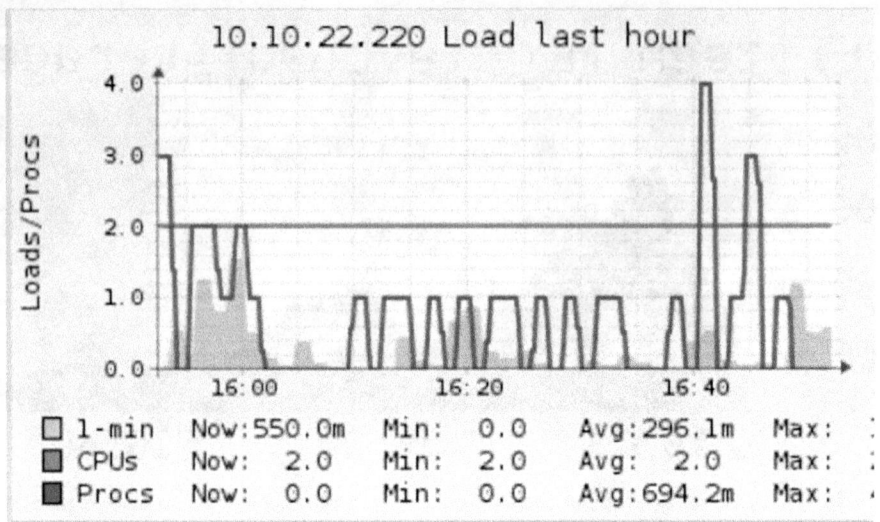

5. FUTURE WORK

- The exactness could be worked on in light of a blend of past insights connect qualities or weighted estimations results without forcing extra above.
- The versatile clock requires really tuning, accordingly, more examination would be important on when more examples are required and when less.
- Besides, more examinations in a genuine climate (for example with genuine traffic) are expected to completely assess the proposed estimation draws near.
- There is part of extension for plan and testing of SDN regulators considering constant actual foundation and traffic conditions.

More than 40 research papers related to research on SDN has been carried out leading to extraction of important information like strengths, weaknesses and gaps. The area being a newly emerging area of research there had been preliminary research on simulation or emulation SDN controllers and their performance analysis using OpenFlow protocol and few other tools available. Looking to the requirement of SDN concept and its future demand, following objectives of the study were finalized.

Objectives

- To create simulation environment for Software Defined Network using Mininet emulator, Floodlight controller and Pox Controller.
- To Design and Implement experimental scenario for network monitoring using SDN environment.
- To Analyze the performance of SDN Controller for variable loads in Real Time.

Experimental Results and Analysis

SDN environment was setup using Linux, Mininet and other tools for creation of network infrastructure scenarios with varying number of switches, hosts, number of links. Tree topology was created and link bandwidth was varied. The performance analysis of SDN controller and the network was done through delay analysis and controller resource utilization under different situations. From the analysis the concept and benefits of SDN controller implementation could be understood. Network monitoring and control becomes easy and fast with the utilization of SDN controller.

REFERENCES

Bosshart, P., Daly, D., Gibb, G., Izzard, M., McKeown, N., Rexford, J., Schlesinger, C., Talayco, D., Vahdat, A., Varghese, G., & Walker, D. (2014, July). P4: Programming Protocol-independent Packet Processors. *Computer Communication Review*, *44*(3), 87–95. doi:10.1145/2656877.2656890

Caesar, M., Caldwell, D., & Feamster, N. (2005). Design and Implementation of a Routing Control Platform. *2Nd Conference on Symposium on Networked Systems Design & Implementation – (Volume 2, pp. 15–28). USENIX Association.

Chen-Xiao, C., & Ya-Bin, X. (2016). Research on load balance method in SDN. *International Journal of Grid and Distributed Computing*, *9*(1), 25–36. doi:10.14257/ijgdc.2016.9.1.03

Feamster, N. Rexford J., & Zegura, E. (2013). The Road to SDN.

Giustiniano, D., Goma, E., Lopez Toledo, A., & Athanasiou, G. (2013). Optimizing TCP performance in multi-AP residential broadband connections via Minislot access. *Journal of computer networks and communications*.

Jehn-Ruey, J. (2011). *Widhi Yahya," Load Balancing and Multicasting Using the Extended Dijkstra's Algorithm in Software Defined Networking*. Springer-Verlag Berlin Heidelberg.

Open Networking Foundation. (n.d.). *Open Networking*. https://www.opennetworking.org/.

OpenFlow Switch Specification. (n.d.). Version 1.4.0. https://www. opennetworking.org /images/stories/ downloads/sdn-resources/onf-specifications/openflow/openflow-spec-v1.4.0.pdf.

Rao, A., Auti, S., Koul, A., & Sabnis, G. (2016). High availability and load balancing in SDN controllers. *Int J Trend Res Dev*, *3*(2), 310–314.

Ryu. (n.d.). Component-based Software Defined Networking Framework. https: // osrg.github.io/ryu/.

Sherwood, R., Gibb, G., Yap, K.-K., Appenzeller, G., Casado, M., McKeown, N., & Parulkar, G. (2009). *Flowvisor: A network virtualization layer. OpenFlow Switch Consortium, Tech. Rep.*

Tennenhouse, L., & Wetherall, D. J. (1996). Towards an Active Network Architecture. *Computer Communication Review, 26*(2), 5–18. doi:10.1145/231699.231701

Zander, J., & Forchheimer, R. (1988). The SOFTNET project: a retrospect. In *Electrotechnics, 1988. Conference Proceedings on Area Communication, EUROCON 88., 8th European Conference on*, (pp. 343–345). 10.1109/EURCON.1988.11172

Chapter 6
Cooperative Rate Splitting (C–RS):
A New Approach for Performance Enhancement of Downlink NOMA

Jyoti Saharan

JCDM College of Engineering, India

Silki Baghla

JCDM College of Engineering, India

Dinesh Kumar Gupta

JCDM College of Engineering, India

ABSTRACT

In a 5G wireless communication system, demands of massive connectivity, ultra-low latency, higher data rates, higher reliability, and many more are fulfilled by NOMA (non-orthogonal multiple access) which is one of the potential candidates that can carry multiple users at the same time and frequency. As NOMA uses the power allocation method to transmit signals, a problem of mismatch between quality of service (QoS) and channel conditions of the users occurs. To resolve this issue, a scheme cooperative rate-splitting non-orthogonal multiple access (C-RS-NOMA) has been proposed in this work to enhance the performance of downlink power-domain NOMA in terms of outage probability, system capacity, spectrum efficiency, and energy efficiency. The performance of the proposed techniques has also been compared with the existing cooperative-NOMA and traditional NOMA techniques. The simulation results showed the improvement in different key parameters of C-RS-NOMA with respect to the existing NOMA and cooperative NOMA schemes.

DOI: 10.4018/978-1-6684-7000-8.ch006

INTRODUCTION

A wireless communication network in which information is carried over a well-defined channel. Each channel has a fixed frequency bandwidth and bit rate. In a cellular mobile communication system, the first wireless network was created over four decades ago in the early 1980s when the 1G mobile network was introduced at that time the demand for more connections grows worldwide which directs the evolution of mobile communication more rapidly. Various networks are developed starting with 1G and going up to 4G all these distinct generations of wireless communication are differentiated based on multiple access schemes used.

For 1G, the frequency division multiple access (FDMA) scheme was used, offering a maximum speed of up to 2.4 Kbps. Then a time division multiple access (TDMA) scheme had been used for 2G mobile communication which offers a maximum speed of up to 50Kbps with general packet radio service (GPRS) and 1Mbps with enhanced data rates for GSM evolution. code division multiple access (CDMA) Technology had been used in 3G wireless communication, offering speeds up to 14Mbps. And for 4G an orthogonal frequency division multiple access (OFDMA) scheme was used that offered speeds up to 1Gbps. All these are the orthogonal multiple access (OMA) Techniques where the spectrum allocation of users was differentiated based on frequency, time, and code. In OMA maximum number of supported users are limited. As demands for wireless communication increase in terms of higher data rates, massive connectivity, lower latency, advanced multimedia functions, and higher spectral efficiency to satisfy these requirements, various enhanced technology for 5G has been proposed such as massive MIMO, mm-wave communication and NOMA (non-orthogonal multiple access) that allows multiple users to share a spectrum at the same time and frequency which is well known in multiple access techniques used for 5G wireless communication and offers increase system throughput. NOMA uses superposition-coded signals at the transmitting side and SIC (Successive Interference Cancellation) at the receiving side, making it possible to decode complete interference. In power-domain NOMA, each user was allotted many different power levels according to their channel conditions so that maximum user fairness exists (Boccardi et al., 2014; Dai et al., 2015). As in the fixed power allocation scheme of downlink NOMA without considering the channel conditions of the user, power levels for far and near users are set on a fixed value. This causes a dissimilarity between the user's service quality and the channel conditions (Joe, 2020).

Other than the NOMA technique, there is one more multiple access was introduced in downlink scenarios which were RSMA (Rate-Splitting Multiple Access) which demonstrates various performance benefits over NOMA (Non-Orthogonal Multiple Access) and SDMA (Space-Division Multiple Access) techniques. As SDMA completely considers interference as noise. As linear-precoded rate-splitting was used at the transmitter of RSMA (Rate-Splitting Multiple Access) with SIC (Successive Interference Cancellation) at the receiving end making it possible to treat a few parts of interference as noise and decode a few parts of interference. And this advantage of RSMA (Rate-Splitting Multiple Access) over SDMA and NOMA makes improvements in QoS (Quality of Service) also reduces complexity and gives enhancements in parameters like Spectral Efficiency (SE) and Energy Efficiency (EE) (Mao, Clerckx, & Li, 2018; Mao, Clerckx, & Li, 2018). RSMA technique simply splits the user signals into two parts, which we can call virtual inputs to splits the signal's rate and enables the user to achieve any rate partition that gives the user flexibility.

This work proposed a C-RS-NOMA (Cooperative-Rate-Splitting Non-Orthogonal Multiple Access) techniques for the future generation mobile communication systems to fulfill the ultra-high requirements of achievable data rates, system capacity, higher SE, and higher EE, and this scheme also solves the problem of dissimilarity between users' quality of service (QoS) and the channel conditions.

LITERATURE REVIEW

In 4G cellular networks, orthogonal frequency division multiple access (OFDMA) was one of the popular methods to provide access to a greater number of customers at the same time. Before this method, FDMA, TDMA, and CDMA techniques were developed time by time to support the increasing number of users. All these techniques of multiple access come under the category of Orthogonal Multiple Access (OMA) having spectrum resources scheduled orthogonally. Because the number of users and their demand for high achievable data rates increased further, OMA-based multiple access techniques become less suitable. So, a new and different multiple access technique NOMA (non-orthogonal multiple access) was developed to address the demands (high capacity, ultra-high connectivity, high throughput, high energy efficiency, and low latency) of a 5G wireless network (Sonia, 2022). Compared to OMA it can carry much more users via a non-orthogonal resource allocation. Power-domain and code-domain multiplexing NOMA are the two main categories for NOMA their basic concepts, key features are explained and compared based on their SE, system throughput, receiver complexity, and many other factors. In (Dai et al., 2015), SoDeMA (software defined multiple access) concepts were defined to support many applications of 5G networks. Power-domain NOMA has been studied to review its fixed power-allocation strategies, user fairness, user pairing, and capacity analysis in NOMA. Performance analysis of NOMA as integrated with other techniques such as cooperative communication, MIMO, and beam-forming has been presented in (Islam et al., 2017). In (Wu et al., 2018), as per the characteristics, different types of NOMA like Scrambling based NOMA, Inter-leaving-based NOMA, spreading based NOMA, and Coding based NOMA have been discussed based on their principles, features, and transceiver block diagrams.

In (, R. & Kizilirmak, 2016), downlink and uplink NOMA schemes were discussed to optimize the system capacity and find out the relation between SE and EE. Also, the impact of imperfect receivers was discussed. For further improvement in the performance of NOMA, the MIMO technique was introduced to make a new MIMO-NOMA technique and analyzed their performance with an existing OMA with a case of fixed power allocation scheme and user pairing (Ding et al., 2016). MIMO-NOMA-based system performance under different modulation schemes was compared with existing NOMA based on different parameters like BER, channel capacity, outage probability, etc (Sonia & Gupta, 2021). In terms of two parameters as overall channel capacity and ergodic sum capacity, MIMO-NOMA performed better compared with MIMO-OMA (Liu et al., 2016). An admission scheme for users was provided for MIMO-NOMA when a greater number of users are grouped into a single cluster and compared with MIMO-OMA showed the superiority of MIMO-NOMA (Zeng et al., 2017). In (Zeng et al., 2018), an evolution from single-carrier NOMA (SC-NOMA) to multi-carrier NOMA (MC-NOMA) was presented. Two MC-NOMA schemes were discussed as PDMA (pattern-division multiple access) and SCMA (sparse-code multiple access) and future research direction was given for MC-NOMA.

NOMA applications were reviewed in different areas like mobile communication networks, wireless sensor networks (WSNs), and in device-to-device (D2D) Communication (Anwar et al., 2019). As demands of future multi-antenna networks like higher throughput, massive connectivity, and many more were needed, a new multiple access design was developed known as RSMA (rate-splitting multiple access) that can partially decode interference and partially treat interference as noise. It provided rate and QoS enhancement and reduction in system complexity. SDMA, NOMA, and RSMA schemes were compared and showed that RSMA was better than both others in terms of spectral and energy efficiency (Mao, Clerckx, & Li, 2018; Mao, Clerckx, & Li, 2018). In downlink RSMA sum-rate maximized by determining the rate allocation for common message and by adjusting the power which was transmitted for the private message to decrease interference and the results showed that RSMA achieved a 19.6% gain of achievable rate compared to NOMA and OFDMA (Yang et al., 2020). A comprehensive study of RSMA was given and compared with other existing MA schemes in 5G. It was shown that RSMA is optimal in several transmission scenarios (Mao, Clerckx, & Li, 2018). In the case of multi-user detection, where user power differentiation was not available, SIC is not applicable directly. So, the rate-Splitting technique was used to control user pairing and power control issues of power-domain NOMA. A similar technique was used with code-domain NOMA to reduce decoding complexity (Zhu, Zhang, Wang et al, 2017). In uplink NOMA schemes, complicated user pairing schemes may cause scheduling overhead so to solve this rate splitting NOMA scheme was proposed. It splits each user signal and successively decoded with the same arriving power and reduces scheduling complexity (Zhu, Wang, Zhang et al, 2017). In (Huang et al., 2017), a novel RS-NOMA scheme was proposed to provide a compromised solution for differentiability between the users' Quality-of-Service (QoS) and channel conditions.

In this work, the C-RS-NOMA technique has been simulated and compared with cooperative NOMA using the MATLAB platform for a better understanding of its basic process followed by a performance evaluation with different parameters.

SYSTEM MODEL FOR C-RS-NOMA

A brief description of NOMA and the Cooperative NOMA system is presented in this section followed by the proposed C-RS-NOMA technique.

a) **System Model for NOMA in Downlink:** In a traditional two-user NOMA system during downlink transmission under a fixed power allocation scheme, a single Base Station transmits two signals non-orthogonally over the same radio channel which means it superimposes the signals at the transmitter stage. Fig 1 shows the downlink NOMA system with 2 users and 1 base station.

Figure 1. System model for basic 2-user downlink NOMA system

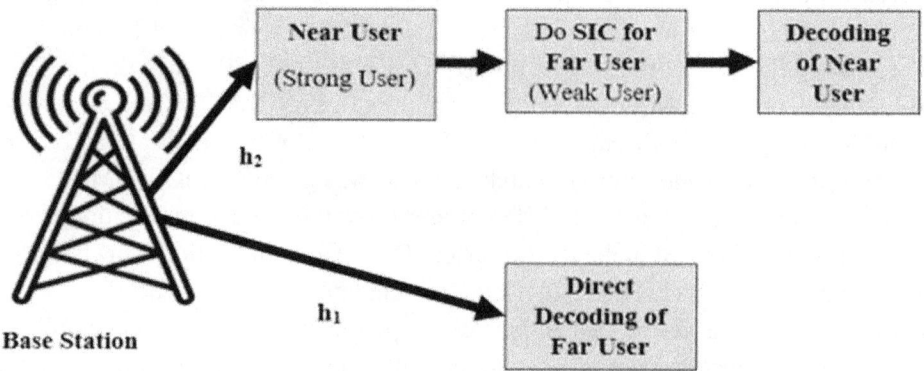

A base station transmits superposition-coded signal x for near and far users by using Rayleigh fading channel shown in eq. 1.

$$x = \sqrt{P}\left(\sqrt{a1} \cdot x1 + \sqrt{a2} \cdot x2\right) \tag{1}$$

Where,

P = Total Transmitted power from Base Station

$x1$ = Far (Weak) User Signal

$x2$ = Near (Strong) User Signal

$a1, a2$ = Power allocation coefficient for far and near user signals respectively

A fixed power allocation scheme was used so here fixed the values of $a1$ and $a2$ irrespective of the channel conditions such that $a1 > a2$ and $a1 + a2 = 1$.

$y1$ is the signal received at the far-user and $y2$ is the signal received at the near-user given in eq. 2 and 3 respectively.

$$y1_{NOMA} = h_1 x + w1 \tag{2}$$

$$y2_{NOMA} = h_2 x = w2 \tag{3}$$

Where, w1 and w2 are the Gaussian noise (AWGN) interference in channel 1 and channel 2 respectively having zero mean and variance of σ^2; h_1 and h_2 are the channel gains of Rayleigh fading channels for far and near-user respectively.

Substitute value of x in eq. 2 and 3, we get eq. 4 and 5

$$y1_{NOMA} = \sqrt{P}\sqrt{a1}x1\left(h_1\right) + \sqrt{P}\sqrt{a2}x2\left(h_2\right) + w1 \tag{4}$$

$$y2_{NOMA} = \sqrt{P}\sqrt{a1}x1\left(h_2\right) + \sqrt{P}\sqrt{a2}x2\left(h_2\right) + w2 \tag{5}$$

- **Far user decoding in NOMA:** Far (Weak) user signal, $x1$ was decoded first because of its dominating power allocation factor ($a1 > a2$). Thus, directly decoding the far user from $y1_{NOMA}$ signal only and considered $x2$ as an interference.

Far user signal-to-noise ratio (SNR) and the rate are given in eq. 6 and 7 respectively.

$$\gamma_{f,NOMA} = \frac{Pa1|h_1|^2}{Pa2|h_1|^2 + \sigma^2} \tag{6}$$

$$R_{f,NOMA} = \log_2(1 + \gamma_{f,NOMA})$$

$$R_{f,NOMA} = \log_2\left(1 + \frac{Pa1|h_1|^2}{Pa2|h_1|^2 + \sigma^2}\right) \tag{7}$$

- **Near user decoding in NOMA:** Near (Strong) user signal decodes $x2$ from $y2_{NOMA}$ signal. As in $y2_{NOMA}$ both $x1$ and $x2$ signals are present and we require only $x2$ to have a lower power allocation factor hence, the $x1$ term is dominating. Thus, Successive Interference Cancellation (SIC) must be performed to decode a strong signal. SIC cancels out the weak signal and then the strong (near) user detects its signal without any interference from another signal.

Near-user signal-to-noise ratio (SNR) and the rate are given in eq. 8 and 9 respectively.

$$\gamma_{n,NOMA} = \frac{Pa2|h_2|^2}{\sigma^2} \tag{8}$$

$$R_{n,NOMA} = \log_2(1 + \gamma_{n,NOMA})$$

$$R_{n,NOMA} = \log_2\left(1 + \frac{Pa2|h_2|^2}{\sigma^2}\right) \tag{9}$$

b) **System Model for Cooperative Downlink NOMA:** Simple two-user cooperative NOMA system in downlink scenario shown in Fig 2.

Figure 2. Two-user Cooperative NOMA System Model

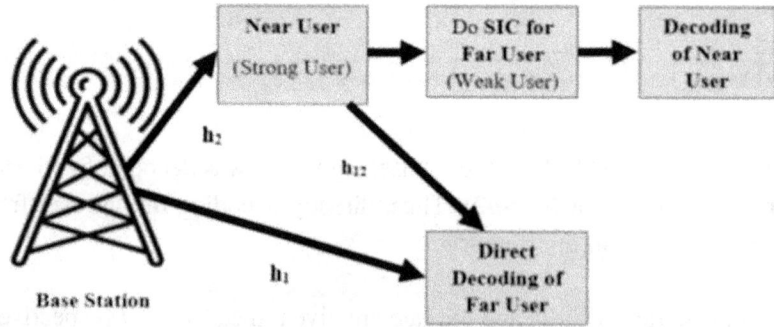

In cooperative communication with the NOMA network, the user with better channel conditions decodes the information for the user with a poor channel (far channel) to the base station. Hence, the strong user acts as a relay for the weak user. Cooperative communication helps to reduce the outage probability of poor channels without any need for additional antennas. Here transmission occurs in two-time slots.

- **Direct Transmission Slot (First-time slot)**

The BS directly transmits the same superimposed signal as in eq. 1 to the near and far user. Then decoding takes place near the user which is required to do SIC before decoding its signal the decoding of far user signal through SIC is needed. But weak user signal decodes by just direct decoding. Eq.10 and 11 gives the data rates for near and far user respectively.

$$R_{n,C-NOMA} = \frac{1}{2}\log_2\left(1 + \frac{Pa2|h_2|^2}{\sigma^2}\right) \tag{10}$$

$$R_{f1,C-NOMA} = \frac{1}{2}\log_2\left(1 + \frac{Pa1|h_1|^2}{Pa2|h_1|^2 + \sigma^2}\right) \tag{11}$$

- **Relaying Slot (Second-time slot)**

In this slot near user just transmits the data to the far user because in the previous slot near user had decoded the far user's data which is directly relayed to the far user. Eq. 12 gives the data rate of the far user at relaying slot end.

$$R_{f2,C-NOMA} = \frac{1}{2}\log_2\left(1 + \frac{P|h_{12}|^2}{\sigma^2}\right) \tag{12}$$

Where the channel between near and far-user was represented by h_{12}.

$$R_{f2,C-NOMA} > R_{f1,C-NOMA}$$

The final achievable rate of the weak signal is given in eq. 13.

$$R_{f,C-NOMA} = \frac{1}{2}\log_2\left(1 + \max\left(\frac{Pa1|h_1|^2}{Pa2|h_1|^2 + \sigma^2}, \frac{P|h_{12}|^2}{\sigma^2}\right)\right) \tag{13}$$

c) **System Model for proposed C-RS-NOMA:** The system model for the proposed scheme with a 2-user downlink scenario is shown in Fig 3. Where near and far users will receive signals from the single base station. The information signal of the far user has been split into two virtual streams of data for downlink transmissions, whereas the near user signal is transmitted without rate-splitting.

Figure 3. System model for C-RS-NOMA with 2 users in downlink transmission

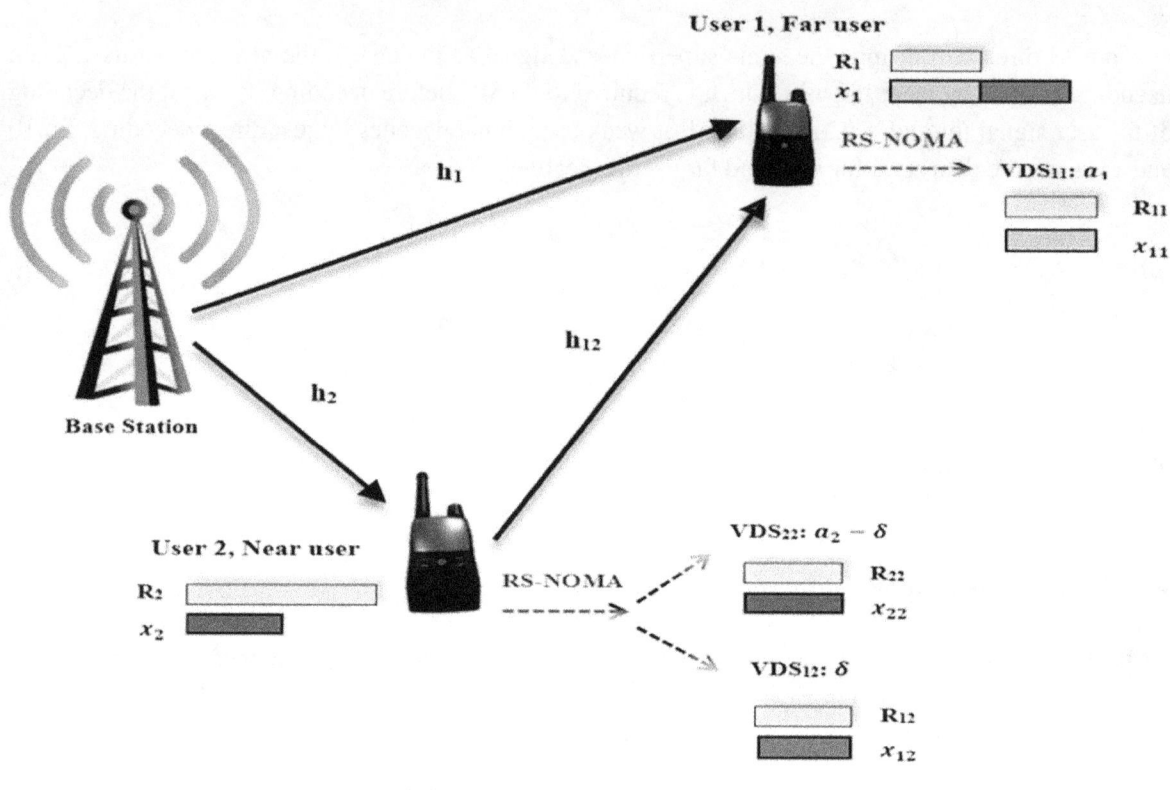

VDS: Virtual Data Stream

As seen from Fig 3. the near user channel condition is quite better than the far user and accordingly rate, R_2 of the strong user is greater than that of the weak user. But in contrast with channel conditions the Quality-of-Service (QoS) of the near user is not that good compared to the far user as depicted by the size of information blocks and the corresponding size of the Rates block. This problem happens in NOMA which is solved by the proposed technique C-RS-NOMA as it can adjust rates according to user requirements. As in a downlink communication system when signal transmits from BS to mobile users the data rates of near users are much higher than that of the far user because near user has a lower power allocation factor due to that near users are decoded after far user detection and near user must perform SIC to first detect Far user signal so that neglect interference from that and then decode its signal due to this process of SIC performed by the near user the rate of the near user was higher than the far user. So, here is the proposed technique of downlink scenario data rate of near user signal splits into two halves to assist information of far user signal that have lower data rate and are not capable to carry whole information signal efficiently. In the rate-splitting, multiple access schemes at the transmitter side original stream of data having N number of users can be split into at most 2N-1 virtual streams of data. So, in our case, we have 2 users and by using a rate-splitting scheme at the transmitter of NOMA a maximum of 3 virtual data streams are possible as shown in fig 3. Three virtual data streams are VDS_{11}, VDS_{12}, and VDS_{22} having information signals x_{11}, x_{12}, and x_{22} respectively. Where x_{22} is the original information signal x_2 of the near user and x_{11} and x_{12} are the split form of the x_1 information signal of the far user.

The BS antenna transmits the superimposed signal to user1 and 2 which is represented with x as shown in eq. no.

$$x = \sqrt{P_s}\left(\sqrt{a_1} \cdot x_{11} + \sqrt{a_2 - \delta} \cdot x_{22} + \sqrt{\delta} \cdot x_{12}\right) \tag{14}$$

Where,

P_s total power transmitted from BS
a_1 and a_2 are the power allocation factors for weak and strong users respectively
δ is the splitting factor

To maintain user fairness must condition are $a_1 > a_2$ and taking $\delta < \dfrac{a_2}{2}$ and $a_1 + a_2 + \delta = 1$.

Hence, according to power allocation factors, the detecting order of the signals must be x_{11}, then x_{22} and at last x_{12} signal is decoded.

The signal received by split far-users VDS_{11}, VDS_{12}, and by near-user VDS_{22} with different channel gains from the Base station transmitter is given in eq. 15,16 and 17 respectively.

$$y_{11} = xh_1 + w_1 = \sqrt{P_s a_1}\,h_1 x_{11} + \sqrt{P_s(a_2 - \delta)}\,h_1 x_{22} + \sqrt{P_s \delta}\,h_1 x_{12} + w_1 \tag{15}$$

$$y_{12} = xh_2 + w_2 = \sqrt{P_s a_1}\,h_2 x_{11} + \sqrt{P_s(a_2 - \delta)}\,h_2 x_{22} + \sqrt{P_s \delta}\,h_2 x_{12} + w_2 \tag{16}$$

$$y_{22} = xh_2 + w_2 = \sqrt{P_s a_1}\,h_2 x_{11} + \sqrt{P_s(a_2 - \delta)}\,h_2 x_{22} + \sqrt{P_s \delta}\,h_2 x_{12} + w_2 \tag{17}$$

Where, w_1 and w_2 are Gaussian Noise (AWGN) interference in channel 1 and channel 2 respectively with zero mean value and variance of σ^2. x_{11} signal detected from far user received signal y_{11}. Similarly, x_{12} and x_{22} are detected from respective y_{12} and y_{22} received signals.

In the cooperative communication concept, the strong user has weak users' data that was relayed to the weak user to aid him. It will reduce the outage probability without any additional antennas. According to the proposed downlink scenario for C-RS-NOMA which has three information signals and x_{12} is the signal that has a stronger channel condition with the base station and detects its signal after decoding both x_{11} and x_{22} signals. This means the x_{12} signal has data from both x_{11} and x_{22} signals which has been relayed to them through channels h_{12} and h_2 respectively. Now the transmission of signal occurs in two-time slots:

i. **Direct Transmission Slot:** In this first half slot, BS transmits the superimposed signal to both users. As per the power allocated to users x_{12} signal does SIC two times to decode firstly x_{22} data then x_{11} data, and then proceeds to decode its data. This x_{12} signal acts as a relay for both other signals.

At the end of the direct transmission slot, the achievable data rates for all three signals are given in eq. 18, 19, and 20.

$$R_{12} = \frac{1}{2} \log_2 \left(1 + \frac{P_s \delta |h_2|^2}{\sigma^2} \right) \tag{18}$$

$$R'_{22} = \frac{1}{2} \log_2 \left(1 + \frac{P_s (a_2 - \delta) |h_2|^2}{P_s \delta |h_2|^2 + \sigma^2} \right) \tag{19}$$

$$R'_{11} = \frac{1}{2} \log_2 \left(1 + \frac{P_s a_1 |h_1|^2}{P_s a_2 |h_1|^2 + \sigma^2} \right) \tag{20}$$

ii. **Relaying Slot:** In this slot, the x_{12} signal already has data of remaining both users because that was decoded in the previous time slot. So, x_{12} just transmits the data to remaining both users.

At the end of relaying slot, the achievable data rates of x_{22} and x_{11} signals are given in eq. 21 and 22.

$$R''_{22} = \frac{1}{2} \log_2 \left(1 + \frac{P_s |h_2|^2}{\sigma^2} \right) \tag{21}$$

$$R''_{11} = \frac{1}{2} \log_2 \left(1 + \frac{P_s |h_{12}|^2}{\sigma^2} \right) \tag{22}$$

Where the channel between the near and far user is represented by h_{12}.

As seen from eq. 19, 20, 21, and 22,

$$R''_{11} > R'_{11} \text{ and } R''_{22} > R'_{22}$$

The final achievable rates of virtual users VDS_{11} and VDS_{22} at the end of both these time slots are given in eq. 23 and 24.

$$R_{11} = \frac{1}{2} \log_2 \left(1 + \max \left(\frac{P_s a_1 |h_1|^2}{P_s a_2 |h_1|^2 + \sigma^2}, \frac{P_s |h_{12}|^2}{\sigma^2} \right) \right) \tag{23}$$

$$R_{22} = \frac{1}{2}\log_2\left(1 + \max\left(\frac{P_s(a_2 - \delta)|h_2|^2}{P_s\delta|h_2|^2 + \sigma^2}, \frac{P_s|h_2|^2}{\sigma^2}\right)\right) \tag{24}$$

d) **Simulation setup:** The proposed C-RS-NOMA technique has been simulated using the MATLAB platform. Table 1 provides the simulation setup used in this new technique to estimate and compare different parameters of performance measures in the proposed techniques with the existing Cooperative NOMA technique. A fixed power allocation scheme has been used in this work and the propagation channel has been assumed as Rayleigh fading channel. Having an Additive White Gaussian Noise (AWGN) which is assumed with zero mean and variance of $\sigma 2$. All these factors are considered for the performance comparison and evaluation also.

Table 1. Simulation Set-Up

Parameter	Values
Techniques Used	Cooperative NOMA, C-RS-NOMA
Power allocation schemes	Fix
Power allocation coefficients	$a1=0.75$, $a2=0.25$, and $\delta=0.12$
Distance of users from BS	$d_1=1000$, $d_2=500$, and $d_{12}=540$
$P_{Circuit}$	40 watts
Bandwidth	10^6 Hz
Path Loss Exponent	4
Fading Channel path	Rayleigh Fading Channel
Noise type	AWGN has zero mean and variance of $\sigma 2$
No. of Monte Carlo Simulations (N)	N=5*10^5
Target rates for both user	Rate1 (far user) = 1, Rate2 (near user) = 2

e) **Simulation results with different Performance Parameters:** Proposed C-RS-NOMA scheme performance has been evaluated with various parameters such as outage probability, the capacity of the channel, SE, and EE. A brief description and comparison with the existing Cooperative NOMA scheme have been shown with their corresponding graphs and tables given as,

i. **Outage Probability:** It can be defined as a probability in which the information rate to channel interference and noise is less than the threshold information rate. Mathematically it is the CDF (Cumulative Distribution Function) of Signal-to-Noise ratio (SNR) as given in eq. 25,

$$P_o(R_{th}) = P(R < R_{th}) \tag{25}$$

Where R is the achievable information rate of the user and R_{th} is the required rate of information Outage probability should be as minimum as possible for better performance of the system.

Outage probability for C-RS-NOMA and C-NOMA for far and near users for variations in Signal-to-Noise ratio is shown in Fig. 4. It has been evaluated that the C-RS-NOMA model provides lesser overall outage probability than that of cooperative-NOMA.

Figure 4. Outage probability for C-RS-NOMA and C-NOMA

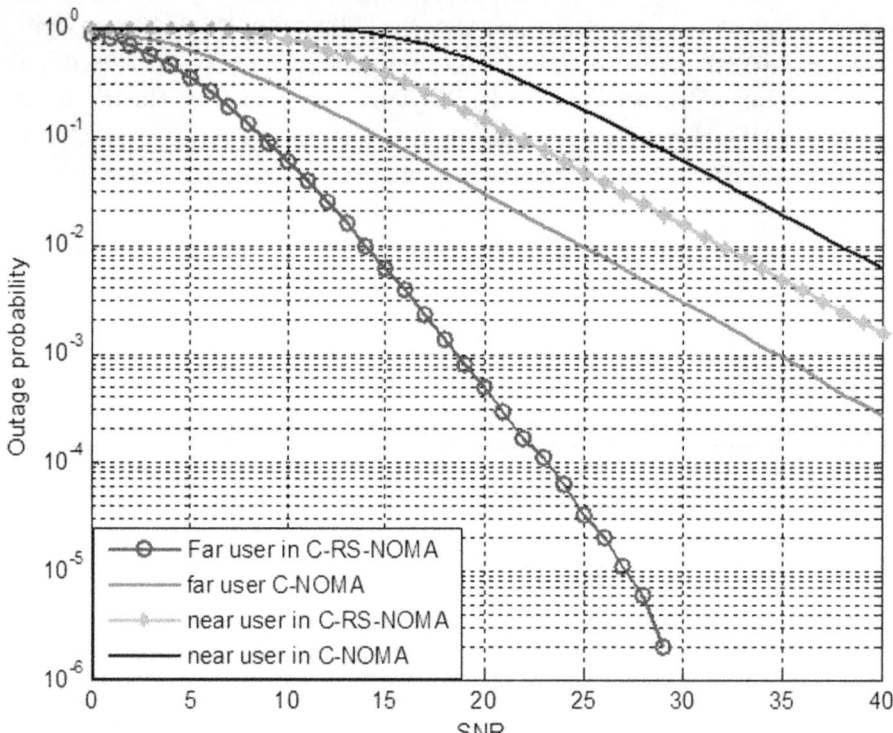

ii. **Channel Capacity:** It can be defined as the maximum amount of data that can be handled by a particular wireless communication channel. According to Shannon Hartley's Theorem, the capacity of a channel has a Bandwidth of B hertz and a Signal-to-Noise ratio (SNR).

$$C = Blog\_2(1+SNR) \text{ in bits/sec} \tag{26}$$

A comparison of the individual far and near user channel capacity between C-RS-NOMA and Co-operative NOMA with the variations in the transmitted power is shown in Fig. 5.

The total capacity for both these techniques with variation in transmitted power is given in Table 2.

From Fig. 5 and Table 2, it can be depicted that the C-RS-NOMA scheme outperforms the existing Cooperative NOMA scheme in terms of channel capacity.

iii. **Spectrum Efficiency (SE):** It can be defined as the maximum amount of data transfer that is being transmitted over a particular bandwidth to maintain the quality of service of transmission in wireless communication. Mathematically, SE is defined as given in eq. 27:

Figure 5. Channel capacity for individual users between C-RS-NOMA and Cooperative NOMA

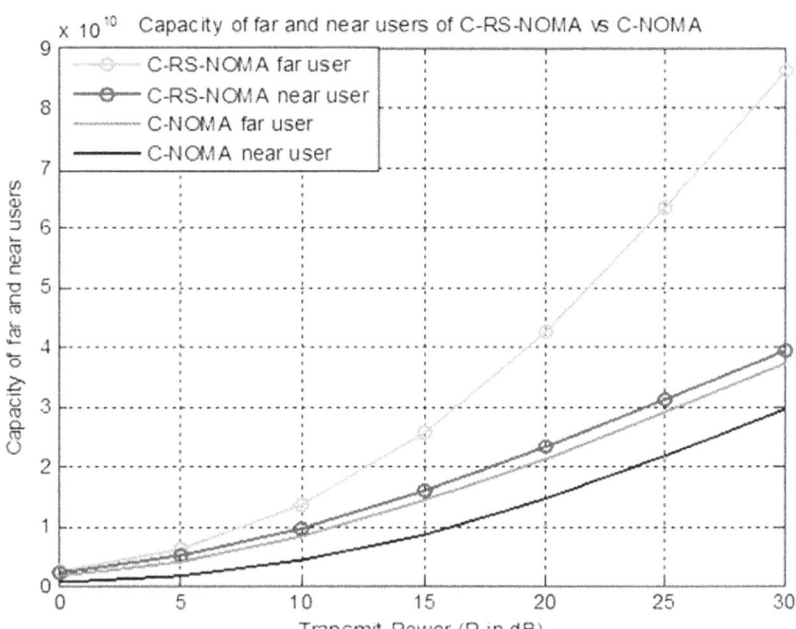

Table 2. Total channel capacity for C-RS-NOMA and Cooperative NOMA in Mbps

Technique/Power(dBm)	0	5	10	15	20	25	30
C-RS-NOMA	0.45	1.11	2.31	4.14	6.57	9.42	12.53
Cooperative NOMA	0.23	0.59	1.26	2.27	3.58	5.06	6.65

$$SE = \frac{R_T}{Bandwidth} \text{ in bps/Hz} \tag{27}$$

Where R_T is the total achievable rate of far and near users in this case it is given in eq. 28.

$$R_T = R_{11} + R_{12} + R_{22} \tag{28}$$

Fig.6 shows the spectral efficiency for C-RS-NOMA and Cooperative NOMA for comparison. It can be observed that the C-RS-NOMA system provides better Spectral efficiency than Cooperative NOMA. SE shows a positive slope with transmitted power.

Table 3 clearly shows that the proposed technique is more spectrally efficient than a Cooperative NOMA.

Figure 6. Spectral Efficiency of C-RS-NOMA and Cooperative NOMA

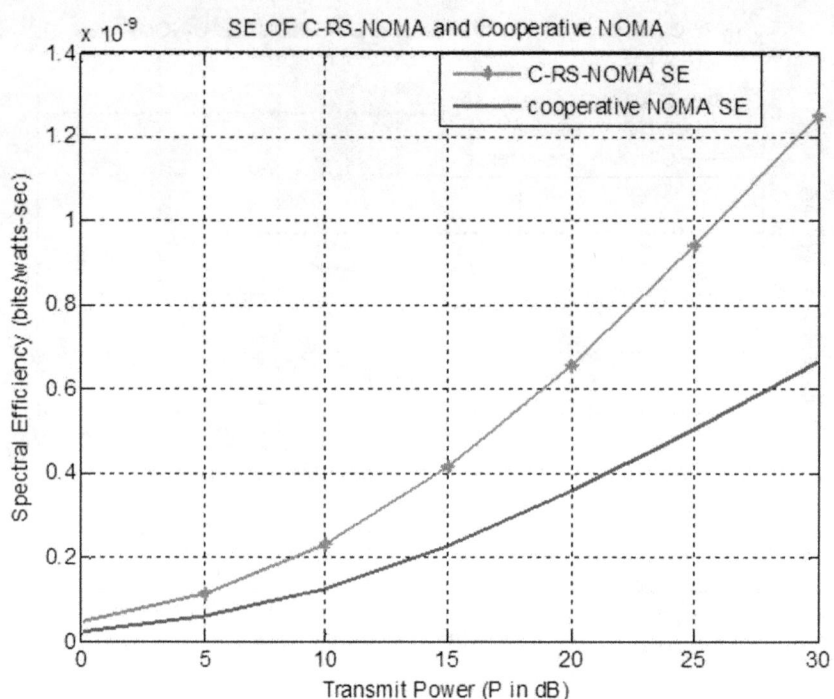

Table 3. Comparison of Spectral Efficiency between C-RS-NOMA and Cooperative NOMA

Technique/Power(dBm)	0	5	10	15	20	25	30
C-RS-NOMA	0.046	0.111	0.230	0.413	0.655	0.939	1.249
Cooperative NOMA	0.023	0.059	0.125	0.227	0.357	0.505	0.663

iv. **Energy Efficiency (EE):** Energy Efficiency is the ratio of the total achievable data/information rate to the total power consumed by the base station and is given in eq. 29 as.

$$EE = \frac{R_T}{P_T} \text{ in bits/joule.} \qquad (29)$$

Here, the total power consumed by the BS antenna P_T is given as in eq. 29:

$$P_T = P_s + P_{Circuit} \qquad (30)$$

Where $P_{circuit}$ is the power which is consumed by the power amplifier circuit at the base station.

Fig. 7 provides the Energy Efficiency for variations in Signal-to-Noise ratio (SNR) being transmitted from the BS. It can be observed that C-RS-NOMA is more energy efficient than a Cooperative NOMA.

f) **Performance comparison of the proposed technique with the existing techniques:** In this section, the performance of proposed C-RS-NOMA techniques has been compared with existing techniques such as Cooperative NOMA(C-NOMA) and NOMA. The effect of variation in transmitted power on sum channel capacity, energy efficiency, and spectral efficiency has been analyzed and compiled in Table 4. The transmitted power has been varied in equal variations from 0dB to 40dB. It was observed that the proposed C-RS-NOMA provides better performance in terms of average sum channel capacity, average energy efficiency, and average spectral efficiency. Simulation results show that in the case of channel capacity, C-RS-NOMA is 65.4% better as compared with NOMA and 47.03% better as compared with cooperative NOMA. Similarly, performance improvement was observed in energy efficiency where the proposed technique turns out to be 65.5% better compared to NOMA and 46.5% compared to cooperative NOMA. Similar results were observed for spectral efficiency. In this case, the performance improvement was detected to be 66% and 48% as compared to NOMA and cooperative NOMA respectively.

Figure 7. Comparison of Energy Efficiency in C-RS-NOMA and Cooperative NOMA

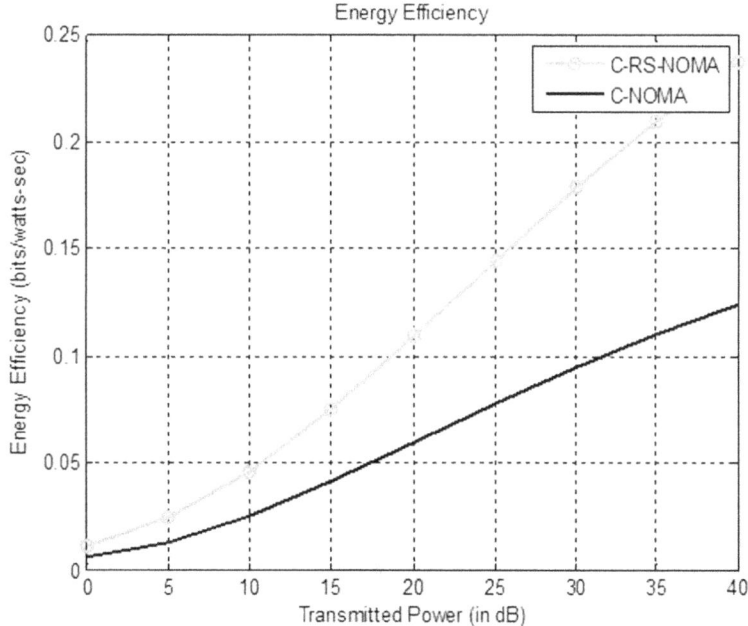

Table 4. Performance comparison of proposed C-RS-NOMA technique with existing techniques

Techniques Parameters	NOMA (existing)	C-NOMA (existing)	C-RS-NOMA (proposed)	Improvement w.r.t. NOMA	Improvement w.r.t. C-NOMA
Average Channel Capacity (in bits/sec)	2.827	4.330	8.175	65.4%	47.03%
Average Energy Efficiency (bits/joule)	0.040	0.062	0.116	65.5%	46.5%
Average Spectral Efficiency (bps/Hz)	2.82	4.30	8.30	66%	48%

The performance of C-RS-NOMA has also been compared with existing techniques based on achievable data rates of far user and is shown in Fig.8. It was depicted that the data rates of the far user for the proposed C-RS-NOMA are quite higher than that of C-NOMA and NOMA.

Figure 8. Comparison of Far user achievable data rates for proposed C-RS-NOMA and existing techniques

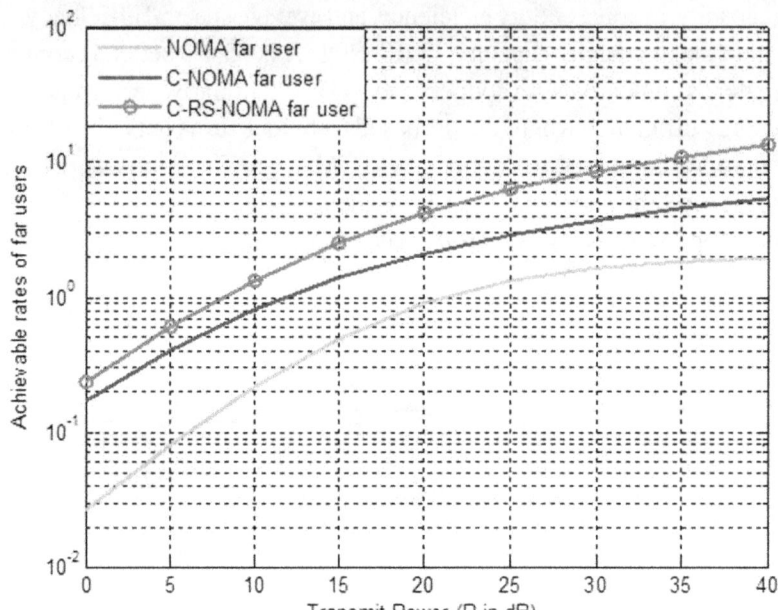

CONCLUSION

In this work, Cooperative Rate-Splitting (C-RS) technique has been integrated with conventional NOMA(C-RS-NOMA) to improve the performance measures of power-domain NOMA in 5G wireless communication networks. The C-RS-NOMA-based system has been modeled for a downlink scenario and its performance has been evaluated for variation in transmitted power and signal-to-noise ratio (SNR). Simulation results show that C-RS-NOMA boosts the quality-of-service (QoS) of conventional NOMA-based systems in terms of various performance parameters such as outage probability, the overall capacity of the channel, spectral efficiency, and energy efficiency. So, the proposed technique can be a better candidate for fifth and beyond fifth generations wireless networks.

REFERENCES

Anwar, A., Seet, B., Hasan, M. & Li, X. (2019). *A Survey on Application of Non-Orthogonal Multiple Access to Different Wireless Networks*. MDPI.

Boccardi, F., Heath, R. W., Lozano, A., Marzetta, T. L., & Popovski, P. (2014). Five Disruptive Technology Directions for 5G. *IEEE Communications Magazine, 52*(2), 74–80. doi:10.1109/MCOM.2014.6736746

Dai, L., Wang, B., Yuan, Y., Han, S., Chih-lin, I., & Wang, Z. (2015). Non-orthogonal multiple access for 5G: Solutions, challenges, opportunities, and future research trends. *IEEE Communications Magazine*, *53*(9), 74–81. doi:10.1109/MCOM.2015.7263349

Ding, Z., Adachi, F., & Poor, H. V. (2016, January). The Application of MIMO to Non-Orthogonal Multiple Access. *IEEE Transactions on Wireless Communications*, *15*(1), 537–552. doi:10.1109/TWC.2015.2475746

Huang, X., Niu, K., Si, Z., He, Z., & Dong, C. (2017). *Rate-Splitting Non-orthogonal Multiple Access: Practical Design and Performance Optimization*. Communications and Networking. 11th EAI International Conference, Chongqing, Chin .

Islam, S. M. R., Avazov, N., Dobre, O. A., & Kwak, K. (2017). Power-Domain Non-Orthogonal Multiple Access (NOMA) in 5G Systems: Potentials and Challenges. In IEEE Communications Surveys & Tutorials, 19(2), pp. 721-742.

Joe. (2020). Does power allocation affect NOMA? *Wireless Communication*. https://ecewireless.blogspot.com/2020/04/how-does-power-allocation-affect-noma.html

Kizilirmak, R. (2016). Non-Orthogonal Multiple Access (NOMA) for 5G Networks. In *Towards 5G Wireless Networks - A Physical Layer Perspective*. IntechOpen.

Liu, Y., Pan, G., Zhang, H., & Song, M. (2016). On the Capacity Comparison Between MIMO-NOMA and MIMO-OMA. *IEEE Access: Practical Innovations, Open Solutions*, *4*, 2123–2129. doi:10.1109/ACCESS.2016.2563462

Mao, Y., Clerckx, B., & Li, V. O. (2018). Rate-splitting multiple access for downlink communication systems: Bridging, generalizing, and outperforming SDMA and NOMA. *J Wireless Com Network*, *2018*(1), 133. doi:10.118613638-018-1104-7 PMID:30996723

Mao, Y., Clerckx, B., & Li, V. O. K. (2018). Energy Efficiency of Rate-Splitting Multiple Access, and Performance Benefits over SDMA and NOMA. *2018 15th International Symposium on Wireless Communication Systems (ISWCS)*, (pp. 1-5). IEEE. 10.1109/ISWCS.2018.8491100

Sonia, D. & Gupta, D. (2021). Performance Enhancement of NOMA: A 5G Candidate. *Thirty Sixth National Convention of Electronics and Telecommunication Engineers on Antenna Design for Efficient Communication and Networking*, (pp. 91-103). The Institute of Engineers, Bhatinda Local Centre, Under the Aegis off Electronics & Telecommunication Divisional Board.

Sonia, S. (2022). Non-Orthogonal Multiple Access Technique in Wireless Communication. In *Emerging Trends in Wireless Communication*, 1-22. Central West Publication.

Wu, Z., Lu, K., Jiang, C., & Shao, X. (2018). Comprehensive Study and Comparison on 5G NOMA Schemes. *IEEE Access: Practical Innovations, Open Solutions*, *6*, 18511–18519. doi:10.1109/ACCESS.2018.2817221

Yang, Z., Chen, M., Saad, W., & Shaikh-Bahaei, M. (2020). Downlink Sum-Rate Maximization for Rate Splitting Multiple Access (RSMA). *ICC 2020 - 2020 IEEE International Conference on Communications (ICC)*, (pp. 1-6). IEEE. 10.1109/ICC40277.2020.9149417

Zeng, J., Lv, T., Liu, R. P., Su, X., Peng, M., Wang, C., & Mei, J. (2018). Investigation on Evolving Single-Carrier NOMA into Multi-Carrier NOMA in 5G. *IEEE Access: Practical Innovations, Open Solutions, 6*, 48268–48288. doi:10.1109/ACCESS.2018.2868093

Zeng, M., Yadav, A., Dobre, O. A., Tsiropoulos, G. I., & Poor, H. V. (2017, October). Capacity Comparison Between MIMO-NOMA and MIMO-OMA With Multiple Users in a Cluster. *IEEE Journal on Selected Areas in Communications, 35*(10), 2413–2424. doi:10.1109/JSAC.2017.2725879

Zhu, Y., Wang, X., Zhang, Z., Chen, X., & Chen, Y. (2017). A rate-splitting non-orthogonal multiple access scheme for uplink transmission. *2017 9th International Conference on Wireless Communications and Signal Processing (WCSP)*, (pp. 1-6). 10.1109/WCSP.2017.8171078

Zhu, Y., Zhang, Z., Wang, X., & Liang, X. (2017). A low-complexity non-orthogonal multiple access systems based on rate splitting. *2017 9th International Conference on Wireless Communications and Signal Processing (WCSP)*, (pp. 1-6). 10.1109/WCSP.2017.8171135

Chapter 7
Compact Dual Fractal Curves-Based Microstrip Patch Antenna for 5G Applications

Manpreet Kaur

Yadavindra Department of Engineering, Guru Kashi College, Punjabi University, India

ABSTRACT

The aim of this work is to design and characterize a compact 5G antenna with an appreciable bandwidth. In the suggested antenna, two fractal curves are applied onto the conducting patch that helped to achieve the targeted operational characteristics. The whole patch is printed on the chosen FR4 material. The modified rectangular patch is placed on the top side, whereas a partial ground is placed on the bottom side. The antenna size observed is 45 mm². The overall adjustments of the specific design parameters are carried out for better S11 characteristics at the fundamental resonance. The operational band stretches from 25.99 GHz to 27.72 GHz with a fractional bandwidth of 6.4%. The designed antenna showed resonance at 26.75 GHz with S11 value -21.00 dB. The proposed antenna possesses good S11 characteristics, appreciable gain, and significant bandwidth at the operational frequency band. Moreover, the influence of variations in the antenna geometry has also been demonstrated for deep understanding.

1. INTRODUCTION TO 5G COMMUNICATION

In the present era, fifth generation (5G) mobile communications technology is growing at a much faster rate (Prasad et al., 2021). This technology is not only able to accommodate the super fast speeds but also satisfy the low data rate requirements of Internet of Things (IoT) applications (Tarpara et al., 2018). It offers higher grade of operational functionality than other configurations of existing communication systems. Additionally it offers groundbreaking solutions that reach across the society. 5G can transfer voice and data concurrently and accurately. A huge growth in the 5G communication sector needs an efficient system with high speed, high throughput, low latency and superior spectrum utilization (Gholb et al., 2017). 5G will reach into every aspect of our business and personal lives, and also maintain a dominant position as the world's foremost cellular technology (Kumawat & Joshi, 2021). It enables a new kind

DOI: 10.4018/978-1-6684-7000-8.ch007

of network that connects everyone and everything to work collectively; therefore it has the capability to expand the mobile ecosystem into new realms (Ozpinar et al., 2020). Same radio frequencies are employed to run 5G systems and in currently available smart phones as well as in widely usable domain i.e. satellite communications. In reality, this technology has the capability to impact every industry, health care, transportation, agriculture, etc (Lota et al., 2017). Moreover, it demands for notable and significant solutions for structured antenna designing along with less degree of device complexity and integration. Antenna designing for 5G communication has received renewed attention from the antenna community. Recently, the interest of researchers in designing compact antennas based on multiple techniques for 5G applications has increased significantly (Hong et al., 2017). Such devices can greatly affect the desired parameters for 5G operation (Kumawat & Joshi, 2021). Compact antenna designs with excellent performance are highly demanded in the available limited space and strong interference scenario (Ozpinar et al., 2020). The choice of an efficient antenna system is considered as a crucial part of all devices and will impose few challenges including size, location and shape (Lota et al., 2017).

2. LITERATURE REVIEW

The purpose of literature review is to profoundly analyze the previous findings of the various researchers. Research on such type of antennas shows a very positive outlook for development. As a stepping stone, different aspects of various types of antennas are examined. Therefore, various antenna designs proposed by distinguished researchers are studied and analyzed. Nowadays, huge development is going on in the field of 5G mobile communications technology. In published literature, several configurations of structures useful for 5G applications have been reported (Tarpara et al., 2018, Gholb et al., 2017 and Kumawat & Joshi, 2021), (Hong et al., 2017, Yu et al., 2018 and Diawuo et al., 2018), Li et al., 2018), and so on. Multi-band patch antennas were discussed in (Mazen et al., 2021). By etching rectangular and circular slots onto the conducting patch, the designed structure showed functionality for Mid-band 5G applications. In (Sree & Nelaturi, 2021), a Multiple Input Multiple Output (MIMO) antenna in which fractal approach was suggested and claimed its functionality for sub 6-GHz applications. The proposed structure was flower shaped and supported by defected ground plane. The designed geometry was also experimentally analyzed. In (Wang et al., 2021), a high isolation monopole antenna functioning at dual bands was demonstrated in an effective way. The authors validated the suggested structure functional for 5G applications. In the design, circular patch was connected with L-shaped branches that are mirror-symmetric. The employed slotted ground was responsible for changes in the distribution parameters associated with the line. A Polyethylene Terephthalate (PET) substrate based modified patch antenna was discussed in (Tighezza et al., 2018) for 5G applications. The whole structure consisted of modified rectangular shaped radiator connected with feedline and a partial ground plane. In (Thi et al., 2021), a low profile, broadband dual-polarized antenna was described. In the structure, coupling occurs between the orthogonal microstrip-lines and the shorted patches. Among the two ports, bandwidth and isolation achieved was 33% and 25 dB, respectively. The broadside gain attained was 7.9-8.5 dBi. In the design, MIMO antenna with eight elements was composed of four antenna pairs (Cheng & Du, 2021). The four shorted radiation patches were used in the radiator of antenna pair. The measured -6 dB working band was from 4.37-5.5 GHz. The acceptable efficiency was higher than 40%. Dolly shaped antenna of size 49 mm² based on Rogers RO3010 was implemented in (Bamy et al., 2021). Parasitic elements have helped to achieve the desired performance. Two operating bands were reported that claimed its functionality for 5G and radar applications. A helical antenna based on Teflon material was projected in (Zeain et al., 2020). It has provided circular polarization and a gain of 11.2 dB. In (Wani et al., 2018), a four port MIMO antenna was implemented on Neltec substrate. Each antenna was loaded with capacitively loaded loops unit cell array. The integration of unit cell array with antenna elements resulted in gain enhancement. Microstrip antennas functional at 28 GHz were realized in (Teresa & Umamaheswari, 2020). Slots used in the geometry resulted in enhanced bandwidth. A lighter remote reconfigurable antenna was recommended for modern 5G networks (Elwi, 2021). Hilbert fractal metamaterial based unit cell was embedded onto the conducting patch. It showed well-controlled bandwidth. Gain spectra were affected greatly by variations in light intensity. A stacked dielectric resonator antenna was suggested in (Sun et al., 2019). In the structure, two thin sheets having high permittivity values along with two hollow slabs were used. They were responsible for four adjacent resonant modes to form wideband. Obtained bandwidth and gain was 54% and 9.2 dBi, respectively.

In the manuscript, dual fractal based antenna design is projected that operates for 5G communication services. It consists of seven Sections. Section 1 deal exclusively with the fundamental concepts of 5G communication and also provides enough motivation to design and implement an antenna for 5G applications. Section 2 presents an elaborated literature review on already suggested antennas for above said services. The detailed design methodology of the geometry is described in Section 3. Deep analysis of the antenna with different changes is effectively narrated in Section 4 of this chapter. Results of projected geometry are covered in Section 5. Section 6 contains the comparison of implemented antenna with published antennas. At last, Section 7 elaborates the conclusion of complete chapter.

3. PROPOSED ANTENNA DESIGN

In this Section, the design methodology of proposed miniaturized antenna is addressed by using an integration of two fractal curves (Kaur & Sivia, 2019). The suggested antenna is based on the principle of electromagnetic theory. The antenna construction consists of a dual fractal curves based radiating patch connected with line feed. Opposite side encloses a partial ground plane. The antenna is etched on a 6 x 7.5 mm^2 FR4 based substrate of 1.6 mm height (Kaur & Sivia, 2019). During the designing process, copper metal is used for the radiator and ground plane. For the construction of patch, Minkowski and Giuseppe Peano fractal curves are applied on two sides of the patch, respectively. On the left side of patch, Minkowski curve is introduced by removing a square whose side length is 1/3th of the original side length. At the middle upper side, Giuseppe Peano fractal is introduced by using two equal sized rectangular blocks (Kaur & Sivia, 2019). For this fractal, an upper block is added and a lower block is removed at the same time. The geometrical dimensions of the antenna are greatly influenced by the chosen substrate material. The used line feed has horizontal dimension i.e. width = 0.75 mm. Additionally, a small rectangular slot is etched from that portion of feed which is joint directly with patch. The area of the partial ground plane size is 41.25 mm^2. The geometrical configuration of suggested dual fractal curves based antenna is properly reported in Fig. 1. The first part i.e. Fig. 1(a) portrays the final design and second part i.e. Fig. 1(b) illustrates the reduced ground. The detailed dimensions of proposed dual fractal curves based microstrip antenna are thoroughly listed in Table 1.

Figure 1. Suggested compact antenna radiating patch

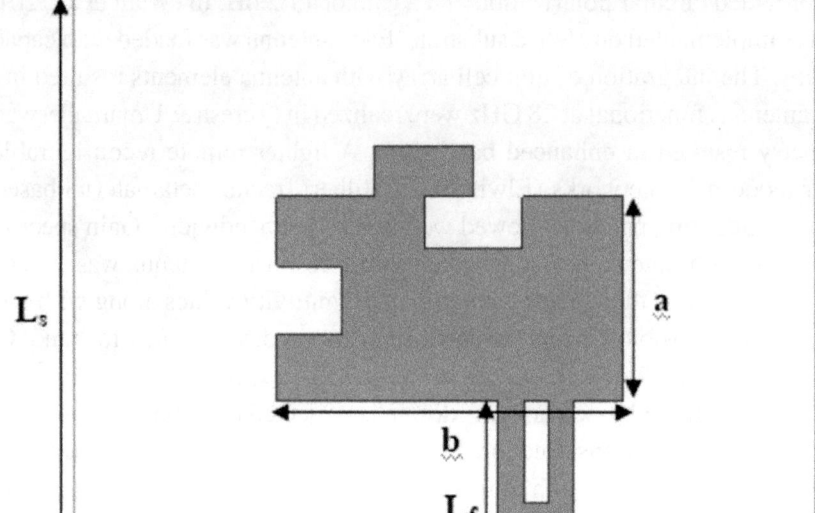

Figure 2. Suggested compact reduced ground

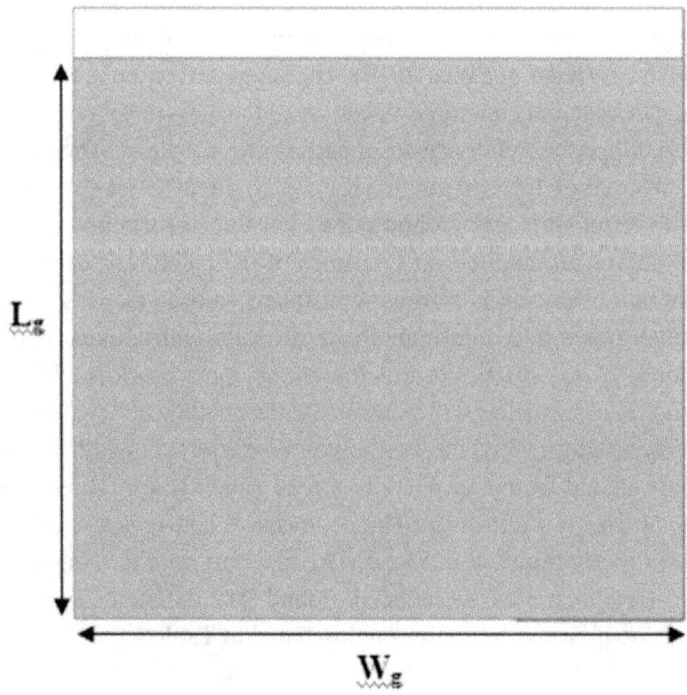

Table 1. Suggested antenna design specifications

S.No.		Parameter	Dimensions (mm)
1.	Substrate	Length (L$_s$)	6
2.		Width (W$_s$)	7.5
3.	Patch	Length (a)	2
4.		Width (b)	3.5
5.	Feedline	Length (L$_f$)	2
6.		Width (W$_f$)	0.75
7.	Ground Plane	Length (L$_g$)	5.5
8.		Width (W$_g$)	7.5
9.	Giuseppe Peano Curve	Length of rectangular block	0.6
10.		Width of rectangular block	1
11.	Minkowski Curve	Length of square	0.67
12.		Width of square	0.67
13.	Etched slot from Feed	Length	1
14.		Width	0.25

4. PARAMETRIC ANALYSIS

To prove the effectuality and reliability of the presented integrated approach, the simplified compact antenna device is simulated at different stages (Kaur & Sivia, 2020). Performance is examined in three different cases which include different fractal curves comparison, with/without introducing slot at the feed and effect of partial/full ground. Results of all forms of structures are briefly compared in terms of S11 value, operational bandwidth in %, gain and VSWR.

4.1 Performance Evaluation With Different Fractal Curves

In Figures 3 & 4, designs with Giuseppe Peano Curve only and with Minkowski Curve only are shown. The final designed antenna exhibits a bandwidth of 1.73 GHz (27.72-25.99 GHz) with dominant frequency at 26.75 GHz. With Giuseppe Peano Curve only, there is only one center frequency i.e. 26.84 GHz. The gain at the observed resonance is 5 dB. With Minkowski Curve only, resonance occurs at 21.45 GHz and 27.96 GHz. Gain values at the respective frequencies are 4.69 dB and 3.21 dB. The simulated S11 characteristics with Minkowski Curve only, Giuseppe Peano Curve only and Final Design are compared and projected in Fig. 5. From this Fig, it can be expounded that suitable results in terms of all desired parameters are achieved in final structure. The VSWR values with Minkowski Curve only, Giuseppe Peano Curve only and Final Design are given in Fig. 6. Simulated results associated with the implemented antenna at different stages including Minkowski Curve only, Giuseppe Peano Curve only and Final Design are given in Table 2.

Figure 3. Design with Giuseppe Peano Curve only   *Figure 4. Design with Minkowski Curve only*

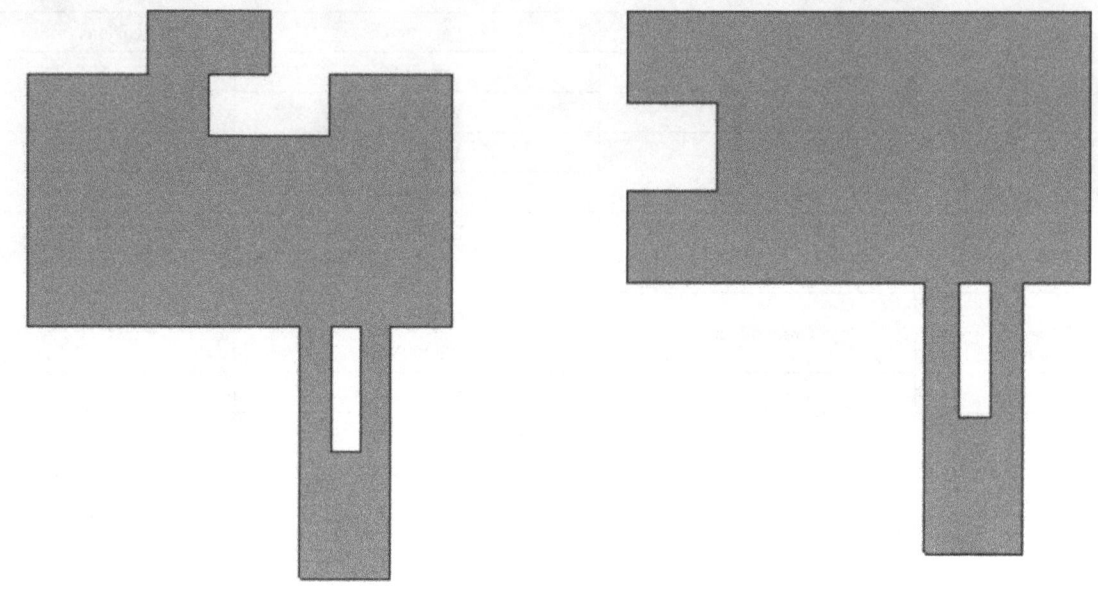

Table 2. Simulated results associated with the implemented antenna at different stages

Antenna type	'F$_L$', GHz	'F$_C$', GHz	'F$_H$', GHz	Functional Bandwidth, GHz	S11, dB	Gain, dB	VSWR
Antenna With Minkowski Curve only	21.00	21.45	22.56	2.1	-16.19	4.69	1.08
	27.16	27.96	28.79	1.63	-12.49	3.21	1.14
Antenna With Giuseppe Peano Curve only	26.05	26.84	27.87	6.7	-27.35	5.00	1.45
Final Design	25.99	26.75	27.72	6.4	-21.00	5.36	1.19

Figure 5. S11 characteristics with Minkowski Curve only, Giuseppe Peano Curve only and Final Design

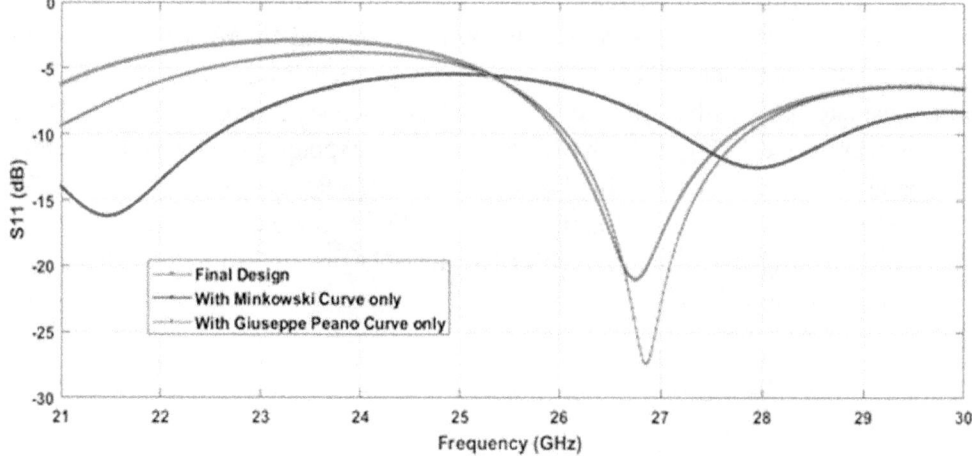

Figure 6. VSWR values with Minkowski Curve only, Giuseppe Peano Curve only and Final Design

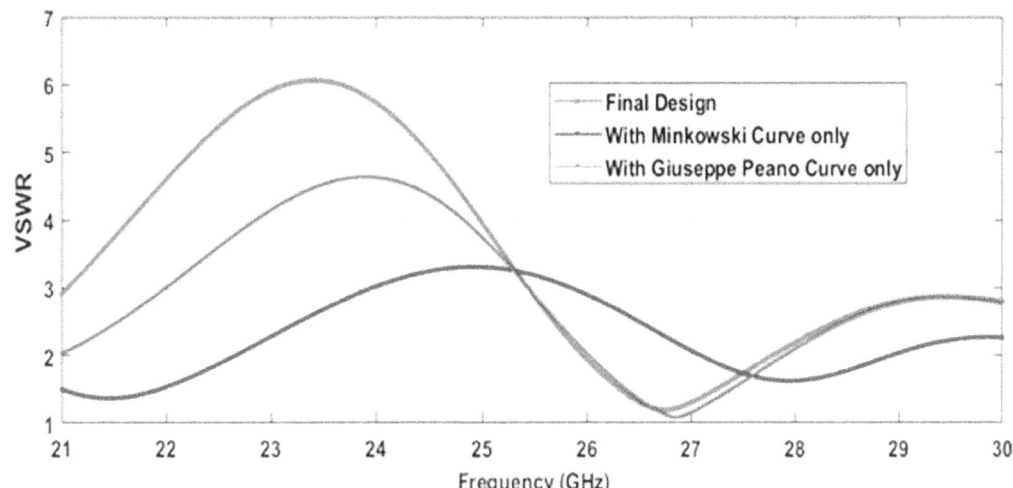

4.2 Performance Evaluation With/Without Slot at the Feed

In this Sub-Section, the effect of introducing slot at the feed associated with the antenna is evaluated. The rectangular slot is introduced at the upper part of the feed. The basic dimensions of the introduced slot are 1 mm and 0.25 mm. The structure designs with slot and without slot at the feed are shown in Fig. 5. S11 characteristics related to the implemented antenna with / without slot at the feed keeping all other parameters constant are delineated in Fig. 6. With this, better impedance characteristics are noted along with enhancement in reflection coefficient, bandwidth and gain. All the simulated results with / without slot at the feed related to the implemented antenna are tabulated in Table 3.

Table 3. Simulated results with / without slot at the feed related to implemented antenna

Antenna type	'F_L', GHz	'F_C', GHz	'F_H', GHz	Functional Bandwidth, %	S11, dB	Gain, dB	VSWR
Final Design with slot at the feed	25.99	26.75	27.72	6.4	-21.00	5.36	1.19
Final Design without slot at the feed	25.96	26.71	27.65	6.3	-20.77	4.25	1.15

4.3 Performance evaluation with Partial/Full ground

In this Sub-Section, performance is examined with Partial ground and Full ground. Among other parameters, there are slight variations except the gain value. The gain achieved with partial ground is 5.36 dB whereas with full ground, the gain attained is 2.54 dB. S11 characteristics related to implemented antenna with Partial Ground and Full Ground are delineated in Fig. 7. Simulated results related to implemented antenna with Partial Ground and Full Ground is given in Table 4.

Table 4. Simulated results related to implemented antenna with Partial Ground and Full Ground

Antenna type	'F_L', GHz	'F_C', GHz	'F_H', GHz	Functional Bandwidth, %	S11, dB	Gain, dB	VSWR
Final Design with Partial Ground	25.99	26.75	27.72	6.4	-21.00	5.36	1.19
Final Design with Full Ground	26.03	26.78	27.88	6.8	-21.07	2.54	1.21

5. RESULTS AND DISCUSSION

In this Section, all the essential parameters are examined. S11 is an important parameter that must be carefully investigated, as it tells whether the antenna is functional or not. The resonance occurs at 26.75 dB and corresponding S11 value is -21.00 dB. With reference to S11 value of -10 dB, the examined 'F_L' and 'F_H' values are 25.99 GHz and 27.72 GHz, respectively. The S11 plot related to the implemented antenna is shown in Fig. 8.

Figure 7. Design with slot at the feed

Figure 8. Design without slot at the feed

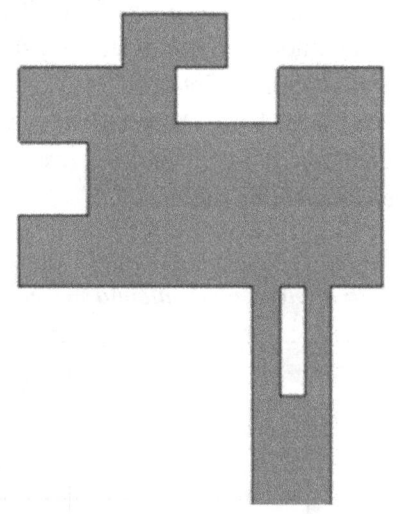

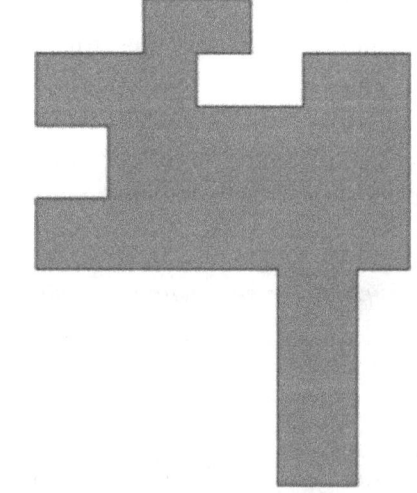

VSWR parameter measures the amount of reflections from the load. For optimum performance of an antenna, its value is less than 2. At the examined resonance, its value is 1.19. The VSWR plot is presented in Fig. 9. With proper adjustments of all geometrical parameters, desired operating band with sufficient bandwidth is achieved. Gain is calculated by taking division of effective radiated output power with input power. The evaluated gain plot is represented in Fig. 10. Further, it is noted that the implemented antenna offers a gain of 5.36 dB at the depicted resonance. Fig. 11 shows that over the entire specified sweep, the simulated peak gain ranges from 4.11 dB to 6.18 dB. The high gain designed miniaturized antenna is the well suitable candidate for the above mentioned applications.

Figure 9. S11 characteristics with / without slot at the feed related to implemented antenna

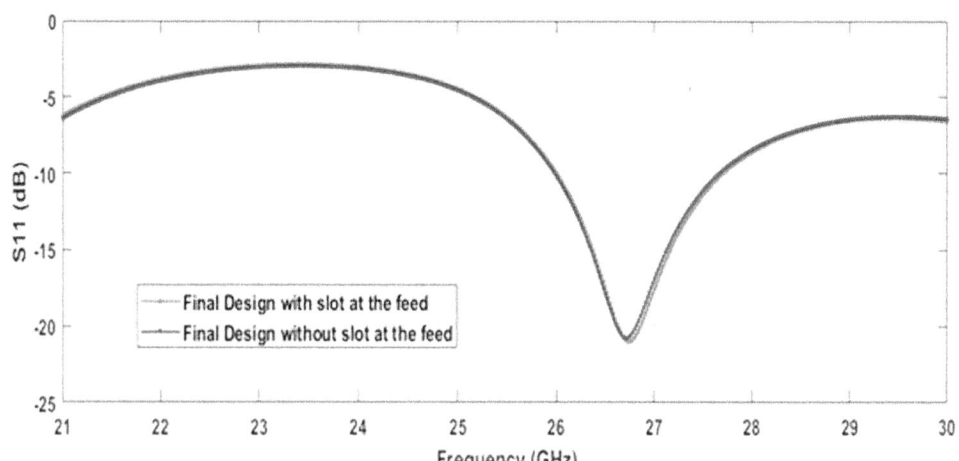

Figure 10. S11 characteristics related to implemented antenna with Partial Ground and Full Ground

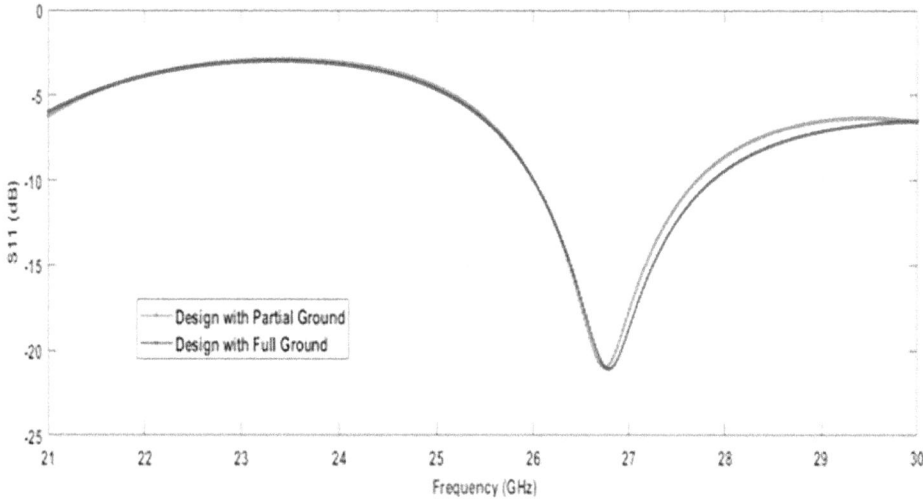

Radiation pattern is observed at the center frequency linked with the functional frequency range. It is the graphical representation of variation in electromagnetic field strength at all equi-distant points from the antenna. The observed radiation pattern at the observed resonance is shown in Fig. 12. The pattern exhibits a arbitrary omni-directional behaviour. Fig. 13 illustrates the current distribution plot. On the designed patch, the computed current distribution is 6.8568e+002 A/m. This value is noted at the realized frequency. The magnitude of surface current density is maximum everywhere except at the centre and boundary edges associated with the patch. Group delay associated with the implemented antenna within the operating range is shown in Fig. 14 and corresponding radiation efficiency is graphically presented in Fig. 15. At the claimed band, group delay is 8.1 x 10^{-10} sec and radiation efficiency is 85%.

Figure 11. S11 characteristics related to the implemented antenna (Final Design)

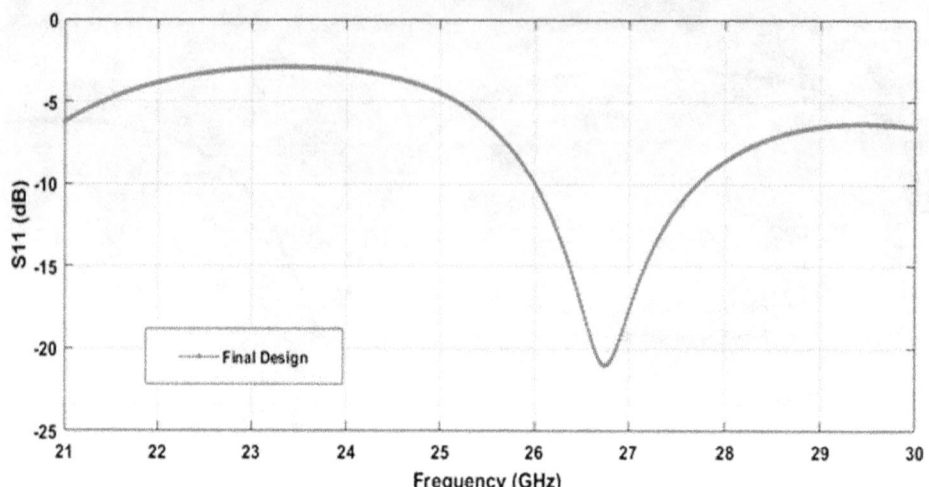

Figure 12. VSWR plot related to the implemented antenna (Final Design)

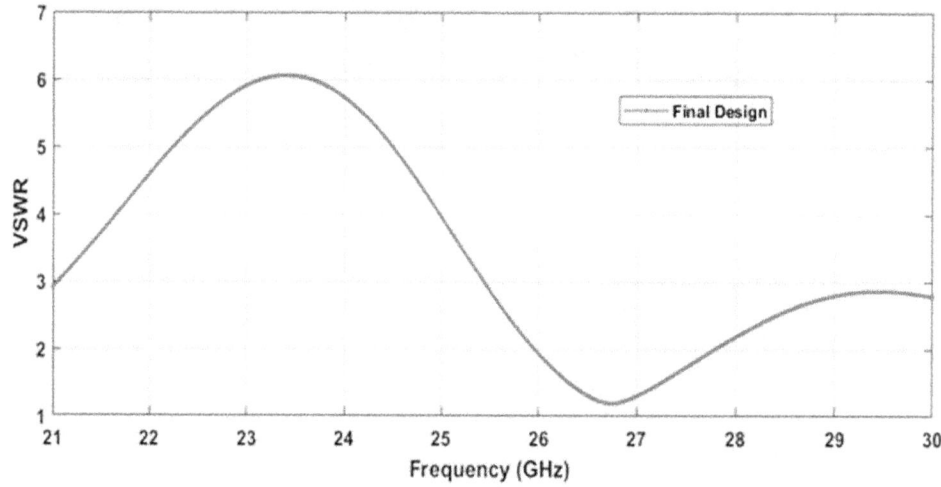

6. COMPARISON WITH FEW PUBLISHED ANTENNAS

A comparison of the suggested antenna design with few formerly designed structures is done thoroughly by taking essential parameters. Table 5 covers the comparison. The chosen parameters are: material / volume, working frequency range, gain, and percentage radiation efficiency. Here, after comparison, it is stated that the proposed dual fractals based antenna exhibiting compact structure with wide operating frequency range is a strong candidate functional for 5G applications.

Figure 13. Gain associated with the implemented antenna

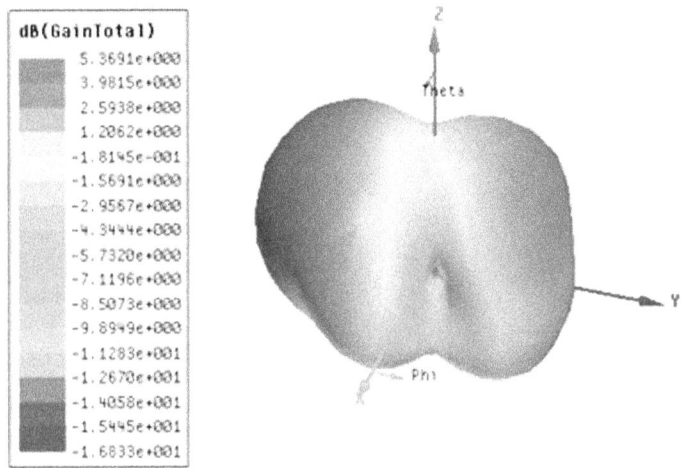

Figure 14. Peak Gain plot associated with the implemented antenna within the operating range

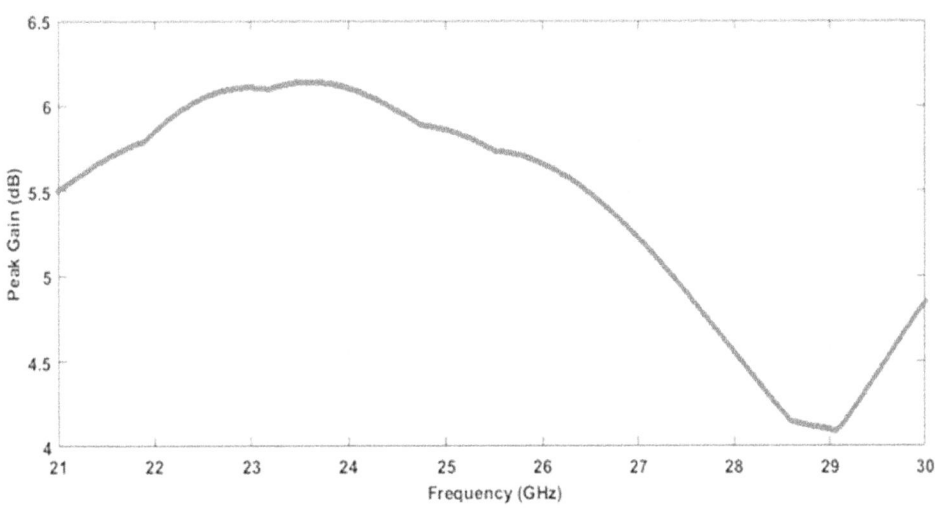

Table 5. Implemented antenna comparison with few formerly designed antennas

Ref. no.	Material/ Volume (mm²)	Working frequencies range (GHz)	Gain (dB)	Radiation efficiency (%)
(Darboe et al., 2019)	Rogers 5880/ 45.47	27.53-28.38	6.62	70
(Jandi et al., 2017)	Rogers 5880/ 361	27.75-28.77	7.54	62
(Kamal et al., 2020)	Duroid RO4003C/ 243	27.03-29.19	8.40	83.51
(Al-Bawri et al., 2020)	Rogers 5880/ 552	24-29.90	13.40	98
(Rusmono et al., 2020)	Rogers 5880/ 621.62	23.89-25.11	---	---
Suggested antenna	**FR4/ 45**	**25.99-27.72**	**5.36**	**85**

Figure 15. Radiation pattern associated with the implemented antenna

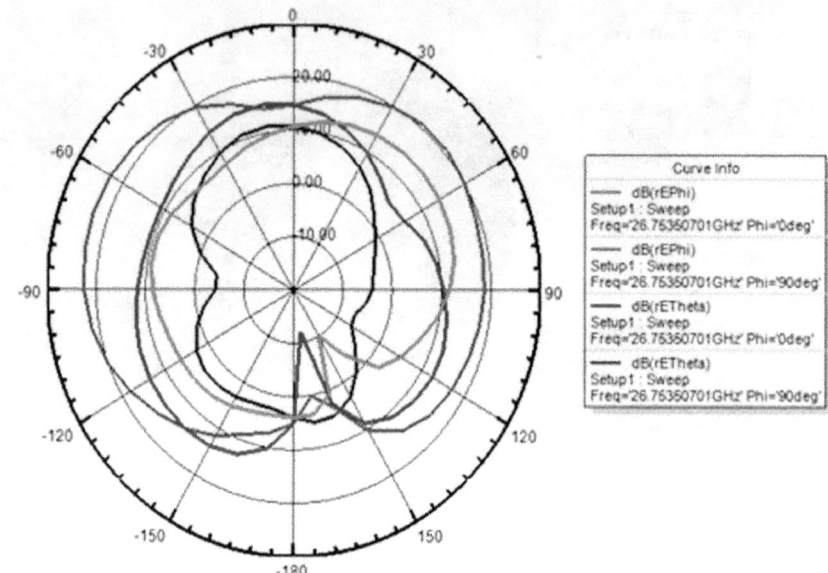

Figure 16. Current distribution on the patch

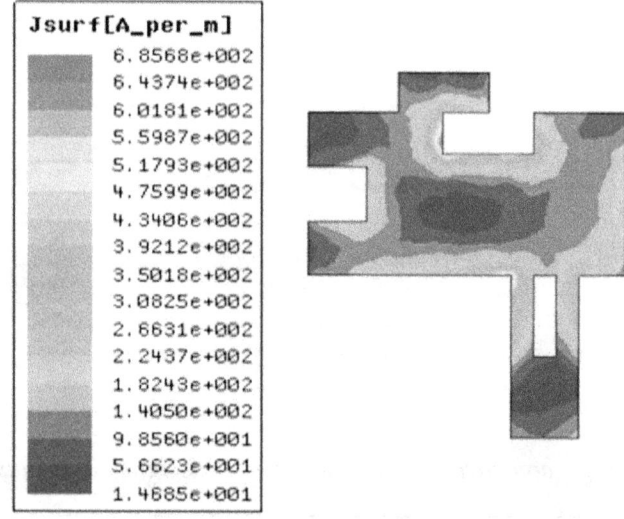

7. CONCLUSION

The proposed antenna is implemented by introducing Minkowski and Giuseppe Peano fractal curves onto the radiating patch. The antenna is conceptualized, designed and evaluated for 5G applications with due considerations. The proposed dual fractal curves based antenna is verified on the basis of essential parameters. The produced results showed that the antenna has a resonance at 26.75 GHz with bandwidth of 1.73 GHz that verifies the -10 dB bandwidth criteria. At the reported resonance, the gain evaluated is admirable. It is claimed that the structure offers good gain and stable radiation characteristics.

Figure 17. Group delay associated with the implemented antenna within the operating range

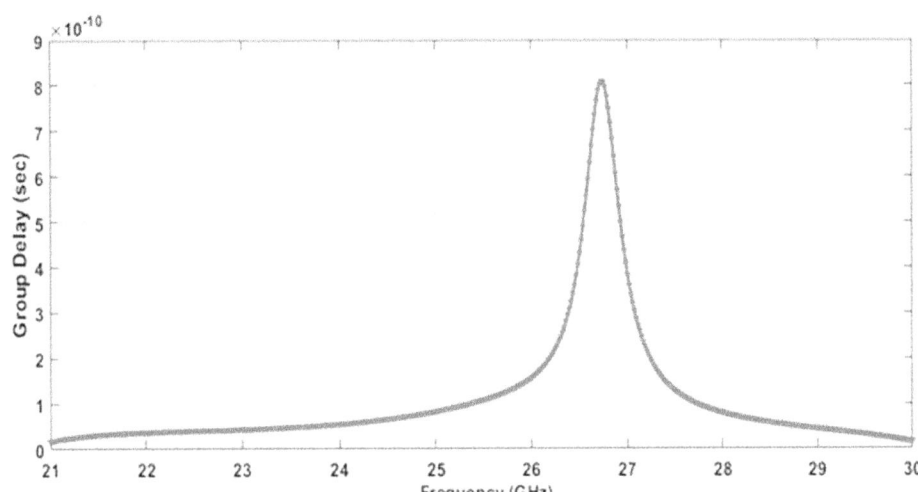

Figure 18. Radiation efficiency associated with the implemented antenna within the operating range

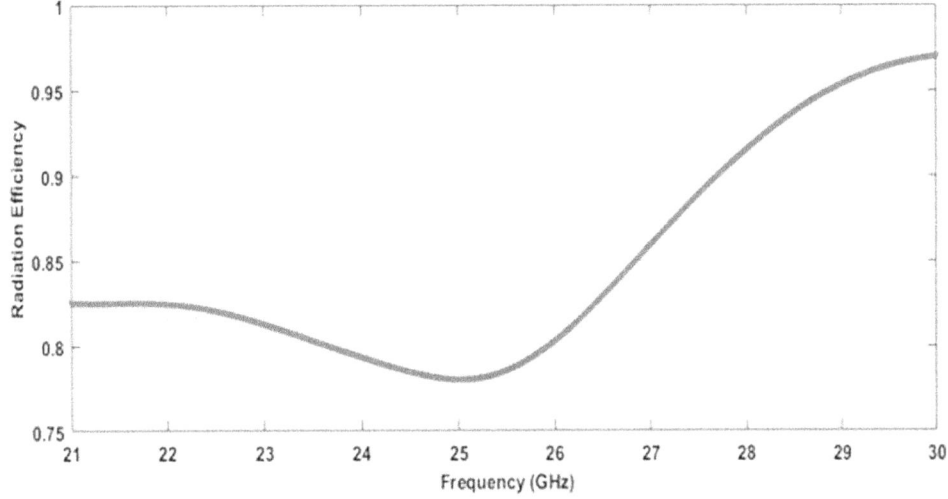

REFERENCES

Al-Bawri, S. S., Islam, M. T., Shabbir, T., Muhammad, G., Islam, M. S., & Wong, H. Y. (2020, October). Hexagonal shaped near zero index (NZI) metamaterial based MIMO antenna for millimeter-wave application. *IEEE Access: Practical Innovations, Open Solutions*, 8, 181003–181013. doi:10.1109/ACCESS.2020.3028377

Bamy, C. L., Mbango, F. M., Konditi, D. B. O., & Mpele, P. M. (2021, April). A compact dual-band Dolly-shaped antenna with parasitic elements for automotive radar and 5G applications. *Heliyon*, 7(4), e06793. doi:10.1016/j.heliyon.2021.e06793 PMID:33948514

Cheng, B., & Du, Z. (2021, July). Dual Polarization MIMO Antenna for 5G Mobile Phone Applications. *IEEE Transactions on Antennas and Propagation*, *69*(7), 4160–4165. doi:10.1109/TAP.2020.3044649

Darboe, O., Konditi, D. B. O., & Manene, F. (2019, June). A 28 GHz rectangular microstrip patch antenna for 5G applications. *International Journal of Engineering Research & Technology (Ahmedabad)*, *12*(6), 854–857.

Diawuo, H. A., & Jung, Y. (2018, July). Broadband Proximity-Coupled Microstrip Planar Antenna Array for 5G Cellular Applications. *IEEE Antennas and Wireless Propagation Letters*, *17*(7), 1286–1290. doi:10.1109/LAWP.2018.2842242

El Gholb, Y., El Amrani El Idrissi, N., & Ghennioui, H. (2017, March). 5G: An Idea Whose Time Has Come. *International Journal of Scientific and Engineering Research*, *8*(3).

Elwi, T. A. (2021, August). Remotely controlled reconfigurable antenna for modern 5G networks applications. *Microwave and Optical Technology Letters*, *63*(2), 2018–2023. doi:10.1002/mop.32505

Hong, W., Baek, K., & Ko, S. (2017, December). Millimeter-Wave 5G Antennas for Smartphones: Overview and Experimental Demonstration. *IEEE Transactions on Antennas and Propagation*, *65*(12), 6250–6261. doi:10.1109/TAP.2017.2740963

Jandi, Y., Gharnati, F., & Oulad, A. (2017). Said, "Design of a compact dual bands patch antenna for 5G applications." *2017 International Conference on Wireless Technologies, Embedded and Intelligent Systems (WITS)*, (pp. 1-4). Semantic Scholar.

Kamal, S., Mohammed, A. S. B., Bin Ain, M. F., Ullah, U., Hussin, R., Ahmad, Z. A., Othman, M., & Rahman, M. F. A. (2020, September). A novel negative meander line design of microstrip antenna for 28 GHz mmwave wireless communications. *Wuxiandian Gongcheng*, *29*(3), 479–485. doi:10.13164/re.2020.0479

Kaur, M., & Sivia, J. S. (2019, February). Minkowski, Giuseppe Peano and Koch Curves based Design of Compact Hybrid Fractal Antenna for Biomedical Applications using ANN and PSO. *International Journal of Electronics and Communications*, *99*, 14–24. doi:10.1016/j.aeue.2018.11.005

Kaur, M., & Sivia, J. S. (2019, March). ANN-based Design of Hybrid Fractal Antenna for Biomedical Applications. *International Journal of Electronics*, *106*(8), 1184–1199. doi:10.1080/00207217.2019.1 582712

Kaur, M., & Sivia, J. S. (2019, November). Giuseppe Peano and Cantor set fractals based miniaturized hybrid fractal antenna for biomedical applications using artificial neural network and firefly algorithm. *International Journal of RF and Microwave Computer-Aided Engineering*, *30*(1), 1–11.

Kaur, M., & Sivia, J. S. (2020, January). ANN and FA Based Design of Hybrid Fractal Antenna for ISM Band Applications. *Progress in Electromagnetics Research C*, *98*, 127–140. doi:10.2528/PIERC19110901

Kumawat, P., & Joshi, S. (2021). A Novel Dual-Band Orthogonal Polarized Elliptical Patch Antenna Array for 5G Applications. *Journal of Scientific Research*, *65*(3), 184–190. doi:10.37398/JSR.2021.650322

Li, T., & Chen, Z. N. (2018, December). Metasurface-based shared-aperture 5G S-/K-band antenna using characteristic mode analysis. *IEEE Transactions on Antennas and Propagation*, *66*(12), 6742–6750. doi:10.1109/TAP.2018.2869220

Lota, J., Sun, S., Rappaport, T. S., & Demosthenous, A. (2017, November). 5G Uniform Linear Arrays With Beamforming and Spatial Multiplexing at 28, 37, 64, and 71 GHz for Outdoor Urban Communication: A Two-Level Approach. *IEEE Transactions on Vehicular Technology*, *66*(11), 9972–9985. doi:10.1109/TVT.2017.2741260

Mazen, K., Emran, A., Shalaby, A. S., & Yahya, A. (2021). Design of Multi-band Microstrip patch Antenna for Mid-band 5G Wireless Communication. *International Journal of Advanced Computer Science and Applications*, *12*(5), 459–469. doi:10.14569/IJACSA.2021.0120557

Ozpinar, H., Aksimsek, S., & Tokan, N. T. (2020, March). A Novel Compact, Broadband, High Gain Millimeter-Wave Antenna for 5G Beam Steering Applications. *IEEE Transactions on Vehicular Technology*, *69*(3), 2389–2397. doi:10.1109/TVT.2020.2966009

Prasad, S., Meenakshi, M., Adhithiya, N., Rao, P. H., Ganti, R. K., & Bhaumik, S. (2021, April). mmWave multibeam phased array antenna for 5G applications. *Journal of Electromagnetic Waves and Applications*, *35*(13), 1802–1814. doi:10.1080/09205071.2021.1917004

Rusmono, E. S., & Marani, T. (2020, July). Design of multiband MIMO antenna for 5G millimeter-wave application. *IOP Conference Series. Materials Science and Engineering*, *852*(1), 012154. doi:10.1088/1757-899X/852/1/012154

Sree, G. N. J., & Nelaturi, S. (2021, May). Design and experimental verification of fractal based MIMO antenna for low sub 6-GHz 5G applications. *International Journal of Electronics and Communications*, *137*(10), 153797. doi:10.1016/j.aeue.2021.153797

Sun, W.-J., Yang, W.-W., Chu, P., & Chen, J.-X. (2019, October). A wideband stacked dielectric resonator antenna for 5G applications. *International Journal of RF and Microwave Computer-Aided Engineering*, *29*(7), e21897. doi:10.1002/mmce.21897

Tarpara, N. M., Rathwa, R. R., & Kotak, D. N. A. (2018, April). Design of slotted microstrip patch antenna for 5G Application. *International Research Journal of Engineering and Technology*, *5*(4), 2827–2832.

Teresa, P. M., & Umamaheswari, G. (2022, September 03). Compact Slotted Microstrip Antenna for 5G Applications Operating at 28 GHz. *Journal of the Institution of Electronics and Telecommunication Engineers*, *68*(5), 3778–3785. doi:10.1080/03772063.2020.1779620

Thi, C. H. L., Ta, S. X., Nguyen, X. Q., Nguyen, K. K., & & C. D-N. (2021). Design of compact broadband dual-polarized antenna for 5G applications. *International Journal of RF and Microwave Computer-Aided Engineering*, *31*(5), e22615.

Tighezza, M., Rahim, S. K. A., & Islam, M. T. (2018, January). Flexible Wideband Antenna for 5G Applications. *Microwave and Optical Technology Letters*, *60*(1), 38–44. doi:10.1002/mop.30906

Wang, W., Wu, Y., Wang, W., & Yang, Y. (2021, June). Isolation Enhancement in Dual-band Monopole Antenna for 5G Applications. *IEEE Transactions on Circuits and Systems II*, *68*(6), 1867–1871. doi:10.1109/TCSII.2020.3040164

Wani, Z., Abegaonkar, M. P., & Koul, S. K. (2018). A 28-GHz Antenna for 5G MIMO Applications. *Progress in Electromagnetics Research*, *78*, 73–79. doi:10.2528/PIERL18070303

Yu, B., Yang, K., Sim, C., & Yang, G. (2018, January). A Novel 28 GHz Beam Steering Array for 5G Mobile Device With Metallic Casing Application. *IEEE Transactions on Antennas and Propagation*, *66*(1), 462–466. doi:10.1109/TAP.2017.2772084

Zeain, M. Y., Abu, M., Zakaria, Z., Al-Gburi, A. J. A., Syahputri, R., & Toding, A. (2020, October). Design of a wideband strip helical antenna for 5G applications. *Bulletin of Electrical Engineering and Informatics*, *9*(No. 5), 1958–1963. doi:10.11591/eei.v9i5.2055

Chapter 8
Sustainable Back–Haul Solution for Beyond 5G Networks:
Survey

Purva Joshi

Information Engineering, University of Pisa, Italy

ABSTRACT

Fifth generation (5G) technology is being incorporated into the infrastructure of many sectors to provide high-quality communication services, smart factories, vehicle-to-vehicle connection, and many other new services. Next-generation networking technology ultra-wideband (UWB) connects many devices and provides location-based services. This chapter reviews 5G enabling technologies and the cellular network as backhaul network interface and solution. Wireless backhauling, which replaces fiber cables for core network communication, is discussed in this chapter. Wireless backhaul sends communications between nodes using radio or microwave frequencies. Wireless backhaul can be installed faster and cheaper than fiber when fiber is too costly or impracticable (which involves trenching). 5G's enhanced connectivity, AI, massive IoT, and other disruptive technologies for essential sectors and public services will be crucial to these solutions. This chapter compares 5G and 6G.

1 INTRODUCTION

In order to successfully manage the ever increasing number of users as well as the ever-increasing demand for bandwidth, future cellular networks of the 5th, Generation, also known as 5G, will combine a variety of radio access technologies (including cellular, satellite, and WiFi, amongst others), as well as different kinds of equipment. 5G networks have one of its key aims to deliver a solution that can manage the ever-increasing demand for bandwidth as well as the in- creased number of users. This is one of the primary goals of 5G. The utilisation of mm-wave frequencies (also known as Millimeter Wave frequencies), huge multi input multi output (MIMO) systems, focused beams, and the construction of super dense network architecture will make this a reality. Wireless back-hauling has arisen as an important component of future networks as a result of the development of these revolutionary technologies and

DOI: 10.4018/978-1-6684-7000-8.ch008

ideas. This makes it feasible to accomplish diversification while keeping higher performance than with other techniques. Back-hauling using wireless links is not a new thing; it was accomplished in networks that predate 5G by making use of microwave connections or relays.

A few years ago, it was predicted that 3G and 4G technologies such as long term evolution (LTE) and high speed packet access (HSPA), would move in that direction. However, initiatives focusing on new 5G and beyond 5G technologies are also on the future. Among these is the Ericsson Startup 5G initiative. The connection between the network that acts as the backbone for other networks and the many other sub-networks is known as the back-haul. Back-haul is also the term that refers to the conveyance of data or network between access points that the general public uses. Back-haul is a term that is used in satellite communication to refer to the process of transporting data to a central location from where it may then be dispersed over a network.

The reason behind backhauling is last-mile aggregation is made possible via wireless back-haul, which improves this link. Direct connection to the internet is possible because these wireless networks can transmit hundreds of data streams and allow for effective and unrestricted throughput for data, video, and voice.

The concept of wireless back-hauls introduce the wireless communication with the advent of mm-Wave frequencies, mMIMO (massive multi input multi output) and beam forming, the criteria for performance are no longer a limiting constraint. However, the issues with inter cell interference can be solved by beam- forming using mMIMO, there are still substantial obstacles that prevent users from realizing the full potential benefit of wireless back-haul. The problem of resource allocation and user association is one that arises frequently in wireless back-hauling, and it has a significant impact on the performance as a whole.

Here, the back-hauling have been done via different mediums so for two types of back-haul network can be developed. When it comes to 5G back-haul, the most contentious question is whether or not to employ a wired or wireless solution. Each one may be advantageous in certain situations and detrimental in others. Depending on the use cases, budget, maintenance needs, and timing, it's possible that certain towns will need to adopt a combination of the two.

There is a sufficient back-haul choice provided by fiber back-haul networks, which is above 10 GBPS and has a latency of hundreds of microseconds. Be- cause of its large capacity and low bit error rate, fiber is quickly becoming one of the most preferred options for back-hauling data traffic. However, a fiber back- haul is not available in all locations, and in the areas in which it is not already present, its installation may be either impossible or exceedingly costly and time- consuming. The use of wireless back-haul eliminates the requirement for cables or underground wires for the delivery of wireless data, making installation more simpler. Wireless back-haul, which may be accomplished by microwave and millimeter wave, appears to be more widely available and simpler to implement.

The main aim behind this survey is to provide the issue and challenges to develop sustainable wireless back-haul solution for 5G and 6G networks. In the next section the overview of 5G and back-haul is explained. After that section 3 illustrates the comparative study of different generation with back-hauling. The issues and challenges for sustainable 5G (Tezergil & Onur, 2022) is demonstrated in section 4. The last section is current state of 5G technology and summary have been presented.

1.1 Objective and Scope

The main objective behind this survey is to compare past research on back- hauling and sustainability. Because of the significance of this subject, there are a great number of books that, from one point of view or another, discuss various aspects of this problem. However, to our knowledge, there are no publications that comprehensively cover all of the different viewpoints that might be taken on this subject. In addition, the majority of the surveys done in this field are already out of date. This chapter sought to give a thorough study on this idea that covered all aspects of it, including the technology that made it possible, potential applications of it in the future, and its current state.

This article provides a complete review of the changes that have taken place prior to 5G wireless back-hauling. Next, paper discuss the fundamental ideas be- hind the 5G wireless back-hauling innovations that are now under construction. In conclusion, considering that the inquiry into this topic carries on to beyond- 5G, the paper demonstrates the works on beyond 5G (Tezergil & Onur, 2022) and 6G networks that may be related with wireless back-hauling. This is because the research on this topic has made more advancements.

1.2 Literature Survey

There are several older studies that cover subjects that are pertinent. Back- hauling is discussed in (Jaber et al., 2016) from a variety of different points of view, such as the evolution of the cellular network prior to the concept of backhauling, the current state of the art that supports this concept, challenges that can be overcome, and configurations that can enable the use of wireless backhaul. 5G networks offer 1000 times more capacity and 10-100 times higher data rates as compared to LTE advance networks (Gupta & Jha, 2015). High connection volumes can be solved via dense deployments of tiny cells, which have a lower transmission power than macro cells (Nie et al., 2017).

The various back-haul options already in use at various frequencies are care- fully examined in (Siddique et al., 2015), along with their advantages, disadvantages, and potential uses. Additionally, other aspects of small cells and their back-hauling are also taken into account, including deployment, adaptability, interference control, de- lay management and signaling overhead. (Bojic et al., 2013) by Bojic et al. examines the wireless small cell back-haul. We compare and contrast wireless and optical mobile back- haul systems.

Ahamed and Faruque present a basic structure of 5G back-haul in (Ahamed & Faruque, 2018), outlining the advantages and disadvantages of each approach, including conventional (using fiber, for example) and wireless (using mmWave or sub 6 GHz) backhaul. The idea of ultra-dense networks and enabling technologies explore the modeling and performance of UDNs (ultra-dense network) is presented by Kamel, Hamouda, and Youssef (Kamel et al., 2016).

Kurt et al. concentrate on HAPS and emphasize its potential for a variety of use-cases (Kurt et al., 2021). The authors of (Kurt et al., 2021) offer a thorough review of radio resource management, handover management, machine learning for HAPS (High Altitude Platform Station), and the most recent advancements in energy and payload systems, as well as how re-configurable smart surfaces may be employed with HAPS.

Jaffry et al. (Jaffry et al., 2020) fully examine the moving networks idea for moving networks, UAVs (Unmanned Aerial Vehicle), and satellites. This paper is currently cur- rent and discusses the ideas around mobile 5G technologies. The system model of ARANs (aerial radio access networks) is described by Dao et al. (Dao et al., 2021), who also examine its transmission propagation, energy consumption, latency and mobility. The development of carrier ethernet made it possible for operators to transmit voice and data over the same network without the use of any additional supporting protocols. This was made possible by the versatility of the new technology (Chia et al., 2009).

Relays were introduced as a new component of the network for LTE and LTE-advanced systems with the intention of extending network coverage and range through the utilisation of wireless back-hauling. A discussion of the relay architectures may be found in the 3GPP technical report 36.806 (3GPP, 2010).). Despite the fact that self-interference and in-band communication reduce the efficiency of the entire arrangement, Gamboa and Demirkol's trials with the LTE self-backhauling strategy indicated improvements in the network's coverage and downlink data rate. (Gamboa & Demirkol, 2018)

Since LTE (Lee et al., 2009), the idea of employing several antennas for communication has been used commercially. Early MIMO systems, however, only used a small number of antennas (less than 10) on both the BS (Base station) and UE (user equipment) sides (Swindlehurst et al., 2014), and the technique was primarily used to take advantage of numerous propagation routes. Theoretically, by serving several users rather than collaborating to serve a single user, this notion may also be utilized to implement spatial multiplexing. Multi-user MIMO was the first name given to this idea (Marzetta, 2010).

2 THE IDEA OF BACK-HAUL AND THE DEVELOPMENT OF THIS CONCEPT

2.1 A Brief Introduction of Back-Hauling

The necessity for broadband and high-speed communication has grown steadily during the recent years. Even Nevertheless, the COVID-19 epidemic has sped up the transition to a more interconnected world. However, this is primarily true for metropolitan areas. In Europe, there is still a problem with access, particularly to the most isolated and rural places. Since they are the least appealing to businesses, they are the most difficult to handle. Operators must accommodate the mobile network's evolving needs as they transition from 4G to 5G. The present and future design of the radio access network is undoubtedly one of the most crucial issues to be addressed (RAN). The RAN (radio access network) serves as the connection between the radio sites and the transport network, and because a RAN may be geographically distant, managing one may be challenging. Any improvement or evolution of the RAN to increase efficiency or lower operating costs requires careful thought because the RAN is where service providers spend the majority of their money.

A back-haul network's significance is usually underrated. For any RAN, back- haul performance in terms of bandwidth and capacity, dependability, and trans- mission latency is crucial since it directly affects the experience of wireless users. No wireless network will ever perform to everyone's satisfaction if the back-haul infrastructure is not consistently operational. Back-haul lines that are not working as planned or that have been set improperly can cause packet loss, excessive latency, carrier jitters, and therefore a terrible user experience.

One example of wireless back-haul is a smartphone that connects to the internet by receiving data from a cell tower or another sort of base station. This type of connection is known as "back-hauling." The wireless back-haul (Dahrouj et al., 2015) is the link that exists between the mobile phone tower and the individual's smartphone.

Wireless back-haul architecture As can be seen in Figure 1, the back-haul network traverses from the cell site all the way to the core network. It begins at the cell site. We will define the main parts of a back-haul network (Dahrouj et al., 2015), as well as a few terms related to this subject, in the paragraphs that follow for the sake of clarity. In a normal wireless network, each access point, also known as an AP, is required to have a wired Ethernet connection in order to provide back-haul to the wired network backbone and, eventually, the Internet. However, connecting Ethernet cables to each AP is not always possible or cost-effective to do so. Under certain circumstances, wireless back-haul from the access point (or another network device, such as a remote IP camera) to the wired network can be accomplished through Wi-Fi itself. This is possible when certain requirements are met. Between the far-off wireless AP and the wireless AP at the network's root that is wired in, there could be a series of many hops. From 4-5 BS/km^2 in 3G to an expected 40-50 BS/km^2 in 5G, the network architecture has been denser with time (Ge et al., 2016).

Figure 1. Wireless back-haul architecture (Source: EURASIP Journal on Wireless Communications and Networking 2015(1)

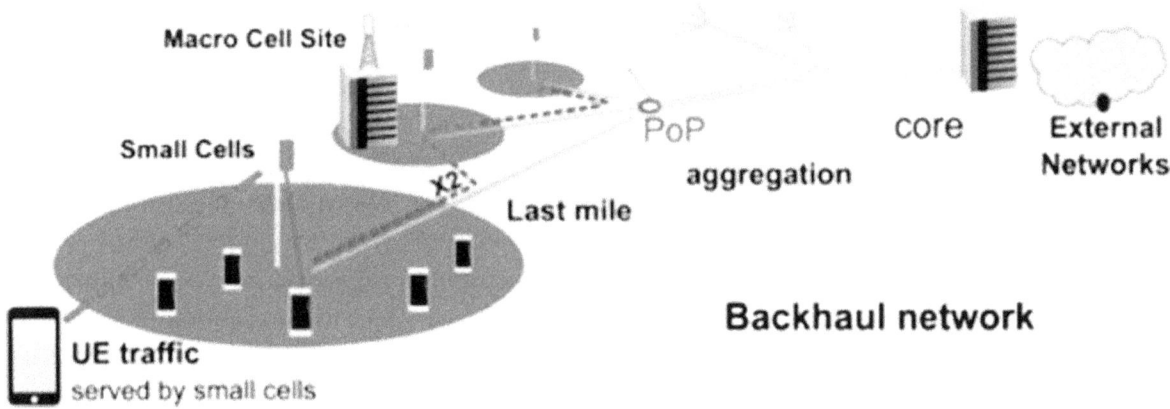

Point-to-point (PtP) and point-to-multipoint (PtM) are the two basic ar- chitectures that are typically utilised by service providers in order to establish a connection for wireless data transmission (PtMP). The PtP back-haul architecture is a radio transmission method that relies on line-of-sight (LOS) and consists of two nodes that can only communicate with one another. Short-range PtP wireless links can connect two places that are just a few hundred meters apart, while long-range PtP wireless links can connect two locations that are tens of miles distant from each other. PtP wireless links can span a wide variety of distances.

PtMP design makes use of a single hub to generate a sector of coverage that is capable of back-hauling numerous stations. A point-to-multi point (PtMP) link is made up of at least one radio from a base station (BS) and many radios from subscriber units (SU). PtMP connections often make use of lower frequency bands that are either licensed or exempt from licensing requirements. These bands enable operations to take place in near-line-of-sight (nLOS) and even non- line-of-sight (NLOS) situations. The simplic-ity of deployment is where PtMP solutions really shine, while the scalability of capacity is where PtP solutions really shine.

Although it is essential to understand the outdoor wireless equipment to create PtP and even PtMP links. As mentioned in figure 2, there are a number of additional components, such as Routers and Switches, that contribute to the overall network operations.

Figure 2. Using a PtMP (Point-to-Multipoint) link and a PtP (Pount-to-Point) link

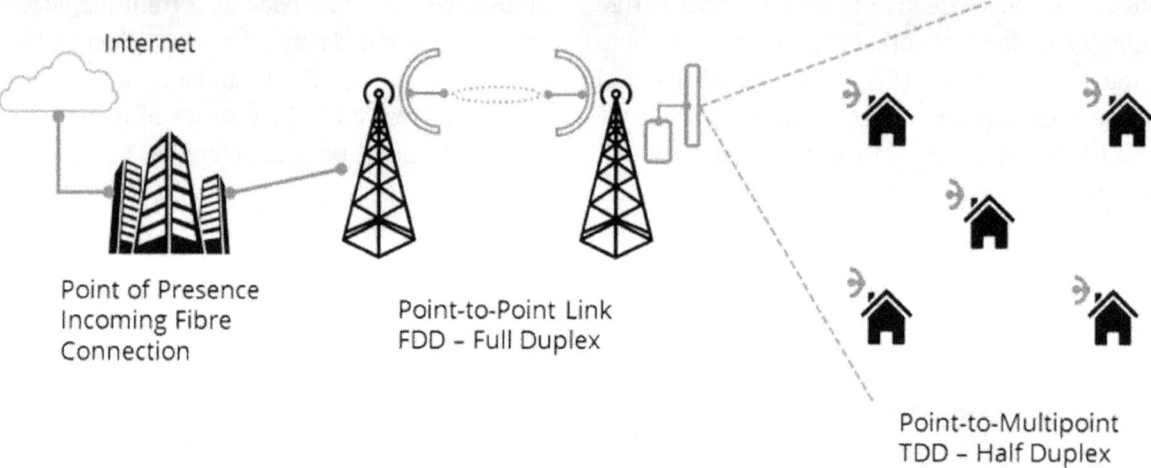

In the beginning, 26-28 GHz frequencies were employed to allow mMIMO and ultra-dense advances in order to achieve higher speeds. For instance, (Coldrey et al., 2014) de- ployed tiny cells in various configurations using 28 GHz frequencies. Another use that utilised those frequencies was point-to-point in-band back-hauling, which was accomplished by utilising the same frequencies as described in article (Taori & Sridharan, 2015). In addition, back-hauling makes use of this frequency band, which sees a lot of utilisation overall. According to (G. Association, 2018), many bands within this frequency range are utilised all over the world for the purpose of wireless back-haul. When it comes to satisfying the KPIs during the initial rollout of 5G networks, the frequency range that was discussed before is the first one that might be used.

In order to establish a PtP link for wireless back-haul, a specific pair of APs that are frequently fitted with integrated directional antennas are configured to operate in Wireless Distribution System (WDS) Bridge mode. This mode allows the APs to function as bridges between two wireless networks. The wired Ethernet interface of the remote access point is used to establish a connection between the WDS Bridge link at the remote end and the remote access point. On one end, a Wi-Fi packet is used to en- capsulate and encrypt a wired Ethernet frame, which is then transported over a wireless network before being de-encapsulated and decrypted.

This wireless connection appears to the rest of the network to be a wired connection. As a result, the WDS Bridge connection maintains the integrity of all wired Layer 2 data, including client MAC (medium access control) addresses, VLANs (virtual local area networks), and other data. Point-to-multipoint WDS Bridge connections are also easily conceivable; however, keep in mind that all distant connections share the link's entire airtime capacity. Figure 3 shows this in detail.

Figure 3. Wireless Back-haul Network Deployment (Source: EnGenius Admin)

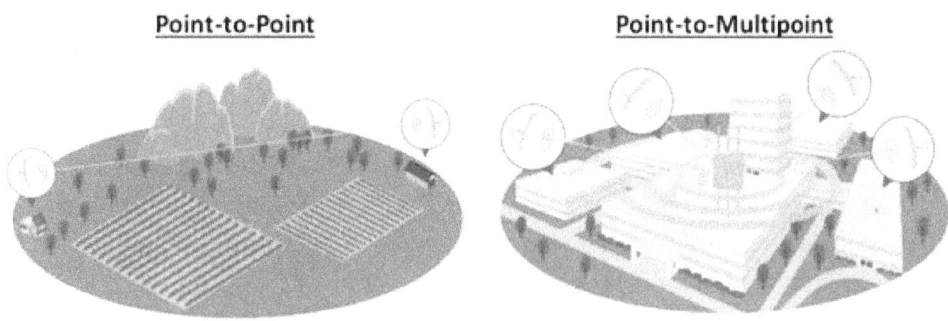

Within the context of a Wi-Fi network, a backhaul link is referred to as a hop. Back-hauling wireless connections to remote access points can be accomplished in a number of different ways. Obviously, there are advantages and disadvantages associated with each potential choice. The most important thing to keep in mind is that each wireless hop adds additional latency, and the total amount of delay increases at an exponential rate as the number of hops increases. In addition, repeaters and mesh inherently perform badly as throughput and user capacity rise, often in a manner that is proportional to the square of the number of hops. The wireless link between an end device and the wireless network's core is provided by RAN. The RAN is often implemented in network equipment form factors and comprises of base station controllers and radios at the distant cell site. To improve performance and cut costs, virtual RANs modify the conventional

RAN design by dividing and centralizing management of wireless operations. Open RAN divides wireless access into components that can come from a range of vendors, such as radio, fronthaul and backhaul (Dahrouj et al., 2015), compute, and radio control. To achieve the claimed performance increases, 5G systems need a lot more cell sites of various sizes than equivalent 4G architectures. Therefore, Open RAN is essential for managing and running brand-new 5G base stations.

3 COMPARATIVE STUDY OF DIFFERENT GENERATIONS

The 5G technology was created through a collaborative effort between a wide variety of businesses and organizations located all over the world, coordinated by the standards body known as 3GPP.

Both of the most recent wireless technologies, 5G New Radio (NR) and Wi-Fi 6, which both debuted on the market around the same time, address the pressing requirement for more improved connection. Each one increases the bandwidth and capacity for existing networks as well as those that will be developed in the future. In addition, both of them are faster than their predecessors (4G/LTE and Wi-Fi 5), more energy efficient in how they utilise that energy, and able to handle a bigger number of devices at the same time than their predecessors were able to support at the same time.

Anyone can read up on how 5G functions and what the requirements are because it is an open, global standard. No matter where you are in the globe or whatever precise gadget you are using, networks and devices will be able to communicate with one another thanks to the worldwide 5G standardized. Since the beginning of the 3GPP, worldwide 5G has been an active member and has had a significant impact on the development of 5G. This is especially true in the areas of ultra-lean radio design, security and subscriber privacy, network slicing, and energy efficiency. In addition, worldwide 5G has been a proactive member. In figure 4, the comparison of different generation mobile networks is pro- vided with various speed limits and also advantages. It is clear that the generations are divided into three categories: 1G, 2G and 3G. First generation was totally analogue and exclusively used for just message audio transmission. Second Generations were developed initially for speech and sluggish data transfer. The development of 2.5G was made possible by the connection of cellular ser- vices with GPRS (General Packet Radio Services). Access to the WAP (Wireless Application Protocol), MMS (multimedia messaging services), and internet communication services such as e-mails and WWW (world wide web) addresses are all provided by this generation of mobile devices.

The initial generation of cellular networks were voice-only analogue systems that could only transmit voice traffic. The packet switched transfer-ring voice protocol is the only one that the 2G standard sup- ports. Because it is a circuit switched cellular network, the 3G cellular network utilises its own gateway in order to translate IP traffic coming from the backbone network. In addition to this, they have their very own communication interfaces and protocols for use within the company. The solution to this conundrum can only be found in 4G networks.

Figure 4. The continual development of specifications for mobile networks

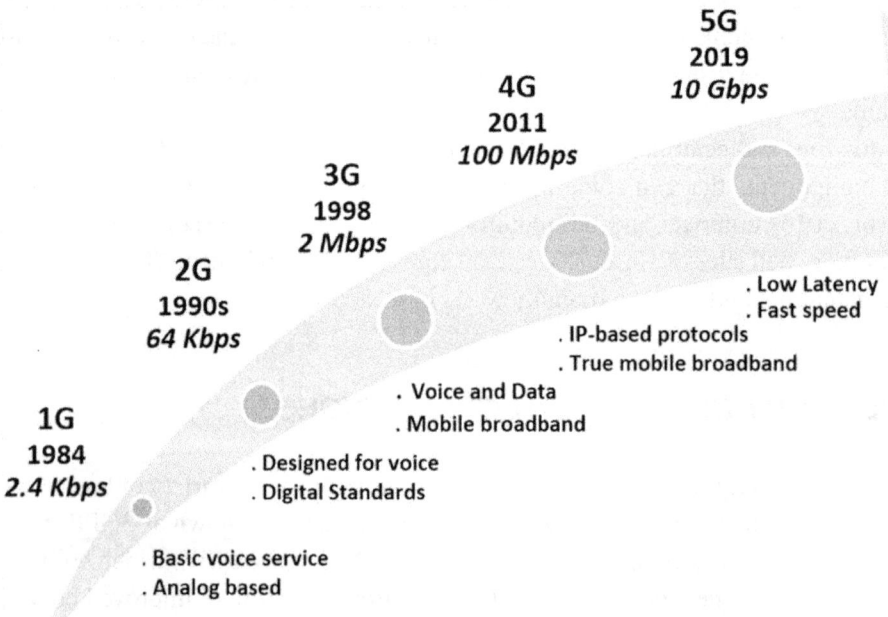

In addition, 3G is lacking in the following features: an end environment; an end-to-end continuous transmission mechanism; an adequate allocation of spectrum; difficulty gradually increasing bandwidth and high data rates for multi- media services; and difficulty roaming across a variety of service environ- ments. The development of fourth-generation (4G) communications is currently under way as a potential new level of solution for these issues.

A cooperative network (CoNet) in 4G is made up of three independent layers: application, connection, and access, which comprise conceptually separate subsystems. Each layer has the potential to be further subdivided into its own distinct sub-layers, as demonstrated in figure 5. In order to facilitate the straight- forward customization of a wide variety of access technologies, relevant techno- logical shifts, and adaptable support for speedy service innovation, the layers that make up an approach need to have interfaces that are clearly defined and be functionally independent of one another. This layer will not be affected in any way by the various transport techniques that are utilised to connect the nodes of the network.

With the fast growth of various wireless communication networks throughout the world, customers' expectations and demand are gradually changing. As a result, the associated wireless networks work several times harder at their capacity limitations. At the end-user level, heterogeneous 4G system interoperability will make it possible for end-to-end connections and service delivery to operate more effectively in a variety of scenarios, as well as make global roaming and dynamic adaptation to local contexts simpler. This will allow for a more seamless user experience overall.

Figure 5. Interoperability is provided through 4G layers (Source: International Journal of Communications, Network and System Sciences)

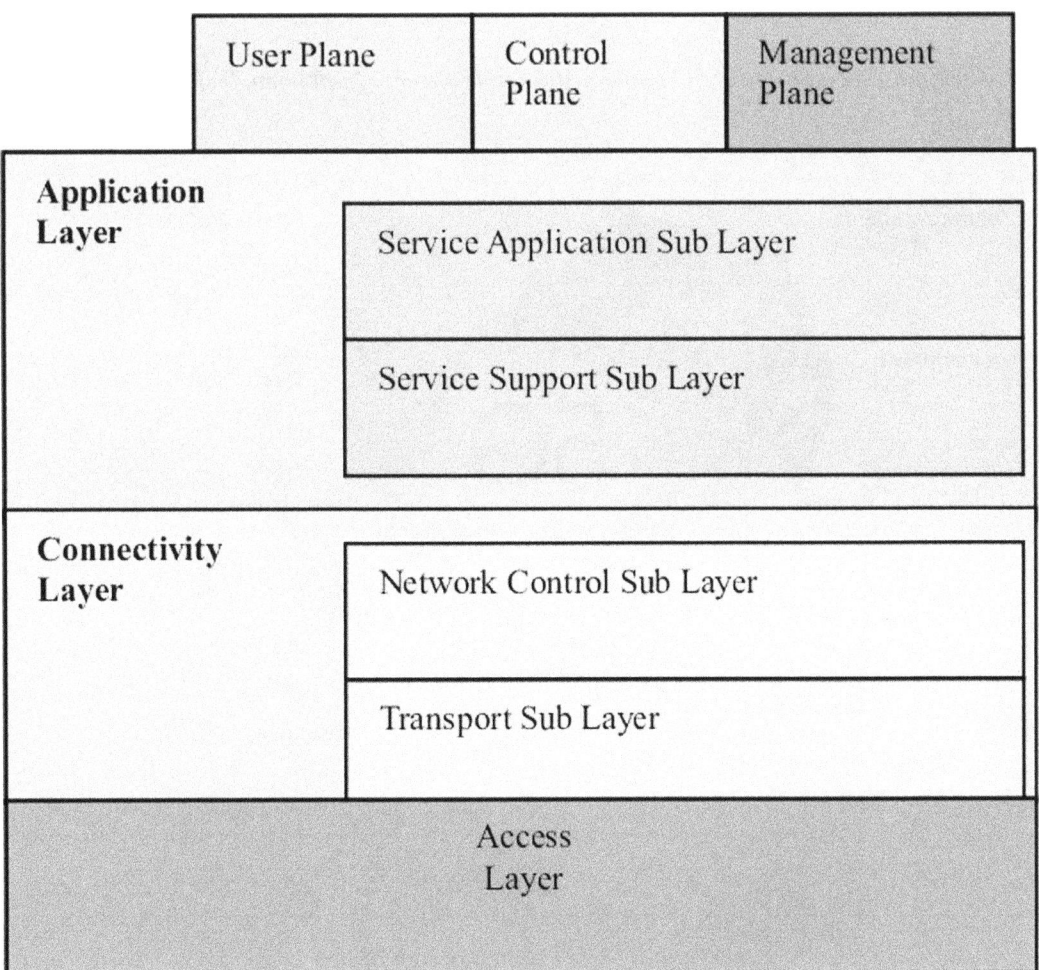

4 ISSUES AND CHALLENGES FOR IMPLEMENTATION OF SUSTAINABLE 5G

There will be an increase in the total number of subscribers as well as the appearance of new types of subscribers as a direct result of the growth of the Internet of Things and the widespread use of wireless telecommunications in a variety of different industries, such as transportation, healthcare, intelligent buildings, and industrial automation. These industries include: intelligent buildings, industrial automation, and transportation. A total of 114 5G commercial networks have launched globally, with 30 networks doing so in the second quarter of 2020, according to Telegeography data. By the end of 2020, this number is predicted to double. Data demand has increased, as has the urgency for the deployment of 5G networks. Figure 6 depicts and summarizes the high expectations of 5G required to realize those scenarios:

As the technology underpinning 5G has advanced, so has our understanding of its complexity. Early capability charts displayed the famous triangle, which emphasized 5G's three main advantages: faster speeds, lower latency, and the ability to host many more IoT connections at once.

The massive internet of things (IoT) that 5G will enable, along with larger data rates and quicker broadband speeds, is still quite promising. Entrepreneurs are pondering the benefits of connected devices, data, automation, and artificial intelligence, as well as the value proposition of 5G and the part it will play in Industry 4.0. Nevertheless, a number of obstacles, like the ones listed below, have been brought up in this following subsections:

Figure 6. The high expectations of 5G implementation (Source: Qualcomm Technologies, Inc.)

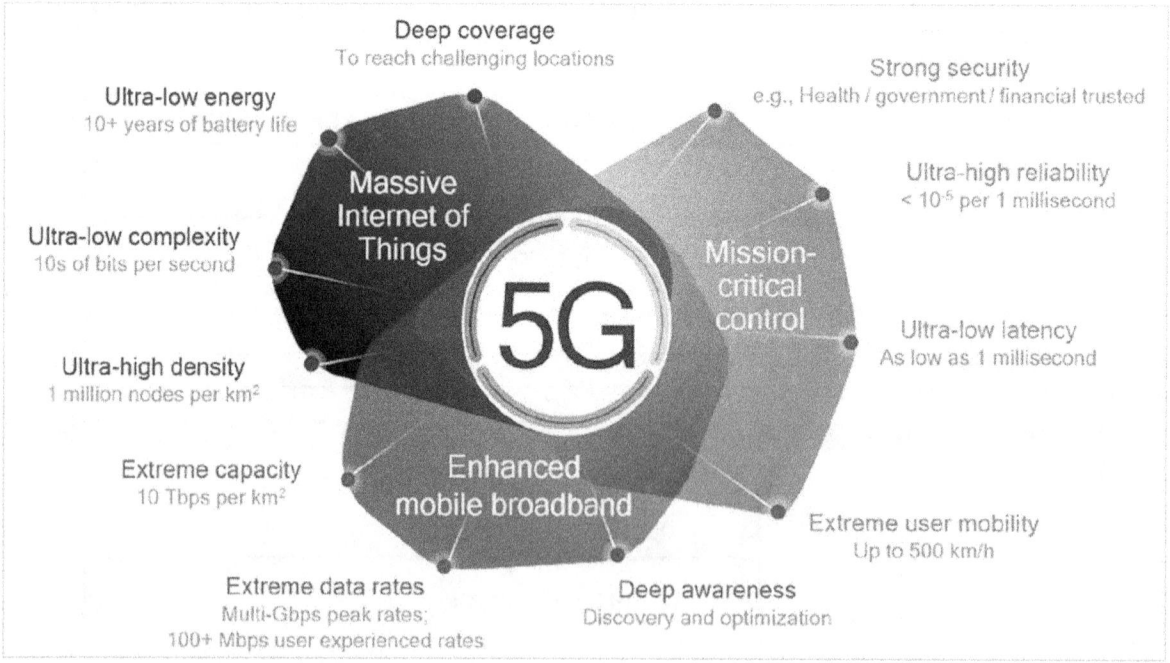

4.1 Spectrum

5G networks run at higher frequencies, almost up to 300GHz, and these bands have expanded capacity in order to give incredibly fast speeds that are 20 times faster than the potential speed of LTE networks. This allows 5G networks to deliver speeds that are exceptionally fast. Despite this, operators continue to express worries regarding the availability and pricing of spectrum band. During the process of continuing to create and operate 5G wireless networks, companies will need to compete with one another for access to higher frequency bands.

Another issue that many countries face is a limited operational spectrum. Spectrum consumption will cost service providers a lot of money every year. 5G necessitates the addition of microwave bands to the present LTE channels in the lower bands (700MHz to 3400 MHz)(Mesodiakaki et al., 2016). Several approaches for releasing more spectrum and expanding 5G networks have already been discussed; spectrum sharing can free up frequency bands that are only infrequently used by opera- tors. 5G NR is intended to accommodate all current spectrum types while also allowing for the adoption of future sharing models.

4.2 The Deployment of Hybrid LTE-NR

In the non-standalone 5G installations that are part of the initial phase of 5G deployments, LTE access serves as the foundation for new radio (NR). LTE eNBs, also known as evolved NodeBs, and 5G gNBs, also known as 5G Next Generation base stations, are responsible for managing control and excess user traffic in addition to LTE traffic. The use of a hybrid LTE-NR configuration that employs the macro LTE layer is beneficial to the introduction of 5G in the mmW (Swindlehurst et al., 2014) spectrum. This design also enhances the mobility of the network between hotspots. This mmW-acquired mobility performs effectively at slow speeds. Despite this, these advantages result from the dual connectivity's (EN- DC) performance, and because of its complexity, testing the 5G network is quite difficult. For data transfer and signaling to the NR, gNB and eNB must be in synchronization.

It is necessary to connect the optical fibre network to thousands of tiny cells and base stations. The fact that 5G networks require a lot more base stations than conventional LTE networks puts a heavy load on network operators. For instance, an operator may initially use Gigabit LTE networks for 4G/5G mobility and coverage while deploying 5G mmWave (Swindlehurst et al., 2014) hotspots for high capacity in crowded metropolitan locations. Examples are the severe oxygen attenuation that affects V-band, the intense rainfall that affects E-band, and the compara- tively high atmospheric attenuation that affects D-band (Mesodiakaki et al., 2016).

4.3 Mobile Network Structure

For the successful implementation of 5G in both the core and RAN networks, IT-related architecture and platforms are crucial. The wireless 5G infrastructure is trickier to understand than it first looks. Due to its feature-rich structure, it is more difficult than ever to manage, necessitating the hiring of skilled personnel with relevant experience.

To cover a large geographic area, the deployment of millions of pieces of support equipment is required. In 5G networks, high-tech equipment is employed to deliver faster data speeds and lower latency. Active and passive back-haul technologies are used in a variety of scenarios to provide faster data transfer.

Passive systems are used to put small cells on utility poles or streetlights. These systems require external fibre connectivity and are made up of a covered unit with antenna modules and other necessary parts.

4.4 Financial Needs

Despite the fact that 5G comes with its own unique set of benefits, the implementation of it will require a significant financial investment. In order to overcome these obstacles with 5G infrastructure, which consists of cables, cell towers, and other similar components, operators are segmenting the deployment process into phase-by-phase deployments while they continue to carry out these tasks on the ground. According to (Taheribakhsh, 2020), another challenge is the high price of the 5G testing equipment that must be purchased in order to conduct an exhaustive analysis of the 5G drive test. This analysis must first include cell site verification as well as an evaluation of user KPIs (Key Performance Indicators), such as speed test, voice and call test, HTTP test, and many other tests, before the technology can be made available to commercial customers. The value of investments in 5G NR network infrastructure, which made up 6% of total wireless infrastructure sales in 2019, is expected to rise by 2020, according to Gartner.

By implementing cutting-edge technologies like SDN (Software-Defined Net- working), NFV (Network functions virtualization), and Micro cells, it is necessary to reconstruct the radio and core networks in order to support unique slicing requirements in the 5G network (Taheribakhsh, 2020). For instance, rolling up a small, dense net- work in an urban environment requires a significant number of resources. On the other hand, because of the expensive capex (capital expenditure), installing micro cell networks (Nie et al., 2017) in rural locations could not be financially viable. Further- more, these expenses are increased by the present network's OPEX (operational expenditure)(Taheribakhsh, 2020). The development of back-haul necessitates investments in fiber deployment, which raises the price of the prior one. In this regard, investment should be taken into account and studied for 5G adoption.

4.5 Guidelines for Radiation

The establishment of regulatory standards is done by regulators with the purpose of ensuring that mobile network carriers build networks to service people in all parts of the country, including rural areas. The testing of 5G was halted because to the stringent radiation regulations that numerous European nations have in place. One of the mobile service providers in Brussels has ambitions to experiment with 5G technology and bring 5G services to the market.

On the other hand, the government prohibited it from doing so on the grounds that there is no way to guarantee that the amount of radiation emission will remain below the acceptable threshold. The government based its decision on the fact that there is no way to guarantee that this. There will be no testing of the new system performed until after these improvements have been completed. The level of electromagnetic fields (EMF) should remain consistent during the rollout of 5G because of networks and devices while we have the necessary ability to do so (Taheribakhsh, 2020). Because of the probability that the deployment of new 5G radios and the installation of smaller cells would result in an increase in the quantity of electromagnetic field (EMF) present in the environment (near areas).

4.6 Safety Issues

Without safe network infrastructure, service security cannot be offered. Traditional networks include components that are separated from one another, whereas 5G networks have virtualized services and shared infrastructure re- sources. In this setting, many virtual network slices are formed, requiring various levels of protection.

In addition, the security heterogeneity that 5G networks present is a brand- new problem that must be taken into consideration. ITU (International Telecommunication Union) service framework claims that 5G supports a number of ser- vices with various needs. These services include Enhanced Mobile Broadband (eMBB) (Taheribakhsh, 2020), URLLC (Ultra-reliable low latency communications), and mMTC (Massive Machine Type Communication). There are varying degrees of safety precautions that are needed for each of them. IoT services, for example, have a low threshold for acceptable levels of se- curity, but URLLC services, such as industrial services, call for a higher level of defense. In this context, it is necessary to have a multi-level architectural security framework in order to enable policies, threat detection, and threat mitigation on a dynamic basis.

5 CURRENT CONDITION OF WIRELESS BACK-HAUL FOR 5G TECHNOLOGIES

5.1 Recent Case Study of Ericsson

In the most recent of their joint efforts in mobile transport, Ericsson and O2 Telefonica have successfully demonstrated 5G wireless back-haul for coverage in rural and suburban areas. This accomplishment was a part of Ericsson and O2 Telefonica's joint endeavour. This significant technological achievement has demonstrated that the firms can transmit data at rates of up to 10 Gbps across a distance of more than 10 kilometers and provide connection that is equivalent to that of fiber optics.

The ability to support the continued build-out of high-performing 5G net- works and enhanced mobile broadband services from urban to suburban and rural areas with microwave back-haul over traditional bands, as demonstrated by this world-first demo, is one of the primary challenges that communications service providers face in accelerating their 5G deployment.

When fiber is not an option, they have developed ground-breaking new, po- tent microwave technologies with Ericsson to back-haul 5G data over great distances in rural locations. We can speed up the implementation of 5G by using this technology to give fiber-like connection through microwave.

The ability to provide such high data rates at lengths greater than 10 kilo- metres, which is equivalent to the cruising altitude of a commercial flight, opens up a whole new spectrum of alternatives for the provision of low-latency, reli- able broadband in locations that are more difficult to access. Such sites have previously been difficult to service due to the high capacity's requirement for enormous bandwidths, which are typically only available in millimetre wave frequency ranges. Historically, this has made it difficult to provide service to such areas (E-band). Because rain has a bigger impact on the E-band than it does on the lower frequency bands, it is more difficult to provide dependable service across extended distances while it is raining.

The back-haul connection utilised the 18GHz frequency band, twin antennas arranged in a MIMO configuration, and commercial Mini-Link radios as its communications infrastructure. Additionally, the link made use of a pre-commercial base-band technique that made it possible to employ MIMO in 2x

112MHz channels. MIMO ensures that scarce spectrum resources are used in the most effective manner possible. In a cross-polar configuration, a bandwidth of 448 MHz would be required to support the same capacity if MIMO weren't used. When compared to fiber deployment, microwave back-haul is typically seen as the alternative that provides more savings in both money and time.

5.2 Case Study of T-Mobile and Nokia

T-Mobile is the fastest and most dependable 5G carrier in the U.S., according to market studies and independent testing, which have been in agreement for years. That shouldn't come as a surprise because it had a significant advantage because it was licensed to use the key middle spectrum, which offers the ideal blend of speed and range. While Verizon could tout much better raw speeds thanks to its early high-frequency mmWave roll-outs.

T-Mobile made the announcement earlier this year that its network suppliers, Ericsson and Nokia, will build a 5G network in 30 cities in 2018, including New York, Los Angeles, Dallas, and Las Vegas, utilising the carrier's 600 MHz, 28 GHz, and 39 GHz spectrum. These cities include: New York, Los Angeles, Dallas, and Las Vegas. New York City, Los Angeles, Dallas, and Las Vegas are among the cities in question. T-Mobile has now put a price tag on these efforts by stating that it will be purchasing approximately $7 billion worth of 5G equipment, with the money being shared evenly between Nokia and Ericsson. This has allowed T-Mobile to put a price tag on these actions.

However, T-Mobile must sign agreements with fiber owners like Zayo in order to backhaul the 5G traffic produced by its cell towers because the firm does not control its own wired network assets.

In Finland, Espoo Nokia has successfully demonstrated a live microwave link using D-Band spectrum, the company said today (130-175 GHz)(Mesodiakaki et al., 2016). D-Band will be used as an ultra-high-capacity extension for 5G back-haul and front- haul in crowded metropolitan areas since it offers substantially more bandwidth than other microwave bands. The research, which was carried out by Nokia using equipment from Nokia Bell Labs, looks at how higher frequencies may accommodate the rising capacity needs of mobile networks.(Source: Press Release Nokia demonstrates live D-Band microwave back-haul connection (13April 2022) As the demand for 5G services rises, this experiment shows how Nokia are continuing to push the envelope to offer best-in-class connectivity. Future high-capacity 5G connection will depend heavily on the D-Band. Nokia team continue to pursue significant improvements that provide coverage and capacity where it is required for their clients.

6 CONCLUSION

Cellular networks typically make use of the technology known as wireless back- hauling, which has been around for a considerable amount of time. Because of 5G and other recent technological developments like mmWave frequencies and beam-forming, wireless back-hauling has reached a new level of development. This is one of the many benefits of the fifth generation of mobile networks. New use cases, such as mobile edge computing, rural connectivity, and effective satellite back-hauling for ubiquitous connection, are made possible as a result of this.

However, in order for wireless back-hauling to be an effective technique, it must overcome certain obstacles. This chapter have briefly discussed case studies and challenges in wireless back-hauling and highlighted the difficulties unique to small cells' multi-hop architecture. This article has now covered the theories un- der consideration for networks that will exist after 5G. The wireless back-hauling will become even more robust thanks to these new technologies, becoming the backbone of the beyond-5G(Tezergil & Onur, 2022) networks.

The potential and technological possibilities of back-haul transport are expanding along with mobile wireless technologies. In-depth descriptions of cutting- edge wireless and wireline back-haul options for 5G are provided in this paper, along with use cases that these systems potentially serve. With the aim of realizing the full potential of 5G technology services and applications for consumers, technological progress is generally still both evolutionary and revolutionary.

REFERENCES

3GPP. (2010). Evolved universal terrestrial radio access (e-utra); relay architectures for e- utra (lte-advanced). *3rd Generation Partnership Project (3GPP), Technical Report (TR) 36.806, 04, version 9.0.0.*

Ahamed, M. M., & Faruque, S. (2018). 5G backhaul: requirements, challenges, and emerg-ing technologies. Broadband Communications Networks: Recent Advances and Lessons from Practice, (vol. 43).

Bojic, D., Sasaki, E., Cvijetic, N., Wang, T., Kuno, J., Lessmann, J., Schmid, S., Ishii, H., & Naka-mura, S. (2013). Advanced wireless and optical technologies for small-cell mobile backhaul with dynamic softwaredefined management. *IEEE Communications Magazine*, *51*(9), 86–93. doi:10.1109/MCOM.2013.6588655

Chia, S., Gasparroni, M., & Brick, P. (2009). The next challenge for cellular networks: Backhaul. *IEEE Microwave Magazine*, *10*(5), 54–66. doi:10.1109/MMM.2009.932832

Coldrey, M., Engstr¨om, U., Helmersson, K. W., Hashemi, M., Manholm, L., & Wallentin, P. (2014). Wireless backhaul in future heterogeneous networks. *Ericsson Review*, *91*, 1–11.

Dahrouj, H., Douik, A., Rayal, F., Al-Naffouri, T. Y., & Alouini, M. (2015). Cost-effective hybrid rf/fso backhaul solution for next generation wireless systems. *IEEE Wireless Communications*, *22*(5), 98–104. doi:10.1109/MWC.2015.7306543

Dao, N.-N., Pham, Q.-V., Tu, N. H., Thanh, T. T., Bao, V. N. Q., Lakew, D. S., & Cho, S. (2021). Survey on aerial radio access networks: Toward a comprehensive 6g access infrastructure. *IEEE Communications Surveys and Tutorials*, *23*(2), 1193–1225. doi:10.1109/COMST.2021.3059644

G. Association. (2018). *Mobile backhaul options: spectrum analysis and recommen- dations.* G. As-sociation.

Gamboa, J., & Demirkol, I. (2018). Softwarized lte self-backhauling solution and its evaluation. In 2018 IEEE Wireless Communications and Networking Conference. WCNC.

Ge, X., Tu, S., Mao, G., Wang, C., & Han, T. (2016). 5G ultra-dense cellular networks. *IEEE Wireless Communications*, *23*(1), 72–79. doi:10.1109/MWC.2016.7422408

Gupta, A., & Jha, R. K. (2015). A survey of 5G network: Architecture and emerging technologies. *IEEE Access: Practical Innovations, Open Solutions*, *3*, 1206–1232. doi:10.1109/ACCESS.2015.2461602

Jaber, M., Imran, M. A., Tafazolli, R., & Tukmanov, A. (2016). 5G backhaul challenges and emerging research directions: A survey. *IEEE Access: Practical Innovations, Open Solutions*, *4*, 1743–1766. doi:10.1109/ACCESS.2016.2556011

Jaffry, S., Hussain, R., Gui, X., & Hasan, S. F. (2020). A comprehensive survey on moving networks. *IEEE Communications Surveys and Tutorials*.

Kamel, M., Hamouda, W., & Youssef, A. (2016). Ultra-dense networks: A survey. *IEEE Communications Surveys and Tutorials*, *18*(4), 2522–2545. doi:10.1109/COMST.2016.2571730

Kurt, G. K., Khoshkholgh, M. G., Alfattani, S., Ibrahim, A., Darwish, T. S., Alam, M. S., Yanikomeroglu, H., & Yongacoglu, A. (2021). A vision and framework for the high altitude platform station (haps) networks of the future. *IEEE Communications Surveys and Tutorials*, *23*(2), 729–779. doi:10.1109/COMST.2021.3066905

Lee, J., Han, J.-K., & Zhang, J. (2009). Mimo technologies in 3gpp lte and lte-advanced. *EURASIP Journal on Wireless Communications and Networking*, *2009*(1), 1–10. doi:10.1155/2009/302092

Marzetta, T. L. (2010). Noncooperative cellular wireless with unlimited numbers of base station antennas. *IEEE Transactions on Wireless Communications*, *9*(11), 3590–3600. doi:10.1109/TWC.2010.092810.091092

Mesodiakaki, A., Kassler, A., Zola, E., Ferndahl, M., & Cai, T. (2016). Energy efficient line-of-sight millimeter wave small cell backhaul: 60, 70, 80 or 140 ghz?" In *2016 IEEE 17th International Symposium on A World of Wireless, Mobile and Multime- dia Networks (WoWMoM)*, (pp. 1–9). IEEE. 10.1109/WoWMoM.2016.7523521

Nie, G., Tian, H., Sengul, C., & Zhang, P. (2017). Forward and backhaul link optimiza- tion for energy efficient ofdma small cell networks. *IEEE Transactions on Wireless Communications*, *16*(2), 1080–1093. doi:10.1109/TWC.2016.2636821

Siddique, U., Tabassum, H., Hossain, E., & Kim, D. I. (2015, October). Wireless backhauling of 5G small cells: Challenges and solution approaches. *IEEE Wireless Communications*, *22*(5), 22–31. doi:10.1109/MWC.2015.7306534

Swindlehurst, A. L., Ayanoglu, E., Heydari, P., & Capolino, F. (2014). Millimeter-wave massive mimo: The next wireless revolution? *IEEE Communications Magazine*, *52*(9), 56–62. doi:10.1109/MCOM.2014.6894453

Taheribakhsh, M. (2020). 5G Implementation: Major Issues and Challenges. *25th International Computer Conference, Computer Society of Iran (CSICC)*.

Taori, R., & Sridharan, A. (2015). Point-to-multipoint in-band mmwave backhaul for 5G networks. *IEEE Communications Magazine*, *53*(1), 195–201. doi:10.1109/MCOM.2015.7010534

Tezergil, B., & Onur, E. (2022). Wireless Backhaul in 5G and Beyond: Issues, Chal- lenges and Opportunities. *IEEE Communications Surveys and Tutorials*, *24*(4), 2579–2632. doi:10.1109/COMST.2022.3203578

Chapter 9
Software Testbed for Space Communication Link Budget Design

M. N. Suma

 https://orcid.org/0000-0001-6112-5526
BMS College of Engineering, India

Sudhindra K. R.
BMS College of Engineering, India

ABSTRACT

Communication link budget is a balance sheet of all gains and losses, where the calculation of link parameters such as received power, path losses, noise power, signal to noise ratio, receiver figure of merit and link margin. There are software available in the market, which can be utilized to design, but are not cost friendly and also have limitations in the form of some predefined data values, which the user cannot alter. In this paper the authors propose a new testbed which is designed to formulate the link budget calculations using Python GUI for space communications link, considering all theoretical parameters needed for computations.

1.1 INTRODUCTION

The area of wireless communication has seen significant advances from last few years to meet the expectation of error free data transmission and high data rates. It is observed that there are several challenges in design of wireless communication system due to various characteristics of radio channel as a physical medium of reliable communication. The transmitted radio wave signal undergoes reflection, refraction, diffraction and scattering causing significant change in amplitude of the signal and intersymbol interference at the receiver. In addition, the wireless devices under mobility pick up signals which are subject to random changes as indicated in ITU_R, P.531-5 (2019). The characterization of wireless channel and calculation of path loss are the key aspects that need to be considered for establishing reliable com-

DOI: 10.4018/978-1-6684-7000-8.ch009

munication link. The link budget is a mathematical tool to calculate path loss between transmitter and receiver. It is computed based on Friss equation and involves the calculation of received power, path losses, noise power, signal to noise ratio, E_b/N_o (Bit Energy per Noise Ratio), receiver figure of merit and link margin and BER(Bit Error Rate) depending on application, operating frequency and operating region (environment). In view of this there is a need to have computational testbed for link parameters avoiding manual calculations as indicated by R-REC-P.618-13,2017, Hoation DAI (2006).

AWGN channel model with line of sight (LOS) communication is best suited for space communication below an altitude of $2x10^6$ Km from earth and considering that low scintillation effect (S4 index less than 0.3) according to ITU-R P.531-5, R-REC-P.618-13,2017.

1.2 TESTBED DESIGN

The Link Budget Testbed is designed using python which has user interface to input necessary parameters for computation for space communication. The block diagram of flow in the design of testbed is shown in figure1.

Figure 1. Flow graph of computation

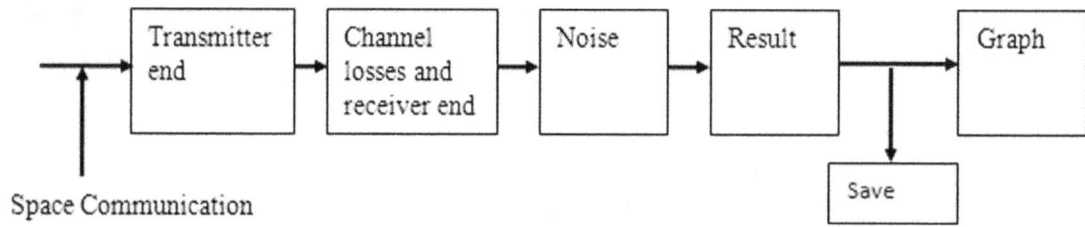

Design considerations

- Radius of the earth is considered to be a uniform 6378.1 km.
- The testbed does not use code gain for link budget calculation
- Testbed supports only QPSK and BPSK given modulation techniques

1.2.1 Operation Procedure for Space Communication

Once the user chooses space communication link budget design, transmitter end window will open, the user has to enter the transmitter end parameters. The ranges of the various parameters can be accessed by clicking the help button present in every window. The channel losses and receiver end window will open, where the all the expected loss in the design and the receiver end parameters can be entered. The noise window will provide option to enter the noise figure, noise temperature of each system, transmission line losses and antenna noise temperature.

Table 1. Graphs supported

Space Communication
Bit rate vs. link margin
BER vs. bit rate
EIRP vs. link margin
BER vs. EIRP
Total losses vs. link margin
BER vs. total losses
Noise temperature vs. link margin
BER vs. noise temperature
Receiver figure of merit vs. link margin
BER vs. receiver figure of merit
Path loss vs. height of satellite from earth(H)
Path loss vs. frequency

Finally, the result window displays all the calculated values. Graphs window displays list of graphs one can plot and analyze the link budget for different values. Result window displays the calculated values. The graphs that are supported in the testbed are indicated in table1.In table 2 operating ranges of communications is indicated.

Table 2. Operation Ranges of Communication

Parameter	Ranges	units
Elevation angle	0-10	degree
Distance	400-37500	Km
Polarization losses	0-3	dB
Scintillation losses	0-6	dB
Misalignment losses	0-3	dB
Fade Margin	0-10	dB
Antenna noise Temperature	10-500	K
Noise Figure	Staring from 0	dB
Gain	Starting from 0	dB
Line Losses	Staring from 0	dB

1.3 LINK BUDGET FOR SPACE COMMUNICATION

Space communication link budget tool is used to calculate the path loss in space communication. The link budget is the compilation of all gains and losses in the space communication link. The network architecture of space communication includes many earth stations which are inter connected through a medium of satellite link. The communication link between ground station and satellite is termed as uplink whereas the link between satellite and ground station is termed as downlink. The transmitter, channel, and receiver parameters for computation of link budget are as shown in table 3. However other parameters included in design is included in table 4.

Table 3. Operation RangesCommunication

Region of communication	Parameters	Comments
Transmitter end	1)Operating Frequency 2)Bandwidth 3)Modulation scheme 4)Power output 5)Transmitter Antenna gain	User inputs
Channel Losses	1)Free space loss 2)Polarization loss 3)Scintillation loss 4)Antenna misalignment loss 5)Atmospheric loss 6)Rain loss	AWGN Channel model is considered ITU-R P.531-5
Receiver Noise	1)Receiver antenna gain and feeder losses 2)Antenna noise temperature 3)Transmission line losses 4)Noise figure of the components	User inputs
Calculations	1)Receiver power 2)System noise 3)Receiver figure of merit 4)BER calculation 5)Eb/N0 6)Signal to noise ratio 7)Link Margin 8)Maximum Channel Capacity 9)Group delay	ITU-R P.531-5

Table 4. Other design parameters Communication

Parameter	Value	Unit
Elevation angle	10	degree
Height of satellite (H)	650	Km
Radius of earth	6378.1	Km
Height of Ground Station (h)	25	M
slant range(d)	2045.341	Km

1.3.1 Design Equations

The Design equations required for link budget computations is given in equations (1) to (4) are indicated in 3GPP TR 38.901(2017), Sklar (2021),

1. Bit Error Rate (BER)

$$BER = Q\left(\sqrt{E_b/N_o}\right) \tag{1}$$

BER is defined as ratio of number of errors that occur in a string of a stated number of bits.

The term E_b represents energy per bit and N_o represents Noise power spectral density.

2. Link Design Equations

The received power at the receiver PR is given by

$$P_R = EIRP + G_R - T_L \qquad (2)$$

T_L - Total losses
G_R - Receiver antenna gain

The term EIRP indicates effective isotropic radiated power. It is defined as maximum power radiated from the antenna after considering antenna gain, transmission loss and transmitted power at the receiver.

3. Channel Losses

The free space path loss is defined as the transmission loss encountered by the signal as it propagates through free space.

$$Free\ Space\ Loss(FSL) = 20\log\left(\frac{4\pi d}{\lambda}\right) dB \qquad (3)$$

where λ indicates wavelength, which is equal to ratio of c by f

c indicates speed of light which is equal to 3×10^8 m/s,
f indicates operating frequency in MHz, d - slant range in kilo meters

where R_E (Radius of Earth) = 6.3781×10^6 m

h - height of the building in ground station (PG Block)
H - Orbital height
α - Elevation angle of ground station antenna

1) Antenna misalignment losses (AML):
    AML are obtained using approximation based on real time data observed in several Ground stations. Generally, it is considered to be 1dB for link budget design.

2) Receiver feeder losses:
    It is the sum of all the losses that occurs across couplers, filters and transmission line of a receiver system.

3)   Rain attenuation:

Defined as the attenuation of the radio waves when they pass through moisture bearing cloud formations or areas in which rainfall increases with the density of the moisture in the transmission path as indicated in Sklar (2017).

Figure 2. Rain attenuation vs frequency

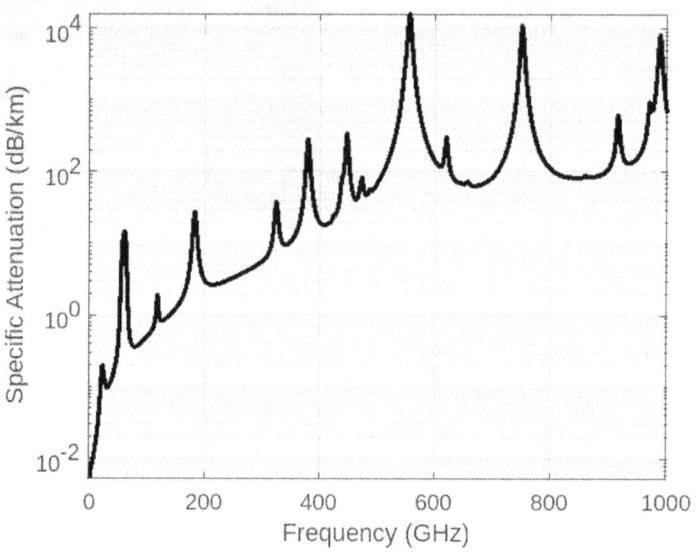

4)   Atmospheric attenuation:

The interaction of the electromagnetic wave signal with the gaseous components of the atmosphere and aerosols results in reduction in distance of propagation which is termed as atmospheric attenuation. The signal absorption and <u>scattering</u> of wave causes atmospheric attenuation.

5)   Gaseous attenuation – The gaseous components in the earth's atmosphere causes reduction in transmitted or received signal level which is termed as Gaseous attenuation as indicated in Roddy(2017), Singh(2017).

6)   Polarization losses:

Loss occurs when a receiver is not matched to the polarization of an incident electromagnetic field. Polarisation of transmit antenna and receive antenna can be vertical or horizontal.

8)   Scintillation losses – It is the result of instantaneous variation of amplitude and phase of the received signal. The occurrence of this loss is due to the physical characteristics of ionosphere.

9)   System Noise Temperature calculation

$$T_{SYS} = TA + ((L_{11} - 1)T_0) + (L_{11}.T_1) + (((L_{22} - 1)T_0)/G_{11})$$
$$+((L_{22}T_2)/G_{11}) + (((L_{33} - 1)T_0/(G_{11}G_{22})) + ((L_{33}T_3)/(G_{11}G_{22}))$$

(4)

Figure 3. Gaseous attenuation vs Frequency

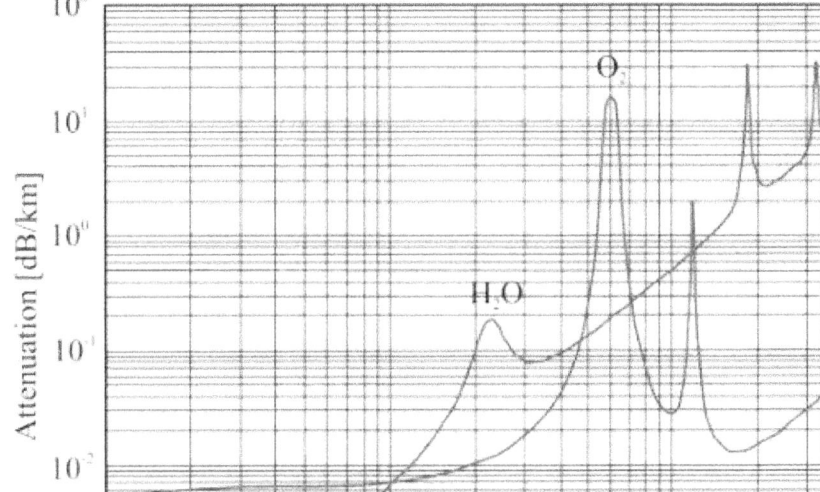

T_A = antenna noise temperature, which is a measure of all the noise that enters into a receiver through its receiving antenna. $T_1, T_2, T_3, G_1, G_2, G_3$ and L_1, L_2, L_3 are the noise temperature, gains of the subsequent blocks and line losses are in units K, dB and dB respectively.

$$T_1 = (F_1 - 1)290 \tag{5}$$

$$L_{11} = 10^{(L_1/10)} \tag{6}$$

$$G_{11} = 10^{(G_1/10)} \tag{7}$$

Where F is the noise figure of the respective blocks. The user can give noise temperature or noise figure as inputs.

Noise figure: Noise figure is calculated as the ratio of total noise output power to the output noise power due to input signal only.

SNR = signal to noise ratio Computation

$$SNR = \frac{P_R}{P_N} \tag{8}$$

where P_R is signal power received, P_N is Noise power

$$\frac{E_b}{N_0} = \frac{((SNR)B_N)}{R_b}$$ (9)

R_b = bit rate ; B_N is Noise Bandwidth

Receiver Figure of Merit
G_R / T_S = antenna gain to system noise temperature ratio (receiver figure of merit).

Link Margin

$$Link\ M\arg in = \left(\frac{E_b}{N_0}\right)(received) - \left(\frac{E_b}{N_0}\right)(required)$$ (10)

Group delay

$$d_t = (40.3)N_T/(Cf^2)$$ (11)

where N_T = ionospheric electron density
c = speed of light
f = operating frequency

Channel capacity
The Channel capacity may be defined using Shannon Hartley capacity theorem. The mathematical relationship of Channel capacity and Bandwidth of channel for a given Signal to noise ratio is given by equation (10)

$$C = B \log_2(1 + SNR)$$ (12)

Where C indicates channel capacity in bits/sec, B indicates bandwidth excluding
Doppler frequency and SNR represent signal to noise ratio as indicated in Singh (2017).

1.3.2 Manual Calculation

In table 5, Sample uplink operation are shown, for which other parameter values required are indicated below in table6 for manual calculations using the formulas.

Table 5. sample up link budget computation

Operating Frequency	Bandwidth	Modulation scheme	BER	E_b/N_o Required	Baud rate
145MHz	5KHz	BPSK	10^{-6}	10.5235 dB	500 symbols/sec

Table 6. Parameters required Communication

Parameter	Value	Units
Transmitter		
Frequency (f)	145	MHz
Baud rate	500	symbols/sec
Bit rate	500	bits/sec
Bandwidth(BW)	5	KHz
	36.989	dB/Hz
Antenna gain (Gt)	10	dBi
Power (Pt)	50	mW
EIRP	-3.0103	dB
Channel Loss		
slant range (d)	2045.342	Km
	63.10766	dB
Free space losses	141.8845	dB
Polarization losses	3	dB
Scintillation losses	6	dB
Rain attenuation	0	dB
Atmospheric attenuation	0	dB
Gaseous attenuation	0	dB
Antenna misalignment losses	1	dB
Receiver		
Receiver feeder losses	1	dB
Total losses (T_L)	151.884452	dB
Receiver Gain(G_R)	1	dBi
Received Power(P_R)	-153.894751	dB
Noise Calculations		
Antenna noise temperature (Ta)	290	K
Other noise temperature (T1)	100	K
System noise temperature (Ts)	390	K
	25.9106	dBK
Noisy Bandwidth (B_N)	38.06768	dBHz
Boltzmann constant (K)	$1.38* 10^{-23}$	$m^2\ Kg\ s^{-2}\ K^{-1}$
	-228.6012	$dBJK^{-1}$
Noise power (P_N)	-164.62292	dB
signal to noise ratio (P_R/P_N)	10.72734	dB
Total electron Content	$10^{\wedge}(16)$	$electron/m^2$
	Computation	
Eb/No received	21.80611	dB
Eb/No required	10.52357	dB

continues on following page

Table 6. Continued

Parameter	Value	Units
Link Margin	11.28254	dB
Group Delay dt	63.97	ns
Gr/Ts- Figure of Merit	0.03859	dB/K

1.3.3 Testbed Result

In the Test bed, windows for transmitter, channel losses and receiver are shown where, parameters are entered as indicated in figure 4, 5, 6. the results obtained has been visualised in figure 7. At the back-end, testbed would take all the input parameters entered and without burden, a quick computation and graphs are obtained.

Transmitter Window:

Figure 4. Transmitter parameter

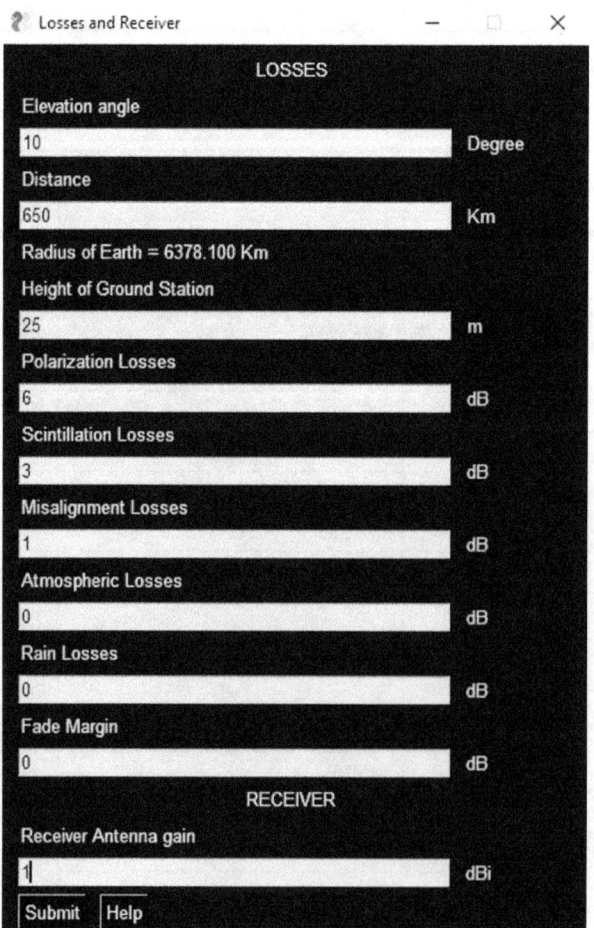

Channel Losses Window:

Figure 5. Channel losses

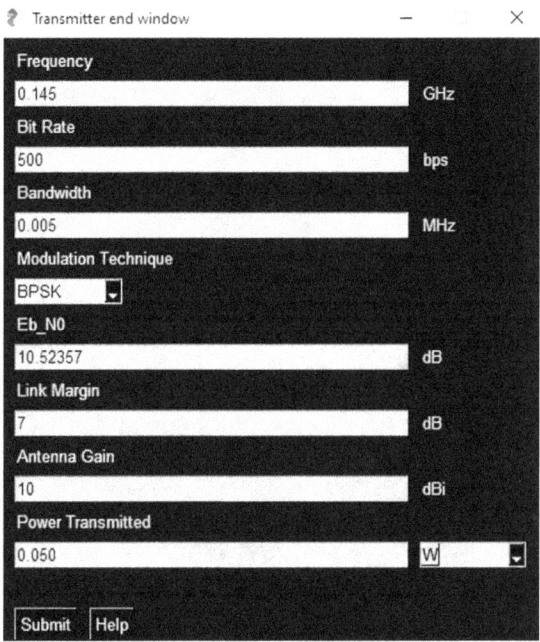

System Noise:

Figure 6. Rreceiver end parameters

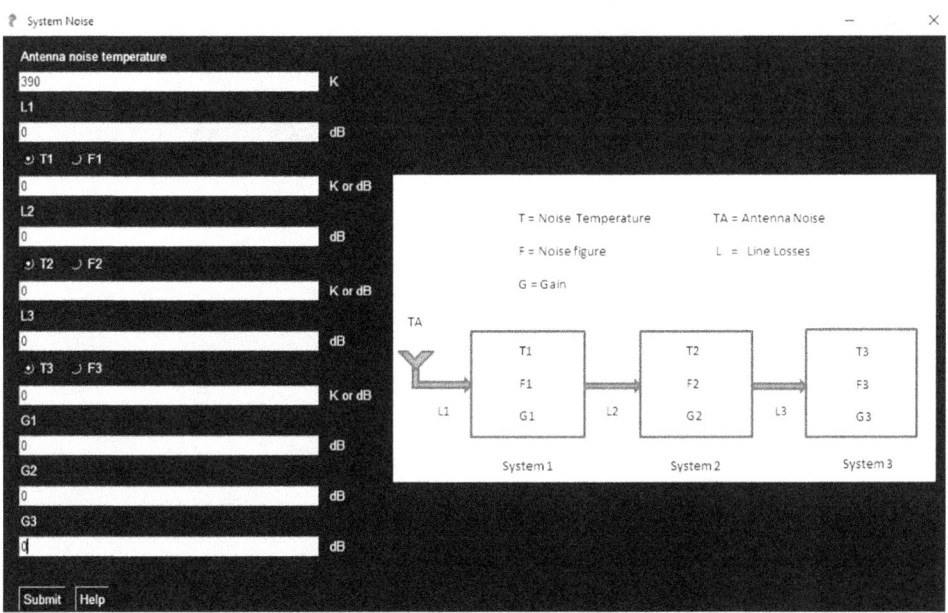

Results window:

Figure 7. Results computed on testbed

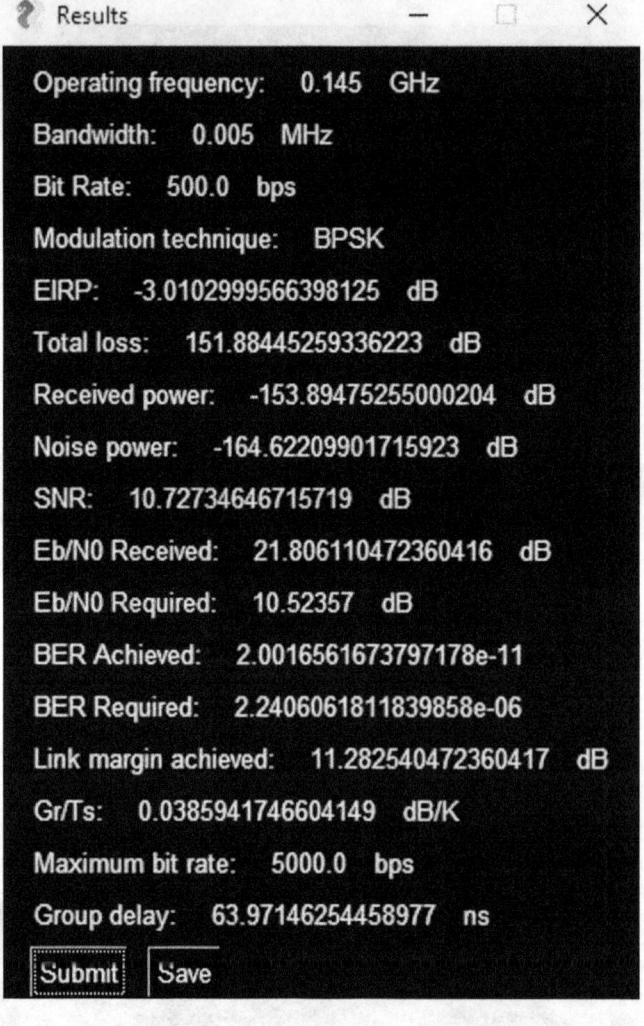

1.4 GRAPHS GENERATED ON TESTBED

The supported graphs that are generated from testbed are shown in figures 8 to figure 19 below.

1) **Bit rate**- Bit rate ranges from 500 to 10000 bps

Figure 8. Bitrate v/s link margin

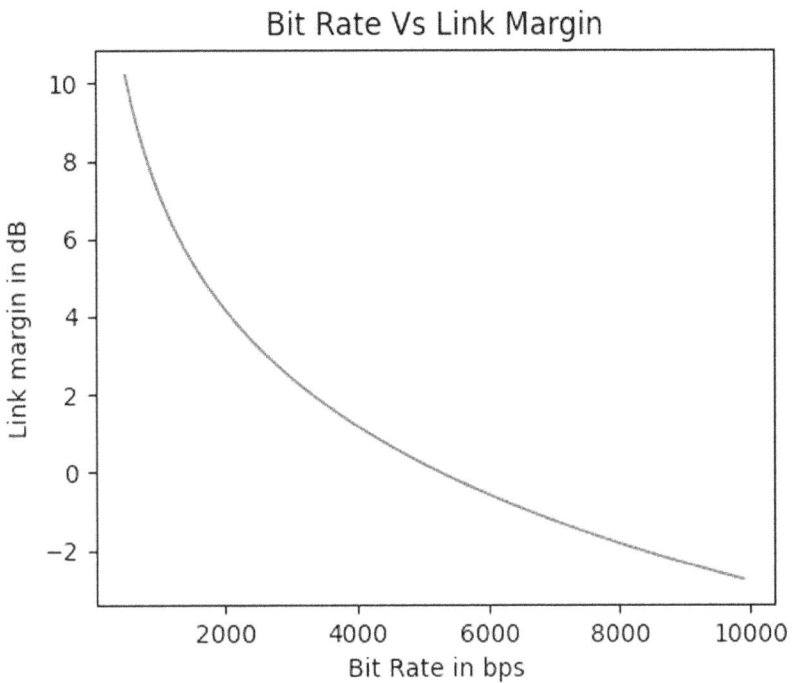

Figure 9. Bit rate vs BER

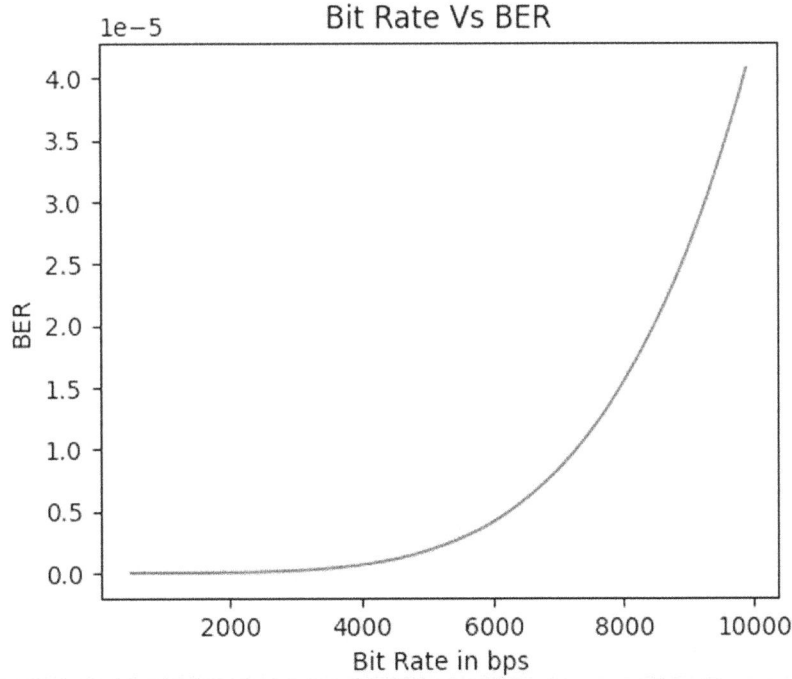

2) **EIRP**- EIRP ranges from -10 to 10 dB

Figure 10. EIRP v/s link margin

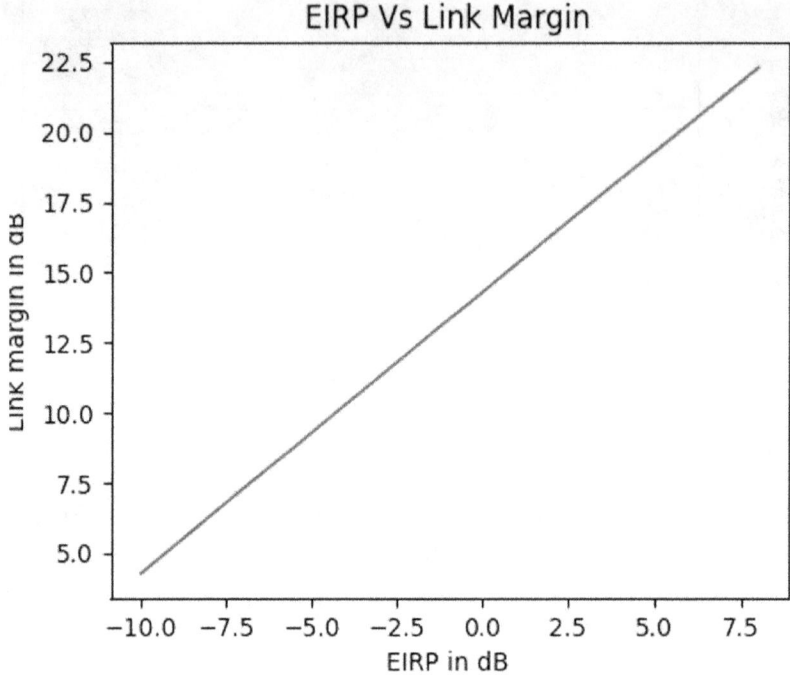

Figure 11. BER vs. EIRP

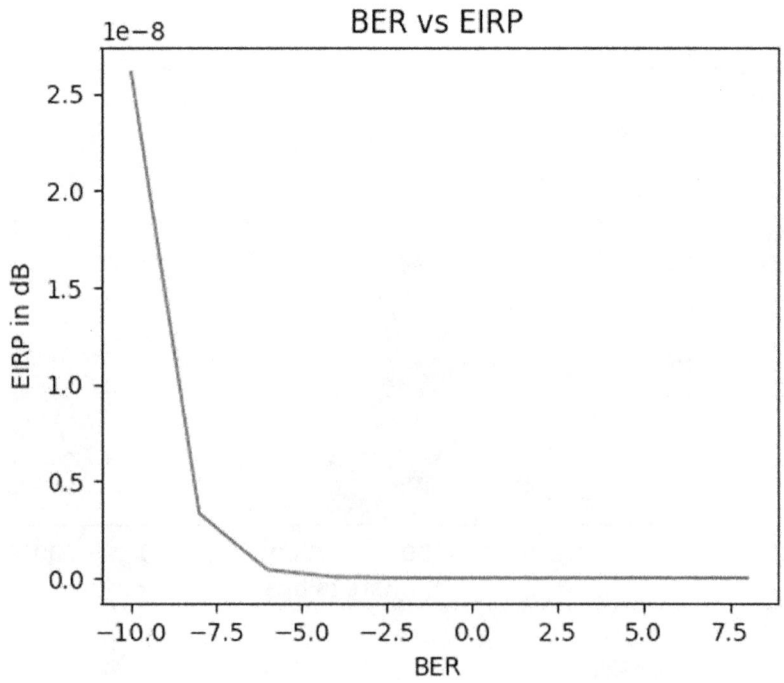

3) **Total losses-** Total losses ranges from 100 to 200 dB

Figure 12. Total losses v/s link margin

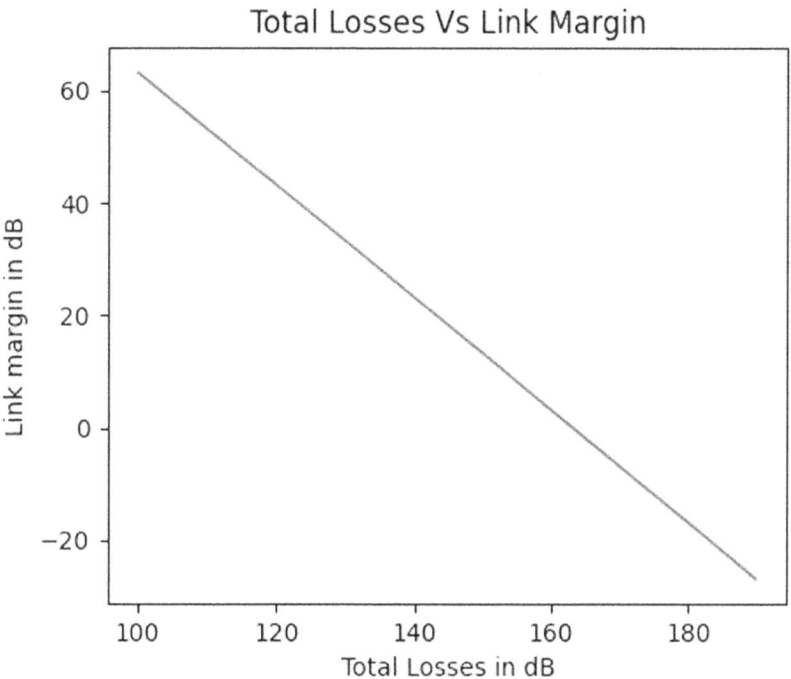

Figure 13. Loses v/s BER

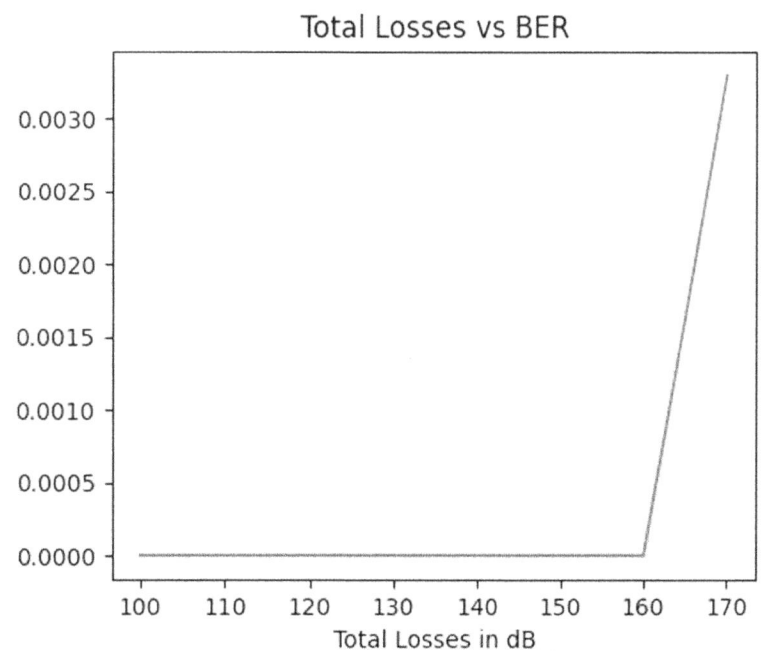

4) **Noise temperature-** Noise temperature ranges from 300 to 500 K

Figure 14. Noise temperature v/s link margin

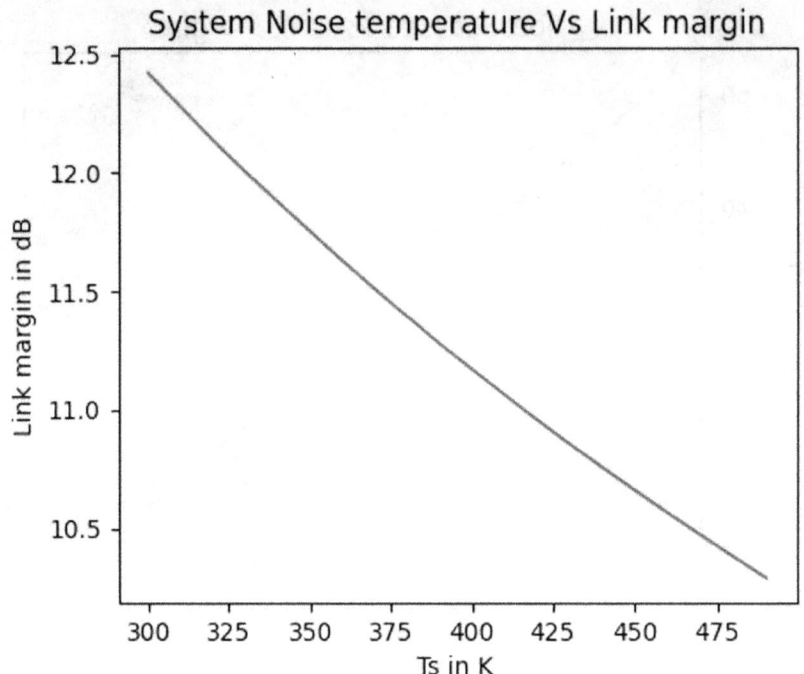

Figure 15. BER vs. Noise Temperature

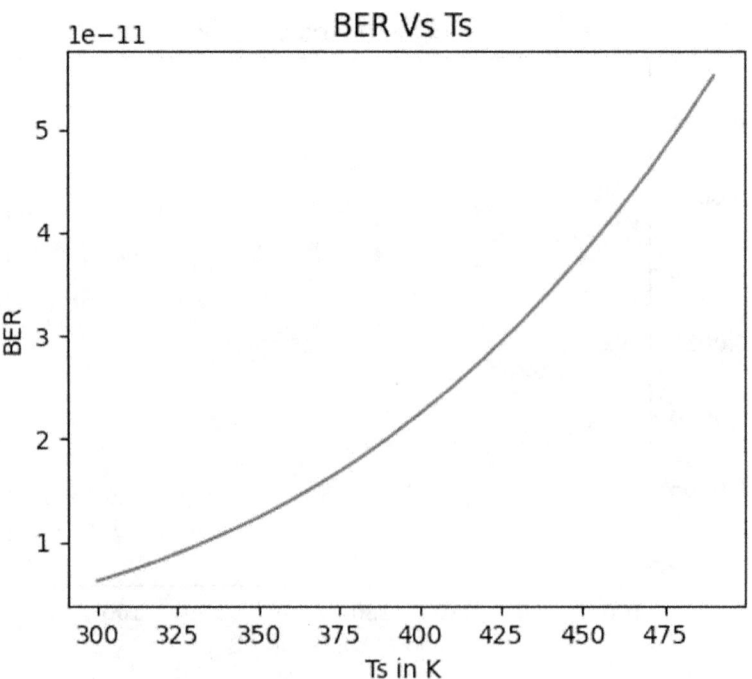

5) **Receiver figure of merit-** Bit rate ranges from -20 to 10 dB/K

Figure 16. link margin v/s receiver figure of merit

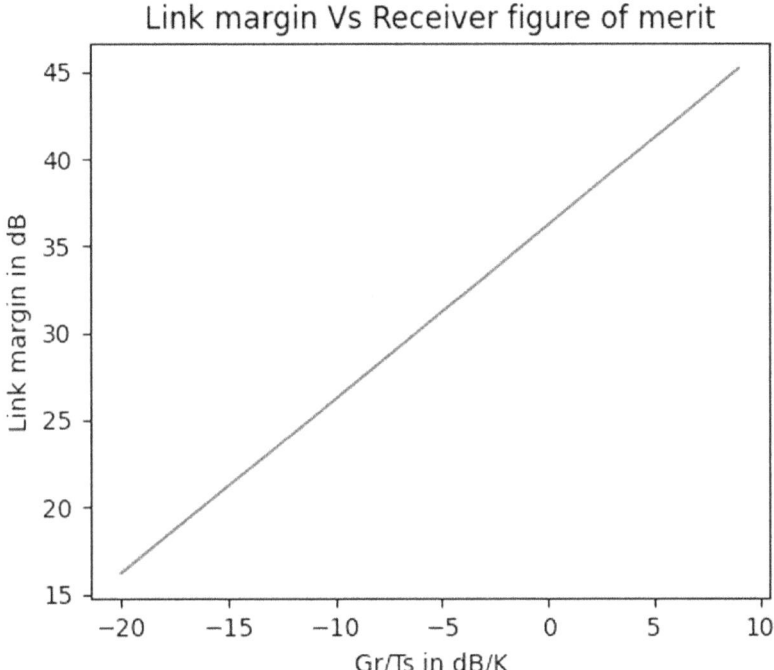

Figure 17. BER vs. Receiver figure of merit

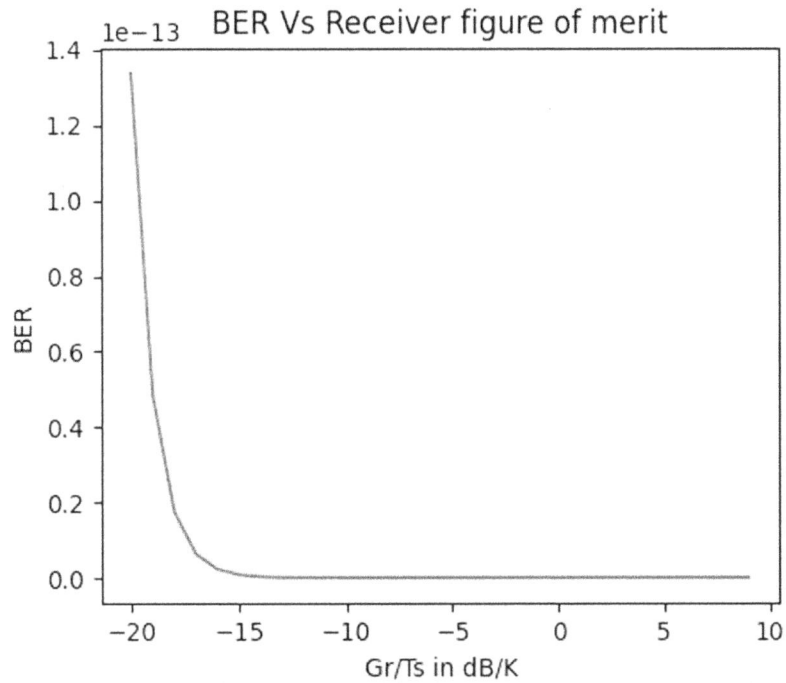

7)   **Path loss-** Height of satellite from earth ranges from 100 to 500 Km

Figure 18. Path loss v/s Height of satellite

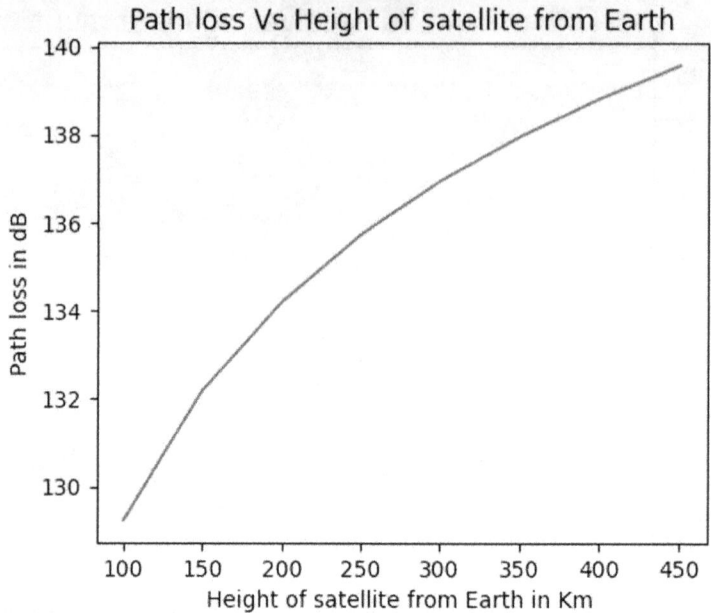

8)   **Path loss vs. frequency-** Frequency ranges from 100 to 1000 MHz

Figure 19. pathloss v/s frequency

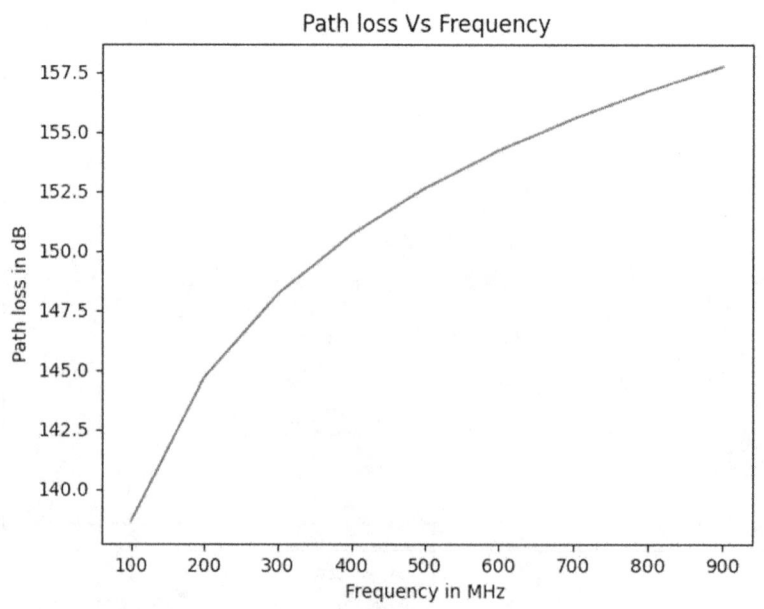

CONCLUSION

A link budget testbed for space communications using PySimpleGUI is developed to provide quick calculations involved in the communication link design. In the subsequent windows, the various transmitter end, channel losses, receiver and noise parameters are entered based on the user's choice within the operation ranges as shown. The results such as BER, link margin, SNR, Eb/N0 and so on are displayed in the next window which allows an option to save the results in the user's system. The results obtained are then used to get different performance graphs as shown in sample graphs. This would provide link budget RF engineers to design Communication system quickly by computing the link budget for any set of parameters and quickly validate through graphs. However, there is scope for further extension on flexibility in operation with more enhancements in user interface.

REFERENCES

Haykin, S. (2014). *Digital communication systems*. Wiley.

International Telecommunication Union (2017). *R-REC-P*, pp. 618-13.

International Telecommunication union. (2019). Rec. *ITU-R*, (August), 531–535.

International Telecommunication Union. (2021). *ITU-R recommendation*. p. 1546.

Link Budget Analysis in Mobile Communication System (2006). Hoation DAI *International conference on communications Technology*. doi:10.1109/ICCT.2006.341977

Roddy, D. (2017). *Satellite Communications* (4th ed.). Mc GrawHill.

Singh, K., Nirmal, A. V., & Sharma, S. V. (2017). A. V. Nirmal, and S. V. Sharma. "Link margin for wireless radio communication link. *ICTACT Journal on Communication Technology*, *8*(3), 3. doi:10.21917/ijct.2017.0232

Sklar, B. and Harris, F. (2022). Digital Communications, Prentice Hall

Study on channel model for frequencies 0.5 to 100 GHz, (2017). *3GPP TR 38.901 version 14.3.0 Release 14,*

Chapter 10
Circularly Polarized Antennas:
An Evident Advancement in Modern Wireless Communications

Utsab Banerjee

 https://orcid.org/0000-0002-8581-8383
MVJ College of Engineering, India

Anirban Karmakar
Tripura University, India

Anuradha Saha
Netaji Subhash Engineering College, India

ABSTRACT

This article is intended to present a comprehensive, technical review of circularly polarized (CP) antennas for an array of applications in various fields of wireless communication, emphasizing on the recent developments in the projected research. The article should also present a comparative study of various approaches reported in the open literature, with an aim to highlight the contribution of CP antenna systems and their chronological development in the domain of wireless communication technology. The primary motive of this literature is to (a) highlight the methodologies used by different attempts to portray and analyze the different aspects in which circularly polarized antennas find their applications in modern day wireless communication, (b) provide a practical viewpoint of the future scope of study, based upon the past and present state-of-art research trends, and (c) provide a conceptual and technical support to present day antenna designers to help the process of enhancement of innovation and multiple system integration.

DOI: 10.4018/978-1-6684-7000-8.ch010

INTRODUCTION

A profoundly crucial and fundamental characteristics related to an antenna in contemporary wireless communication systems is polarization, as discussed in (Balanis, 2005). An antenna's polarization is defined as the locus that is traced out by the tip of the electric field vector of the wave it is radiating or receiving as a function of time. The orientation of the electric field often determines an antenna's polarization. The senses of polarization typically encountered are the linear, and the circular polarization (CP) as described in (Balanis, 2005). The direction of an antenna's electromagnetic field, radiated in a plane that is perpendicular to the orientation of an electromagnetic wave. Co-polarization, for instance, occurs when an antenna that is vertically polarized can expeditiously transmit and receive only a vertically polarised wave and vice versa (Ludwig, 1975). The reciprocity attribute of the antenna causes its transmission and reception behaviour to be comparable. Cross-polarization, which causes a significant signal loss, occurs when a transmitting antenna is horizontally polarised while a receiving antenna exhibits vertical polarization, or vice versa (Ludwig, 1975). In contrast, the antenna in circular polarization radiates electromagnetic radiation in a spiralling, circular pattern that includes the azimuthal, elevation, and all planes in between. Because the circular pattern of the field radiated by the transmitting and/or receiving antennas are definitely coherent to the radiation pattern of the incident wave, circular polarisation renders the orientation of the transmitting and receiving antennas unimportant. Circularly polarised antennas are becoming paramount components of contemporary wireless communication systems due to their unique features, which include immunity to the Faraday rotation effect and multipath rejection, thereby preventing signal interference, as encountered in linear polarisation.

Figure 1. Electric field vector carving (a) Linear, (b) Circular and (c) Elliptical Polarizations

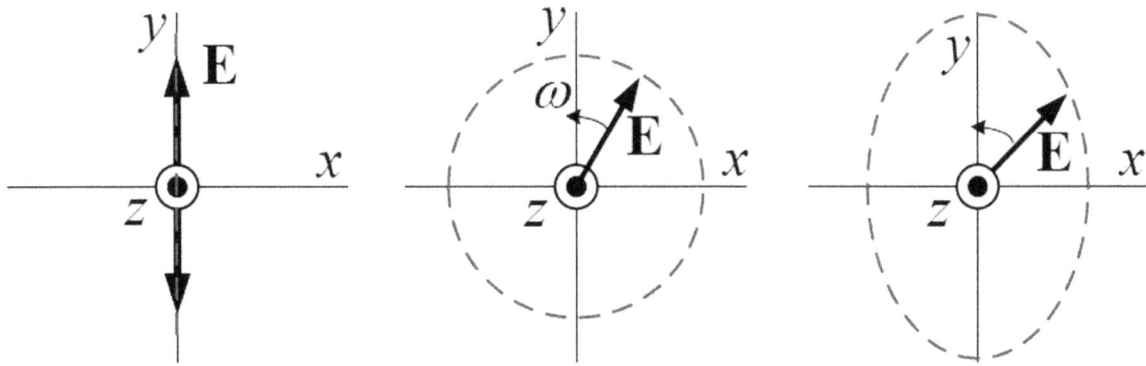

Figure 1 shows the orthogonal field components in the case of circular polarization are approximately of same amplitude, spatially orthogonal, and in phase quadrature. As a result, as the resultant electric field vector evolves over time, its tip produces a circular locus.

The tip of the electric field vector shows a linear oscillation, in this case along the x- or y-axes, to generate linear polarization, as seen in figure 1.

Although the quadrature phase difference between the two field components is preserved for elliptical polarization, as observed in figure 1(c), the amplitude of the field components is not the same as in the case of CP, depicted in figure 1. (b). Thus, this type of polarization is justifiable to be unique situation of elliptical polarization. As a matter of fact, both Linear and Circular polarizations are special cases of elliptical polarization. CP is technically defined by the Axial Ratio (AR) in (Balanis, 2005), that can be defined to be the ratio of the major to minor axes of the polarization ellipse and is represented in decibels as:-

$$\text{Axial Ratio (dB)} = 20\log_{10}\left[\frac{E_{x_0}}{E_{y_0}}\right] \qquad (1)$$

This equation portrays that $1 \leq AR < \infty$. An ideal CP is realized when $AR = 1$ or 0 dB, indicating that both the axes are of equal lengths. This is virtually impossible to obtain practically. Therefore, a frequency range across which AR £ 3dB is accepted for practical applications

$$\text{Axial Ratio Bandwidth} = \frac{f_2 - f_1}{f_{min}} \qquad (2)$$

Where, f_1 and f_2 are the terminal frequencies for $AR \leq 3dB$ band and f_{min} is the frequency at which the minimum value of AR occures. Figure 4 depicts the polarization ellipse along with the major and minor axes. The highest magnitudes of the electric field along the said axes are E_{x0} and E_{y0}, which are mentioned in (1). These values are the same for circular polarization. As a result, circular polarization is a particular instance of elliptical polarization. The tilt angle (τ), which specifies the orientation of the ellipse, is another parameter that is shown in Figure 4. It is the measure of obliqueness of the major axis of the ellipse with reference to the vertical axis. The largest electric field magnitudes along the x- and y-axes are indicated by E_{x0} and E_{y0}, respectively., whereas OA and OB stand in for the major and minor axes of the polarization ellipse, respectively.

A primitive, planar microstrip patch antenna usually does not emit circularly polarized waves on its own; therefore, some modifications to the patch antenna are required to generate the circular polarization, as discussed in (Row, 2004). This article goes into great detail about the many methods researchers have taken to create circular polarization in the parts that follow. Each categorized description includes a comparison of the many antennas described in the literature, the underlying techniques, and the gradual growth of design trends. The number of feeding points required to produce circularly polarized waves classifies circularly polarized microstrip antennas. The most commonly applied feeding networks consists of dual or multiple feed and single feed, as pointed out in (Banerjee, 2020).

Dual feed architecture: Dual feed is a natural alternative to generate circular polarization in microstrip antennas since phase quadrature between the perpendicular fields that are radiated is a necessary prerequisite in doing so. Figure 3 illustrates how the two feed locations were selected to be spatially orthogonal to one another, as presented in (James, 1989). Signals with identical magnitude and quadrature phase stimulate the microstrip patch antenna with the aid of an external polarizer. The substrate is 45 mm wide and 55 mm long. The feeding strips are 3.5 mm wide and 18 mm and 16.5 mm long, respectively. The gap g is assumed to have a dimension of 1 mm. Dual feed architecture is discussed in (Li, 2005). In general, dual-fed CP antennas result in a relatively wide operating bandwidth. But in order to get the desired CP behavior, this comes at the expense of elevated design complexity.

Figure 2. Polarization Ellipse and its parameters (Balanis, 2005)

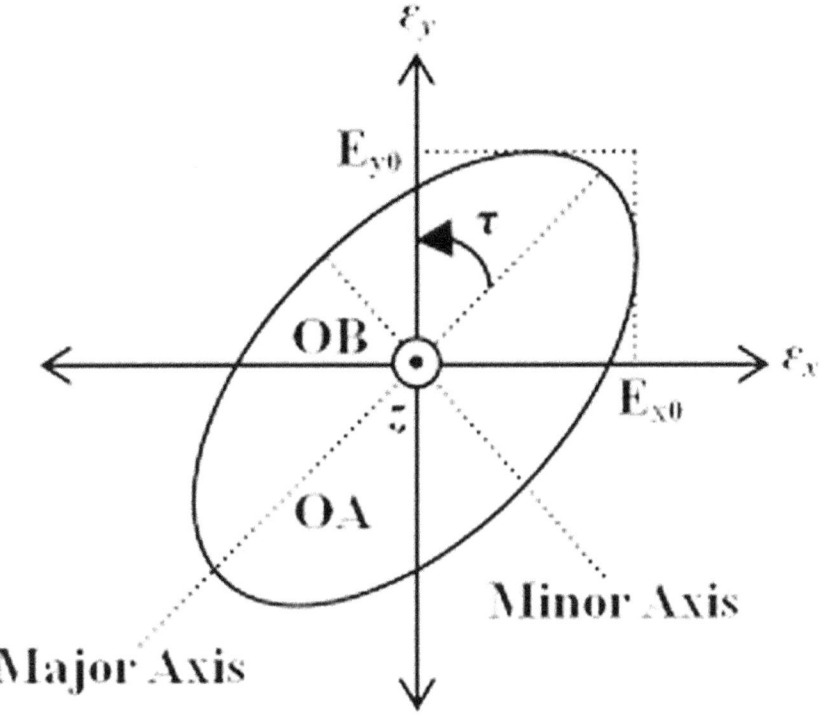

Figure 3. A dual-fed CP patch antenna (James, 1989)

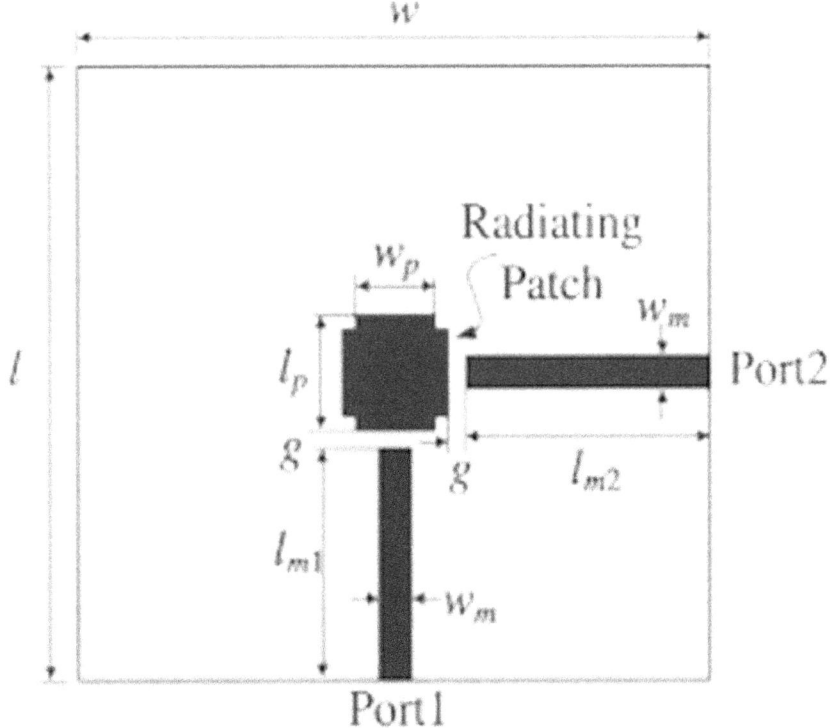

Single feeding structures: As illustrated in Figure 4, single fed microstrip antennas boast of a simple structure, fabrication is usually easy, low cost, and show compactness in their structure. It eliminates the need for complex hybrid polarizers, which are difficult to implement in practice, discussed in (Haneishi, 1982). Single fed circularly polarized microstrip antennas are among the simplest antennas capable of producing circular polarization, discussed in (Heidari, 2009). In order to realize circular polarization with a only one feed, a couple of degenerate resonant components, with identical amplitude and quadrature phase difference should be exited. Because basic shapes of microstrip antennas produce linear polarization, some design changes are required to produce circular polarization. The field is splitted into two orthogonal modes with equal magnitude and a 90° phase shift using perturbation segments. On account of which, the criteria for generating CP are satisfied. An obvious benefit of single-fed CP antennas is their easy architecture. However, because single fed CP antennas have a limited operating bandwidth, some modifications to enhance the operational bandwidth, while sustaining structural simplicity is needed. The design's acceptability is dependent on the mentioned mechanisms and optimal results.

Figure 4. A Coaxial-fed CP Antenna with a central cross slot (Heidari, 2009, reproduced courtesy of The Electromagnetics Academy)

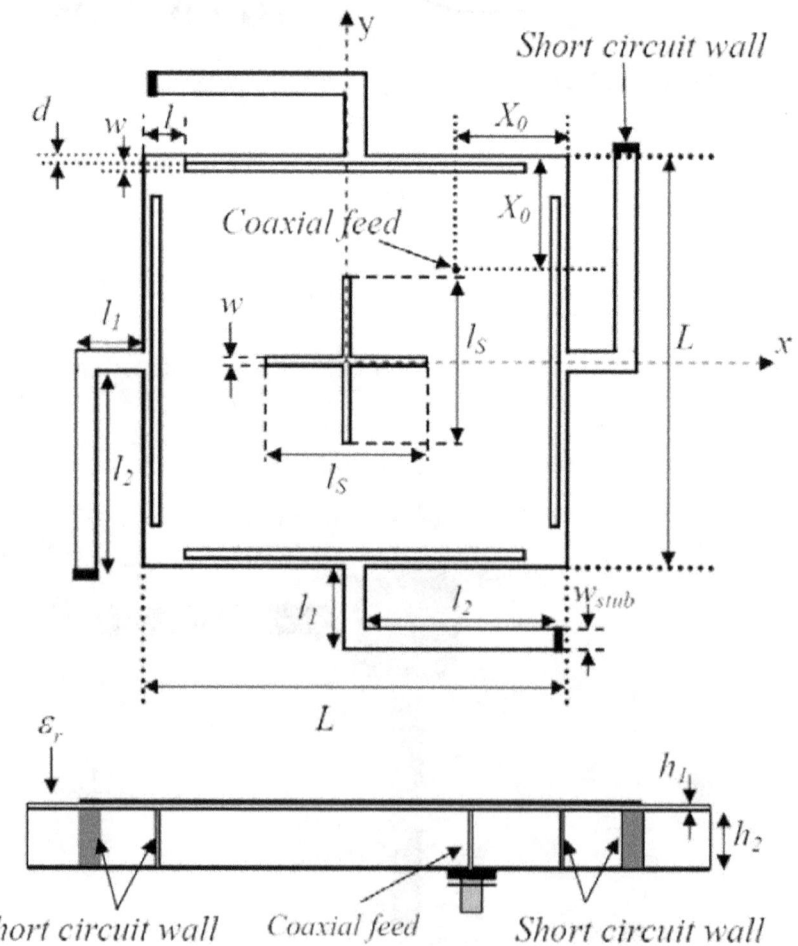

In general, circularly polarized antennas can be constructed in a variety of ways. The introduction of perturbation in the antenna is one method for generating CP waves, discussed in (Ravipati, 1999). To trigger a couple of mutually perpendicular field components having identical amplitude and 90° phase shift.

Due to their simplicity, ease of fabrication, low cost, and compact structure, single fed architectures are frequently employed in contemporary microstrip antenna designs. Several methods were employed to obtain circular polarization in a single feed microstrip antenna. Few are covered in this section. The antenna in (Haneishi, 1982) depicts a single-fed CP patch antenna constructed on a Teflon fibre glass panel with parameters 'h' = 1.2 mm, 'W_x' = 27 mm, and 'W_y' = 27 mm. The circular disc radiating patch has a radius of 10.5 mm. Circularly polarized antennas, based upon dielectric resonators, fractal based CP antennas, and other CP antenna types have also been the subject of review investigations in the past, as seen in (Ullah, 2017). However, it is uncommon in the current literature to find a thorough and scientific illustration of CP antennas in a culmination, together with the infinite possibilities of their utilities. This article aims to fill the gap within the volume of research documents.

SUBDIVISION OF CIRCULARLY POLARIZED ANTENNAS

Overall, circularly polarized antennas could be categorized in to various divisions, depending on a multitude of facets. This is exhibited in the following Figure 5: -

Figure 5. Categorization of CP Antennas depending upon various factors

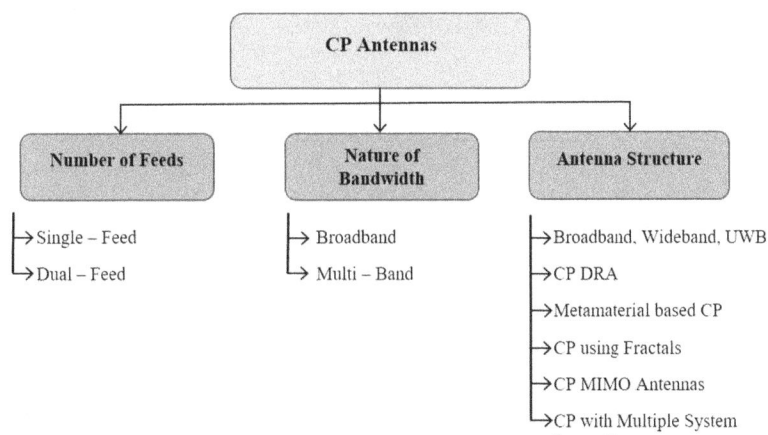

1. Circularly Polarized Wide Band Antennas

Single feed, discussed in (Sudha, 2004), and dual or multiple feed, discussed in (Li, 2005) are the two prominent methods for realizing circular polarization, as was pre-postulated. Despite having a simple structural design, the majority of single feed CP antennas have extremely narrow CP (3-dB) axial ratio (AR) bandwidths and impedance bandwidths. In order to sustain this bulk data transmission, significant majority of today's high data rate communication across communicating systems requires additional bandwidth. As a result, single-fed antennas cannot be utilized in wide band applications. Dual-feed CP antennas, on the other hand, have a wider axial ratio bandwidth but entails more design complexity. An appreciable design of a CP antenna is discussed in (Sitompul, 2019), shown in Figure 6 (a – d). Microstrip line and coplanar waveguide are the most appealing feeding methods for wideband CP slot antennas (CPW). Various different types of wide-slot antennas with various shapes are reported with improved CP radiation. These comprise an L-shaped slot antenna with a 46.5% AR bandwidth, an asymmetric aperture antenna employing a CPW feed, a circular slot antenna fed by microstrip line, and a square slot antenna fed by a CP waveguide with a 49% AR bandwidth.

Figure 6. A microstrip-fed CP Antenna exhibiting (a) top view (b) back view (c) Impedance bandwidth and (d) Axial Ratio Bandwidth (Sitompul, 2019, reproduced courtesy of The Electromagnetics Academy)

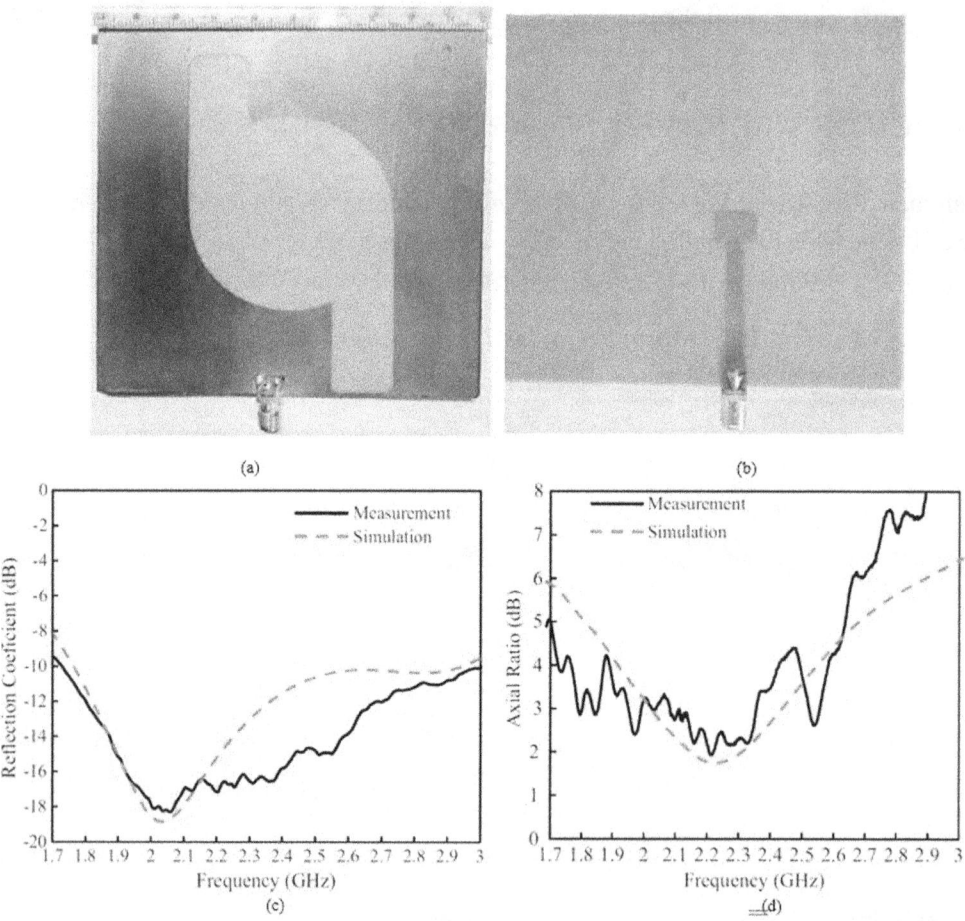

There has recently been an increase in demand for ultra-wideband (UWB) antenna technology for wireless applications with a short range and a high data rate. As a result, research into UWB equipment and antenna design has become more important. These antennas often attempt to cover the 3.1 to 10.6 GHz bandwidth that the Federal Communications Commission (FCC) recommended in their study from 2002, as mentioned in (Fu, 2014), which is known as the Ultra-Wideband (UWB) spectrum. Monopole

UWB radiators are commonly used because of their light weight, wide return loss bandwidth, decent radiation characteristics, ease of construction, and cost efficiency. But, linear polarization is used in the majority of UWB monopole antenna designs. Because of its low transmitting power, UWB technology is limited in terms of distance. The application of circular polarization (CP) in Ultra-wideband technology is beneficial for obtaining distance gain as well as the previously mentioned merits. Ultra-wideband CP antennas find huge applications in defense applications such as Electronic countermeasure systems, UWB RADAR, and medical imaging applications. Some design techniques have been investigated in order to obtain UWB CP, discussed in (Banerjee, 2019). In (Jou, 2009), the authors generate a resonant mode for broadband impedance bandwidth using an inverted - L shape in the ground plane and trigger two orthogonal resonant modes with same amplitude and phase quadrature. Furthermore, a bevel in the radiator patch enhances the return loss bandwidth. The frequency varied from 2.17 to 8.47 GHz, not being able to cover the complete UWB frequency range, and the impedance bandwidth increased in the second design compared to the first design. A CPW-fed monopole antenna that presents a fairly wideband circular polarization characteristics is described in the research in (Banerjee, 2019). The close proximity coupling of the monopole, a parasitic strip on the upper edge, and an engineered parasitic Hilbert shaped strip results in a very broad impedance bandwidth, enough for encompassing the C-band (4–8 GHz).

In (Jhajharia, 2018), a novel geometry of a wideband circularly polarized antenna with an asymmetric meandered-shaped monopole is discussed. The antenna's wideband characteristics are realized by two asymmetric meandered-shaped orthogonal edges of the monopole and defected ground structure with oblique edge. In terms of return loss bandwidth (nearly 102.04% spanning through 2.4 - 7.4 GHz) and 3 dB axial-ratio (AR) bandwidth (approximately 37.5% within the frequency cap of 4.9 - 6.9 GHz), the proposed prototype performs decently. The obtained functional bandwidth of the radiator engulfs a number of narrowband communication spectra.. This work distinctly explains that the strategically tailored defect in the ground plane is primarily responsible for the wide return loss bandwidth. Moreover, an appreciable axial ratio bandwidth can also be obtained, as demonstrated.

The literature in (Chaudhary, 2020) details a novel, planar wideband circularly polarized MIMO antenna with pattern diversity across the entire axial-ratio operating bandwidth for 4G and Sub-6 5G bands (1.35 - 2.75 GHz). The design, shown in Figure 7, is comprised of a couple of tri-branch planar inverted-F antenna (PIFA), connected by a ground T-stub to attain circular polarization, with significantly high isolation. It also sustains circular polarization across the LTE band. The MIMO antenna's axial ratio bandwidth (ARBW) is 1.08 GHz (1.47 - 2.55 GHz), its return loss bandwidth (RLBW) is 1.4 GHz (1.35 - 2.75 GHz), and its isolation exceeds 13.4 dB. With a well-orchestrated design and prominent MIMO performance, the antenna has been reasonably adjudged to be suitable for applications in handheld communication devices.

A comparative analysis on the various methods investigated to present a significant return loss bandwidth with circular polarization is discussed.

2. Circularly Polarized Dielectric Resonator Based Antennas (CP DRA)

A certain class of radiators, known as Dielectric Resonator Antennas (DRAs) has great promise for high frequency applications. Richtmyer published the first thorough study proving that dielectric resonators can also be employed as effective radiators in (Richtmyer, 1939), even though these resonators were initially applied as filters and oscillators, or energy storing devices. Long et al. conducted the initial theoretical and experimental study of a cylindrical DRA in (Long, 1983). Since then, DRA technology

Figure 7. A planar, inverted, F - antenna antenna generating Circular Polarization (Chaudhary, 2020, reproduced courtesy of The Electromagnetics Academy)

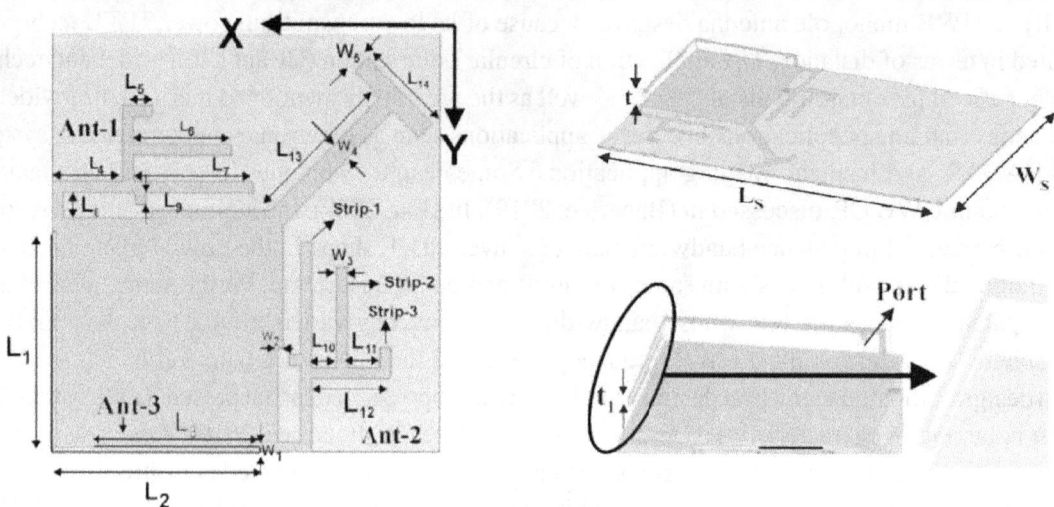

Table 1. A comparison of various CP Antennas with broad impedance and axial ratio bandwidths

Reported papers	Antenna volume (mm³)	Type of feed	Return loss Bandwidth	Axial Ratio Bandwidth
(Nasimuddin, 2011)	60×60×1.6	Single CPW feed	107%	68%
(Yang, 2008)	82×82×1.524	Single L-shaped feed	57%	46.5%
(Sitompul, 2019)	60×60×0.8	Single CPW feed	51.4%	48.8%
(Pourahmadazar, 2011)	60×60×0.8	Single CPW feed	132%	32.2%
(Oraizi, 2013)	21.66×10.83×9.5	Single Aperture feed	36.15%	33.6%
(Yong-Xin, 2009)	180×180×0.8	Proximity coupled hybrid feed	38%	29.7%

has advanced significantly and is now of particular interest to contemporary wireless communication researchers. Essentially, a DRA is a block of dielectric material (ideally one with a high dielectric constant) that has been adhered to the upper face of a coplanar monopole radiator on top of a substrate. (of a permittivity that is comparatively less). The presence of a ground plane is however, optional.

The paper in (Agarwal, 2017) presents the design and experimental validation of a state-of-the-art elevated gain and broadband modified dielectric resonator antenna (DRA). The DRA and an elemental slot are combined with a small rectangular notch in the proposed antenna architecture, effectively enlarging the impedance bandwidth by fusing the slot and DRA resonances. Both antennas are excited concurrently by an inverted T-shaped feed line. It enables the articulation of different slot antenna and DRA resonant modes. From 1.67 to 6.7 GHz, the designed antenna has a 120% impedance bandwidth. A reflector placed beneath the antenna at an optimal distance improves antenna gain. Following the optimization of the vertical dimension and measurements of this reflector, the gain of the antenna is escalated from 2.2 dBi up to about 8.7 dBi around the frequency of 1.7 GHz.

A thorough analysis of the modes and radiation characteristics of a variety of DRs was published by (Mongia, 1997), which included precise functional detailing of the antenna. The majority of DRA research volume has been documented to have a linearly polarized radiation pattern from their inception. The operation is thus frequently constrained by linear polarization, in spite of the inherent advantages of a DRA, such as superior gain and radiation efficiency due to minimal metallization. This potential difficulty was taken into account, and as a result of ongoing research, circularly polarized DRA was introduced (CP DRA). The possibility to sustain multiple radiating modes concurrently in DRAs is judiciously explored in the design and development of CP DRAs.

In (Varshney, 2020), the authors have recommended to use a two-port MIMO dielectric resonator antenna (DRA) with CP characteristics. The proposed architecture, as seen in Figure 8, enables an isolation better than 15 dB over the operative passband, even when the distance within the radiators is minimized. The antenna has a 3-dB axial ratio bandwidth and a 10-dB impedance bandwidth of 34.85% and 4.55%, respectively. By computing the parametric variables, the presented radiator's MIMO performance is demonstrated. The designed antenna can be considered to be suitable for C- band satellite communication applications.

Figure 8. A two port fed Circularly Polarized DRA (Varshney, 2020, reproduced courtesy of The Electromagnetics Academy)

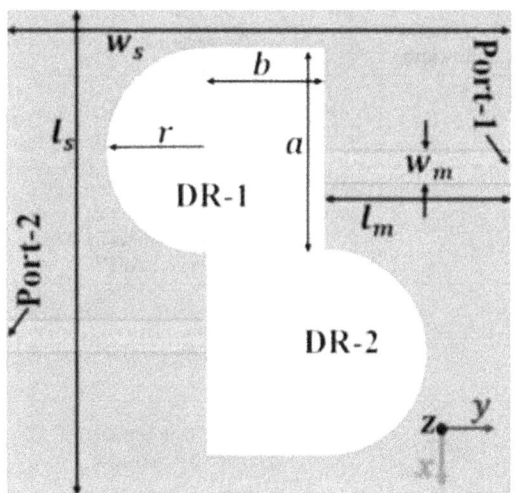 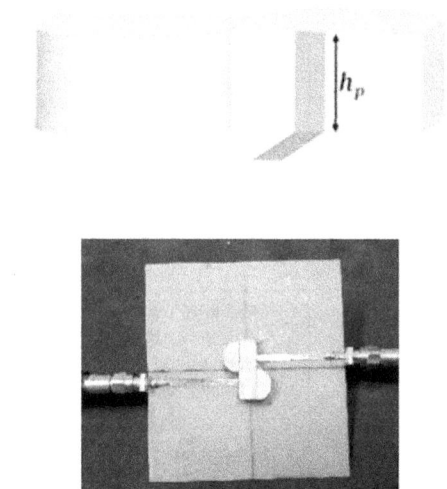

An extensive investigation on the numerous procedures employed for producing CP DRAs has been carried out.

It is observed that, a multiple-point fed CP DRAs and progressive rotation approach leads to the creation of a feeding network that is comparatively more complicated and challenging to control in an optimum manner, but, if handled effectively by sound mathematical modelling and simulation methods, may yield attractive results.

Multi-point fed DRAs, while architecturally complicated, can have greater axial ratio bandwidths than single point fed CP DRAs, those are easy in construction but often have limited bandwidths. A cuboidal

DRA stimulated with vertical strips and coupled with a phase quadrature network was described by the authors of (Bin, 2009). This excitation led to the simultaneous generation of several circularly polarized modes (TE_{111} and TE_{113}) in the rectangular DR, boosting the axial ratio bandwidth. Despite the fact that the theoretical resonant frequency of the higher order TE_{113} mode was found to be different from that of the fundamental mode, the resonance of these two modes took place close to one another. In order to create a phase quadrature at the output ports, two quarter wave transformers were responsible for splitting the incoming signal into two distinct routes before transmitting it. A 48% axial ratio bandwidth is produced by this arrangement, and the antenna exhibits stable principal beam radiation qualities in the broadside direction.

The antenna components are activated in a sequential 90° phase shift at each subsequent feeding point owing to effective engineering of the feeding network. Existing literature is populated by a number of research articles describing the application of the sequential rotation feeding strategy for producing circular polarization in DRA.

According to a number of experimental studies, inhomogeneous dielectric resonators with particular geometries or altering the dielectric characteristics of typical dielectric resonator shapes in the azimuth direction can both be used to further elevate the CP DRA performance.

Table 2. A comparison of various Circularly Polarized Dielectric Resonator based Antennas (CP DRAs)

Reference Number	DR Dimension[each component in (mm)]	Feed Type	Impedance Bandwidth	Axial Ratio Bandwidth	Observations
(Esselle, 1996)	14×3.7×3.24	50Ω microstrip line through a single aperture	1.4%	3%	Structurally simple, but application is limited due to a very narrow bandwidth.
(Leung, 2000)	100×100×11.6	Dual conformal strip feed	13.7%	20%	A bulky overall structure is created by a sophisticated feeding system that uses a 90° hybrid coupler but allows for direct integration with MMICs.
(Leung, 2001)	Radius of the resonator = 20 Height of the resonator = 20	Coaxial feed with a perturbed annular slot	18%	3.4%	The authors point out that a significant gain is obtained.
(Chair, 2006)	18.34×9.66×5.2	Microstrip line fed, slot coupled	36.6%	10.6%	Novel structure with appreciable radiation characteristics.

3. Fractal Based Circularly Polarized (CP) Antennas

The primary characteristic attribute of the fractal geometry can be considered to be its "self-similarity", that is preserved across several orders of its progressive iteration, as is evident in Figure 9 (a – c).

A radiator that uses a fractal, self-similar geometry to increase the perimeter of a material (such as a patch or a monopole) that may accept or radiate electromagnetic energy while preserving a specific overall contour or volume is known as a fractal antenna.

The main characteristic of such fractal antennas, also known as multilevel and space-filling curves, is their recurrence of a pattern over multiple scale orders, or "iterations." Because of this, fractal antennas are typically relatively small in size. These antennas are expected to display wideband or multiband radiation and have practical uses in recent developments in wireless communications, such as telephonic transmission. The response of a fractal antenna is significantly different from that of conventional antenna designs since it can operate with good to exceptional performance at numerous frequencies at once. Standard antennas often only function properly at the resonant frequency intended for their application since they must be physically sized for it.

Figure 9. Various fractal geometries like (a) Koch Snow flake (b) Sierpinski Carpet and (c) Hilbert Fractal (Mandelbrot, 1982)

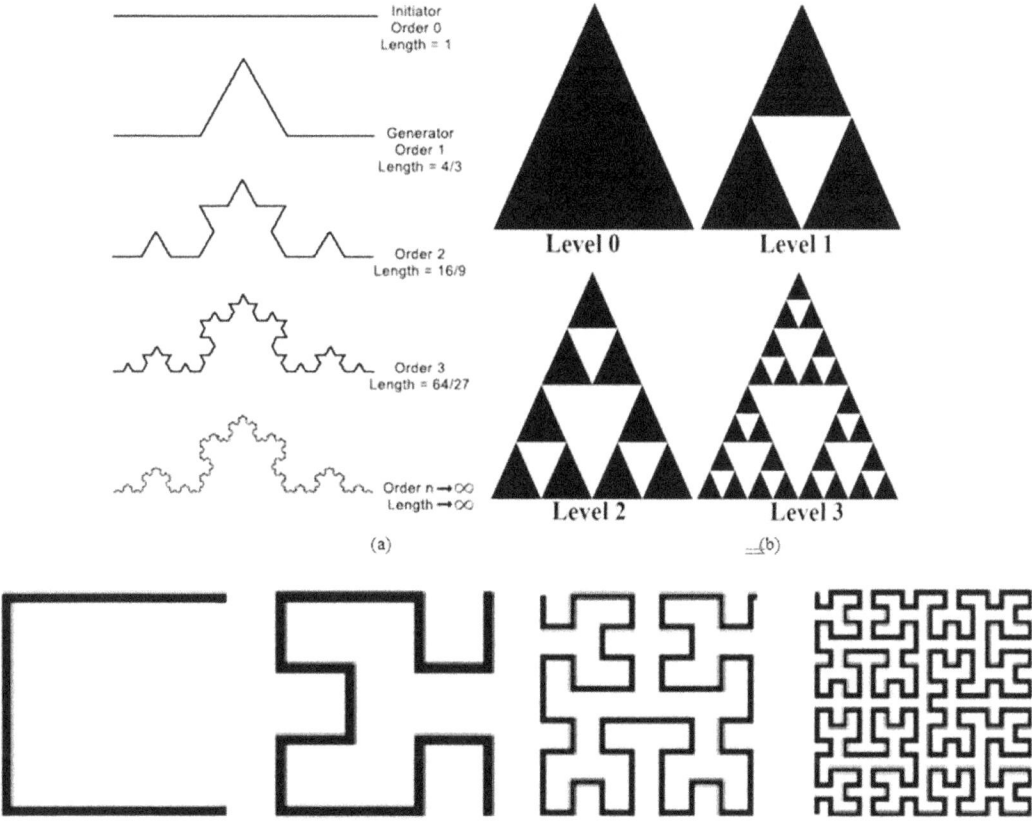

In (Gupta, 2021), a design for a single-feed circularly polarized microstrip antenna is presented, and is shown in Figure 10 (a – c). With an intention to enhance the return loss bandwidth, gain, and axial ratio bandwidth, the structure utilizes the idea of an E-shape patch with a slanted fractal defective ground structure. A fractal in the form of an angled E is etched onto the ground plane to produce the obtained return loss bandwidth, axial ratio bandwidth, and gain. The designed antenna exhibits a 12.7% impedance bandwidth at the band's center frequency of 3.47 GHz and a 2.39% 3-dB AR bandwidth at the band's center frequency of 3.626 GHz. The calculated maximum gain for the antenna is 8.1 dBi. The authors claim that the antenna might be appropriate for many modern communication applications.

The method put out in (Prajapati, 2015) entails substituting various fractal curves for the square patch's straight sides. A couple of modes, orthogonal in space are triggered by this tiny difference in electrical length in the patch's couple of mutually orthogonal orientations. It is possible to build small CP fractal microstrip antennas by effectively dimensioning the fractal curves. Achieving structural miniaturization is always achievable in addition to CP generation.

In (Heydari, 2016) a fractal Jerusalem cross (JC) slot antenna is designed for CP. In order to create circular polarization, a unique slot form called a broken cross is needed to change the surface current distribution of the antenna. In order to lower the antenna and ensure symmetrical formation while obtaining circular polarisation, a T-shaped feed line is used. Impedance matching of the antenna is enhanced by a triangular slot in the feed line. Additionally, the coupling to the ground plane may be fairly regulated by varying the length of the feed line, indicating that the perfect matching length has been achieved. The Jerusalem Cross (JC) on the ground plane has finally been incorporated.

In (Oraizi, 2012), a new square and Giuseppe Peano fractals are used to introduce a multiband CP antenna. It is intended for use with the IEEE802.11b/g (one of the WLAN bands) at 2.4 to 2.484 GHz, and the fourth generation (4G) of mobile communication systems at 4.6 to 5.2 GHz. The square fractal microstrip patch antenna's miniaturization and multi-banding characteristics are also examined and reported.

Figure 10. Figure showing (a) geometry of a fractal based CP Antenna and the (b) S_{11} and (c) Axial Ratio Bandwidth (Gupta, 2021)

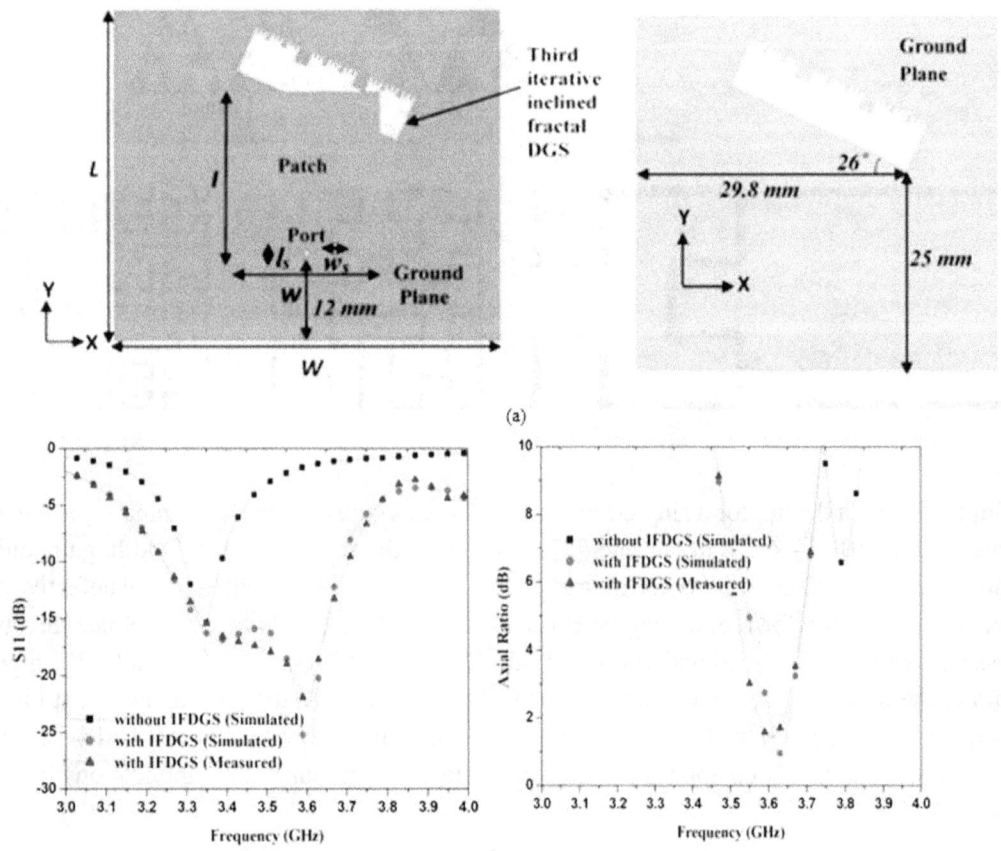

4. Metamaterial based Circularly Polarized (CP) Antennas

In order to provide electromagnetic properties that are rare or challenging to achieve in nature, meta-materials are structural units that are often built with innovative or artificial components. In the recent years, antenna design has included this idea. A family of antennas, known as metamaterial antennas uses metamaterials to increase their performance or perform innovative functions. They fall under the categories of resonator-type antennas and leaky-wave antennas. Meta-resonators and/or metamaterial loadings may also serve as the foundation for metamaterial-based antennas. These antennas offer notable advantages such compact physical dimensions, affordable price, broad bandwidth, and high radiation efficiency.

Compact antennas made of metamaterial are suggested as a way to control the near field boundary conditions, thereby facilitating antenna miniaturization while preserving appreciable radiation characteristics. Researchers have learned how to get around the constrictive efficiency-bandwidth restriction for small antennas thanks to metamaterial-based antennas.

In (Park, 2011), it is suggested to use a cylindrical DRA fed by a coaxial probe to achieve an axial ratio bandwidth of 0.4% and an impedance bandwidth of 0.9%. The arched branches are another way that the horizontal polarization is accomplished. The gain attained is minimal, about − 0.4 dBic.

Table 3. A comparison of various fractal based Circularly Polarized Antennas

Cited papers	Antenna Dimension (mm³)	Type of feed	Nature of fractal employed	Return loss Bandwidth	AR Bandwidth	Observations
(Wei, 2017)	45×45×3.18	Single probe feed	Fractal DGS	2%	0.4%	The structure is complex because of the third order iterative Fractal DGS. Additionally, the maximal gain of 2.2 dB and operational bandwidth are both rather low.
(Gupta, 2021)	41×41×1.6	Coaxial probe feed	Inclined Fractal defected ground structures	12.7%	2.39%	The effect of the defected ground structure, based upon an inclined fractal geometry has been exploited, and decent results were obtained.
(Joy, 2018)	55×55×1.6	Single probe feed	Minkowski Fractal	3.7%	2.8%	The authors refer the antenna to be suitable in Satellite communication and GPS applications.
(Prajapati, 2015)	90×90×3.2	Microstrip line fed	Koch curve Fractal with Defected Ground Structure	9.27%	2.1%	Antenna gain is high, at roughly 5.5 dB. It has uses in satellite communication and L-band communication.

Circularly polarized antennas are now frequently designed using metamaterial-based antennas. The literature (Blanco, 2015) describes a number of metamaterial–based CP antennas, some of which employ complementary Split-ring resonators (CSRR), 22 mushrooms, etc. It is seen that these described antennas have a limited impedance bandwidth, a limited axial ratio bandwidth, weak gain, a rather complex design, a large dimension, etc.

In (Wang, 2021), miniature circularly polarized (CP) antennas for RFID applications are proposed and constructed using dispersion-engineered metamaterial transmission lines (TLs). The composite right-handed transistor (CRLHTL), which stands out for having a novel, relatively affordable double-layered

3D structure, used for the initial research and development of this invention. The implementation of one CP antenna and one polarization-reconfigurable radiator using CRLH-TLs follows. Comparatively to the conventional right-handed (RH) TL-based antennas, miniaturization is obtained by decreasing the dispersion curve to a smaller frequency. This antenna has a 10 dB bandwidth of 4.8%. The antenna can be considered to be worthy of 902–928 MHz UHF band RFID applications. The antenna and the results are shown in Figure 11 (a – e).

Figure 11. Figure showing (a) perspective view (b) Top view (c) Fabricated prototype (d) S_{11} Bandwidth and (e) Axial Ratio bandwidth of a Metamaterial based CP Antenna (Wang, 2021)

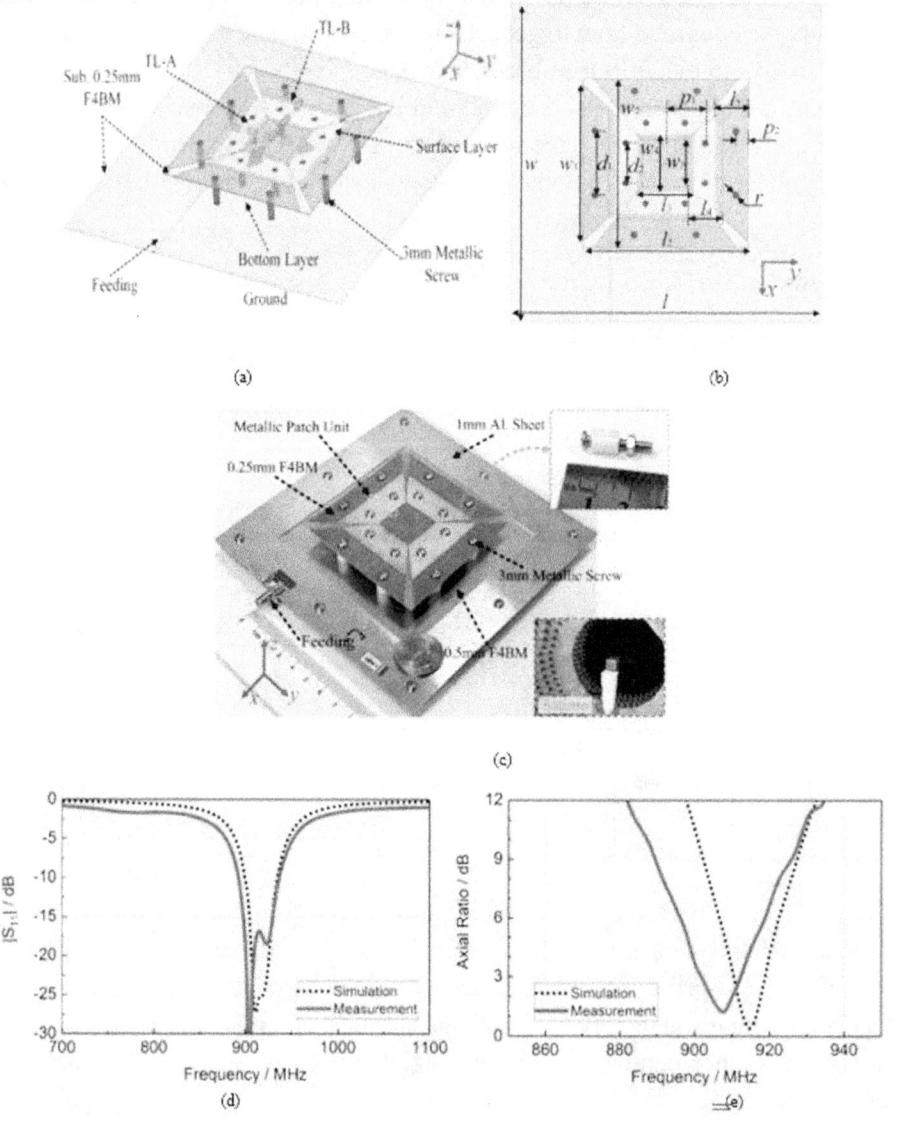

In (Wang, 2020), a metamaterial inspired compact CP antenna is demonstrated for RFID applications. The focal concept is the application of the dispersion curves of Composite Right/Left Handed transmission lines. On account of this, the effective electrical size of the antenna has been reported to be 0.24 $\lambda_0 \times 0.24 \lambda_0 \times 0.03 \lambda_0$. The return loss bandwidth of the radiator stands at 8.1%, while the axial ratio bandwidth stands at 4.8%. Maximum gain of the radiator is measured to be 5.4 dBi. The authors have demonstrated that the proposed antenna is having huge potential in terms of dynamic miniaturization, thin profile, and attractive radiation properties.

In (Ameen, 2019), a Metamaterial inspired wideband CP antenna is discussed. The antenna uses rotated V-shaped Metasurfaces to make the system suitable for small satellite applications. The design concept is based on composite metamaterials, along with modified coplanar waveguide based feeding technique. The unit cell has a dimension of $0.088 \lambda_0 \times 0.19 \lambda_0$, at a central frequency of 2.4 GHz, while the overall antenna dimensions are $0.37 \lambda_0 \times 0.37 \lambda_0 \times 0.18 \lambda_0$, at 2.5 GHz. Parametric measurements of the designed prototype show an acceptable return loss bandwidth of 860 MHz (33.06%), while a considerable AR bandwidth of 411 MHz (15.92%). An enhanced gain of 5.76 dBic is recorded within the functional bandwidth of the antenna. As it is known that the antennas that find themselves useful for on-board satellite communication applications, need to have a decent gain. Having that consideration as motive, and successfully being able to implement that, this antenna is hence well suitable for satellite communication applications.

In order to contribute to the rapidly growing and sophisticated field of circularly polarized antennas, a comparative study is presented.

Table 4. A comparison of various Metamaterial based Circularly Polarized Antennas

Cited papers	Antenna dimensions(in mm)	Type of feed	Impedance Bandwidth	Axial Ratio Bandwidth	Observations
(Park, 2011)	Exterior Diameter = 56 Interior Diameter = 40 Substrate profile = 3.175	Coaxial probe feed	0.9%	0.4%	The structure is extremely complex, and the gain realized is only about -0.4 dBic, which is also quite low.
(Ko, 2013)	60×60×3.17	Coaxial probe feed	Lower band = 2.9% Upper band = 0.6%	Lower band = 0.7% Upper band = 0.18%	The suggested model achieves a very short operational bandwidth. Furthermore, for each band, radiation efficiencies of 82% and 65%, respectively, are reported.
(Wang, 2021)	150×150×10	Microstrip line and slot coupled feed	4.8%	1.21%	As compared to the structural complexity involved, in terms of CRLH TL, metallic screws and MIM capacitor loading, the achieved bandwidth could have been more enhanced
(Ameen, 2018)	20×20×2.1	Coaxial probe feed	10.86%	2.54%	The authors assert that the innovative aspect of the suggested antenna is its small size, low profile, and broad bandwidth.

5. Multiband Circularly Polarized Antennas Having Novel Structures and Band Specific Applications

Multiband CP antennas with structural compactness and large bandwidth attract considerable interest because of the increased system requirements of contemporary wireless communication applications. Multiband antenna technologies with circularly polarized radiation patterns are also becoming more popular to accommodate various particular bandwidth ranges. An antenna with an annular patch and a "T"-shaped slit in (Li, 2012) displays decent axial ratio bandwidths at various frequency bands. Although multiband, it has been shown that the axial ratio bandwidths at the various working bands are rather small. Artificial Magnetic Conductor can considerably increase the CP operational bandwidth while reducing the physical dimensions of the antenna. However, this design strategy complicates the antenna structure, making it inappropriate for use in integrated devices. This issue can be solved successfully by using parasitic patches in addition to the radiating patch. A C-shaped parasitic patch-enhanced dual CP broadband antenna is presented in.

A low-profile dual band antenna for wearable applications that are both indoor and outdoor is suggested in (Paracha, 2019). This antenna reduces its size by 15% compared to other electromagnetically coupled CP antenna designs, reported in (Torres, 2015) and by roughly 63.86% when compared to a conventional CP slotted patch antenna. Complimentary Split Ring Resonator (CSRR) is used in this antenna (Paracha, 2019) to both reduce size and enable CP characteristics in the lower band. Additionally, a small, multi-band AMC unit cell is employed to increase the antenna's gain and front-to-back ratio (FBR). Gain was improved by up to 3.47 dBi as a result, and the FBR was increased by up to 21.88 dB in the former band and up to 22 dB in the latter band, which is pretty substantial.

In (Altaf, 2018), a brand-new slanted D-shaped monopole antenna is suggested with enhanced double-band and double-sense circular polarization. By fusing two inverted C-shaped and slanted I-shaped structures that are capacitively coupled to one another, the tilted D-shaped design is created. Over the entire operational band, cross-polarization discrimination is demonstrated to be above 20 dB, which defends the purity of circular polarization.

In (Kumar, 2019), by using a dual-band, circularly polarized antenna with microstrip line feeding and polarization reconfigurability for WLAN applications, a solution to the previously discussed problem was found. In one of the bands, the antenna radiates RHCP, while in the other, LHCP. On the feed line, there are four PIN diodes the switching states of which are changed to modify the nature of polarization. This is apparently the debut method to produce a double-band, double-sense CP reconfigurable antenna that has been reported. The obtained axial ratio bandwidths in the two bands are, respectively, 16.6% and 5.7%.

In (Wang, 2012), a further simple design for a single-fed, triple band annular slot antenna is described. Using an L-shaped series step impedance feed setup, the antenna is fed. The two annular slots etched out of the two slots can have their dimensions altered to tailor the antenna's frequency ratio. The measured 3 dB axial ratio bandwidths in the three operational bands are 7.3%, 1.2%, and 5.5%, respectively. Furthermore, the antenna consistently achieves gains of above 4 dBi across all the bands.

In (Lu, 2017), a square-ring tag antenna with circular polarization was suggested. The antenna's axial ratio bandwidth is around 36 MHz (902 MHz - 938 MHz), which covers the American, European, and Taiwanese UHF RFID frequency bands (902 MHz - 928 MHz). In addition to having a simple structure, the antenna is approximately 64% reduced in size than other traditional tag antennas.

In (Ferreira, 2017), a small, dual circularly polarized, cavity-backed ring slot antenna is suggested. The feeding network, which is made up of a few T-shaped capacitive feed structures, is coupled to a miniature hybrid branch line coupler, making the design complicated. Antenna matching also benefits from the capacitive feed network. Measured axial ratio bandwidth is 4.8%, while measured impedance bandwidth is 12.0%. A steady gain of up to 7.4 dBi is obtained because of the cavity backing, which is rather appreciable. Additionally, it appears that the planned antenna has a substantially better front-to-back (FBR) ratio.

The authors of (Karthikeyan, 2017) describe a brand-new broadband dual circularly polarized microstrip fed monopole antenna. The radiating monopole has a remarkably distinctive shape; it initially appears as a square monopole and later endured further modification by the addition of bevels to the lower edges of the monopole. On account of this, the monopole adopts the geometry of an asymmetric hexagon. Two feed lines that are orthogonal to one another feed the antenna, while the overall symmetry of the structure ensures polarization diversity. The axial ratio bandwidth is 80.7%, while the resulting impedance bandwidth is 149%. The antenna can be treated as a highly potential candidate in WLAN (5 GHz, 5.2 GHz, and 5.8 GHz), WiMAX (5.5 GHz), and some C-band applications in various multipath scenarios.

A sophisticated design strategy to achieve circular polarization of a microstrip antenna, excited through unbalanced vertical feed lines is provided by the authors in (Thakur, 2006). In order to accomplish circular polarization, the design additionally includes Defected Ground Structures (DGS) underneath the imbalanced feed lines. In practice, the DGS is used to adjust the impedance of the feed line in order to maximize the CP bandwidth. By utilizing DGS, the proposed structure eliminates the design challenges, that are associated with using hybrid couplers with diagonally balanced feed lines. While the axial ratio bandwidth is just 1.9%, the obtained return loss bandwidth is only 7.3%.

In (Karmakar, 2019), a compact, CPW-fed, triple band CP ultra-wide band monopole antenna is reported. To achieve an ultra-wide bandwidth of 112%, the design incorporates a rectangular stub with modified I-shaped and E-shaped slots on the right side of the ground plane and monopole, respectively (3.1–11 GHz). The antenna's overall dimension is $(27 \times 27 \times 1)$ mm^3.

Table 5. Comparative Analysis of various multi band Circularly Polarized Antennas having structural novelty

Reference	Overall Dimension (mm³)	Type of Feed	Return Loss Bandwidth	Axial Ratio Bandwidth
(Ding, 2015)	25×25×1	Microstrip line	87.7%	65.2%
(Yang, 2015)	50×50×0.5	Multi-port probe feed	19.6%	15.3%
(Xiong, 2018)	60×60×0.8	CPW feed	117.3%	Double band:- 32,2% and 3.8%
(Paracha, 2019)	70.4×76.14×3.11	Y-shaped Electromagnetic coupled Feed line	Double band:- 1.84% and 0.736%	0.83%
(Abed, 2018)	18×18×0.8	Double microstrip line feed	Double band:- 4%; 21.5%	Double band:- 4%; 5.1%
(Karmakar, 2019)	27×27×1	CPW feed	112%	Triband:- 6.09%; 15.35%; 16%

The proposed antenna's surface current distribution for the second CP band's central frequency of 5.52 GHz is shown in Figure 12(b). At 0° phase, the triangular stepped ring resonator, upper rectangular ring resonator, and modified "T" and "I"-shaped slots are where the current density predominates. The dominant currents on the surface of the modified "T"- and "I"-shaped slot are ignored since they are influenced by equivalent components of one another, which affect the horizontal and vertical currents on that surface. Therefore, it is clear from the analysis that when the phase progresses from 0° to 90°, the dominant current vector is rotating in a clockwise direction, as depicted in figure 12(c), which results in LHCP response from the +Z direction and RHCP response from the Z direction. The creation of a small printed Ultra Wideband monopole antenna with coupled tri-band circular polarization characteristics that can enjoy the benefit of independent control by altering the antenna design parameters is what makes this structure novel. The antenna structure, and the results are shown in Figure 12 (a – d).

Figure 12. Figure showing (a) geometry of a multi-band CP Antenna and its Surface current distribution at (b) 5.52 GHz at 0°phase (c) 5.52 GHz at 90° phase (d) Axial Ratio Bandwidths (Karmakar, 2019)

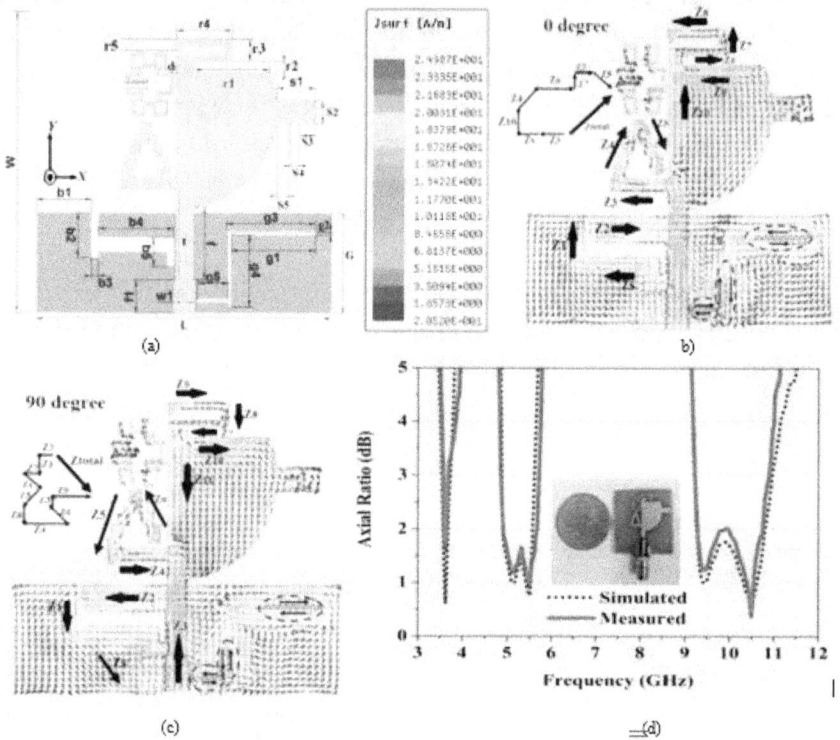

CONCLUSION

A thorough analysis of the various uses for circularly polarized antenna systems in contemporary wireless communications has been provided in this article. Although there is an unending list of noteworthy study in this topic, with due respect for the efforts of all the scholars who continuously strive in this domain, the discourse is kept to a minimum for the review's brevity. In order to get over the drawbacks that linearly polarized systems inevitably have, like loss from mismatch in polarization, vulnerability to multipath and fading effects, Faraday rotation, and bad weather, among others, the review demonstrates that circularly polarized antennas are the inevitable choice. However, majority of the traditional CP antennas featured narrowband characteristics. This sparked attention in the quest for potential solutions for boosting the bandwidth. Few concepts are highlighted in this article. Additionally, it was found that attempts to improve bandwidth frequently resulted in bulky overall structures that rendered the design inappropriate for multi-level system integration. The goal of research is presently focused to reduce the antenna volume while maintaining CP radiation. As a result, many design strategies emerged, including the use of CP DRAs, the incorporation of fractal shapes, and metamaterial loading. In this article, a few design strategies are discussed. Additionally, it has been demonstrated that DRAs with notably less metallization can improve gain of an antenna and operational efficiency, however maintaining circular polarization responsiveness. Ultimately, it has been stated that the use of circularly polarized antennas in specific narrow band applications has ignited the creation of double or multi-band antennas.

REFERENCES

Abed, A. T., Singh, M. S. J., & Islam, M. T. (2018). Compact fractal antenna circularly polarized radiation for wi-fi and WiMAX communications. *IET Microwaves, Antennas & Propagation*, *12*(14), 2218–2224. doi:10.1049/iet-map.2018.5213

Agrawal, S., Gupta, R. D., Parihar, M. S., & Kondekar, P. N. (2017). A wideband high gain dielectric resonator antenna for RF energy harvesting applications. *AEÜ. International Journal of Electronics and Communications*, *78*, 24–31. doi:10.1016/j.aeue.2017.05.018

Altaf, A., Seo, M., & Tilted, A. (2018). D-shaped monopole antenna with wide dual-band dual-sense circular polarization. *IEEE Antennas and Wireless Propagation Letters*, *12*(12), 2464–2468. doi:10.1109/LAWP.2018.2878334

Ameen, M., & Chaudhary, R. K. (2019). Metamaterial-based wideband circularly polarised antenna with rotated V-shaped metasurface for small satellite applications. *Electronics Letters*, *55*(7), 365–366. doi:10.1049/el.2018.7348

Balanis, C. A. (2005). *Antenna Theory: Analysis and Design* (3rd ed.). John Wiley & Sons, Inc.

Banerjee, U., Karmakar, A., & Saha, A. (2020). A review on Circularly Polarized Antennas, trends and advances. *International Journal of Microwave and Wireless Technologies*, *12*(9), 922–943. doi:10.1017/S1759078720000331

Banerjee, U., Karmakar, A., Saha, A., & Chakraborty, P. (2019). A CPW-fed compact monopole antenna with defected ground structure and modified parasitic hilbert strip having wideband circular polarization. *AEÜ. International Journal of Electronics and Communications*, *110*, 152831. doi:10.1016/j.aeue.2019.152831

Blanco, D., Rajo-Iglesias, E., Maci, S., & Llombart, N. (2015). Directivity enhancement and spurious radiation suppression in leaky-wave antennas using inductive grid metasurfaces. *IEEE Transactions on Antennas and Propagation*, *63*(3), 891–900. doi:10.1109/TAP.2014.2387422

Chair, R., Yang, S. L. S., Kishk, A. A., Lee, K. F., & Luk, K. M. (2006). Aperture fed wideband Circularly Polarized Rectangular Stair-shaped Dielectric Resonator Antenna. *IEEE Transactions on Antennas and Propagation*, *54*(4), 1350–1352. doi:10.1109/TAP.2006.872665

Chaudhary, P., Kumar, A., & Yadav, A. (2020). Pattern diversity MIMO 4G and 5G wideband circularly polarized antenna with integrated LTE band for mobile handset. *Progress in Electromagnetics Research M. Pier M, 89*, 111–120. doi:10.2528/PIERM19111202

Chen, W.-S., Wu, C.-K., & Wong, K.-L. (2001). Novel compact circularly polarized square microstrip antenna. *IEEE Transactions on Antennas and Propagation, 49*(3), 340–342. doi:10.1109/8.918606

Ding, K., Gao, C., Yu, T., & Qu, D. (2015). Broadband C-shaped circularly polarized monopole antenna. *IEEE Transactions on Antennas and Propagation, 63*(2), 785–790. doi:10.1109/TAP.2014.2380437

D.S., C., & Karthikeyan, S. S. (2017). A novel broadband dual circularly polarized microstrip-fed monopole antenna. *IEEE Transactions on Antennas and Propagation, 65*(3), 1410–1415. doi:10.1109/TAP.2016.2647705

Esselle, K. P. (1996). Circularly polarized higher order rectangular dielectric resonator antenna. *Electronics Letters, 32*(3), 150–151. doi:10.1049/el:19960171

Farswan, A., Gautam, A. K., Kanaujia, B. K., & Rambabu, K. (2016). Design of Koch Fractal Circularly Polarized Antenna for Handheld UHF RFID reader Applications. *IEEE Transactions on Antennas and Propagation, 64*(2), 771–775. doi:10.1109/TAP.2015.2505001

Ferreira, R., Joubert, J., & Odendaal, J. W. (2017). A compact dual-circularly polarized cavity-backed ring-slot antenna. *IEEE Transactions on Antennas and Propagation, 65*(1), 364–368. doi:10.1109/TAP.2016.2623654

Fu, S., Kong, Q., Fang, S., & Wang, Z. (2014, May). Broadband circularly polarized microstrip antenna with coplanar parasitic ring slot patch for L-band satellite system application. *IEEE Antennas and Wireless Propagation Letters, 13*, 943–946. doi:10.1109/LAWP.2014.2323113

Gupta, S., Patil, S., Dalela, C., & Kanaujia, B. K. (2021). Analysis and Design of Fractal Defected Ground Based Circularly Polarized Antenna for CA band Applications. *International Journal of Microwave and Wireless Technologies, 13*(4), 397–406. doi:10.1017/S1759078720001142

Han, R., Zhong, S. S., & Liu, J. (2014). Broadband Circularly Polarized Dielectric Resonator Antenna fed by wideband Switched Line Coupler. *Electronics Letters, 50*(10), 725–726. doi:10.1049/el.2014.0809

Haneishi, M., Nambara, T., & Yoshida, S. (1982). Study on ellipticity properties of single-feed-type circularly polarised microstrip antennas. *Electronics Letters, 18*(5), 191–193. doi:10.1049/el:19820132

Heidari, A. A., Heyrani, M., & Nakhkash, M. (2009). A Dual – band Circularly Polarized stub loaded microstrip patch antenna for GPS applications. *Progress in Electromagnetics Research, 92*, 195–208. doi:10.2528/PIER09032401

Heydari, S., Jahangiri, P., Arezoomand, A. S., & Zarrabi, F. B. (2016). Circular polarization fractal slot by Jerusalem cross slot for wireless applications. *Progress in Electromagnetics Research Letters, 63*, 79–84. doi:10.2528/PIERL16070802

Ittipiboon, A., Roscoe, D., Mongia, R. K., & Cuhaci, M. (1994). Circularly polarized dielectric resonator antenna. *Electronics Letters, 30*(17), 1361–1362. doi:10.1049/el:19940968

James, J. R., & Hall, P. P. (1989). *Handbook of Microstrip Antenna*. Peter Peregrinus Ltd.

Jhajharia, T., Tiwari, V., Yadav, D., Rawat, S., & Bhatnagar, D. (2018). Wideband Circularly Polarised Antenna with an asymmetric meandered-shaped monopole and defected ground structurefor wireless communuication. *IET Microwaves, Antennas & Propagation, 12*(9), 1554–1558. doi:10.1049/iet-map.2018.0092

Jou, C. F., Wu, J. W., & Wang, C. J. (2009). Novel broadband monopole antennas with dual-band circular polarization. *IEEE Transactions on Antennas and Propagation, 57*(4), 1027–1034. doi:10.1109/TAP.2009.2015827

Joy, S., Natarajamani, S., & Vaitheeswaran, S. M. (2018). Minkowski fractal circularly polarized planar antenna for GPS application. In *Proceedings of the 8th International Conference on Advances in Computing and Communication* (ICACC), 143, 66–73.

Karmakar, A., Chakraborty, P., Banerjee, U., & Saha, A. (2019). Combined triple-band circularly polarized and compact UWB monopole antenna. *IET Microwaves, Antennas & Propagation, 13*(9), 1306–1311. doi:10.1049/iet-map.2018.5459

Kumar, P., Dwari, S., Saini, R. K., Mandal, M. K., & Dual-Sense, D.-B. (2019). Dual-Band Dual-Sense Polarization reconfigurable circularly polarized antenna. *IEEE Antennas and Wireless Propagation Letters, 18*(1), 64–68. doi:10.1109/LAWP.2018.2880799

Kumar Mongia, R. K., & Ittipiboon, A. (1997). Theoretical and experimental investigations on rectangular dielectric resonator antennas. *IEEE Transactions on Antennas and Propagation, 45*(9), 1348–1356. doi:10.1109/8.623123

Laisné, A., Gillard, R., & Piton, G. (2002). Circularly polarized dielectric resonator antenna with metallic strip. *Electronics Letters, 38*(3), 106–107. doi:10.1049/el:20020075

Leung, K. W., & Mok, S. K. (2001). Circularly polarized dielectric resonator antenna excited by perturbed annular slot with backing cavity. *Electronics Letters, 37*(15), 934–936. doi:10.1049/el:20010658

Leung, K. W., Wong, W. C., Luk, K. M., & Yung, E. K. N. (2000). Circularly Polarized Dielectric Resonator Antenna excited by dual Conformal Strips. *Electronics Letters, 36*(6), 484–486. doi:10.1049/el:20000453

Li, B., Hao, C.-X., & Sheng, X.-Q. (2009). A dual-mode quadrature-fed wideband circularly polarized dielectric resonator antenna. *IEEE Antennas and Wireless Propagation Letters, 8*, 1036–1038. doi:10.1109/LAWP.2009.2030700

Li, G. H., Zhai, H. Q., T. L., & Liang, C. H. (2012). A Compact antenna with broad bandwidth and quad-sense circular polarization. *IEEE Transactions on Antennas and Wireless Propagation, 11*(7), 761–794.

Li, T. W., Lai, C. L., & Sun, J. S. (2005). Study of Dual-Band Circularly Polarized Microstrip Antenna. In *Proceedings of the European Conference on Wireless Technology*, (pp. 79 – 80). Paris, France: IEEE.

Liang, W., Jiao, Y. C., Luan, Y., Tian, C., & Dual-Band, A. (2015). A Dual-Band Circularly polarized complementary antenna. *IEEE Antennas and Wireless Propagation Letters, 14*, 1153–1156. doi:10.1109/LAWP.2015.2392787

Liang, Z., Li, Y., & Long, Y. (2014). Multiband monopole mobile phone antenna with circular polarization for GNSS application. *IEEE Transactions on Antennas and Propagation, 62*(4), 1910–1917. doi:10.1109/TAP.2014.2299821

Long, S., McAllister, M., & Liang, S. (1983). The resonant cylindrical dielectric cavity antenna. *IEEE Transactions on Antennas and Propagation, 31*(3), 406–412. doi:10.1109/TAP.1983.1143080

Lu, J. H., & Chang, B. S. (2017). Planar compact square-ring tag antenna with circular polarization for UHF RFID applications. *IEEE Transactions on Antennas and Propagation, 65*(2), 432–441. doi:10.1109/TAP.2016.2633162

Ludwig, A. C. (1975). The definition of Cross – polarization. *IEEE Transactions on Antennas and Propagation, 21*(1), 116–119. doi:10.1109/TAP.1973.1140406

Mak, K. M., & Luk, K. M. (2009, October). A circularly polarized antenna with wide axial ratio beamwidth. *IEEE Transactions on Antennas and Propagation, 57*(10), 3309–3312. doi:10.1109/TAP.2009.2029370

Mandelbrot, B. (1982). The fractal geometry of nature, 1186–1189. W. H. Freeman and Company

Nakano, H., Nogami, K., Arai, S., Mimaki, H., & Yamauchi, J. (1986, June). A spiral antenna backed by reflector a conducting plane. *IEEE Transactions on Antennas and Propagation, 34*(6), 791–796. doi:10.1109/TAP.1986.1143893

Nasimuddin, C. Z. N., Chen, Z. N., & Qing, X. (2011, October). Symmetric aperture Antenna for Broadband Circular Polarization. *IEEE Transactions on Antennas and Propagation, 59*(10), 3932–3936. doi:10.1109/TAP.2011.2163757

Oraizi, H., & Hedayati, S. (2012). Circularly polarized multiband microstrip antenna using the square and Giuseppe Peano fractals. *IEEE Transactions on Antennas and Propagation, 60*(7), 3466–3470. doi:10.1109/TAP.2012.2196912

Oraizi, H., & Pazoki, R. (2013, March). Wideband circularly polarized aperture-fed rotated stacked patch antenna. *IEEE Transactions on Antennas and Propagation, 61*(3), 1048–1054. doi:10.1109/TAP.2012.2229378

Paracha, K. N., Abdul Rahim, S. K. A., Soh, P. J., Kamarudin, M. R., Tan, K. G., Lo, Y. C., Islam, M. T., & Low Profile, A. (2019). A Low Profile, Dual-band, dual polarized antenna for indoor/outdoor wearable application. *IEEE Access: Practical Innovations, Open Solutions, 7*, 33277–33288. doi:10.1109/ACCESS.2019.2894330

Park, B. C., & Lee, J. H. (2011). Omnidirectional circularly polarized antenna utilizing zeroth-order resonance of epsilon negative transmission line. *IEEE Transactions on Antennas and Propagation, 59*(7), 2717–2721. doi:10.1109/TAP.2011.2152337

Pourahmadazar, J., Ghobadi, C., Nourinia, J., Felegari, N., & Shirzad, H. (2011). Broadband CPW-Fed circularly polarized square slot antenna with inverted-L strips for UWB applications. *IEEE Antennas and Wireless Propagation Letters, 10*(May), 369–372. doi:10.1109/LAWP.2011.2147271

Pouyanfar, N. (2013). Broadband square-slot circularly polarized antenna for WiMAX and WLAN applications. *Microwave and Optical Technology Letters, 55*(9), 2191–2195. doi:10.1002/mop.27805

Prajapati, P. R., Murthy, G. G. K., Patnaik, A., & Kartikeyan, M. V. (2015). Design and testing of a compact circularly polarized microstrip antenna with fractal defected ground structure for L-band applications. *IET Microwaves, Antennas & Propagation*, *9*(11), 1179–1185. doi:10.1049/iet-map.2014.0596

Rao, P. N., & Sarma, N. V. S. N. (2008). Fractal boundary circularly polarized single feed microstrip antenna. *Electronics Letters*, *44*(12), 1710–1711.

Ravipati, C. B., & Shafai, L. (1999). A wide Bandwidth Circularly Polarized Microstrip Antenna Using a Single Feed, In *Proceeding of the IEEE Antennas and Propagation Society International Symposium*, 1 (pp. 244–247). 10.1109/APS.1999.789126

Reddy, V. V., & Sarma, N. V. S. N. (2014). Single feed circularly polarized poly fractal antenna for wireless applications. *International Journal of Computer and Information Technology*, *8*(11), 1710–1713.

Reddy, V. V., & Sarma, N. V. S. N. (2014). Triband circularly polarized Koch fractal boundary microstrip antenna. *IEEE Antennas and Wireless Propagation Letters*, *13*, 1057–1060. doi:10.1109/LAWP.2014.2327566

Richtmyer, R. D. (1939). Dielectric resonators. *Journal of Applied Physics*, *10*(6), 391–398. doi:10.1063/1.1707320

Row, J. S., & Ai, C. Y. (2004). Compact Design of Single-Feed Circularly Polarised Microstrip Antenna. *IEEE Electronics Letters*, *40*(18), 1093–1094. doi:10.1049/el:20045602

Rui, X., Li, J., & Wei, K. (2016). Dual-band dual-sense circularly polarized square slot antenna with simple structure. *Electronics Letters*, *52*(8), 578–580. doi:10.1049/el.2015.4499

Sitompul, P. P., Sri Sumantyo, J. T., Kurniawan, F., Santosa, C. E., Manik, T., Hattori, K., Gao, S., & Liu, J. Y. (2019). Circularly Polarized Circularly-Slotted-Patch Antenna with Two Asymmetrical Rectangular Truncations for Nanosatellite Antenna. *Progress in Electromagnetics Research, C*, 90, 225–236. doi:10.2528/PIERC18120503

Sudha, T., Vedavathy, T. S., & Bhat, N. (2004). Wideband single-fed circularly polarized patch antenna. *Electronics Letters*, *40*(11), 648–649. doi:10.1049/el:20040407

Thakur, J. P., & Park, J.-S. (2006). An advance design approach for circular polarization of the microstrip antenna with unbalance DGS Feedlines. *IEEE Antennas and Wireless Propagation Letters*, *5*, 101–103. doi:10.1109/LAWP.2006.872425

Torres, A. E., Marante, F., Tazón, A., & Vassal'lo, J. (2015). New microstrip radiator feeding by electromagnetic coupling for circular polarization. *AEÜ. International Journal of Electronics and Communications*, *69*(12), 1880–1884. doi:10.1016/j.aeue.2015.09.016

Ullah, U., Ain, M. F., & Ahmad, Z. A. (2017). A review of wideband circularly polarized dielectric resonator antennas. *China Communications*, *14*(6), 65–79. doi:10.1109/CC.2017.7961364

Varshney, G., Singh, R., Pandey, V. S., & Yaduvanshi, R. S. (2020). Circularly polarized Two-Port MIMO dielectric resonator antenna. *Progress in Electromagnetics Research M. Pier M*, *91*, 19–28. doi:10.2528/PIERM20011003

Wang, L., Guo, Y. X., & Sheng, W. (2012). Tri-band circularly polarized annular slot antenna for GPS and CNSS applications. [Early access]. *IEEE Antennas and Wireless Propagation Letters*, 1–1. doi:10.1109/LAWP.2012.2200869

Wang, Z., Dong, Y., & Itoh, T. (2021). Metamaterial-based, miniaturised circularly polarised antennas for RFID application. *IET Microwaves, Antennas & Propagation*, *15*(6), 547–559. doi:10.1049/mia2.12064

Wang, Z., & Yuandan, D. (2020). A Dual Band Circularly Polarized Ring Antenna based on Composite Right and Left Handed Metamaterials. *IET Microwaves, Antennas & Propagation*, *10*(8), 363–375. doi:10.2528/mia2.08292

Xiong, X., Li, X., Zhang, W., & Zhang, H. (2018). Enhance dual-band circularly polarized broadband antenna by using parasitic patch. *IET Microwaves, Antennas & Propagation*, *12*(13), 2085–2088. doi:10.1049/iet-map.2018.5186

Xu, R., Li, J., Qi, Y. X., Guangwei, Y., & Yang, J. J. (2017). A design of triple-wideband triple-sense circularly polarized square slot antenna. *IEEE Antennas and Wireless Propagation Letters*, *16*, 1763–1766. doi:10.1109/LAWP.2017.2674677

Xu, R., Li, J.-Y., Liu, J., Zhou, S.-G., Xing, Z.-J., & Wei, K. (2018). A design of dual-wideband planar printed antenna for circular polarization diversity by combining slot and monopole modes. *IEEE Transactions on Antennas and Propagation*, *66*(8), 4326–4331. doi:10.1109/TAP.2018.2836670

Yang, S. S., Kishk, A. A., & Lee, K. F. (2008, June). Wideband circularly polarized antenna with L-shaped slot. *IEEE Transactions on Antennas and Propagation*, *56*(6), 1780–1783. doi:10.1109/TAP.2008.923340

Yang, W., Che, W., Jin, H., Feng, W., & Xue, Q. (2015). A polarization-reconfigurable dipole antenna using polarization rotation AMC structure. *IEEE Transactions on Antennas and Propagation*, *63*(12), 5305–5315. doi:10.1109/TAP.2015.2490250

Yong-Xin, G., Lei, B., & Xiang Quan, S. (2009, August). Broadband circularly polarized annular-ring microstrip antenna. *IEEE Transactions on Antennas and Propagation*, *57*(8), 2474–2477. doi:10.1109/TAP.2009.2024584

Chapter 11
Impact of Call Drop Ratio Over 5G Network

Jay Kumar Pandey

https://orcid.org/0000-0003-4086-5730

Shri Ramswaroop Memorial University,India

Shahanawaj Ahamad

University of Hail, Saudi Arabia

Vivek Veeraiah

Adichunchanagiri University, India

Nishchal Adil

Rungta College of Engineering and Technology, India

Dharmesh Dhabliya

Vishwakarma Institute of Information Technology, India

Ashok Koujalagi

https://orcid.org/0000-0002-0195-3976

Godavari Institute of Engineering and Technology (Autonomous), India

Ankur Gupta

https://orcid.org/0000-0002-4651-5830

Vaish College of Engineering, India

ABSTRACT

The 5G network is the main topic of this investigation. 5G is expected to be far more advanced than 4G since it makes use of three distinct bands of the network spectrum. The acronym "5G" refers to the latest generation of wireless communications. With 5G, communications will improve all across the world. The purpose of this research is to examine the consequences of the increasing call drop ratio in 5G networks. In other words, if a user's current session is interrupted, they will need to make a new connection to continue using the service. One area where 5G excels over its predecessors, 4G (and LTE), is in reducing latency. Also, there would be fewer lost calls for individuals utilizing VoIP because network uptime will have risen significantly. Reduced call failure rates lead to happier customers.

DOI: 10.4018/978-1-6684-7000-8.ch011

1 INTRODUCTION

5G improves global connection. This study examines the call drop ratio on 5G networks. A call drop indicates a user's ongoing session is dropped, necessitating a new connection to resume services. 5G offers better latency than 4G (and LTE). Increased network uptime means fewer lost VoIP calls. Fewer dropped calls improve customer satisfaction (Aldmour et al., 2017). In section 1 the call drop ratio and key performance indicator optimization for call drop ratio are considered. Then Radio-Induced Call Drops is elaborated with hardware execution failure. Then evaluation of wireless toward 5G along with most compelling aspects is presented. Section also considers optimum 5G and performance indicators of 5G such as throughput, deployment, mobility, connected devices, energy efficiency, data volume, latency and reliability. Section 2 is considered research related to 5G network. In the field of mobile cellular networks, there have been several studies that have investigated the 5g network. A few of the authors focused their attention on the percentage of dropped calls on 5G networks. Section 3 is focused on the challenges and issues faced by 5G networks (Anand et al., 2018).

Those research faced issues related to call drop, performance, error rate and accuracy. Section 4 is presenting the process flow of overall work where WSN, localization, machine learning and call drop ratio related research work are studied and problems such as call drop, performance, error, accuracy are identified. Then a novel approach for mobile node localization has been proposed that is making use of machine learning and considering call drop ratio. Finally the performance and accuracy evaluation has been made in order to find the reliability of work. Section 5 is presenting the simulation results for optimized and non optimized dataset. Section 6 is considering conclusion while section 7 is focused on future scope (Al-Maitah et al., 2018).

1.1 Call Drop Ratio

The percentage of phone conversations that are disconnected mid-conversation due to technical difficulties is known as the dropped-call rate (DCR) in the field of telecommunications. Typically, this proportion is expressed as a share (or percentage) of total calls.

$$\text{Call drop rate} \frac{\text{number of call drop times}}{\text{number of call setup success times}}$$

The number of call setup success times+1: After the Alerting message is received.

1.2 KPI (Key Performance Indicator) Optimization for Call Drop Ratio

The percentage of calls lost over LTE is a crucial key performance indicator. Since the advent of VoLTE, this KPI has taken on increased significance in LTE, and as a result, every network is working to enhance it. A call drop in LTE is the interruption of an active session, necessitating the establishment of a new connection before further use is possible. Upon inspection of the Context Release message, the eNB will recognize this as an anomalous release for reasons indicated by the cause code (Anioke et al., 2015).

1.3 Radio-Induced Call Drops

a)   DL RLC Retransmissions

The most common decrease that can be traced back to radio issues is the result of RLC retransmissions. If in the downlink, a network only permits a maximum of 16 RLC retransmissions, the eNB will try resending the message 16 times at the RLC layer before declaring an RLF if the UE still cannot decode it or sending an acknowledgment. If a vendor has initiated a UE Context Release, an unusual release and call drop could occur at this time. While the eNB may "peg" a radio-induced call loss and make room for new traffic during DL retransmissions, it cannot do so during UL RLC retransmissions. When a UL RLC retransmission occurs, however, the eNB is typically ignorant that the UE has reached RLF circumstances until the UE sends an RRC Reestablishment Request. Problems in coverage and quality are mostly to blame (Babu et al., 2022). Optimizing the physical environment around the radio transmitter and receiver is the most effective method for solving this problem. Increasing the value of the RLC retransmission threshold may potentially help with this problem (Dhanya et al., 2016).

b)   Handover Execution Failure

Take the case of an unsuccessful handover attempt by UE. The UE will initiate an RRC Reestablishment with the reasonable value of Handover failure when the amount of time provided by T304 has passed (Dighriri et al., 2017). If this RRC Reestablishment attempt also fails and the UE is still unable to establish a connection, the source eNB will release the context once the internal timeout has elapsed. A new version will be released soon since the X2 Reloc Overall Timer is about to run out. In this case, the failure of the call is considered to be a Handover failure. Two common reasons for this failure are a cell radius that is too narrow to fit the UE's position and a target cell that is too far away for the UE to reach when the handover begins. Therefore, the handover fails because the destination cell could not correctly decode the UE's specialized RACH. The solution is to either raise the target cell's cell radius or down-tilt it so that it does not go beyond its intended range (Eluwole et al., 2018). Signals readily tend to get reflected across the water; therefore, this can also occur in regions with vast water bodies. Handovers should be prevented to these cells or made more challenging by adding offsets for them.

c)   Drop Due to No Response

A signaling message, such as RRC Reconfiguration, sent by an eNB requires a reply from the user equipment (UE). The eNB initiates a release if the UE fails to send an RRC Reconfiguration Complete message before the eNB's internal timer expires. Given the often high value of this timeframe, this decrease stands out. If you see drops of this sort, you may want to double-check that the eNB's internal timer is not set to an absurdly low value. Keep in mind that RRC Reconfiguration for Mobility Command requires the UE to react to the destination cell rather than the source cell, thus this method cannot be used in that scenario (Gaur et al., 2016)

1.4 5G

As the name implies, 5G is the fifth generation of mobile network technology. Using three different parts of the network spectrum, 5G is set to represent a significant improvement over the previous generation, 4G. The acronym "5G" stands for "Fifth Generation." It's equipped with cutting-edge capabilities that might be used to address a wide range of everyday issues (Khan et al., 2018). It's helpful for the government because it streamlines administration; it's useful for students because it makes cutting-edge educational resources accessible 24/7, and it's helpful for the general public because it makes it possible to access the internet from just about everywhere.

1.5 Evolution of wireless towards 5G

It's no surprise that 4G has replaced 3G. Increased demand for mobile phone service necessitated the development of a more robust and reliable network infrastructure. We are on the cusp of a transition like the one that brought us from 3G to 4G LTE and it's resulting in substantially faster connections. 5G, the fifth generation of mobile networks, is expected to radically alter the landscape of wireless connection (Kliks et al., 2018). Just like 3G made smartphones possible, 5G will pave the way for a plethora of new technologies. Three of the most compelling aspects of 5G are as follows:

a)   Lower Latency

The response time for every smartphone command is not quick. Even though modern smartphones are lightning-quick, they still have some latency. Sometimes it takes a few seconds for a new browser window to appear, and sometimes programs take a while to fully load or don't work as intended. This is because our communication services are sent by signals, and these signals must travel across multiple carriers. The term for this delay is "latency," and if it's too great, it might hinder your device's connection to the Internet. Latency issues can impact even the Hushed app. To combat this, 5G was developed. The delay will be measured in milliseconds, not seconds (Morgado et al., 2018). It will be able to send messages far more quickly and completely than its predecessors. For example, 4G networks frequently broadcast insignificant data snippets. The security of 5G networks will improve (Veeraiah et al., 2022).

b)   VoIP (Voice over Internet Protocol)

VoIP is a type of cloud-based telephony. With VoIP, rather than using the traditional phone company's network, all communications take place online. Those using Voice over IP phone numbers can only make and receive calls while they are connected to the Internet. They require minimal latency and a reliable connection. Call quality and connection stability will suffer on networks with high latency. The reliability of VoIP calls on 4G networks was significantly higher than on 3G or LTE networks, but it is expected to improve even more on 5G. If VoIP services can become more consistent, users may anticipate reduced latency, improved audio quality, and fewer service disruptions. Better global connection for users of telephony applications like Hushed is one of the promises of 5G. At the moment, many individuals only feel comfortable making VoIP calls via WiFi, but with 5G, they will be able to access the internet even in places where 4G doesn't exist (Sabella et al., 2016).

c) The Internet of Things

5G will bring unprecedented network stability and coverage expansion. There may be as many as a million gadgets per square mile that it can support! To increase their usefulness, commonplace objects may now transmit and receive data. A term for this is "the Internet of Things." It is no longer necessary to wait for the morning traffic report since your automobile may exchange information with other vehicles and the GPS system in real-time. As the temperature outside drops, the smart thermostat will automatically regulate itself. The mirror in the hallway tells you to get an umbrella since it's pouring outside. A building site where everyone works from afar. Some of these predictions seem plausible now, while others, like flying automobiles and jetpacks, do not. However, with 5G, it's feasible to do all of those things in theory (Singh et al., 2017).

1.5.1 Optimum 5G

The term "5G NR B1 Thresholds" describes this cutoff. Most existing 5G deployments don't operate in isolation from one another (NSA). Because of this, the mobile device must establish an LTE connection before it can access the 5G network. Connections between 5G nodes and the LTE site should be set up (frequency and cell relations). If a mobile device is 5G-capable, the LTE base station will transmit a B1 Measurement Control to it. The purpose of this B1 measurement control message is to have the mobile device check its 5G signal strength to see if it is inside a 5G coverage area. Mobile devices will submit their measurements to the eNB if they are inside 5G coverage. A 5G node for the mobile device will be added by the eNB via X2 signaling after the measurements are complete (SgNB Addition procedure). When the 5G NR leg is finally installed, the mobile device will connect to it using the RACH method. An uplink handshake is a method of establishing synchronization between mobile devices. To determine the TA (Timing Advance) value, the mobile transmits a RACH preamble, and the gNB responds with RAR (Tarkaa et al., 2018).

1.5.2 Performance of 5G

When determining if a 5G network lives up to its promise, eight performance indicators must be taken into account.

Figure 1. Performance indicators of 5G

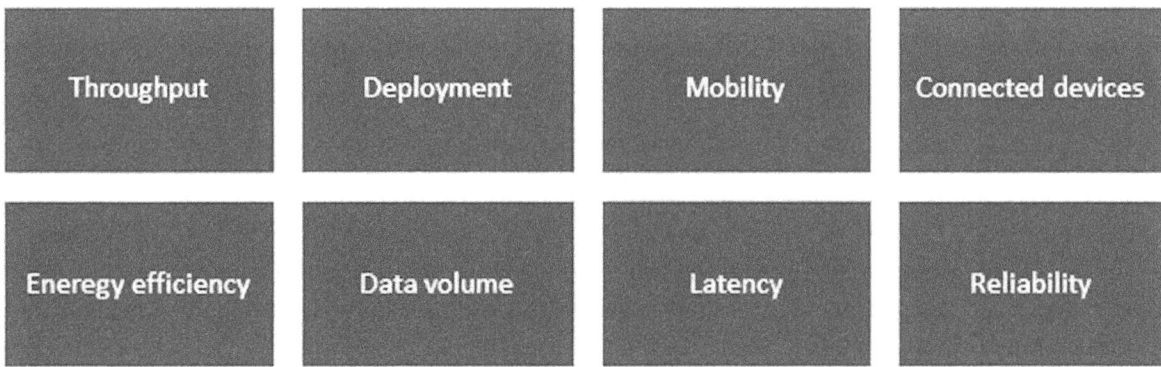

a)   Throughput

There have been multiple firms who have tested the limits of its award-winning 4G network, reaching speeds of up to 1.45 Gbps in 4G LTE Advanced with six-channel carrier aggregation and 953 Mbps in a real-world situation using 4G LTE Licensed Assisted Access (LAA). In theory, 5G should be able to offer speeds multiple times faster than current 4G, opening the door to uses like intelligent video, remote diagnostics, and mobile command centers for live audio and video. One day, users of 5G networks will be able to take advantage of peak data speeds of 10 Gbps. Many carriers' 5G Home services can achieve downlink speeds of 600–800 Mbps and uplink speeds of 250 Mbps, according to third-party tests.

b)   Deployment

As a result of network virtualization, services and applications may be deployed without the need to set up new infrastructure. As a result, the normal time needed to implement a service will drop from six months to just 90 minutes. New features and enhanced security (Anand et al., 2018) may be implemented rapidly thanks to the reduced time required for deployment.

c)   Mobility

Devices traveling at speeds of up to 500 kilometers per hour (310 miles per hour) will still be able to maintain a connection because of advancements in 5G networking. To keep people connected in fast-moving cars and trains, they have tested 5G network handoff mechanisms.

d)   Connected devices

By 2030, there will be more linked gadgets than people on Earth. Up to one million devices per square kilometer will be supported by 5G. Multiple carriers predict that by 2020, 5G Ultra-Wideband would be able to support tens to hundreds of times as many connected devices per square kilometer as 4G. For urban applications like 5G-powered smart lighting, remote security monitoring, intelligent rail, and smart parking systems, this is fantastic news.

e)   Energy efficiency

Several operators see sustainability as a high priority. The energy needed to run a 5G network will be only 10% of what it is now. Furthermore, with 5G, sophisticated operations may be performed within the network, close to the end user. Consequently, less processing power and energy will be required in the device used by the end user.

f)   Data volume

5G is intended to sustain densities of up to 10 terabits per second per square kilometer. As a result, a 5G network can support a huge number of users at once while still transporting a big amount of data. Thus, the high speeds and low latency of 5G service will be available to consumers in densely populated regions such as airports, stadiums, and cities.

g)   Latency

Because of its very low end-to-end latency, 5G will make data transmission speeds quicker than the blink of an eye by a factor of several orders of magnitude. A common lag time on modern 4G LTE networks is around 50 ms. Multiple operators predict that by 2020, the 5G Ultra-Wideband will have a reaction time of less than 10 ms from beginning to finish.

h)   Reliability

The most dependable 4G LTE service in the country is provided by many different companies. They are using the same level of skill and attention to detail in the design and construction of the 5G network.

1.6 Applications of 5G

This section provides details regarding the application areas of 5G:-

1.6.1 Manufacturing

Industrial control and automation systems, planning and design systems, and field devices will all be profoundly affected by 5G, which will have far-reaching consequences for the manufacturing industry as we know it. The use of this technology might help businesses in several ways, including reduced costs, quicker production times, fewer injuries, and increased output.

a)   **Flexibility in operations** – Humans oversee and control operations in real-time, while connected robots with latency as low as 1 ms take charge. Think about how much easier and more efficient things would be! A robot in the factory communicates with its counterpart in the warehouse to control stock and alter production. The need for routine equipment maintenance is highlighted by a second example; the potential applications are staggering. The most cost-effective method of providing IoT services (Anand et al., 2021; Kaur et al., 2023; Tripathi et al., 2023), which may yield a return of 1.5 times the initial investment, involves slicing the 5G network into many virtual networks suited to individual devices and functions.

b)   **Productivity boost** – Productivity advantages from digitalization are magnified the more complicated a manufacturing process is. Increasing productivity by just 1% may have a significant impact on the economy, perhaps adding billions of dollars to GDP. Increased data gathering, more automated processes, and real-time monitoring for greater precision are just a few ways in which 5G might increase industrial output. For instance, 5G has enabled 75%-time savings in the process design phase for MTU Aero Engines. In addition, enabling the usage of augmented reality may assist in the improvement of staff capacities by facilitating the remote training and assistance of hundreds of workers.

c)   **Real-time predictive maintenance** – Manufacturers might lose millions of dollars due to broken machinery if production has to be stopped. To avoid unexpected breakdowns, 5G will make it possible to do predictive maintenance in real-time. To monitor critical equipment characteristics like acceleration, temperature, humidity, etc., and issue timely alarms, 5G networkscanbroadcast4K videos, and IoT data.

d) **Employee health and safety** – The incorporation of 5G sensors into wearables like smart helmets would allow workers to be monitored and led safely through potentially dangerous situations.

e) **Cost savings** – Supply chain integration, predictive maintenance, and increased productivity all contribute to a more cost-effective manufacturing facility. The expenses associated with maintaining an organization's physical infrastructure are cut down further by the fact that less computer power is needed on-site.

However, 5G's actual worth resides in the creativity it may unleash in the manufacturing industry, rather than in its capacity to enhance traditional measurements of industrial performance. New avenues of profit and methods of doing business may be made possible. Manufacturers, for instance, can offer their idle time to others as a manufacturing-as-a-service option provided they have a precise understanding of when their equipment is in operation or undergoing maintenance.

1.6.2 Agriculture

In the future, farms will rely less on pesticides and more on data. Using information collected by sensors planted strategically throughout fields, farmers can zero in on exactly which plots are thirsty, sick, or in need of pest control. Health monitoring for animals may potentially develop as wearables become more affordable and 5G makes it simpler to build networks including huge numbers of IoT devices. Farmers can safely use fewer antibiotics if they have access to more precise health data.

1.6.3 Retail

The quality of the shopping experience for consumers is paramount for 5G retail applications. The shelves of today's stores may not be the norm in the future. Just picture a store that functions more like a showroom, where you can virtually "browse" for products and then "buy" them without ever leaving your seat. Similarly, 5G might be used in stores for real-time stock management and inventory control. Alterations to the conventional checkout line, such as cashier-less establishments that only monitor what you put in your shopping basket, are also possible.

1.6.4 Healthcare

With 5G, healthcare providers and patients will have better communication than ever before. A patient's symptoms might trigger an alarm from a wearable device, such as an internal defibrillator that notifies a team of ER cardiologists to prepare for an incoming patient and provides a full record of data acquired by the device.

1.6.5 Logistics

Stock management is time-consuming, labor-intensive, and costly in the logistics and transport industry. With 5G, it may be possible for more cars to talk to each other and to the infrastructure around them. With 5G, it will be much simpler to keep track of and guide large fleets. An augmented reality system that detects and highlights possible dangers without taking a driver's eyes off the road might be used to aid in navigating.

1.6.6 Autonomous Vehicles

Autonomous cars are cutting-edge technology that will soon be available to the public. Machine Learning's techniques and application areas are continuously being evaluated (Jain et al. 2022). However, the advent of 5G and its accompanying speed, low latency, and expanded deployment, can make this fantasy a reality. The vehicle-to-everything network is the primary motivation for autonomous vehicle communication. In response to the detection of certain environmental cues, the vehicle will take appropriate action without any input from the driver. To deal with pedestrians, roadblocks, road signs, and so on, the vehicle must be able to send and receive data in milliseconds.

1.7 Call Drop Over 5G Network

In comparison to its predecessors, 4G (and LTE) networks, 5G networks have significantly reduced latency. 4G's reduced latency is a major drawback that might degrade call quality. Since most contact centers must deal with a high volume of calls simultaneously, 4G might degrade call quality significantly. Calls made over VoIP services may see fewer lost calls due to the network's improved reliability. Improved customer satisfaction can be directly attributed to a reduction in call failures.

1.7.1 RACH Drops

- Disruptions in 5G connections may be due in part to a phenomenon called random access memory corruption, or RACH.
- Accessibility in other technologies is intriguing. However, 5G NR (NSA) calls will be terminated if the RACH fails.
- This is because a 5G NR RACH only takes place after the 5G NR leg has been successfully set up, i.e. after an ENDC setup has been completed successfully.
- If the RACH fails, the UE will switch to LTE, and the call will be dropped.
- When a UE's RACH attempts have all failed, it sends SCG Failure Information to the anchor eNB, which releases the SCG. The gNB has detected a 5G Drop.

1.7.2 Uplink RLC Drops

- Data retransmissions are another potential cause of 5G call drops.
- If the UE has had a configurable number of transmission or reception failures with the gNB, the UE will begin an SCG Failure Information towards the anchor eNB, leading to SCG Release. On the gNB, this is also represented as a 5G Drop.

1.7.3 Poor Quality Drops

- The connection adaption channels are continuously monitored by the UE to ensure they are functioning properly.
- This means that the UE will constantly be aware of the SSB channel's SINR since it will be constantly monitoring it.

- Quotes occur when the SINR is so low that the UE's projected BLER for PDCCH is more than 10%. If the SSB period is 20 milliseconds, then a Court will occur after 20 milliseconds. (3GPP 38133)
- Accordingly, if N310 is 3, then 3xN310 would take around 60 ms.
- To begin the T310 timer, the UE must detect three consecutive N310 (preconfigured) events. A UE will stop the T310 timer if it can maintain the link (N311) before the duration ends.
- An estimated PDCCH BLER of less than 2% corresponds to a signal-to-noise ratio (SNR) of N311 (3GPP 38133)
- After T310 has elapsed if the UE is still unable to maintain the link, RLF will be initiated.

1.7.4 Downlink RLC Drops

- SCG Release towards the anchor eNB is also initiated by the gNB if the UE has not successfully received data from the gNB a configurable number of times. There will be a 5G drop on the gNB as a result of this.
- The UE reports a problem with the uplink by sending an SCG Failure to the anchor eNB, while the gNB reports a problem with the downlink by detecting a drop and sending an SgNB Release to the anchor eNB.

2 LITERATURE REVIEW

There are several researches in area of mobile cellular network that considered 5g network. Some of the author focused on call drop ratio over 5G networks.

Aldmour et al. (2017) presented wireless broadband tools and their evolution toward 5g networks. This book examines the historical and contemporary technological means by which networks have met rate and traffic needs, and it speculates on the potential role those means may play in future networks. In particular, the next fifth generation (5G) must overcome several opposite problems related to capacity, spectrum, energy, connection, performance, complexity, and cost. Current and historical solutions were contrasted with the suggested instruments presented to solve these difficulties and the manner they operate to accomplish so. Methods including millimeter waves, huge arrays of antennas with various inputs and outputs, barriers between inside and outside, ultra-dense cooperative networks, and greater dependence on users' terminals are also considered. Differences in structure and equipment between present and future 5G are highlighted.

Al-Maitah et al. (2018) explained a hybrid approach to call admission control in 5g networks. Complex challenges in science, technology, and everyday life are often tackled with the help of artificial intelligence systems. Management, optimization, and prediction in the field of communication systems make extensive use of artificial intelligence methods. For 5G mobile communications, supporting a large number of users and providing a high-quality service were two of the most critical requirements for a network's architecture. The quality of service that was delivered depends greatly on the effectiveness of call admission control.

When it comes to mobile cellular networks, Anioke et al. (2015) zeroed emphasis on methods to reduce call drops during handovers. Because mobile devices may be moved from one cell to another, calls can be transferred across cells. If the handoff fails, the call may be disconnected. Techniques for minimizing call drops during handoffs help ensure that both incoming and outgoing calls are treated fairly. To reduce the number of dropped calls, this article investigates what happens when retrial queues and handover priority systems are used together. The results reveal that the number of dropped calls during handovers was reduced to a minimum when call drop-reduction approaches are used.

Dhanya et al. (2016) provided GSM as a standard to describe protocols for 2G networks. An essential key parameter indicator (KPI) of a GSM network, the Call Drop Rate (DCR) is closely related to consumer satisfaction. There is a call drop if the traffic channel is released unexpectedly after being effectively occupied. When the Bit Error Rate is too high, calls might disconnect unexpectedly (BER). In practice, several currently available wireless systems implement a minimum BER before disconnecting a call. The goal of this effort is to lower the percentage of calls that are dropped because of bad BER. To boost the overall system performance, a new signal processing subsystem has been implemented in the receiver. Parts for creating signal-noise mixes, as well as for centering and whitening, were added to the block. Specifically, an independent component analysis (ICA) system was used to compute an un-mixing matrix and conduct automatic identification for determining the differences between the various signals. Inverse correlation analysis (ICA) was a method for unmixing linear mixtures of sources. Separating a multivariate signal into additive subcomponents using this approach was a computational technique based on the assumption of the statistical independence of the individual signals.

Dighriri et al. (2017) looked measurement and classification of smart systems data traffic over 5g mobile networks. This study chapter provides a new model for 5G radio resource data traffic aggregation based on making optimal use of the smallest unit of PRB by combining the data from several M2M devices. In addition, the smart city use case depends on a novel 5G network slicing paradigm, where network slices would differentiate smart systems data flow based on QoS needs. The QoS for data traffic slices will be evaluated using the OPNET. Classes of users who utilize FTP, VoIP, and video are all represented in the simulated 5G data traffic. Popular, sensitive, and high data flow are the three M2M communication "slices" used to classify the circumstances. The results will reveal how the suggested models affect the Quality of Service of 5G data traffic across M2M connections.

Eluwole et al. (2018) reviewed from 1g to 5g, what next. There have been periods of very slow data rates and periods of extremely fast data rates (in the startling magnitude of 1 Tbps), both of which are theoretically feasible with 5G technology. This review article summarises the development from 1G to 5G. Trends in technology as a result of this development are also discussed, along with some of the most pressing problems that experts in the area and tech-savvy people must solve in today's rapidly changing environment. As an added bonus, they also provide a discussion of some possible future directions for R&D.

Call drops are a serious problem in India, and Gaur et al. (2016) provided a comprehensive overview of the problem and potential solutions. It was reasonable to assume that in this day and age of wireless communications; telecom service providers would deliver only the highest quality service. However, this was not the case, and in India, the telecom industry was grappling with a Call Drop Improvement. Nuisance called Call Drops. Call drop rates in India are far higher than the TRAI's recommended threshold. Erlang B Formula, as well as other techniques of proof utilized in other scientific works, such as the Poisson Probability Function with a discrete variable, has been examined to show the mathematical importance of these approaches. Evidence gathered from some sources highlights the seriousness of the situation. Methods like cell splitting and sectoring lead to makeshift gentle handover procedures, which in turn reduce call loss rates. The term "call drops" was defined and several basic but effective solutions are proposed in this study.

Khan et al. (2018) focused on defeating the downgrade attack on identity privacy in 5g. To prevent users' personal information from falling into the hands of IMSI catchers, 3GPP Release 15 (the initial 5G standard) implements privacy protections. Public-key encryption is the foundation of these safeguards. Latter facilitates LTE pseudonym recovery in the event of synchronization failure. When a user's device connects to 5G, this method causes the pseudonyms in both the user's device and the home network to be brought into sync. To protect users' identities, our systems simply need adjustments to the user's gadgets and local area network, since they make use of already LTE and 3GPP Release 15 communications. Furthermore, legal interception calls for minimal providing network patching.

Kliks et al. (2018) looked at perspectives for resource sharing in 5g networks. A commercial launch of 5G networks was scheduled for 2020. However, many problems remain, and standards must be established before the typical end-user may reap their benefits daily. It is expected that 5G technology would be able to better meet the ever-changing demands of its users via increased speed, improved service quality, and more. Since the 5G network's implementation must prioritize economy and adaptability, it was a natural match for virtualization principles that promote the pooling of physical resources for the benefit of many network operators, and applications. In this study, they explore the workings of a model that is expected to arise in such a complex network environment and offer an overview of these notions that emerged from our interactions between academic researchers and active network builders.

Morgado et al. (2018) introduced 5G networks which have made great strides in recent years. Several potential technologies might pave the way for the 5G mobile system age, and they were all now under investigation. The ultimate goal is to develop a mobile network that is inherently adaptable and significantly outperforms conventional mobile systems in every measurable way. Realizing the potential of 5G, will involve the combined efforts of many groups and individuals, including government agencies, standardization organizations, industry forums, mobile network operators, and equipment manufacturers. In this paper, they compile data on 5G from a variety of sources so that they can do two things: I get a bird's-eye view of the technology and (ii) survey the technologies that will make 5G a reality. The progress made on 5G so far, as seen by these various stakeholders, will pave the way for exciting new services and applications to run over next-generation wireless networks.

Sabella et al. (2016) introduced energy management in mobile networks towards 5G. Mobile networks have evolved since the introduction of the first generation of devices, and projections for the next decade show that both the network infrastructure and the traffic it carries will expand. Even while newer systems may be more energy efficient than older ones, the deployment of newer systems on top of existing ones would always raise energy usage. More consumption equals higher expenses, which in turn means lower profits for business owners and a bigger carbon imprint for the whole planet. However, operators would be unable to completely abandon older systems in favor of newer ones because of the continued use of outdated terminals. Because of this, operators must conduct a thorough evaluation of the energy efficiency of their 2G, 3G, and 4G networks by considering the long-term effects of factors like traffic growth, shifts in prevailing paradigms and business models, the advent of next-generation networks (5G, etc.). In this book chapter, we examine the operator's viewpoint on energy efficiency and how it relates to the sustainable growth of mobile networks to 5G. The scope of the research will include all facets of network deployment, including but not limited to equipment development, energy-efficient feature introduction, and associated standardization efforts. (Kaushik et al., 2021)

Singh et al. (2017) reviewed optimizing call drops in the cellular network using an artificial intelligence-based handover schema. Congestion and call drops are commonplace during peak hours on wireless cellular networks, necessitating smart handover management methods for a variety of calls to provide fair load balancing and resource distribution. As it is, the cellular network service provider is up against certain challenges in terms of load balancing and resource sharing for mobile consumers. As a result, the different load balancing and sharing schemes suggested for cellular communication have not proven effective in reducing the occurrence of dropped calls during handovers. In this study, they put forward a smart plan for dividing up work during transitions between nodes. The simulation results demonstrate that the proposed approach was capable of optimizing call drops under significant peak traffic loads through handover management that involves load balancing and sharing.

Tarkaa et al. (2018) focused on a comparative analysis of drop-call probability due to handover and other factors. The term "handover" describes this stage. If the handoff process doesn't go smoothly, the call might get disconnected. However, it has been discovered that electromagnetic causes, erratic user behavior, and abnormal network response were the main causes of call dropping in an established network. These results show that cellular network performance research might benefit from including estimates of the drop-call probability caused by handover and other variables. Therefore, the probability of calls being dropped owing to handover and the other variables are estimated and compared in this study by first using operational data acquired from a GSM network. As the likelihood of a dropped call during handover was higher than the probability of a dropped call for any of the other six MSC regions, the research shows that the GSM network requires more improvement.

Table 1. Literature survey

Author/Year	Title	Methodology	Limitation
Aldmour /2017	The Development of Wireless Broadband Tools for Future 5G Networks.	5G Networks	There is a lack of performance
Al-Maitah / 2018	Call Admission Control in 5G Networks: A Hybrid Approach	5G Networks	There is a lack of performance
Anioke /2015	Methods for Reducing the Frequency of Handoff-Induced Call Losses in Wireless Networks	Call Drop	Lack of technical work
Dhanya /2016	Reducing the Frequency of Missed Cellular Calls.	Call Drop	There is not performed in future
Dighriri /2017	Methods for Quantifying and Categorizing Data Traffic in 5G Mobile Networks for Smart Systems	5G Networks	There is less technical work
Eluwole /2018	What comes after 1G, 2G, 3G, 4G, and 5G networks?	5G Networks	The performance of this research is very low
Gaur /2016	The danger and potential solutions to the increasing frequency of call failures in India are discussed.	Call Drop	There is no implication in future
Khan /2018	Protecting user data against a downgrade attack in 5G networks.	Privacy, 5G Networks	There is considering less technical work
Kliks /2018	5G network resource sharing perspectives.	5G Networks	There was no use of a clustering mechanism
Morgado /2018	Regulatory, standards, and business considerations for 5G technology.	5G Networks	Lack of technical work
Sabella /2016	Cellular networks' energy management for 5G.	5G Network	The scope of this research is very less
Singh /2017	A Smarter Handover Schema for Cellular Networks Based on Artificial Intelligence	Call Drops	Lack of accuracy
Tarkaa/2018	Probability of Call Dropping During Handover and Other Factors Analyzed Side by Side.	Call Drops	Lack of accuracy

3 PROBLEM STATEMENT

Numerous studies in the field of WSN have successfully localized nodes. However, the problem is the rising call drop ratio brought on by node mobility. However, to carry out node localization in an effective way, several researchers examined optimization mechanisms and other machine-learning methodologies. However, the reach of these studies is constrained. Call drop difficulties, performance concerns, error rate, and correctness all need further improvement.

4 PROPOSED WORK

The proposed study took into account research in the fields of call drop ratio, localization, WSN, and machine learning. After then, the issue was discovered. These problems include call drop difficulties, performance challenges, and accuracy and error rate problems. The suggested unique technique uses machine learning to localize mobile nodes while also taking call drop rates into account. The performance, accuracy improvement, and comparison of error rate and accuracy between the suggested model and the standard model have all been evaluated.

Figure 2. Research Methodologies

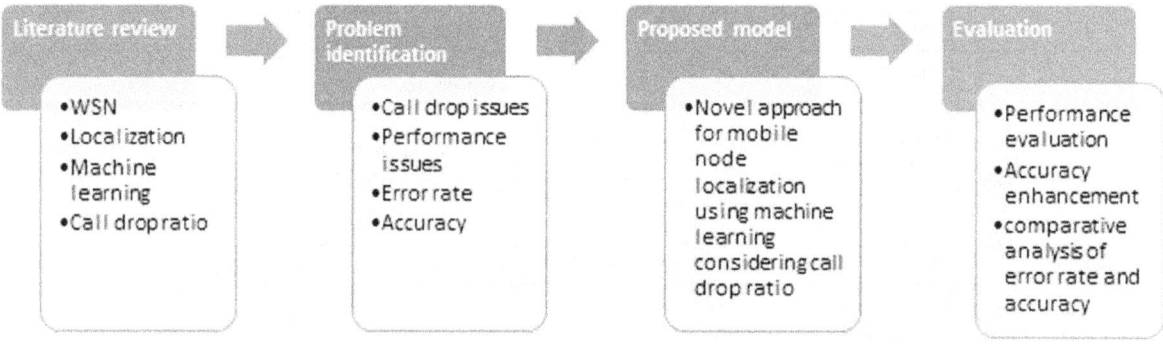

In the proposed approach, the nodes are configured in WSN and taken into account for localization. The current suggested work localizes nodes to determine call drop rates for associated nodes.

Create a dataset and optimize it for training and testing after obtaining the call drop rate. Currently, research is adjusting the batch size, epoch, and hidden layer to initialize machine learning models. Next, the suggested job is to conduct training and testing. The study then does node categorization based on high, low, and moderate call drop ratios. In the end, the assessment is done using the accuracy parameters. Accuracy parameters used in research are recall value, precision and f1-score.

Figure 3. Process flow of proposed work

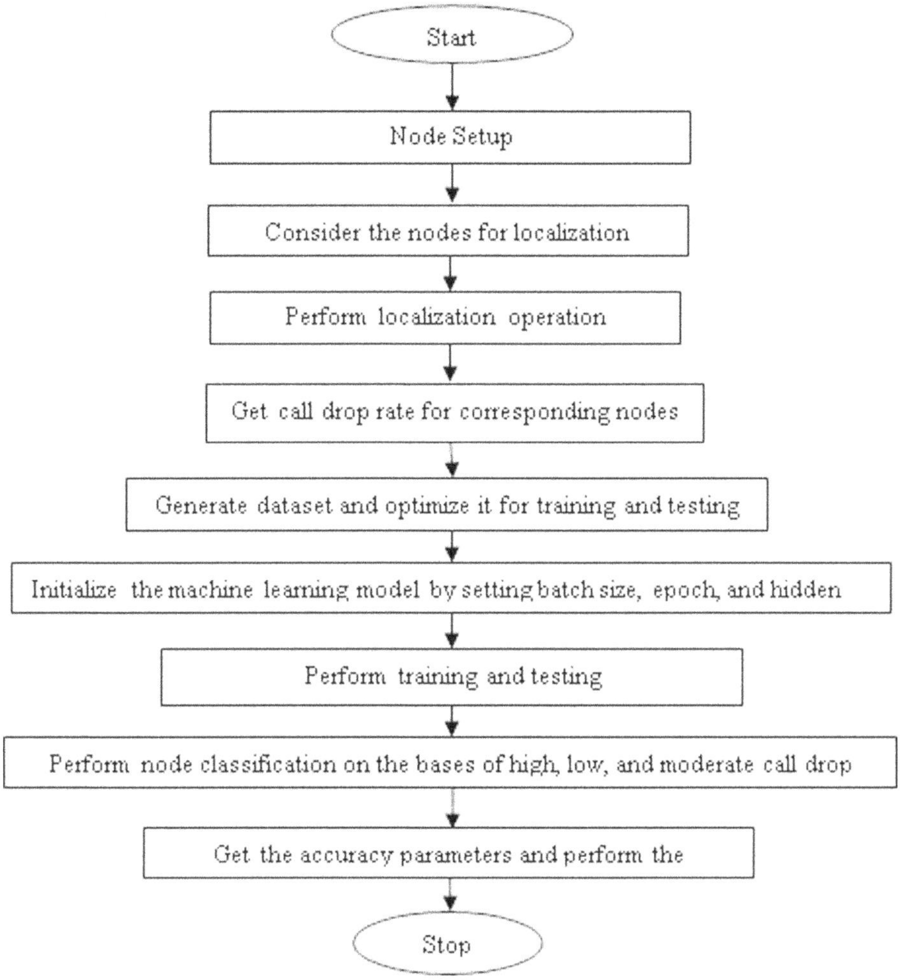

5 RESULTS AND DISCUSSION

Research in the disciplines of call drop ratio, localization, wireless sensor networks (WSN), and machine learning (Garg & Anand, 2011; Khan & Anand, 2022) were taken into consideration for the suggested study. The problem wasn't identified until much later. These issues include difficulties with call drops, performance concerns, accuracy and error rate issues, and other similar issues. The novel approach that was recommended makes use of machine learning to pinpoint mobile nodes, and it also takes into consideration the rates at which calls are dropped. It has been determined whether or not the recommended model performs better than the standard model, whether or not it improves accuracy, and whether or not it has a lower error rate than the standard model. The nodes in the WSN are given configurations and are taken into consideration for localization in the technique that has been presented. In the work that is now being recommended, nodes are localised such that call drop rates for connected nodes may be determined. After determining the call drop rate, the next step in the proposed research is to optimize the dataset for training and testing (Raghavan et al., 2022). When it comes to initializing machine learning models, research is now focusing on modifying the batch size, epoch, and hidden layer. The next job that has been recommended is one that involves training and examination. The next thing that the research performs is classify nodes according to whether they have high, low, or moderate call drop rates. In the end, the evaluation is carried out with the help of the accuracy parameters. Recall value, precision, and f1-score are some examples of accuracy measures that are used in research. This section is considering the accuracy parameters for node classification with and without optimization. Table 2 is presenting the case of non-optimization and Table 3 is presenting its accuracy. Table 4 is presenting the case of optimization and Table 5 is presenting its accuracy.

Table 2. Confusion matrixes for node classification without optimization

	High call drop	Moderate call drop	Minimum call drop
High call drop	901	49	50
Moderate call drop	40	912	48
Minimum call drop	37	34	929

Results
TP: 2742
Overall Accuracy: 91.4%

Table 3. Accuracy parameters for node classification without optimization

Class	n (truth)	n (classified)	Accuracy	Precision	Recall	F1 Score
1	978	1000	94.13%	0.90	0.92	0.91
2	995	1000	94.3%	0.91	0.92	0.91
3	1027	1000	94.37%	0.93	0.90	0.92

Table 4. Confusion matrixes for node classification with optimization

	High call drop	Moderate call drop	Minimum call drop
High call drop	949	25	26
Moderate call drop	20	956	24
Minimum call drop	19	21	960

Results
TP: 2865
Overall Accuracy: 95.5%

Table 5. Accuracy parameters for node classification with optimization

Class	n (truth)	n (classified)	Accuracy	Precision	Recall	F1 Score
1	988	1000	97%	0.95	0.96	0.95
2	1002	1000	97%	0.96	0.95	0.96
3	1010	1000	97%	0.96	0.95	0.96

5.1 Comparison Analysis

Table 6 is presenting the comparison analysis of accuracy, Table 7 is considering precision and Table 8 is considering recall value while Table 9 has considered f1-score.

1.   Accuracy

Taking into account the data in table 6, we can now show the precision of the filtered dataset in comparison to the unfiltered one in figure 4.

The precision of previous work and proposed work are taken for class 1, class2, and class 3 and shown in table 7. It is observed that the Precision of the filtered dataset for the unfiltered dataset.

Table 6. Comparison analysis of accuracy

Class	Node classification without optimization	Node classification with optimization
1	94.13%	97%
2	94.3%	97%
3	94.37%	97%

Figure 4. Comparison analysis of accuracy

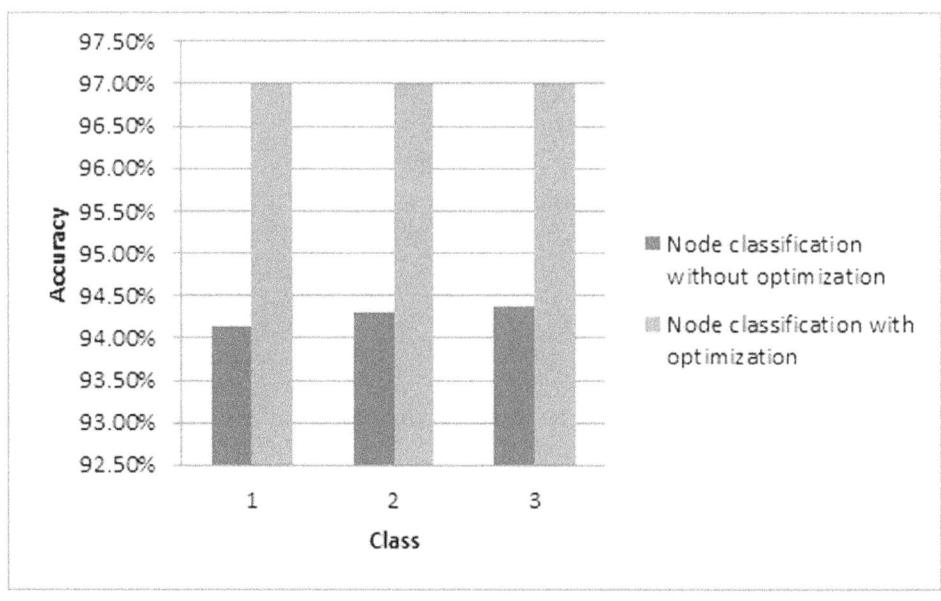

2.   Precision

Considering table 7, figure 5 is drawn to visualize the precision of the filtered dataset for an unfiltered dataset.

The recall value of previous work and proposed work are taken for class 1, class 2, and class 3 and shown in table 8. It is observed that the Recall value of the filtered dataset for the unfiltered dataset.

Table 7. Comparison analysis of precision

Class	Node classification without optimization	Node classification with optimization
1	0.90	0.95
2	0.91	0.96
3	0.93	0.96

Figure 5. Comparison analysis of precision

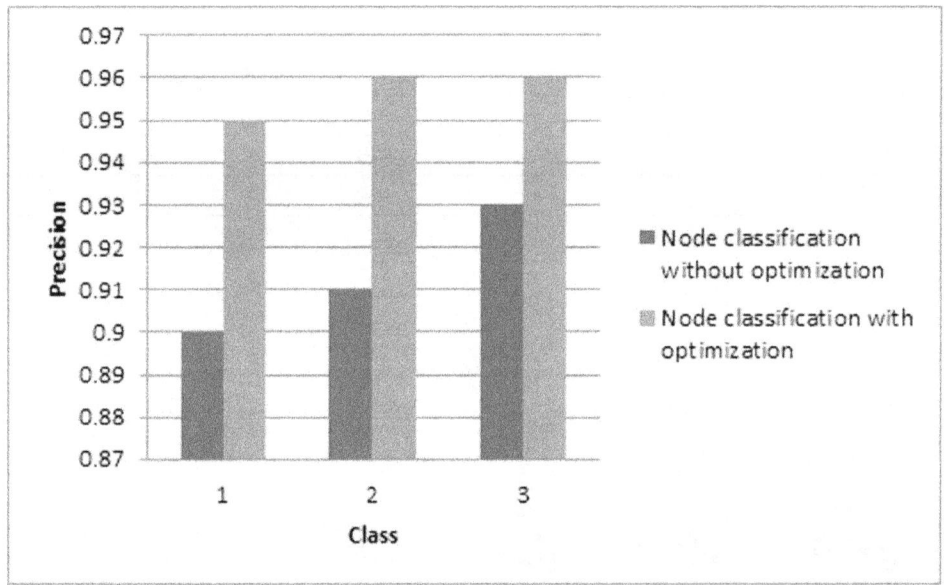

3.   Recall Value

Considering table 8, figure 6 is drawn to visualize the recall value of the filtered dataset for the unfiltered dataset.

F1-Score of previous work and proposed work are taken for class 1, class2, and class 3 and shown in table 9. It is observed that the F1-Score of the filtered dataset is for an unfiltered dataset.

Table 8. Comparison analyses of recall value

Class	Node classification without optimization	Node classification with optimization
1	0.92	0.96
2	0.92	0.95
3	0.90	0.95

Figure 6. Comparison analysis of recall value

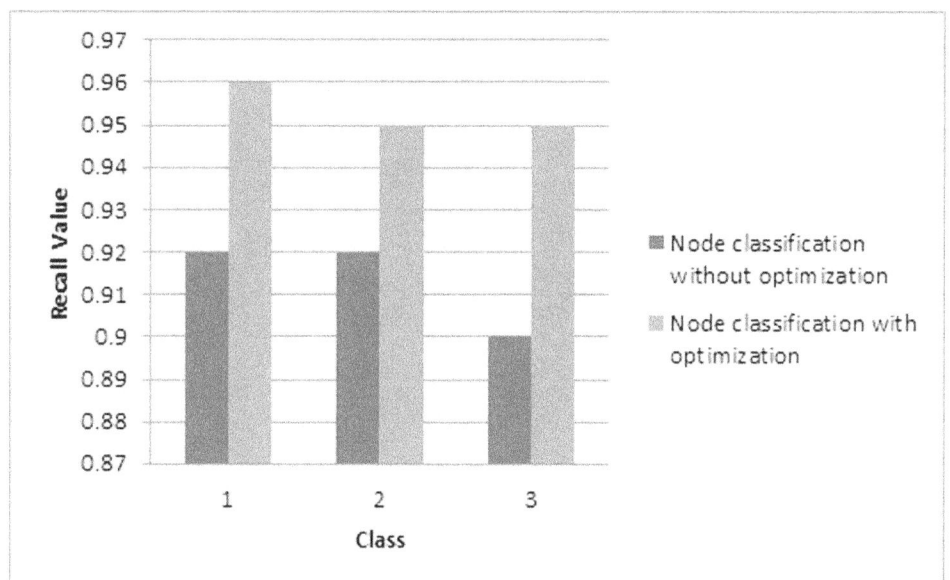

4.   F1-Score

Considering table 9, figure 7 is drawn to visualize the F1-Score of the filtered dataset for the unfiltered dataset.

Table 9. Comparison analyses of f1-score

Class	Node classification without optimization	Node classification with optimization
1	0.91	0.95
2	0.91	0.96
3	0.92	0.96

6 CONCLUSION

According to the simulation's findings, node classification with optimization is more accurate than without optimization. Recall value, f1-score, and precision are superior accuracy metrics in the case of the suggested optimal strategy. As a result, the suggested hybrid strategy is more adaptable, scalable, and technically possible. (Gupta et al., 2021)

Figure 7. Comparison analyses of f1-score

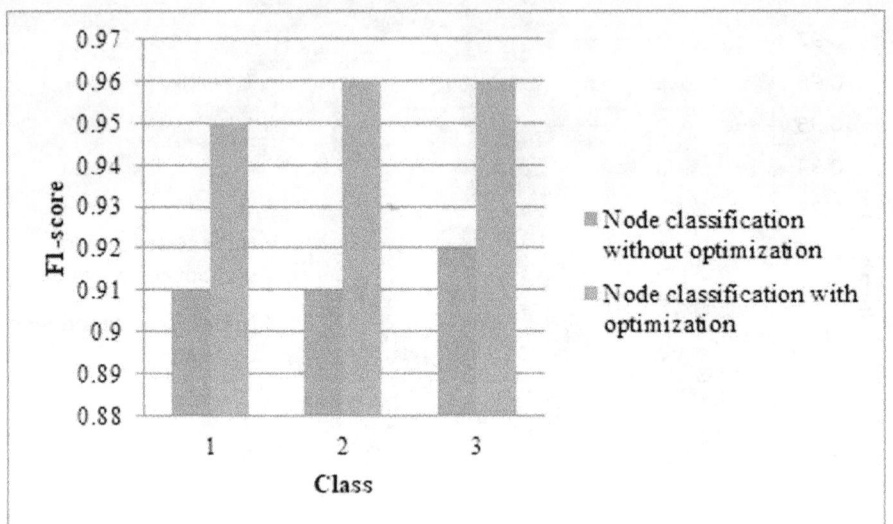

7 FUTURE SCOPE

Scalable and adaptable machine learning has been made possible by the proposed methodology. System dependability has increased thanks to an optimized strategy. Such a technology would be able to provide an effective underwater transmission option. Such clever strategies might control the call drop percentage during the localization of mobile nodes in underwater wireless sensor networks (Chibber et al., 2021; Meivel et al, 2023).

REFERENCES

Al-Maitah, M., Semenova, O. O., Semenov, A. O., Kulakov, P. I., & Kucheruk, V. Y. (2018). A hybrid approach to call admission control in 5G networks. *Advances in Fuzzy Systems*, *2018*, 1–7. doi:10.1155/2018/2535127

Aldmour, I. (2017). Wireless Broadband Tools and Their Evolution Towards 5G Networks. *Wireless Personal Communications*, *95*(4), 4185–4210. doi:10.100711277-017-4058-x

Anand, R., Shrivastava, G., Gupta, S., Peng, S. L., & Sindhwani, N. (2018). Audio watermarking with reduced number of random samples. In *Handbook of Research on Network Forensics and Analysis Techniques* (pp. 372–394). IGI Global. doi:10.4018/978-1-5225-4100-4.ch020

Anand, R., Sindhwani, N., & Saini, A. (2021). Emerging Technologies for COVID-19. *Enabling Healthcare 4.0 for Pandemics: A Roadmap Using AI, Machine Learning, IoT and Cognitive Technologies*, 163-188.

Anioke, C., Nnamani, O., & Ani, C. (2015). *Call Drop Minimization Techniques for Handover Calls in Mobile Cellular Networks*, 1–5. UNN. http://www.dspace.unn.edu.ng/handle/123456789/5866

Babu, S. Z. D. (2022). Analysation of Big Data in Smart Healthcare. In M. Gupta, S. Ghatak, A. Gupta, & A. L. Mukherjee (Eds.), *Artificial Intelligence on Medical Data. Lecture Notes in Computational Vision and Biomechanics* (Vol. 37). Springer. doi:10.1007/978-981-19-0151-5_21

Bansal, R., Gupta, A., Singh, R., & Nassa, V. K. (2021). Role and Impact of Digital Technologies in E-Learning amidst COVID-19 Pandemic. *2021 Fourth International Conference on Computational Intelligence and Communication Technologies (CCICT)*, (pp. 194-202). IEEE. 10.1109/CCICT53244.2021.00046

Chibber, A., Anand, R., & Arora, S. (2021, September). A Staircase Microstrip Patch Antenna for UWB Applications. In *2021 9th International Conference on Reliability, Infocom Technologies and Optimization (Trends and Future Directions)(ICRITO)* (pp. 1-5). IEEE. 10.1109/ICRITO51393.2021.9596108

Dhanya, D., & Sankar, P. (2016). Call Drop Improvement in the Cellular Network. International Journal of Advanced Research in Electrical. *Electronics and Instrumentation Engineering*, *5*(2), 1117–1123. doi:10.15662/IJAREEIE.2016.0502025

Dighriri, M., Lee, G., & Baker, T. (2017). *Measurement and Classification of Smart Systems Data Traffic Over 5G Mobile Networks,* 1–363. Springer. doi:10.1007/978-3-319-60137-3

Dushyant, K., Muskan, G., Gupta, A., & Pramanik, S. (2022). Utilizing Machine Learning and Deep Learning in Cyber security: An Innovative Approach. In M. M. Ghonge, S. Pramanik, R. Mangrulkar, & D. N. Le (Eds.), *Cyber security and Digital Forensics*. Wiley. doi:10.1002/9781119795667.ch12

Eluwole, O. T., Udoh, N., Ojo, M., Okoro, C., & Akinyoade, A. J. (2018). From 1G to 5G, what next? *IAENG International Journal of Computer Science*, *45*(3), 413–434.

Garg, P., & Anand, R. (2011). Energy efficient data collection in wireless sensor network. *Dronacharya Research Journal*, *3*(1), 41.

Gaur, P. (2016). A review of Menace of Call Drops in India and possible ways to minimize it. International Journal of Mathematical. *Engineering and Management Sciences*, *1*(3), 130–138. doi:10.33889/IJMEMS.2016.1.3-014

Gupta, A. (2019). Script classification at word level for a Multilingual Document. *International Journal of Advanced Science and Technology*, *28*(20), 1247–1252. http://sersc.org/journals/index.php/IJAST/article/view/3835

Gupta, A. (2020). An Analysis of Digital Image Compression Technique in Image Processing. *International Journal of Advanced Science and Technology*, *28*(20), 1261–1265. http://sersc.org/journals/index.php/IJAST/article/view/3837

Gupta, A., Kaushik, D., Garg, M., & Verma, A. (2020). Machine Learning model for Breast Cancer Prediction. *2020 Fourth International Conference on I-SMAC (IoT in Social, Mobile, Analytics and Cloud) (I-SMAC),* (pp. 472-477). IEEE. 10.1109/I-SMAC49090.2020.9243323

Gupta, A., Singh, R., Nassa, V. K., Bansal, R., Sharma, P., & Koti, K. (2021) Investigating Application and Challenges of Big Data Analytics with Clustering. *2021 International Conference on Advancements in Electrical, Electronics, Communication, Computing and Automation (ICAECA)*, (pp. 1-6). IEEE. 10.1109/ICAECA52838.2021.9675483

Gupta, N., Khosravy, M., Patel, N., Dey, N., Gupta, S., Darbari, H., & Crespo, R. G. (2020). Economic data analytic AI technique on IoT edge devices for health monitoring of agriculture machines. *Applied Intelligence*, *50*(11), 3990–4016. doi:10.100710489-020-01744-x

Jain, S., Sindhwani, N., Anand, R., & Kannan, R. (2022). COVID Detection Using Chest X-Ray and Transfer Learning. In *International Conference on Intelligent Systems Design and Applications* (pp. 933-943). Springer. 10.1007/978-3-030-96308-8_87

Kaur, J., Sindhwani, N., Anand, R., & Pandey, D. (2023). Implementation of IoT in Various Domains. In *IoT Based Smart Applications* (pp. 165–178). Springer. doi:10.1007/978-3-031-04524-0_10

Kaushik, D., & Gupta, A. (2021). Ultra-secure transmissions for 5G-V2X communications. *Materials Today: Proceedings*. doi:10.1016/j.matpr.2020.12.130

Kaushik, K., & Garg, M. Annu, Gupta, A. & Pramanik, S. (2021). Application of Machine Learning and Deep Learning. In M. Ghonge, S. Pramanik, R. Mangrulkar and D. N. Le (eds.) Cyber security: An Innovative Approach, in Cybersecurity and Digital Forensics: Challenges and Future Trends. Wiley.

Khan, J. A., & Anand, R. (2022). A review on healthcare data privacy and security. *Networking Technologies in Smart Healthcare: Innovations and Analytical Approaches, 79*.

Khan, M., Ginzboorg, P., Järvinen, K., & Niemi, V. (2018). Defeating the downgrade attack on identity privacy in 5G. Lecture Notes in Computer Science (Including Subseries Lecture Notes in Artificial Intelligence and Lecture Notes in Bioinformatics), 11322 LNCS, (pp. 95–119). Springer. doi:10.1007/978-3-030-04762-7_6

Kliks, A., Musznicki, B., Kowalik, K., & Kryszkiewicz, P. (2018). Perspectives for resource sharing in 5G networks. *Telecommunication Systems*, *68*(4), 605–619. doi:10.100711235-017-0411-3

Meivel, S., Sindhwani, N., Valarmathi, S., Dhivya, G., Atchaya, M., Anand, R., & Maurya, S. (2023). Design and Method of 16.24 GHz Microstrip Network Antenna Using Underwater Wireless Communication Algorithm. In *Cyber Technologies and Emerging Sciences* (pp. 363–371). Springer. doi:10.1007/978-981-19-2538-2_36

Morgado, A., Huq, K. M. S., Mumtaz, S., & Rodriguez, J. (2018). A survey of 5G technologies: Regulatory, standardization and industrial perspectives. *Digital Communications and Networks*, *4*(2), 87–97. doi:10.1016/j.dcan.2017.09.010

Müller, C. F., Galaviz, G., Andrade, Á. G., Kaiser, I., & Fengler, W. (2018). Evaluation of Scheduling Algorithms for 5G Mobile Systems. *Studies in Systems, Decision, and Control*, *143*, 213–233. doi:10.1007/978-3-319-74060-7_12

Pandey, B. K. (2022). Effective and Secure Transmission of Health Information Using Advanced Morphological Component Analysis and Image Hiding. In M. Gupta, S. Ghatak, A. Gupta, & A. L. Mukherjee (Eds.), *Artificial Intelligence on Medical Data. Lecture Notes in Computational Vision and Biomechanics* (Vol. 37). Springer. doi:10.1007/978-981-19-0151-5_19

Pathania, V. (2022). A Database Application of Monitoring COVID-19 in India. In M. Gupta, S. Ghatak, A. Gupta, & A. L. Mukherjee (Eds.), *Artificial Intelligence on Medical Data. Lecture Notes in Computational Vision and Biomechanics* (Vol. 37). Springer. doi:10.1007/978-981-19-0151-5_23

Raghavan, R., Verma, D. C., Pandey, D., Anand, R., Pandey, B. K., & Singh, H. (2022). Optimized building extraction from high-resolution satellite imagery using deep learning. *Multimedia Tools and Applications*, *81*(29), 1–15. doi:10.100711042-022-13493-9

Sabella, D., Rapone, D., Fodrini, M., Cavdar, C., Olsson, M., Frenger, P., & Tombaz, S. (2016). Energy management in mobile networks towards 5G. In *Studies in Systems* (Vol. 50). Decision and Control. doi:10.1007/978-3-319-27568-0_17

Shukla, A., Ahamad, S., Rao, G. N., Al-Asadi, A. J., Gupta, A., & Kumbhkar, M. (2021). Artificial Intelligence Assisted IoT Data Intrusion Detection. *2021 4th International Conference on Computing and Communications Technologies (ICCCT)*, (pp. 330-335). IEEE. 10.1109/ICCCT53315.2021.9711795

Singh, A., Singh, S. P., Tripathi, U. N., & Mishra, M. (2017). Optimizing Call Drops in Cellular Networks using Artificial Intelligence based Handover Schema. *Ijarcce*, *6*(1), 286–290. doi:10.17148/IJARCCE.2017.6155

Sreekanth, N., Rama Devi, J., Shukla, A., Mohanty, D. K., Srinivas, A., Rao, G. N., Alam, A., & Gupta, A. (2022). (2022). Evaluation of estimation in software development using deep learning-modified neural network. *Applied Nanoscience*. doi:10.100713204-021-02204-9

Tarkaa, N., & Mom, J. (2018). Comparative Analysis of Drop-Call Probability Due to Handover and Other Factors. *International Journal of Innovative Research in Science, Engineering and Technology*, *7*(7), 8029–8040. doi:10.15680/IJIRSET.2018.70707069

Tripathi, A., Sindhwani, N., Anand, R., & Dahiya, A. (2023). Role of IoT in Smart Homes and Smart Cities: Challenges, Benefits, and Applications. In *IoT Based Smart Applications* (pp. 199–217). Springer. doi:10.1007/978-3-031-04524-0_12

Veeraiah, V., Ahamad, G. P. S., Talukdar, S. B., Gupta, A., & Talukdar, V. (2022) Enhancement of Meta Verse Capabilities by IoT Integration. *2022 2nd International Conference on Advance Computing and Innovative Technologies in Engineering (ICACITE)*, (pp. 1493-1498). IEEE. 10.1109/ICACITE53722.2022.9823766

Veeraiah, V., Khan, H., Kumar, A., Ahamad, S., Mahajan, A., & Gupta, A. (2022). Integration of PSO and Deep Learning for Trend Analysis of Meta-Verse. *2022 2nd International Conference on Advance Computing and Innovative Technologies in Engineering (ICACITE)*, (pp. 713-718). IEEE. 10.1109/ICACITE53722.2022.9823883

Veeraiah, V., Kumar, K. R., Lalitha, K. P., Ahamad, S., Bansal, R., & Gupta, A. (2022). Application of Biometric System to Enhance the Security in Virtual World. *2022 2nd International Conference on Advance Computing and Innovative Technologies in Engineering (ICACITE)*, (pp. 719-723). IEEE. 10.1109/ICACITE53722.2022.9823850

Veeraiah, V., Rajaboina, N. B., Rao, G. N., Ahamad, S., Gupta, A., & Suri, C. S. (2022).Securing Online Web Application for IoT Management. *2022 2nd International Conference on Advance Computing and Innovative Technologies in Engineering (ICACITE),* (pp. 1499-1504). IEEE. 10.1109/ICACITE53722.2022.9823733

Verma, A., Gupta, A., Kaushik, D., & Garg, M. (2021). Performance enhancement of IOT based accident detection system by integration of edge detection. *Materials Today: Proceedings*. doi:10.1016/j.matpr.2021.01.468

Chapter 12
Performance Enhancement of Health-Tech Applications Using Agile Methodology

Vikas Goyal

Malout Institute of Management and Information Technology, India

Geetanjali Goyal

Malout Institute of Management and Information Technology, India

Hritik Ranjan Nanda

Malout Institute of Management and Information Technology, India

ABSTRACT

HealthCare is an application which connects the patients/users and health care professionals registered on the HealthCare application, acting as merely intermediary. The data, which is provided by the users/patients or from any health care professional, belongs simply to them. Here a person is given access with some credential, non-exclusive license to use the HealthCare Services as described therein. The authors are using Flutter and Firebase as the technology to develop the App and WebSite. The authors are using Flutter technology, which is provided by Google for creating front-end/back-end, along with crafting beautiful designs (FutureLearn,); moreover, it is a compiled environment for mobile, web, and desktop by using a single codebase. Furthermore, the authors are using Google's Firebase technology that is provided by Google as a back-end service that helps in the development of software, which lets the developers to make iOS, Android, and web applications. It basically provides the tools which helps keep a tab on investigation, addressing, fixing application bugs, making marketing strategies, and product testing.

DOI: 10.4018/978-1-6684-7000-8.ch012

I. INTRODUCTION

HealthCare - As we know that health sector is under high pressure not only from national political point of view as well as from the overgrowing marketplace, for the management of patient/user services more effectively. Any change for the reconstruction in the field of health care may result in a drastic positive revolution, resulting in shifting of overall hospital management philosophy. Then the main objective of hospitals and clinics would be to reduce stress, high tenure, growth in patients admissions and hick in cases load (Perez, 2021). Most challenging part for hospitals and clinics would be to gave a quality of health services in the manner of effectiveness and cost-efficient in the real world. Which include getting the patients cured and out of the hospital as soon as possible. The reputation of the hospitals and clinics will totally depend on their success and how they are responding towards the problems which are being asked by the patients/users (Esper et al., 2020). Whereas the hospitals and clinics are integral part for a social, economical and for a medical organisations, The main goal of "HealthCare" is to provide a complete health related services to the entire population, which are health-giving and preventive for the patients/users. whose output of is the provide In-Home and Out-Home facilities. Due to increased pressure it was not possible to provide quality facility and services.

Looking on that many of the startups and companies came forward to solve the problem of intense pressure on the Health services. Name of some such companies are: **MFine** (Mathew, 2019), **Practor** (Practo,), **DocsApp** (DocsApp,), **IMG** (1mg,), **Sastasundar** (Flipkart Health,), **CureJoy** (Tracxn,) and **Call Health (CallHealth,)**.

But still there are a major drawback in these lineup. And those problems are:

- Display the incorrect result of the Patients Condition.
- Increased cost of the treatment for the Patients/Users.
- Lack of Information about the Patients.
- Patients' Online Treatment Through Technology.
- Time-Consuming in process in each sector.
- No quick service in the time emergency.
- No dedicated system for the in time medical notification.
- All health related facilities are not available on the single platform.
- Security concern.
- No direct interaction with the Health Care Providers.

This all is making the Health-Tech softwares unreliable to use and believe. So to come up with this we created "HealthCare" (Zanotto et al., 2021).

HealthCare is an application, that connects the Patients/Users and Health Care Professionals registered on HealthCare application, acting as merely intermediary.

First step is to download the app from app store and play store, then HealthCare application starts with the registration process Once it is completed by the Users/Patients or the Health care providers, then the HealthCare works on their device as an application.

In case of any problem any type of user or HCP are free to connect with the HealthCare team through different social media panels and call centre.

Figure 1. How HealthCare work

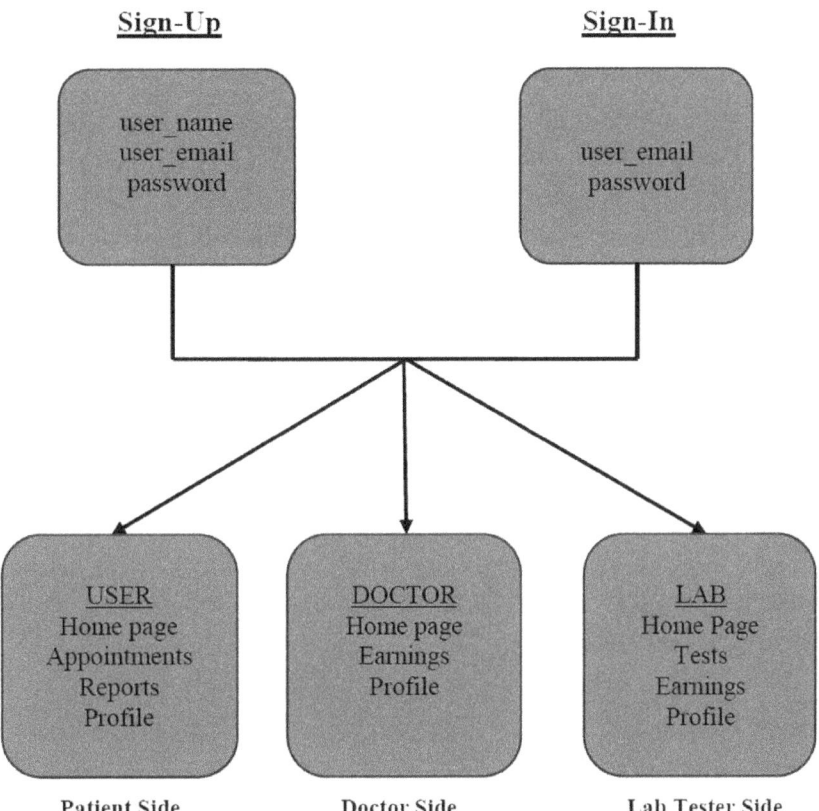

After registering as a User/Patient you will automatically allows HealthCare team to use your data if they need. Our system synchronize the users/patients data in a secure environment to make and keep a backup copy so users/patients can store and retrieve their data at any point of time, from any place and from any device that lets them do so.

If Health Care Providers and Users/Patient wants to enter, access, modify, correct, update, or delete their profile information and reject the processing of their data from the Application so they can do so.

Moreover by using HealthCare application Users/Patients and Health care providers can communicate with each other by the concept of client-service model even more advice, monitor and access data.

Users have direct & free access to the physical health related information through video content.

Also, users can book for the physical trainer for the physical health.

1. The Environment and Materiality of HealthCare Conditions

Terms of Services and Privacy Policies together create a legal documentation (the "Agreement") between User/Patient and HealthCare. Although the relationship in between visiting and using the Application and Services leads to the better use of all the functionality of HealthCare. Thus, it is more reliable and easy to use for each and every type of user. This all terms are related to the governmental legations, thus each and every service is legal (Ford, 2022).

2. Relevance Algorithms

Healthcare related algorithms for Health care providers is totally an interconnected system that provides information regarding Health care providers, profile, and their experience of practice on HealthCare application. These lists of HCP does not reflect any type of constant aim of ranking or endorsement by HealthCare. HealthCare is not responsible for any type of changes in the relevance of Health care providers on searched output, which may or may not occur time to time.

The health care providers list is based on an automated and inter-connective methodology of various conditions which includes comments and feedback that was made by users/patients. All such criteria vary time to time, in order to improve the HealthCare algorithms. Healthcare will under no circumstances be responsible for the accuracy and relevance of Health care provides listing on the HealthCare platform.

3. Security

HealthCare has been created with the modern technology like *"Flutter and Firebase"* which help. The HeathCare by providing the security from Google, Apple and many more. This helps the user to create the secured account by using email authentication. Once the account is created each user is given an encrypted HCID which is basically HealthCare ID, this id makes the details of users to be stored in form of encrypted data which is totally hack free.

4. Records

HealthCare can provide end-user with a free service called as 'Records' on our application 'HealthCare'. There are two types of information available in your records:

User-made: Information uploaded by users/patients or information that was given by Health care providers during the time of consultation with the HealthCare environment, for example: book appointment, a medicine order you gave.

Created by cure: Health records created by doctor at the time of interaction with users/patients.

Problems Solved by HealthCare

As our main logic is the service from the nearby so thats by the services are quick and fast. So the medicines are delivered fast, lab testers comes to the door step in no time delay and many more.

- Each and every record is double verified and are clearly shown in the mobile interface of the users.
- Each and every product and service are charged at their base price and in many of the cases discounts are also available.

HealthCare in very unique in the case of providing services because on the single platform you get access to the doctor consultations where after consultation, you can directly order the medicines or can book any type of lab testing. And the reports of the respective are also shown in the report's sections. Also, you get the facility of physical exercise guide, along with that problems like hair fall and skin are separately solved. HealthCare is connected with the security of google.

- Direct online or offline interaction with HealthCare provides.
- HealthCare has a special section for ayurvedic. You can directly get the access to all the ayurvedic medicines and doctors.
- Patients have the easiest and most reliable UI/UX experience (Swiggy, 2023).

How HealthCare is Different

During research we conclude that there are many ongoing Health-Tech startups in India, serving to solve the above mentioned problems. Below in Tab.1.1 we show that how our project/startup is different from others. Along with the features in the table HealthCare have many more features which make it very unique and easy to use. Those feature lets the users to get all the health-related services at the same platform.

Those features include:-

- Fast & Easy appointment from HCP
- Hospital & Clinic Connectivity
- E-Laboratory
- Online Pharma
- E-Gym
- Skin care
- Hair Care
- Ayurvedic
- Easy Support Chat

Table 1. HealthCare is unique in comparison with others

	HealthCare	Practo	mFine	Pharmeasy
Consultancy	✓ (Available)	✓ (Available)	✓ (Available)	✗
Order Medicine	✓ (Available)	✓ (Available)	✓ (Available)	✓ (Available)
Lab Test	✓ (all types)	✓ (only blood test)	✓ (only blood test)	✓ (all types)
Ambulance	✓ (Available)	✗	✗	✗

Approach Used

To develop the project, we use "Agile" methodology for managing the application that we are developing.

Agile methodology is a latest method for any type of project management and software/application development which helps organisations, teams, and groups to provide value to their end customers, users faster with fewer complexities. It works on sprint basics that divides all the task based on discrete schemas and help to wind-up all the work in an efficient manner. It provides testing approach at the last stage of sprint.

Agile methodology helps us to run the sprints which show us the proper time of development, phase of development done, phase of development left, phase of development going on. Fig 1.2 shows our approach for the Agile in HealthCare.

Agile is a methodology that helps to complete a work in a flow without any consequences. It is not illustrated for the development approach or for set of ceremonies. Moreover, agile is a group of methods that illustrate a commitment of tight feedback chain and constantly upgrading. The actual Agile methodology did not describe two-week iterations or a size of the ideal team. It simply lay out a set of essential values that put people first.

Figure 2 HealthCare Agile Methodology

II. LITERATURE SURVEY

We gone through many research articles and research paper to get into the conclusion to solve the above mentioned problem statements by creating a Health- Tech startup named-**"HealthCare"** . Research paper we gone through are:-

- "Analysis of impact of technological innovation on healthcare services" - by **José Figueiredo** from the Universidade Fernando Pessoa, Praça 9 de Abril, 349, 4249-004 Porto, Portugal. (Int. J. Behavioural and Healthcare Research, Vol. 1, No. 3, 2009)

Under this article author explained about, how Technological innovation is essential driver of performance in most of service areas. Also, this article focuses on the evidence of relationship like this is much more communicated in healthcare services / sectors. Furthermore, the concept of **José Figueiredo** about technological innovation is very clear in healthcare sector, which is very broad when new medical equipment, new pharma products, new types of forms for contact with patients and new work procedures are being taken. So, the study about technological innovation contributes to the clarifies that technological innovation in healthcare is very important. It also expands the idea about, how to evaluate the impact of technological innovation in different areas of the healthcare environment. The output identification of cause of technological innovation and its relationship with different types of health related services which may lead to a much more understandable relationship between technological innovation and its performance.

- "Technology and the Future of Healthcare" - by **Harold Thimbleby** at J Public Health Res. 2013 Dec 1 (Published online 2013 Dec 1. doi: 10.4081 / jphr.2013.e28)

Healthcare changes drastically due to technological developments or changes in many sectors, from anaesthetics and antibiotics to magnetic resonance imaging scanners and radiotherapy. In future technological innovation and development are going to keep changing and transforming healthcare sector, yet technologies (drugs and treatments, new and advance devices, new media support for healthcare, etc) which will develop ideas for different innovation, human factors will remain one of the stable limitations of breakthroughs. No judgment can predict everyone; instead, this article explores view of the future to see how to think and how to see more clearly about how to get where we want to go.

III. IMPLEMENTATION OF HEALTHCARE

HealthCare works on the modular approach in which each and every section of the HealthCare app / application is being divided into different modules to make it easy and effective while understanding and implementing any type of feature (Sullivan et al., 2018). In HealthCare application we used Flutter and firebase for each and every development, which is being balanced by Atlassian (uses Agile methodology). Figure 3, shows the entire flow of implementation.

In the above figure 3, *main.dart* file which is the main file of the implementation process is being shown. In this file we can clearly see how we have linked each and every file of the modules are connected with the *main.dart*. Using all the 56 modules, which is being developed in flutter, all the features are being implemented by the latest technology using GetX property of routing. Wherein, routing is the property of connecting different files with the route without calling the module itself.

Whereas, in the left side of the figure shows how each and every features are being created and furthermore how they are divided in sub modules. For example, in *Page* module we further have Chat, doctor, gym_trainer and many more.

Figure 3. Module structure

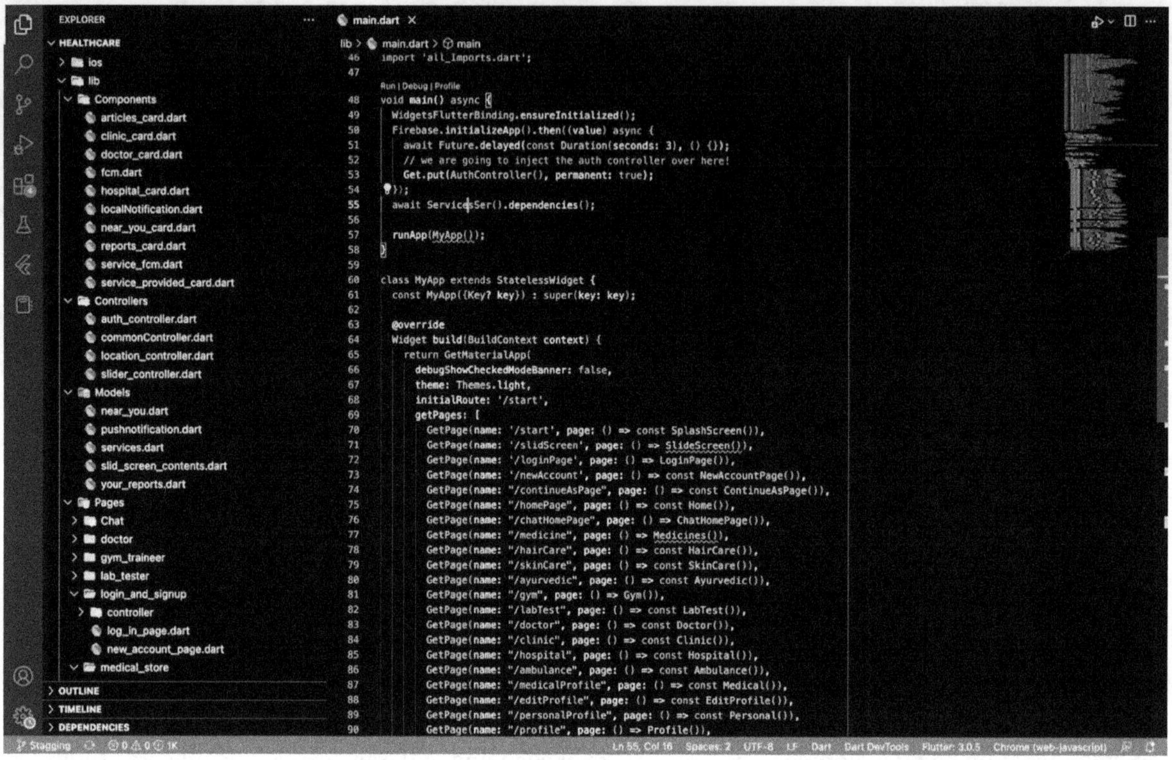

Moreover, we have total of 56 module section for the every images, icons & code which are being used in the HealthCare. And those files are being managed under *pubspec.yaml* where route of all images and icons are present. And in this file, we have the directory of each and every packages, which are being mentioned in this file and then a command "*flutter pub* get" is being run on the command line to install that in the HealthCare application.

As it is being developed by Flutter so it is feasible to run this on a IOS Devices. But it is not possible directly, so to run that on IOS devices we need to install *Podfile.lock* to make it compatible for IOS environment. Also, IOS environment needs more security concerns so before creating and converting any code for IOS device we should have Apple developer membership.

IV. FUNCTIONALITY

HealthCare is concerned to provide the quality and fast services to all the users, which makes it different from other Health-Tech Applications (Vainieri et al., 2019). All the features in HealthCare are interlinked to each other due to the interactive feature of the technology which we use for the development of this Application. This allows the users to use the services anytime and anywhere. *Fig.4* shows all the features which are provided to the patients/users by HealthCare.

Figure 4. HealthCare User Interface

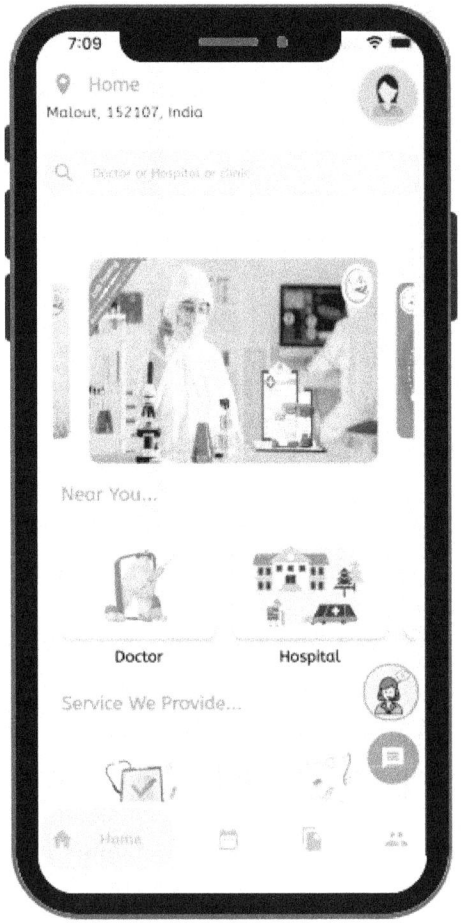 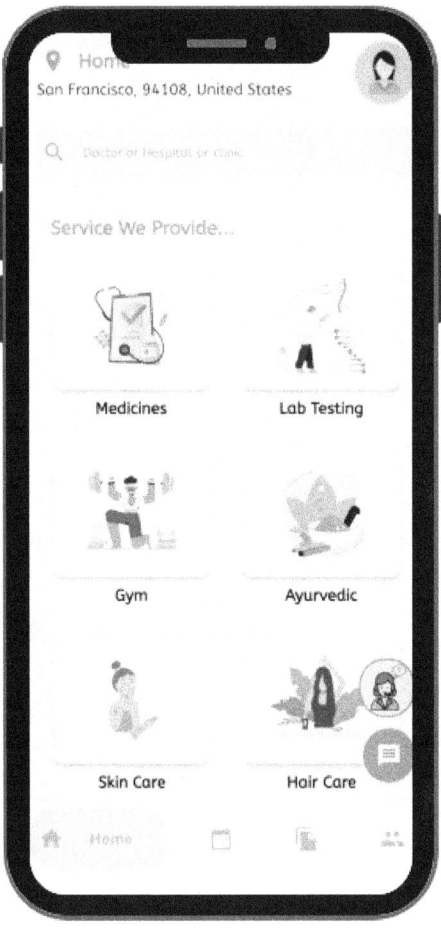

HealthCare includes the features like:-

- Fast & Easy appointment from HCP
- Hospital & Clinic Connectivity
- E-Laboratory
- Online Pharma
- E-Gym
- Skin care
- Hair Care
- Ayurvedic

1. **Fast & Easy appointment-** With just a click we can make appointment from the nearest doctor in your location. Appointment can be in any form of your choice, which can be either online and offline. Once your appointment is approved from the doctor you get a meeting link from the safest meeting platform from the google. This make your communication end-to-end encrypted and secure (Leggat et al., 2010).

Figure 5. Doctor Appointment

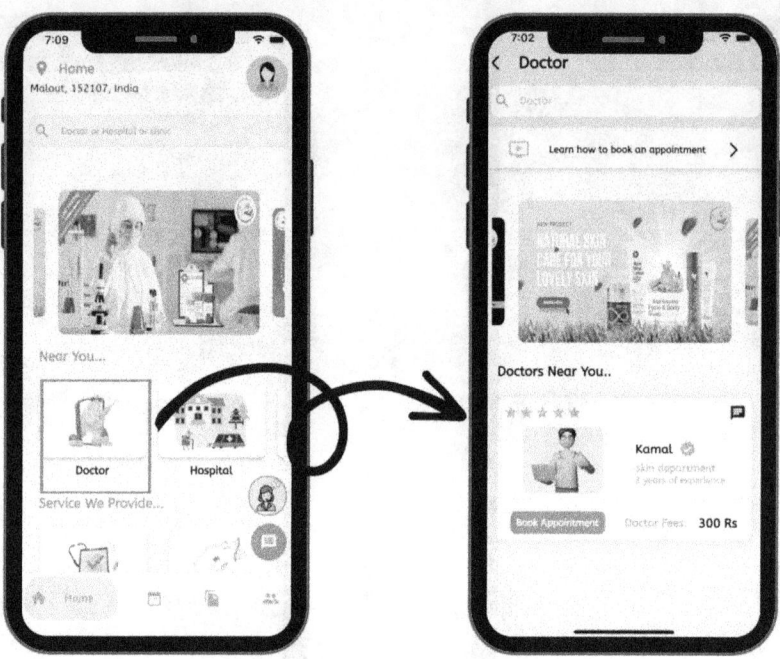

2. **E-Laboratory-** E-lab service is a feature that is used to understand and provide them a clear view of what they are current facing the inner body problems through experimentation and conclusion. This provides an interface for the users to see all the lab tester near them, which allows them to book the appointment from them to make the test for them by giving them the flexibility to collect the sample from them or the user can reach the lab by himself. This make the HealthCare different from others as the service will be very fast due to the near by services. Also the reports for the testing will also be delivered to their app within a very quick time. Moreover multiple tests can be booked by the user at a time. Also the user has the feature to upload the prescription prescribed by the doctor and according to that the team of the testers can come for the indoor service for the sample collection. Once the appointment from the lab tester is confirmed then the user has to pay with the most secured gate way which is being used in the Healthcare and it has the mode of payment like UPI, net banking, debit/credit card, etc. Fig 4.2 shows the best of the E-Laboratory feature in the HealthCare and how one can benefit from it is clearly being shown.

3. **Hospital & Clinic Connectivity-** You will get a full fledged access to all the near by Hospitals and clinics which helps the users / patients to get connectivity with full of ease and satisfaction. They can connect with hospitals and clinics in both the mode of connectivity, which are offline and online.

4. **Online Pharma-** Many Health-Tech applications provide the feature of online pharmacy. But the pinch point is that those applications provide the delivery with the time gap of either two or three days. But the unique feature of HealthCare is that user is shown the Medical Stores near to their location, which enables the HealthCare to deliver the medicines within a day because the medicines will be picked from the nearer locations.

Figure 6. E-Laborarty

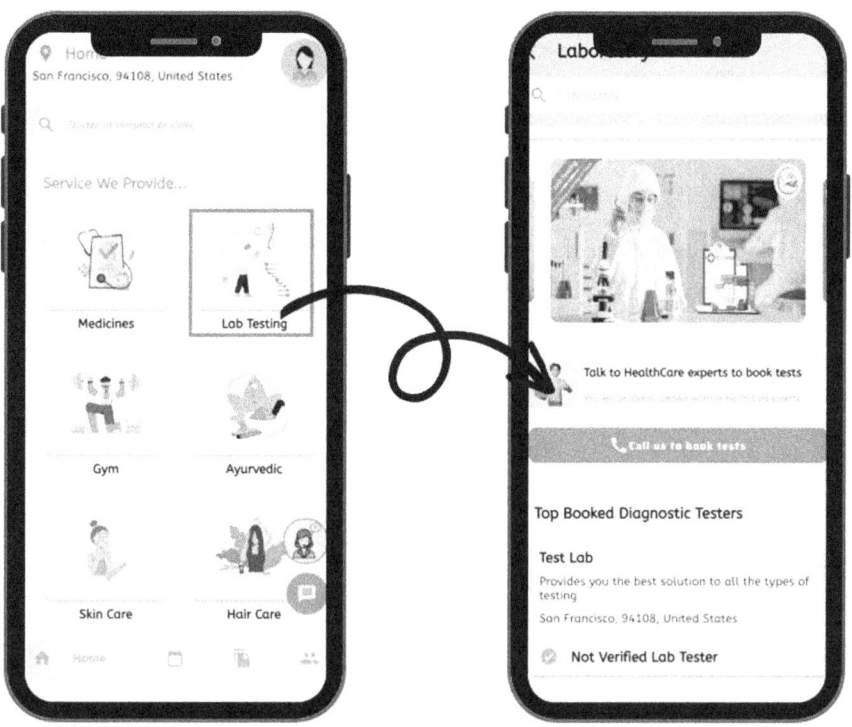

Figure 7. Connection with Hospitals & Clinic

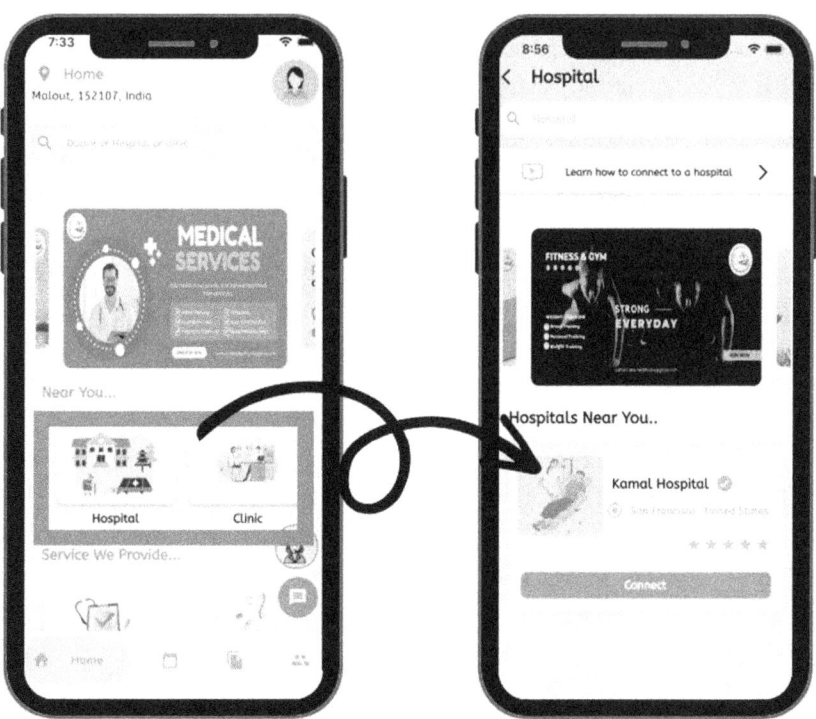

Figure 8. Online-Pharma

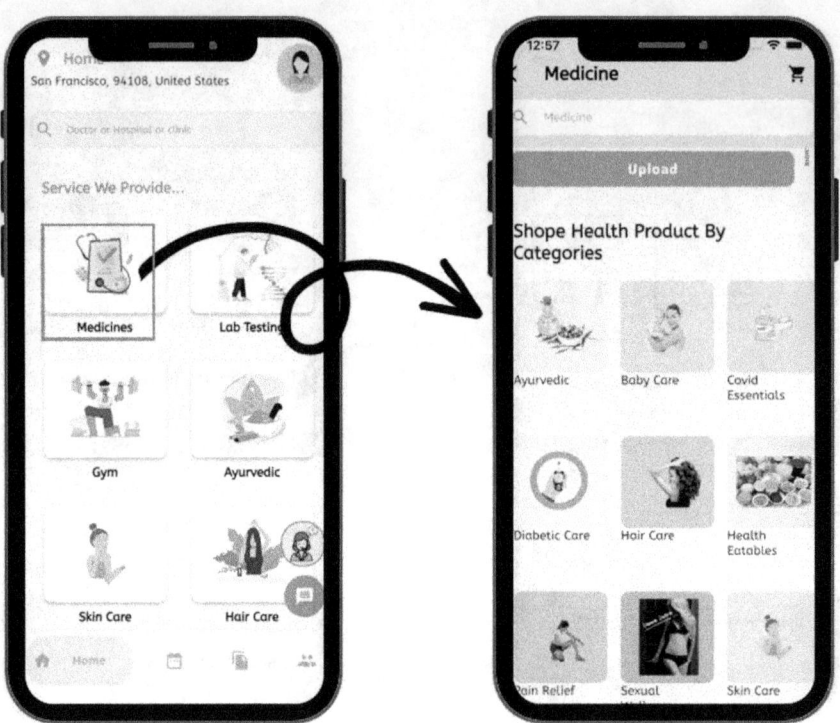

5. **E-Gym-** Now a days all the persons are associated with their physical health. Each and everyone wants to be fit and tight but in the world of busyness no one has the time for going specially to gym or any exercise centre all they need is the in home alternative for this. So HealthCare comes with the feature of E-Gym where they can have excess to all the types of exercise videos according to your body level. If someone is in the beginner level then they can start from that or if they are in intermediate level or in professional level then they can access the service according to that.

6. **Skin Care-** Skin care apps like HealthCare always guides a users/patients in improving and keeping a daily check on their skin health and providing a better and healthy way to make their skin much more healthy. As we know that in the entire human body skin is one of the most sensitive part and widely distributed organ in the entire body. In he current edge-era skin is most cared organ by any being as this is the only portion of the body which is being interacted with the outer world. So that is why all take special steps for the nourishment of skin. HealthCare app provides a special feature of skin care that provide information about list of skincare conditions, skincare daily routines, professional made skincare plans for users with specific skin conditions, and how to face them to improve their skin. Even HealthCare allows users/patients to buy skincare and beauty products. Also HealthCare provides the nearby doctors for the skin care which can easily be appointed and you can get prescribed, also you get the entire interface of medicines related to skin.

Figure 9. E-Gym

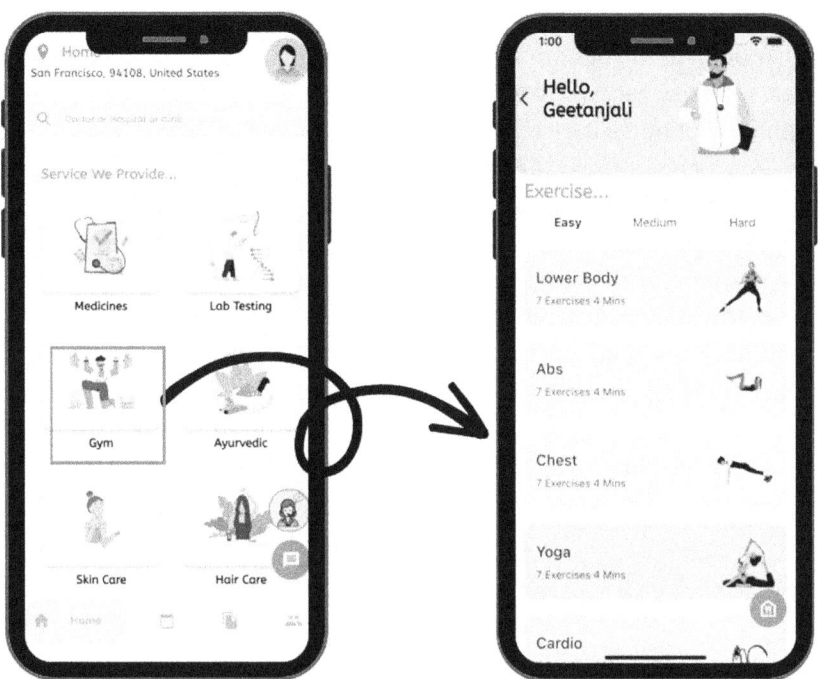

Figure 10. Skin Care

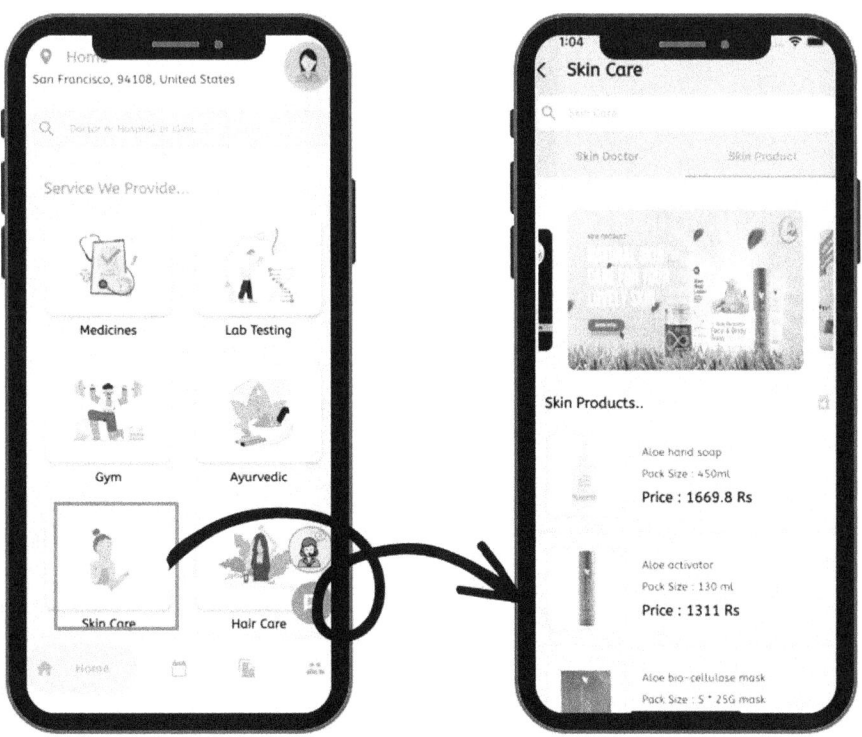

7. **Hair Care-** Hair Care is the another most cared part in terms of human nature. Each and every human wants a hair care routine with hygiene which involves the hair which grows from the scalp, pubic and other body hair. These routines differ according to an individual's culture, the physical characteristics, what one eats and how he spends his day. Hair of humans may be shaped, coloured, cut, shaved, plucked or might have removed with some activities like waxing, sugaring and threading. Hair care treatments are basically being given by salons, barbershops and spas. And items are available in market for daily home use. Thus HealthCare is up with the special feature of Hair Care where you get the ease to access all the hair care specialist for the better care of your hair. Our hair care professionals provide the best solutions for your daily care or style goals, hair problems and any queries regarding hair care. Even you have the access to all the hair care products which can be delivered to the door step with in a day. What makes it most unique is the access to the free videos which help the individual to watch and learn how to properly take cake of the hair. This allows all the near by service provider to get a good business as it is most trendy part to get the hair care in an app. Even the UI used in the HealthCare is quit simple make it more effective to use and get the relevant services easily. The image show it very clearly.

Figure 11. Hair Care

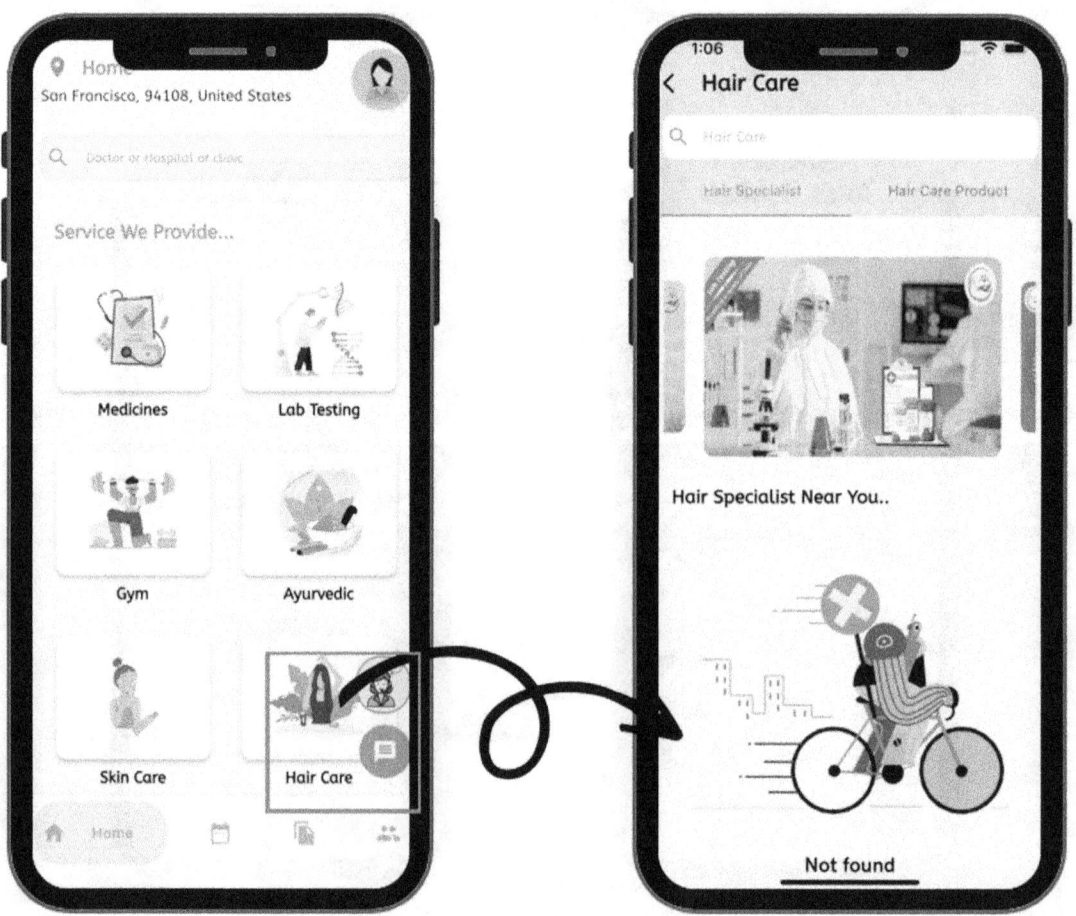

8. **Ayurvedic-** It is a natural system of medicine developed in India from more than 3,000 years ago. The term Ayurveda is stated from the words ayur which means life and veda which means science of knowledge. So, it represents as life of knowledge. It is derived on the thinking that any health problem is due to an out of balance or more pressure in a once's life, It give rise to many lifestyle inventions and environmental therapies to maintain a balance between the body, thinking, sole, and the lifestyle. Ayurveda therapy starts with an inner purification body process, that is advanced by a better diet plan, natural remedies, therapy, yoga, exercise and meditation. HealthCare app provides a special feature of ayurvedic section that provide information about list of body conditions, ayurvedic daily routines, nature-made ayurvedic medicines for users/patients with specific cure, and how to solve them to improve their body condition with ayurvedic treatment. Even it allows you to buy ayurvedic products which nourishes your body. Also HealthCare provides the nearby Ayurvedic doctors for users which can easily be appointed and can get prescribed, also you get the entire interface of medicines related to ayurveda. Items which are used in Ayurvedic treatment has natural herbs, minerals, or other metals that may be harmful if not used in a proper way or without the guidance of a professional doctor, so HealthCare comes with the properly trained HCP for the medicine recommendations.

Figure 12. Ayurvedic

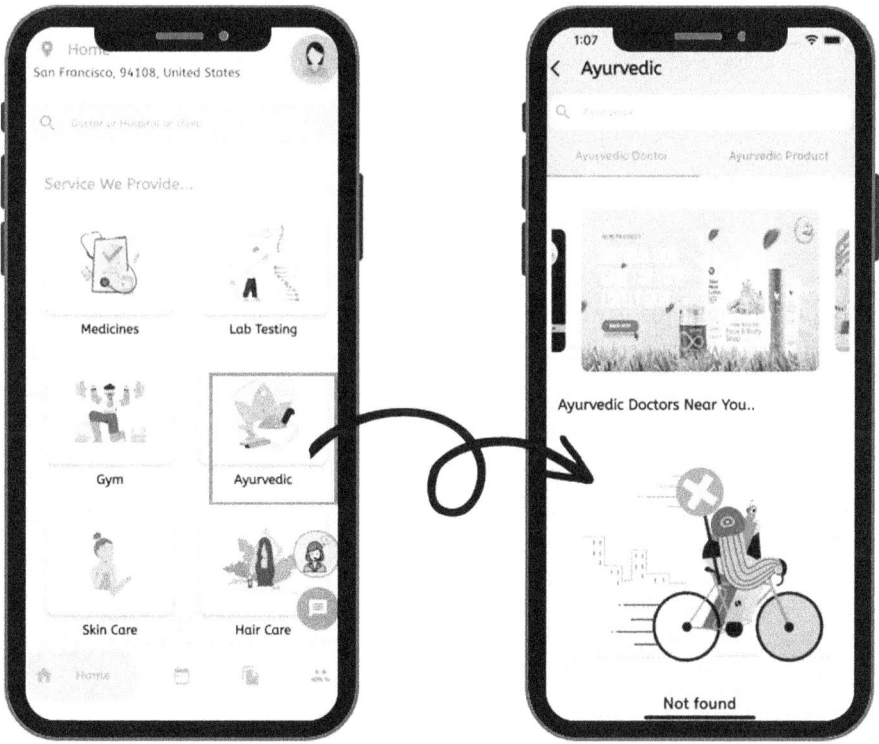

V. HEALTHCARE MANAGEMENT WEBSITE

Management Website is basically website which help us to manage the entire user/patients and HCP in our platform through which we can verify each and every user and HCP (Issel, 2020). So that there are no fake users on our health care platform with cost effectiveness (Waldman et al., 2010). Healthcare management website starts with an authentication screen in which only HealthCare members who are providing the services can login. Once the Authentication processes is complete then the main screen for the management of user's patient, hospitals, etc., opens up through which each and every thing can be verified by checking out their certificates and ID's. Currently management website is proving the features like:

1. **Content Entry**
    ◦ Medicine
    ◦ Add Video
    ◦ Post
    ◦ Notifications
2. **User Management**
    ◦ Patients
    ◦ Doctors
    ◦ Hospitals
    ◦ Clinic
    ◦ Lab Testers
3. **Support Chat**- (One to One)

Content Entry

It is a section in management website where content for the HealthCare environment is added. Content which is added on the management website is directly reflected on the HeathCare environment. The feature enables us to add, update and delete any data according to the current requirements of the users/ patients. Content includes the addition of any new medicines, any change in sliding banners, addition of any new exercise video and most importantly we can schedule the notification at our flexible time (Lewis et al., 2019).

User Management- It is a section in which HealthCare team verifies the doctors, lab testers and many more for the authenticity (Lewis et al., 2019). From here any user can be banned or deleted in case of any malfunctioning.

Support Chat- It is section for the CRM to maintain the balance on the enquires and other types of support services, which are requested by the users. Even the CRM gets the access to all the contact numbers to which they can call and make the support system more effective. This is controlled by the firebase and its feature of firebase messaging, which lends the user to get proper notifications from CRM.

Figure 13. Content Entry(Management Website)

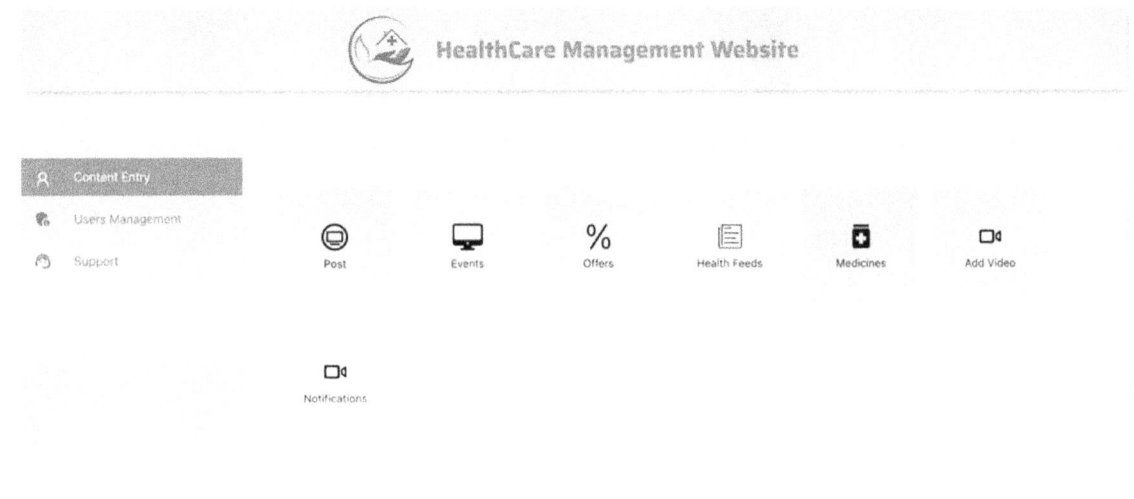

Figure 14. User Management

How HealthCare is Unique as an Application?

This Figure below show how HealthCare is Unique as an application in comparison with others, the constraints included are Security, UI/UX, Relevance Algorithm, Records and Environment. And the rating is done from the value out of 25 and the graphing is done on the basics of the general reviews of the population around us.

Figure 15. Support Chat

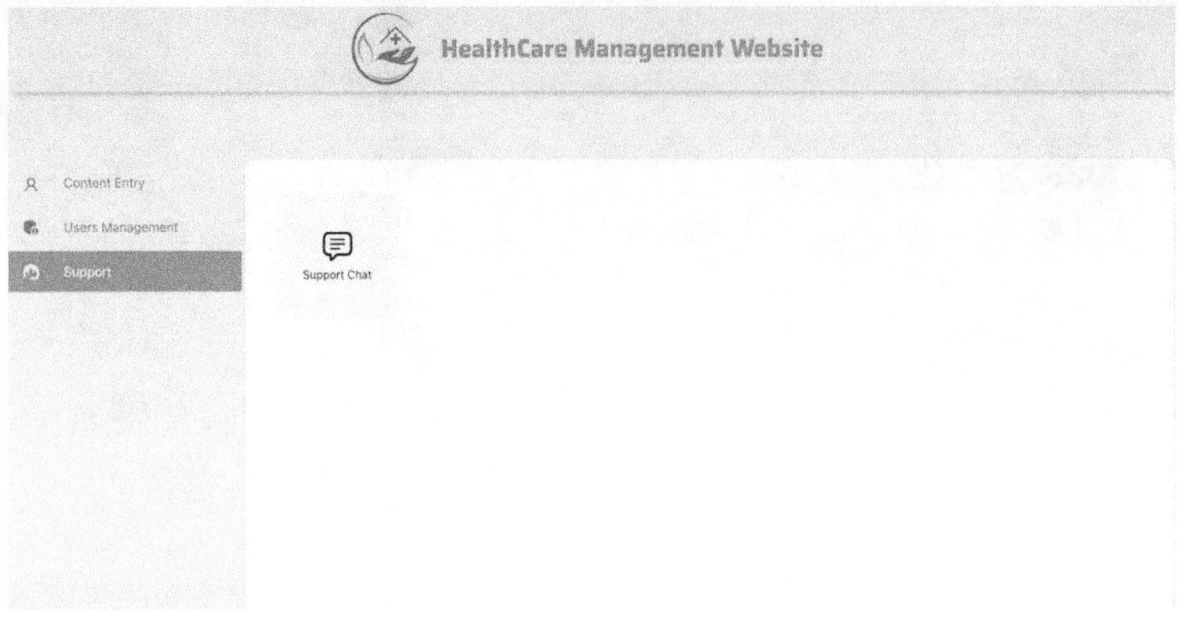

Figure 16.

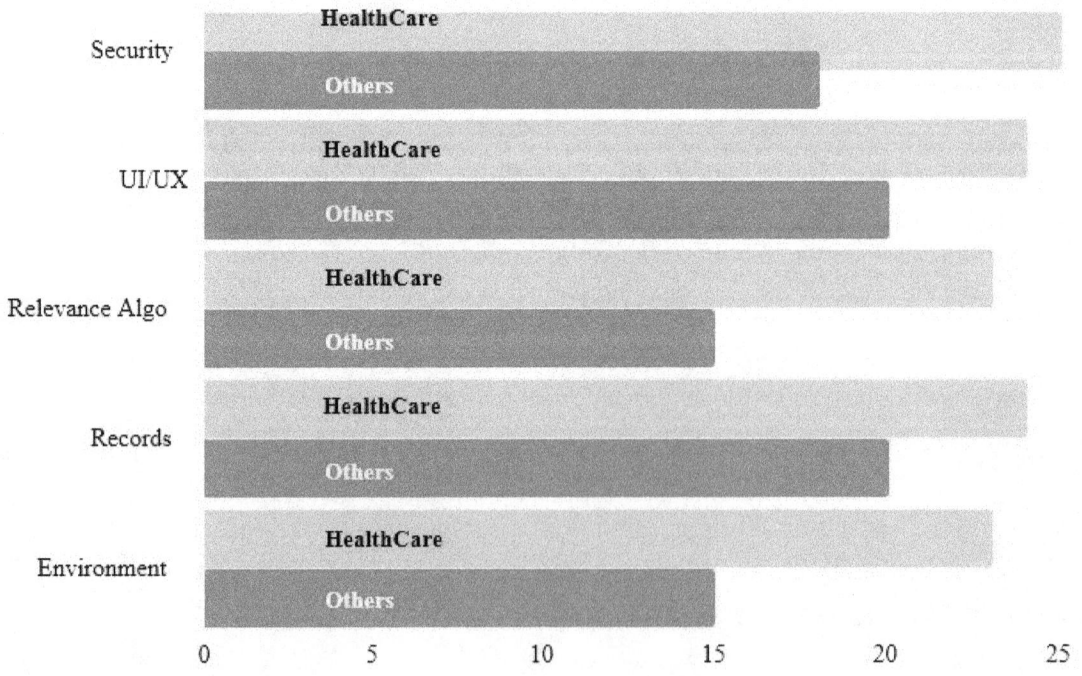

VI. CONCLUSION

HealthCare is a unique application which provides a quite interesting features which make it different from many other Health-Tech applications in the market. Main thing which make it different is that it works on the basis of current location. User gets the access of all the near by service around them, which enables them to get fast and effective services either in online or offline mode. Thus, the services like online pharma, lab testing and many more can be provided with in couple of hours.

REFERENCES

1mg. (n.d.). *Online pharmacy India: Buy medicines from India's Trusted Medicine Store*. 1mg. https://www.1mg.com/

CallHealth. (n.d.). *Everything about health*. Call Health. https://www.callhealth.com/

DocsApp. (n.d.). *DocsApp is Now MediBuddy*. MediBuddy. https://m.docsapp.in/

Esper, G., Sweeney, R., Winchell, E, Duffell, J., Kier, S, Lukens, H., & Krupinski, E. (2020). Rapid Systemwide Implementation of Outpatient Telehealth in Response to the COVID-19 Pandemic. *Journal of Healthcare Management, 65*(6), 443-452. doi:10.1097/JHM-D-20-00131

Flipkart Health. (n.d.). *Health*. Flipkart Health. https://www.sastasundar.com/index.php/user/login/

Ford, E. (2022). Creating Your Personal Board of Directors. *Journal of Healthcare Management, 67*(5), 303-305. doi:10.1097/JHM-D-22-00160

FutureLearn. (n.d.). *Online courses and degrees from top universities*. FutureLearn. https://www.futurelearn.com/

Issel, M L. (2020). Value Added of Management to Health Care Organizations. *Health Care Management Review, 45*(2), 95. doi:10.1097/HMR.0000000000000280

Leggat, S. G., Bartram, T., Casimir, G., Stanton, P. (2010). Nurse perceptions of the quality of patient care: Confirming the importance of empowerment and job satisfaction. *Health Care Management Review, 35*(4), 355-364. doi:10.1097/HMR.0b013e3181e4ec55

Lewis, V. A., Schoenherr, K., Fraze, T., Cunningham, A. (2019). Clinical coordination in accountable care organizations: A qualitative study. *Health Care Management Review 44*(2), 127-136. doi:10.1097/HMR.0000000000000141

Mathew, S. (2019, March 19). Mfine app: Connects patients to doctors instantly. *The Hindu*. https://www.thehindu.com/sci-tech/health/mfine-app-60-seconds-to-a-doctor/article26568179.ece

Perez, J. (2021). Leadership in Healthcare: Transitioning From Clinical Professional to Healthcare Leader. *Journal of Healthcare Management 66*(4), 280-302. doi:10.1097/JHM-D-20-00057

Practo. (n.d.). *Video consultation with doctors, Book doctor appointments, order medicine, diagnostic tests*. Practo. https://www.practo.com/

Sullivan, J. L., Adjognon, O. L., Engle, R. L., Shin, M. H., Afable, M. K., Rudin, W., White, B., Shay, K., & Lukas, C. V. (2018). Identifying and overcoming implementation challenges: Experience of 59 noninstitutional long-term services and support pilot programs in the Veterans Health Administration. *Health Care Management Review, 43*(3), 193-205. doi:10.1097/HMR.0000000000000152

Swiggy. (2023). Swiggy App for design. [Application].

Tracxn. (n.d.). *Technology + human-in-the-loop for deal discovery*. Tracxn. https://tracxn.com/

Vainieri, M, Ferrè, F, Giacomelli, G., & Nuti, S. (2019). Explaining performance in health care: How and when top management competencies make the difference. *Health Care Management Review, 44*(4), 306-317. doi:10.1097/HMR.0000000000000164

Waldman, J. D, Kelly, F., Arora, S., Smith, H. L. (2010). The shocking cost of turnover in health care. *Health Care Management Review, 35*(3), 206-211. doi:. doi:10.1097/HMR.0b013e3181e3940

Zanotto, B., Etges, S., da Silva, A., Marcolino, Zago, M., & Polanczyk, C. (2021). Value-Based Health-care Initiatives in Practice: A Systematic Review. *Journal of Healthcare Management, 66*(5), 340-365. doi:10.1097/JHM-D-20-00283

Chapter 13
Optimization of Multi-Objective Emperor Penguin Handovers Over Dissimilar Networks

V. V. Satyanarayana Tallapragada
https://orcid.org/0000-0002-8764-9982
Mohan Babu University, India

M. Venkatanaresh
Mohan Babu University, India

N. Gireesh
Mohan Babu University, India

M. Naresh
Matrusri Engineering College, India

ABSTRACT

Media independent handover (IEEE 802.21 MIH) services are an area of particular focus for IEEE. The primary objective of IEEE 802.21 MIH is to streamline the handover procedure, resulting in a more reliable and less time-consuming handover service. There are two distinct kinds of handoffs, called horizontal and vertical, respectively. Among these, the performance of vertical handover (VH) can be improved through parameter optimization. In order to achieve the best possible outcome and reduce the rate of handover failure, careful parameter selection is required. Despite the fact that many options exist for VH management optimisation processes, many current works only consider one or two parameters for VH optimisation. Therefore, the ideal handover solution that emerges is inferior in terms of reliability, responsiveness, and precision. So, a technique called multi-objective emperor penguin handover optimisation (MOEPHO), which takes into account nearly all network characteristics for VH optimisation is discussed in this chapter.

DOI: 10.4018/978-1-6684-7000-8.ch013

1. INTRODUCTION

Each of the various forms of communication technology that have emerged has been developed separately to cater to certain data, coverage, or portability needs. There are benefits and drawbacks to using these various forms of communication technology. There is currently no technology that can meet the needs of a sizable user base while also allowing for high bandwidth, high mobility, and a huge service area. An approach based on establishing interconnections between devices allows for this such that the benefits of integrating different technologies result in minimizing their shortcomings. The term "heterogeneous networks" stems from combining multiple telecommunication technologies to leverage their additional properties. The flexibility of heterogeneous networks allows users to hop between several access points, each of which may offer a somewhat different set of features and capabilities. In addition, by distributing traffic across many access technologies, heterogeneous networks can help alleviate mobile network congestion. Rather than focusing on just one service or application, heterogeneous networks offer a variety of services, which allows not only the presence of different kinds of traffic within the network but also the capacity for a single network to accommodate all applications without sacrificing the quality of service. Because of their rapid and massive rollout, mobile networks will be composed of a collection of heterogeneous systems operated by separate companies and linked together via access networks. Mobile terminals will function as network multi-interfaces, allowing them to seamlessly switch between different communication systems. Users should be able to seamlessly switch between networks on this heterogeneous infrastructure. Networks need to be built on top of already-existing infrastructure by linking already-deployed networks.

Telecommunications companies have made substantial investments in third-generation networks, and customers now have several options from which to pick. Today's telecommunications infrastructure is based entirely on heterogeneous systems, allowing customers to seamlessly switch between networks without disrupting their connections. Simple switching techniques between networks are required to guarantee an operational connection in every service zone. Moreover, a mobile terminal capable of functioning on different access technologies is consequently required. Therefore, it is possible to categorize the needs of various networks into two major categories. First is that the operators are tasked with ensuring interoperability and facilitating changeover procedures across various access technologies; secondly, a set of metrics concerns the efficiency with which endpoint devices may access and use network services. Figure 1 shows a heterogenous network (Naresh M et al., 2020).

Radio waves are transmitted and received for wireless communication between mobile users. A centralized or decentralized medium access control (MAC) system can manage radio spectrum access. Figure 1 depicts a typical mobile user communication paradigm, which employs a centralized MAC mechanism in which a mobile user connects with other users via a base station. This form of network is sometimes referred to as an infrastructure network. In this setup, the base station is in charge of dividing up available wireless bandwidth across different mobile devices. The advantage of this method of allocating network resources is that it allows for the management and planning of mobile user interference from the perspective of the wireless system as a whole, allowing for the throughput, dependability, speed, and fairness of the network as a whole to be optimized. The potential for latency experienced by mobile users interested in direct communication is a significant downside of this network design.

Figure 1. Heterogeneous Network

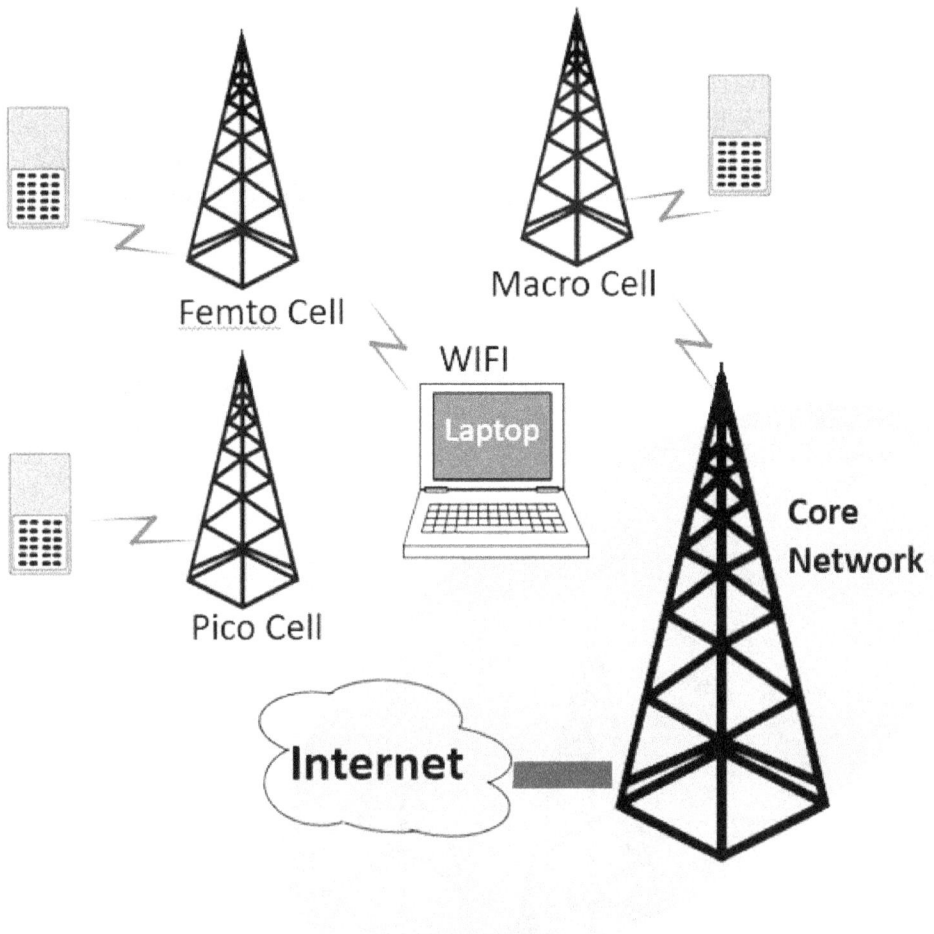

Longevity in an ad hoc wireless sensor network is directly proportional to the network's ability to conserve energy. Every node in an ad hoc wireless sensor network consumes some amount of energy, and the rest is sent between nodes. Increasing the energy efficiency for interactive communication in a network of clusters is another goal of this work that aims to extend the ad hoc wireless sensor network (WSN) lifetime. Because the nodes that make up a clustered network are selected at random, there is a chance that the cluster hubs will be placed in an unreliable order. All the relay stations talk to all the cluster hubs, which talk to all the nodes. For situations where the base station's range is inadequate, it may rely on intermediary cluster heads (CHs) to relay messages between the BS and CHs. Within the WSN, this kind of dialogue is known as interactive communication (Anand, R et al., 2022).

Figure 2 depicts the distributed media access control method used by ad hoc networks. Ad hoc networks are mobile users interested in talking with one another, and the network's resources are distributed according to the wireless channel circumstances they experience. The allocations are normally made from a pool of available wireless channels. Ad hoc networks can function autonomously without the assistance of a pre-existing network or the network infrastructure (base station). Ad hoc networks' advantages include decreased latency compared to infrastructure networks and independence from the base station location. Ad hoc networks have lesser throughput than infrastructure networks and are more vulnerable to interference from nearby mobile users. Heterogeneous network (Het Net) development is crucial for capacity improvement in wireless networks to properly manage the increasing influence of mobile traffic demand. Systems having Micro, Macro, Femto, and Wi-Fi layers are made possible by this. A combination of high data demand and quick internationalization has led to a meteoric rise in mobile data usage. Certain mobile terminals provide high-quality internet access to their most connected users. When moving from one Access Point (AP) to another, handover is crucial in maintaining a constant internet connection. Unlike the more conventional Vertical Handover (VH), which keeps data sessions inside the same link layer, Horizontal Handover (HH) moves sessions between different APs. Despite the availability of many reference collections, most need to be more flexible to accommodate different networks. Handover approaches can be implemented in both IEEE 802 and non-IEEE 802 networks thanks to the media-independent procedures provided by IEEE 802.21(Kunarak S et al., 2021), (Shiwei G et al., 2021), (Naresh M. et al., 2020), (Yu H et al., 2019).

Figure 2. Adhoc Network

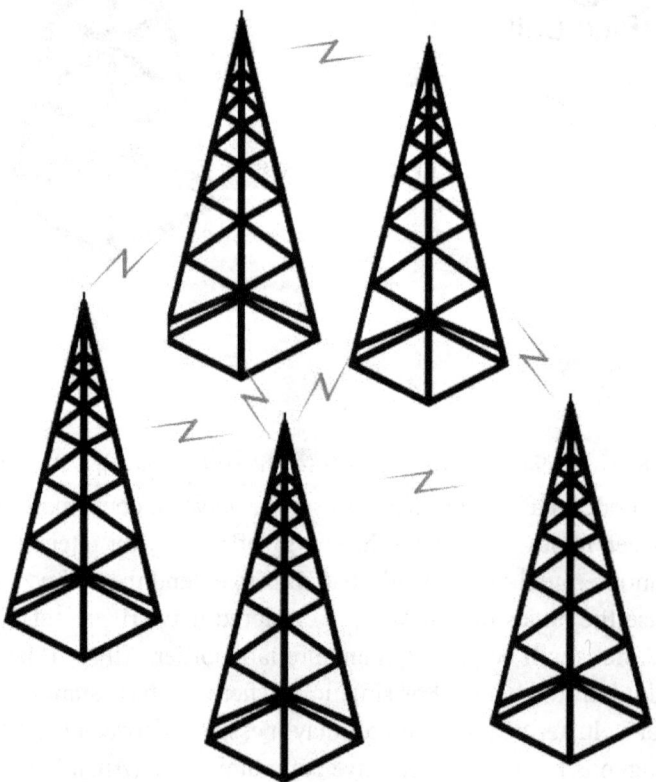

Heterogeneous network (Het Net) advancement is critical for capacity improvement in wireless networks (Bagubali A et al., 2018) due to the increasing effect of mobile traffic demand. It allows systems with Micro, Macro, Femto, and Wi-Fi layers to coexist. The massive data demand and the quickening pace of globalization are driving mobile data usage. For the most connected customers, a few mobile terminals provide reliable internet service (Mahajan, P. 2018). Handover is critical in maintaining continuous internet service by seamlessly shifting user connections from one Access Point (AP) to another (Stamou A et al., 2019). Data sessions can be moved across Access Points (APs) with equal link layers using Horizontal Handover (HH), while the standard Link Layer Method is used in Vertical Handover (VH) (Bhatt, M. C et al. 2018). Despite the availability of several standard groups, most lack adaptability for different kinds of networks (Saeed, R. A. 2019). IEEE 802.21 provides media-independent techniques to support handover approaches in IEEE 802 and non-IEEE 802 networks (Aljeri N., & Boukerche A 2018, Khan, B. M., & Bilal R 2018, Ahmed N., & Rikli, N. E. 2018).

Existing publications have proposed various optimization (Anand, R., & Chawla, P. 2022) strategies for VH management in heterogeneous networks. To lower the failure rate of handoffs, parameter selection for optimization is crucial, as the optimal outcome is sensitive to the parameters. The VH optimization was often just done based on one or two factors in much prior research. Therefore, the best possible handoff solution will not improve upon the baseline regarding reliability, speed, or precision. This motivated a VH optimization technique that takes into account all aspects of a network, such as power usage, connection cost, bandwidth, latency, throughput, MN velocity, BER, BER, SNR, RSS, and data traffic, to produce the best possible outcome for a reliable handover.

2. RELATED WORK

In order to make decisions in heterogeneous systems while keeping context in mind, a Neural Network based Whale Optimization Algorithm (WOA-NN) (Parambanchary, D., & Malleswara Rao, V. 2020) is presented. Initially, RSS properties were observed to compile a NN model training library. This way, trained networks can handle handover problems by making RSS predictions. Additionally, WOA has been used to verify that NN has indeed learned the correct behavior of RSS. The primary goal of implementing handoff between two cellular networks is to provide the user with uninterrupted service. Two basic rules, the horizontal handoff and the vertical handoff are used to initiate the handoff between two networks. In reality, a connection between homogeneous networks is indicated by a horizontal handoff, but a vertical handoff shows a connection between heterogeneous networks. Vertical handoff is when a handoff is made between two cells that use distinct technologies. The movement of a node between various wireless access networks is another way of putting it. Since the node hops between networks, its IP address and underlying technology may evolve as it does so. As a result, the IP address and network interface are the primary targets of this procedure. When talking about WLAN, the WiFi network is one of the unsung technologies. Initially, only a select few could take advantage of its services. However, now that the network's performance is at its peak, so are the available services. The WiFi network consists of terminals (clients) and base stations (Access Points; APS) and transfers data at a rate of 108 Mbps.

Nonetheless, a collision formation situation can arise because WiFi data is transmitted across the air. Therefore, several radio packets have a significant impact on data transfer. In comparison to WiFi, WiMAX is a more modern network that provides both faster transfer rates (70 Mbps) and a wider coverage area. WiMax's value comes from its speed and range, superior bandwidth, and reduced interference. More trust can be placed in data transmitted over WiMAX's two channels (the uplink and the downlink). The uplink is used for data transmission from the base station to the user, while the downlink is used for data transmission from the user to the base station.

An intelligent decision-maker for handover management in a heterogeneous network was proposed using a fuzzy-based energy-efficient architecture (Coqueiro, T et al. 2019). Fuzzy system decision-making prioritizes the maximization of battery life, minimization of energy usage, and enhancement of Quality of Experience (QoE). When a mobile device is in the range of two or more networks, it must decide which one to connect to, even when the device lacks any form of mobile support or intelligence that could guide it in making this selection. No resource exists, not even technological norms, that can help with decisions like network selection and setup. At first, the RSSI is used to determine which base station or access point a mobile device can connect to. Since a mobile device with low battery capacity could be linked to a network that raises its energy consumption and, as a result, shortens its battery life, the lack of support for decision-making while picking the network might cause the user problems.

Similarly, a user with a long-lasting battery could be linked to a crowded network with inadequate bandwidth, negatively impacting the user's program. The ping-pong handoff is an example of how user mobility can degrade the quality of experience when users with high mobility do too many handoffs. Because of its mobility, a mobile device will be linked to a new network even if it will only spend a short amount of time within the region of coverage of this new network, requiring it to reestablish the connection with the point of contact or previous base station.

A refined version of the Modified Grey Relational Analysis (MGRA) for VH was created by Chattatte et al. (Chattate et al., 2019). MGRA's Always Suitable Connection (ASC) approach uses the Analytic Hierarchy Process (AHP) and Fuzzy AHP to identify a good network. Similarly, a network classification method, an Enhanced Fuzzy Technique, is presented (Chattate I et al., 2019) for Order of Preference by Similarity to Ideal Solution (EFTOPSIS) to be used during VH. EFTOPSIS uses a weight-updating algorithm that combines Fuzzy AHP and a ranking approach to achieve precise categorization. As a result, top networks were chosen, resulting in optimal performance. Prior research has only managed VH based on a handful of critical factors, such as RSS, power consumption, and latency. Our review of the relevant literature leads us to conclude that a more than precise handover with improved QoS is needed. Thus, in our study, we develop an algorithmic method of multi-objective VH, which considers all network factors for optimization, allowing for precise handover with good quality of service. A novel network selection method is created for a pervasive network based on an improved version of the Fuzzy TOPSIS algorithm to prevent service disruptions when transitioning between different types of networks. Fuzzy AHP is used to assign relative importance to each criterion in the evaluation, and Fuzzy TOPSIS is used as an improved ranking approach. Our new FE-TOPSIS approach is based on many criteria. Thus we compared it against the traditional F-TOPSIS method to see how well it performed. Experiments conducted in a simulated environment demonstrate that the quality of service (QoS) is significantly enhanced compared to the conventional technique. This algorithm for making vertical handover decisions can quickly and easily pinpoint the optimal candidate access network.

3. HANDOVER TECHNIQUES

The range of base stations in cellular networks is restricted. Hence handover is required to ensure continuity of service for mobile devices, which are typically in motion. The term "handover" or "hand-off" refers to a process in telecommunications and wireless communication in which a cellular transmission (voice or data) is handed over from one base station (cell site) to another without disrupting the cellular connection. Creating data sessions or connecting phone calls between mobile devices continuously on the move requires handover, which is a crucial part of implementing mobile transmission. Two types of handoffs are soft and hard(Duong et al., 2020).

In a hard handover, the source cell of the channel is disconnected from the network before the target cell's channel is activated. As the name implies, this handoff occurs when the connection to the source is severed before or simultaneously with establishing the connection to the destination. The term "break before making handover" describes this time period(Stamou A et al., 2019). During a gentle handover, the source cell is kept and continues to operate in parallel with the target cell. Here, the link to the target cell is made before the original one is severed. With soft handover, many channels can be connected in tandem to improve service quality (Shao S et al., 2020).

The last two decades have seen a clear paradigm shift in telecommunications and the usage of the Internet and its allied services. Up to the turn of the millennium, progress was mostly focused on technology, and user demands were sometimes overlooked. Several significant steps were taken by the modern mobile Internet evolution to improve the quality of wireless data and network services for its consumers. Third-generation (3G) cellular networks are the pinnacle of mobile network evolution, bringing unprecedented data rates and a vastly improved quality of life for users. After the 3G standard, the following generation of networks optimized and improved data speeds and services to provide users with more freedom of movement (Shayea et al., 2020).

Beyond 3G (B3G) and 4G are common names for these types of systems. They rely heavily on heterogeneous networking technologies to provide mobile users with more options when accessing multi-service networks that offer a wide variety of services, from IP-based real-time multimedia to location-aware navigation services to seamless Internet connections via heterogeneous wireless networks. Next-generation all-IP networks need to be built to make it possible for all relevant services to work regardless of the underlying network technologies, which is essential for achieving network, service, and application convergence. Service providers will benefit from this since they can efficiently supply their network services to end customers regardless of the type of network or the capabilities of the terminals they use. There is a growing demand for more wireless services, such as higher bandwidth, throughput, and lower packet loss, and all-IP-based apps are pushing this demand.

Since users need to be able to move between different next generation networks (NGNs), mobility management with quality of service (QoS) support is an essential issue in NGNs. In contrast to wired data networks, wireless cellular networks do not rely on physical connections between users. The term "handover" is commonly used to describe this procedure. The academic community has classified this procedure into three broad categories: horizontal, vertical, and diagonal handovers.

A horizontal handover occurs when a mobile node (MN) switches between two cells that employ the same technology. By masking the IP address change, as in Mobile IP, or by dynamically delivering up-to-date information on the altered IP address, as in mobile stream control transmission protocol, intra-cell (intra-domain) handover allows running services to continue uninterrupted (mSCTP). A vertical handover is a transition between two access technologies. When a user roams into an adjacent cell, all of their terminals' connections must be transferred to a new base station, a process known as Inter-cell (Inter-domain, Inter-RAT1) handover (BS). The primary goal of vertical handover is to keep services functioning during an IP address change and any other changes to network interfaces or quality of service that may occur.

The Wi-Family is based on IEEE standards, and many groups are now collaborating to establish wireless technologies that can communicate with one another via diagonal handovers. The diagonal handoff is a cross between the horizontal and the vertical. Suppose the mobile node passes over cells with a common underlying technology (say, Ethernet). In that case, the user can seamlessly transition from the Wi-XX network to the Wi-YY network while maintaining the necessary quality of service (QoS). The IEEE 802.21 working group developed this phrase when two different types of networks must use the same frequency range (typically, IEEE networks and broadcast networks or downlink-only networks). It is also known as a "media-neutral handoff" (MIH)[11]. Figure 3 shows a typical handover mechanism in which a mobile terminal moves from one Base Transceiver Station (BTS) to another within the given time frame. It is an assumption that it is moving in this case.

The initial step is to identify the power level and signal. The signal level becomes weak as it moves to the cell border, making the system vulnerable to a handover. Figure 4 shows a typical vertical handover mechanism. The yellow dashed line shows the vertical handover. The accurate Vertical Hand Over (VH)) is designed and developed so that the mobile will have good service continuity, over the side having network discovery, network selection, and better security. The other issues are also taken care of are the device's power management and QoS issues. The VHO process is divided into three phases. The first phase is the information-gathering phase. During this stage, data is gathered not just about the network but also about the mobile devices, access points, and user preferences that make up the system. Many terms describe this stage, including handover information collecting, system discovery, system detection, handover initiation, and network discovery [62,48,43]. During this stage, data is gathered for the decision-making process during handoff. Typically, the following data is gathered, viz., Throughput, cost, packet loss ratio, handoff rate, Received Signal Strength (RSS), Noise Signal Ratio (NSR), Carrier to Interference Ratio (CIR), Signal to Interference Ratio (SIR), Bit Error Ratio (BER), distance, location, and QoS parameters are all examples of what neighboring network links make available. Battery life, available storage space, transfer rates, and connection quality are a few indicators of a mobile device's overall health. Information on the user's requirements(Lahby M et al., 2018), (Ahmed A et al., 2013).

The next phase of the handover is where decisions are made crucial. Preparation for the handoff can refer to selecting a system, a network, or even a handoff. During this stage, the When and the Where of the handover is determined based on the data collected. Choosing the finest network that meets our needs for the switch hinges on two decisions: when to make the switch and where to make the switch.

Assuming we all utilize the same networking technology, the When of a handover is typically determined by RSS numbers, but the Where is immaterial (horizontal handover). When dealing with heterogeneous networks, the answers to these problems become quite nuanced. Let us make the most informed choices possible. We need to analyze the data in light of a wide range of criteria gleaned from various data sources (such as the network, mobile devices, and individual users). The parameters involved in a handoff are weighted and evaluated using a Vertical Handover Decision Algorithm (VHA) for each criterion. Fuzzy logic, neural networks, and pattern recognition are a few methods that can be used to analyze cross-layer multi-parameters.

The two most common kinds of handoffs are handoffs inside the same network (HH) and between networks (VH). In HH, the RSS is used to make the handover decision, but in VH, the asymmetrical behavior of the networks means that the RSS evaluation alone is insufficient. Without VH support, the handoff service cannot proceed as planned. Therefore, the suggested works focus on optimizing VH across different parameters utilizing the Emperor Penguin Algorithm to ensure a seamless handover service in a network with a mix of technologies.

Figure 3. Typical Handover Mechanism

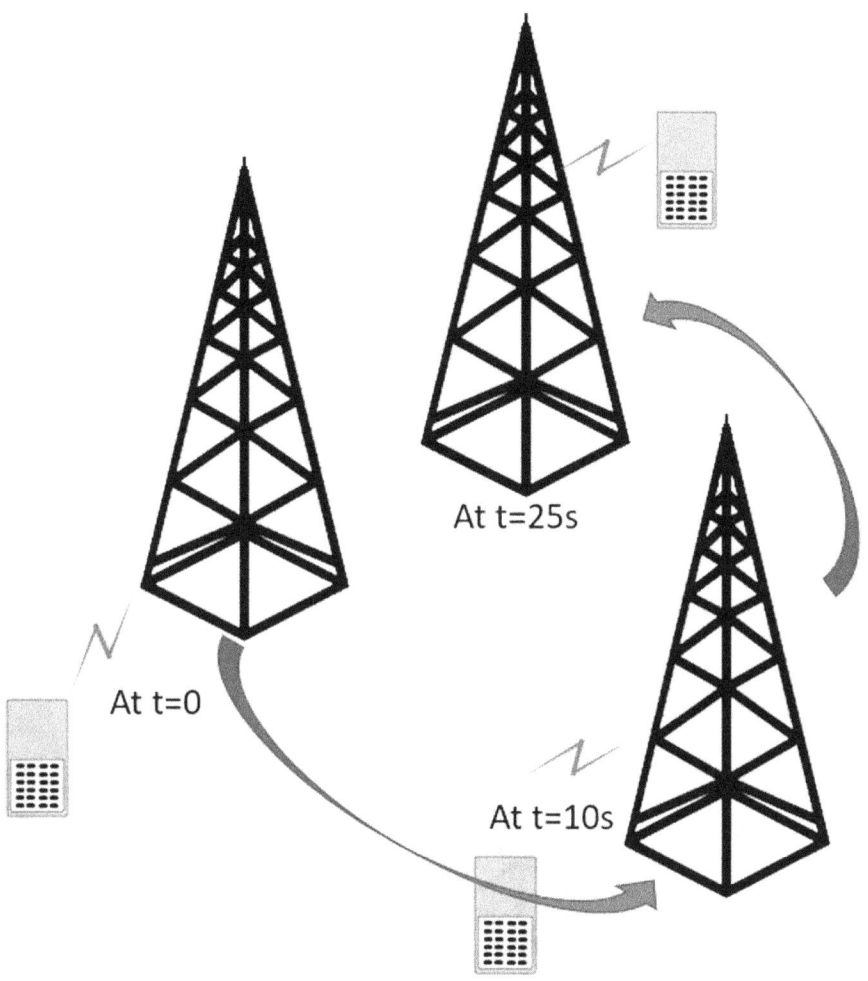

Finding the ideal network in which to effect a seamless handover is a crucial part of VH optimization. During the discovery phase, scanning is typically used to continuously examine available networks. However, in a heterogeneous environment where devices sport a variety of interfaces, a scanning-based approach to network discovery is ineffective. This causes problems with service quality under heterogeneous VH, such as increased power consumption, poor network choice, and increased latency. The huddle boundary generation stage of EPO is used to successfully manage these challenges MOEPHO[1]. As part of the network discovery process, the total utility (μt_{otal}) of several parameters are added together to form the huddle border region, and this region's equation (1) need to be satisfied.

$$\mu_{total} = \begin{cases} \dfrac{\partial \mu}{\partial \mu_i} \geq 0 \\ \lim_{\mu_i \to 0} \mu = 0 \quad ; \forall i = 1, 2, \dots n \\ \lim_{\mu_1 \dots \mu_{n-1} \to 0} \mu = 1 \end{cases} \tag{1}$$

(Naresh, M et al., 2020) Where, μi is utility of it^h parameter and n is number of parameters considered for network discovery(Naresh, M et al., 2020).

Combination of n with their weights and network negotiation factor is liable for cost function calculation as represented in equation (2).

$$\mu_{\cos t} = \sum_L \prod_i E^L_{n_i,i} \sum_i \left(\left(\mu_{\cos t} \right)^L_i \right) \left(W^L_i \right) N \left(\mu^L_{n_i,i} \right)$$ (2)

(Naresh, M et al., 2020) Where, $E^L_{n_i,i}$ is network negotiation factor which should be either 1 or ∞. For instance, $E^L_{n_i,i}$ becomes ∞, if delay requirements are not satisfied by a network for certain applications. $N\left(\mu^L_{n_i,i} \right)$ denotes normalized utility and W^L_i is weighting function of application (L) in ith network (n$_i$) by means of ith parameter.

Figure 4. Vertical Handover

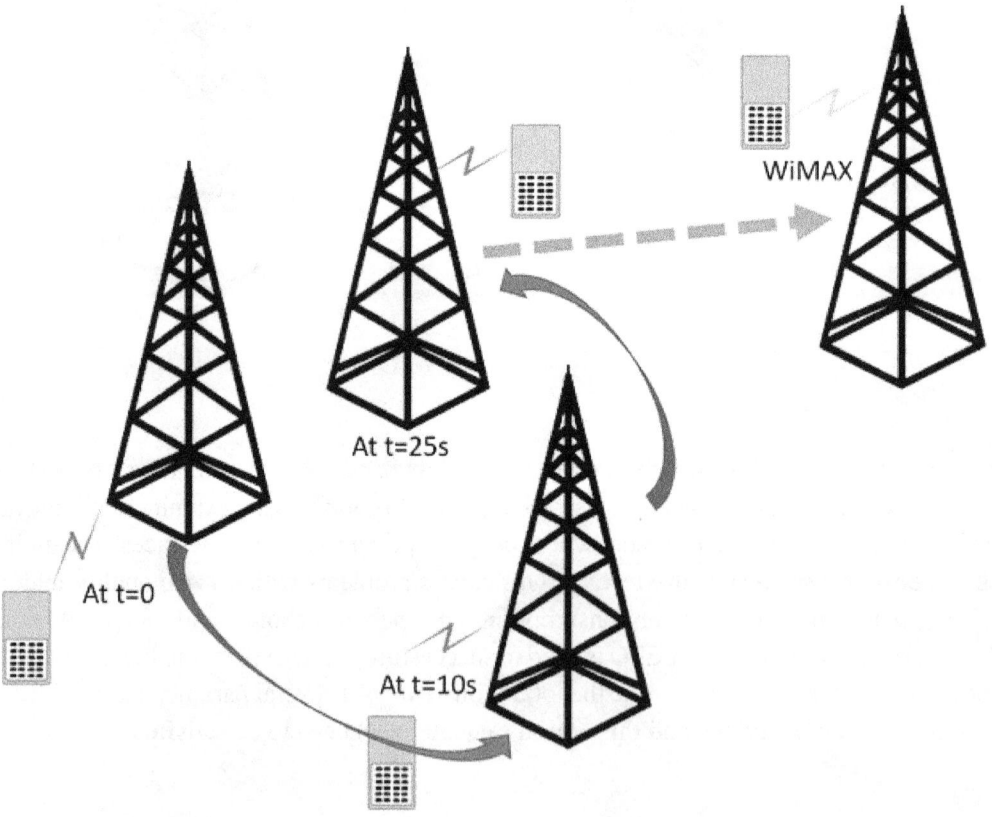

Expression (3) illustrates how the objective function of VH optimization can be used to compute multi-criteria decision making. Each iteration will take into account the current solution's optimization parameters in order to update those parameters. The current optimal solution and the optimization parameters of the previous iterations are compared and recorded as $(\vec{V})$.

$$Objective\ Function\ (\eta) = \max\left(\frac{\left(\beta \times T \times \varepsilon_{snr} \times B\right)}{\left(Y \times \varepsilon_{ber} \times v \times d \times P_c \times \mu\right)}\right) \tag{3}$$

(Naresh, M et al., 2020) Where

$$\vec{V} = \left\| \left(\left(\sqrt{\gamma.e^{-f/\gamma} - e^{-f}}\right)^2 \times \eta(\vec{f}) - rand\ (\overrightarrow{\eta_{vec}(f)})\right)\right\| \tag{4}$$

(Naresh, M et al., 2020) Where, $\eta(\vec{f})\ \&\ \overrightarrow{\eta_{vec}(f)}$ represents optimal solution and its position vector during current iteration respectively. Identification $\vec{V}$ at every iteration helps to enhance parameter updation procedure in order to find optimal solution for VH in IEEE 802.21. Table 1 shows the Algorithm for MOEPHO.

Table 1 Algorithm for MOEPHO(Naresh, M et al., 2020)

Step 1:	Initialize optimization parameters.
Step 2:	Define the objective function
Step 3:	Generation of Huddle boundary
Step 4:	Optimization parameter computation
Step 5:	Evaluation of Objective function
Step 6:	Check for termination condition. If not met, goto step 3 by updating the optimization parameters
Step 7:	Terminate by providing the optimal solution.

4. RESULTS

Results are plotted using various simulation parameters listed in table 2, and the research assessment of MOEPHO for IEEE 802.21 is carried out with the assistance of Network Simulator 2 (NS2). For MOEPHO implementation, an area measuring 600 x 600 mts containing 25 MNs and 4 BSs is considered. Every base station has a coverage area of 250 meters, and the total number of packets that each node may generate and transmit is 50 per second. The time it takes to transmit each packet is 0.02 seconds, and the entire process of putting the recommended plan into action takes 20 seconds. The simulation analyzes factors such as RSS, energy consumption, PDR, packet loss, throughput, handover time, handover accuracy, and handover failure rate. In this section, the following parameters are compared with the existing benchmark techniques viz., WOA NN(Parambanchary D et al., 2018), Adaptive Cross-Layer Design (ACLD) (Al Emam F.A. et al., 2018), Fuzzy Intelligent Decision Making (FIDM) (Coqueiro T et al., 2019), and Novel Type 2 FLC (Saeed M et al., 2018).

Table 2. Network Simulation Parameters(Naresh, M et al., 2020)

Network	IEEE 802.21
Number of Base Stations	4
Number of MNs	25
Number of Packets Generation	50 packets/second
Packet Size	1000 bytes
Initial energy	25J
Radio Range	250 m
Number of Iterations	50
Simulation Time	20 seconds
Area	600×600m
Packet Interval	0.02 seconds
Rate of Transmission	50 packets/second

It is proposed to test the simulation of the MOEPHO by initializing nodes and transmission of packets. Figure 5 shows four BSs, each of which is represented by a unique color: red for BS0 (node id = 4), blue for BS1 (node id = 5), pink for BS2 (node id = 6), and green for BS3 (node id is 7). At this early stage, BS0 is linked to six nodes identified by the identifiers 9, 14, 20, 18, 22, and 15. Similarly, 6, 8, and 5 nodes are connected to BS1, BS2, and BS3, respectively. Six nodes with ids of 32, 12, 24, 13, 21, and 11 are linked to BS1. Similarly, BS2 is connected to 8 nodes with ids 25, 10, 17, 23, 28, 27, 26, and 31, and BS3 is connected to 5 nodes with ids 30, 16, 19, 8, and 29. After the traditional VH approach, Figure 5 depicts the MOEPHO-based VH optimization. Six nodes (ids 15, 18, 14, 21, 20, and 27) are linked to Base Station 0 (BS0), four nodes (ids 22, 11, 12, and 32) are linked to Base Station 1 (BS1), eight nodes (ids 24, 25, 10, 31, 16, 30, 17&13), and seven nodes (ids 9, 8, 30, 29, 23, and 28&19) are linked to Base Station 2 (BS2). As more nodes are connected throughout the handover process, BS2 and BS3 experience extremely high levels of data traffic. As a result, it could cause increased energy costs, handoff delays, and inaccurate data. So, in Figure 6, it can be observed that MOEPHO's optimized version of VH. Six nodes (15, 18, 14, 21, 20, and 9) are connected to BS0, and node 9 chooses BS0 as the optimal option, relieving BS3 of its load in MOEPHO. Also, six nodes (with identifiers 22, 11, 12, 13, 24, and 32) are linked to BS1, and in this case, nodes with identifiers 24 and 13 choose BS1 as the optimal approach to alleviate service problems at BS2. The node with the id of 27 has discovered that BS2 is the optimal solution, and it is now connected to six other nodes (ids 25, 10, 31, 16, 30, and 17). At this point in the process, BS3 is connected to six nodes (ids 8, 30, 29, 23, 28, and 19). As a result of our proposed approach, VH will be a very precise, nearly instantaneous, and somewhat energy-efficient procedure.

Figure 5. Initialization of parameters in the network(Naresh, M et al., 2020).

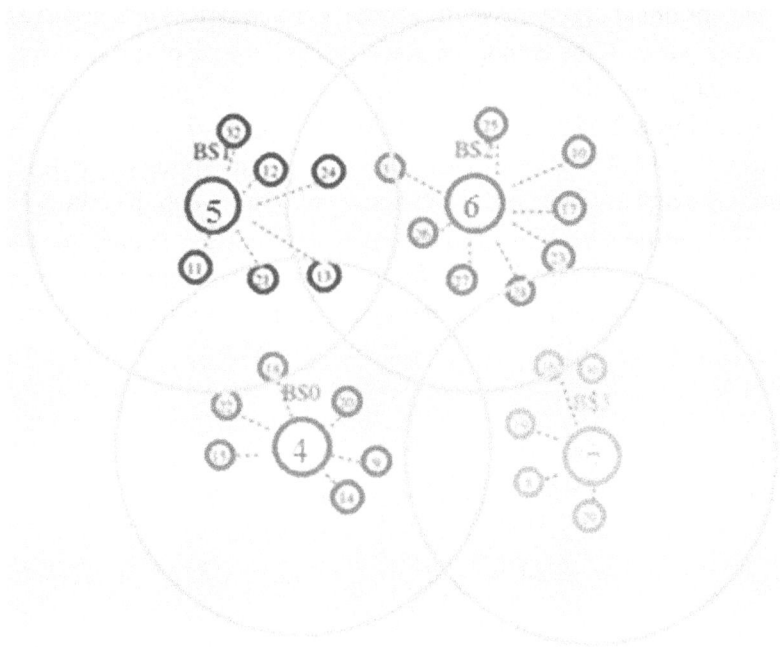

Figure 6. Status of the network after handover optimization using MOEPHO (Naresh, M et al., 2020)

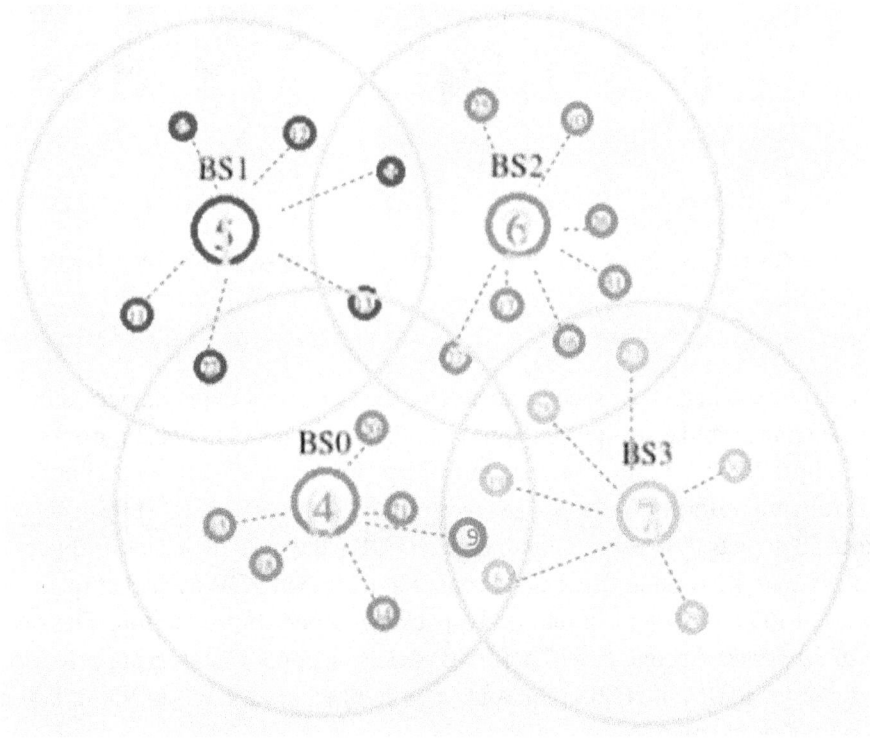

Figure 7 shows the results of comparing the energy used by different methods during VH. The initial energy stored in each node gradually decreases due to the VH technique's operating behavior. VH operations require much more power as the number of nodes grows because of the electricity they waste. The MOEPHO uses less energy than the current systems, as shown in figure 7. To finish the VH process with 25 nodes, MOEPHO uses 7.2 J of energy, while WOA-NN uses 14 J, ACLD uses 15 J, FIDM uses 21 J, and type 2 FLC uses 24 J. The RSS characterizes the BS's reception capabilities regarding signal strength, radio conditions, and distance. RSS might go up or down when the number of nodes changes, depending on the node's position. Because each BS can receive a strong signal, maximizing RSS results in a high data rate. As may be shown in Figure 8, MOEPHO receives significantly more signals than the benchmark techniques. This new technique outperforms prior efforts such as WOA-NN (120KdBm), ACLD (190KdBm), FIDM (280KdBm), and type 2 FLC, exhibiting a 570 KdBm RSS value for 25 nodes (490KdBm).

Figure 7. Consumption of Energy by the nodes in different techniques

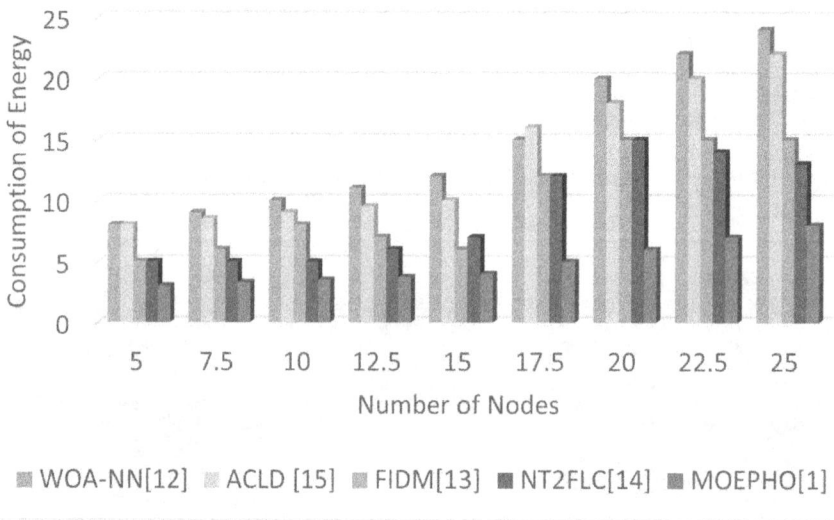

Throughput, expressed in bits per second, measures the amount of data transferred in a specific time. The amount of data transferred grows proportionally with the number of nodes. Figure9 shows MOEPHO throughput of 20 bps, 26 bps, 32 bps, 34 bps, and 42 bps for 5, 10, 15, 20, and 25 nodes, respectively. The suggested work outperforms WOA-NN by 74.86%, ACLD by 47%, FIDM by 40%, and type 2 FLC by 19.2%. Figure 10 depicts the Packet Delivery Rate (PDR), the ratio of successfully delivered packets to the total packets sent. Increasing data traffic causes a decline in PDR as packet rates increase. In the proposed effort, the PDR is 99% for a rate of 10 packets/second in production. The numbers drop to 93% for a rate of 30 packets/second and 87% for 50 packets/second. Considering prior art, the PRD for type 2 FLC is 84%; for FIDM, it is 80%; for ACLD, it is 72%; and for WOA-NN, it is 65% at a rate of 50 packets/second of production.

Figure 8. Received signal strength for various techniques with varied number of nodes

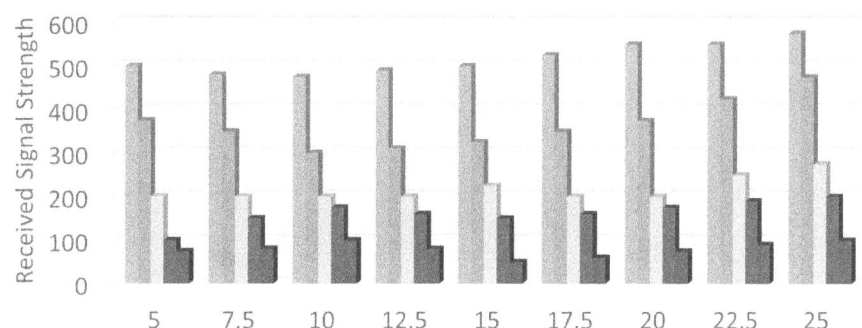

Figure 9. Throughput of various techniques with varied number of nodes

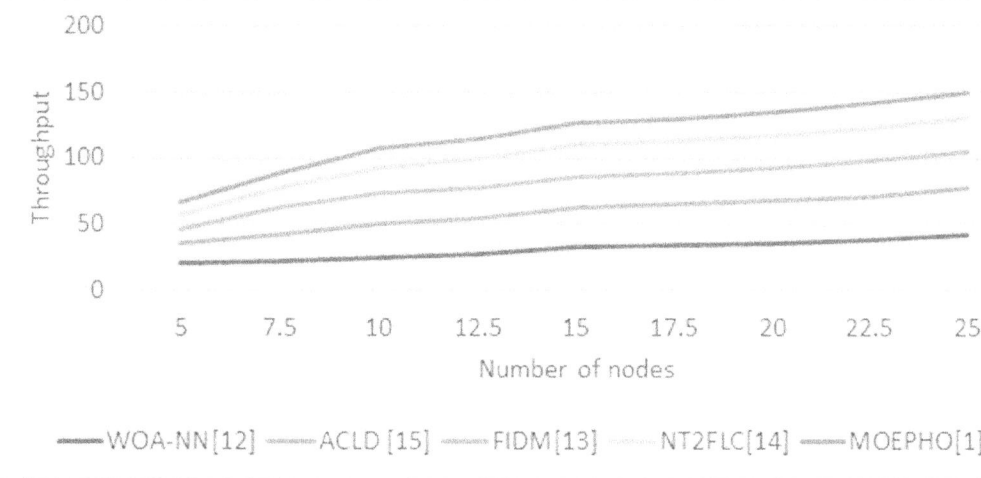

Table 3 shows packet loss performance, and Table 4 shows the handover failure rate. Since transferring a large number of packets results in packet drops because of insufficient reception capacity, packet loss grows in tandem with an increase in packet creation. At a packet generation rate of 10, 20, 30, 40, and 50, MOEPHO shows packet losses of 5%, 10%, 19%, 27%, and 30%. WOA-NN, on the other hand, fares poorly compared to other approaches, with packet loss percentages of 39%, 39%, 43%, 45%, and 58% for packet generation rates of 10, 20, 30, 40, and 50, respectively. The rate at which a call is lost while moving a network from one base station to another is known as the handover failure rate. Table 3 displays the effectiveness of the failure rate concerning the number of nodes. MOEPHO outperforms WOA-NN by as much as 72% when it comes to reducing the failure rate at 25 nodes and by as much as 66% when the number of nodes is 5. The failure rate is reduced by 58% and 44%, respectively, using MOEPHO compared to ACLD and FIDM. Furthermore, MOEPHO has a 23% lower failure rate than type 2 FLC.

Figure 10. packet delivery ratio for varied number of nodes for various techniques

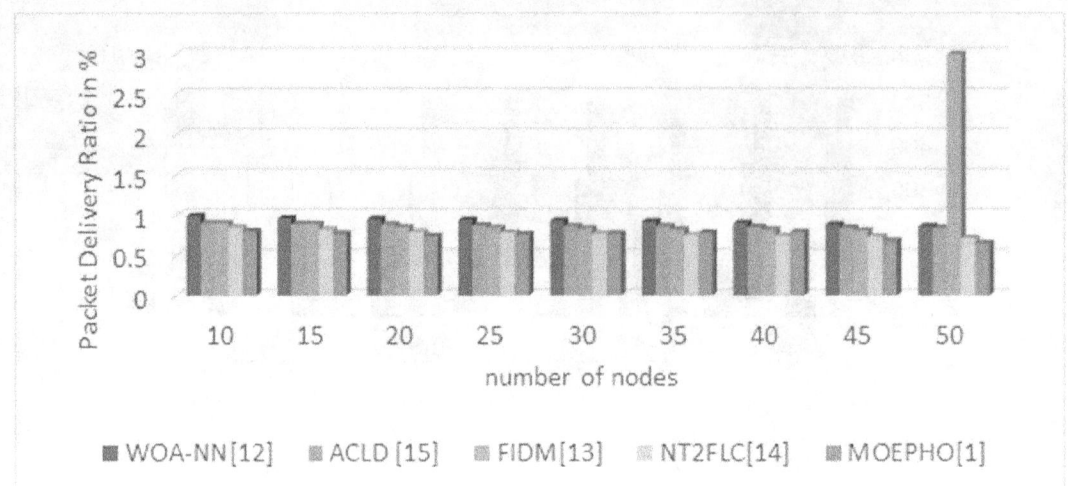

Table 3. Packet loss in % for various number of nodes for different techniques.

Packet Rate(Packets/sec)(X)	Packet Loss(%)				
	WOA-NN[12]	ACLD [15]	FIDM[13]	NT2FLC[14]	MOEPHO[1]
10	0.32	0.4	0.21	0.1	0.4
15	0.34	0.4	0.24	0.14	0.6
20	0.4	0.4	0.26	0.2	0.1
25	0.42	0.4	0.32	0.26	0.16
30	0.44	0.4	0.38	0.34	0.2
35	0.44	0.42	0.38	0.35	0.24
40	0.46	0.43	0.38	0.4	0.28
45	0.52	0.46	0.42	0.42	0.28
50	0.58	0.48	0.44	0.42	0.3

Handover delay, measured in milliseconds, is the time it takes for one BS to hand off control of a connection to another in (ms). Figure 11 shows the handover delay for a range of the node. As more and more devices join the network, there will inevitably be more requests for service from each base station (BS), which can cause delays. The performance of existing comparable approaches is worse than MOEPHO. For example, WOA-NN performs 51% worse than MOEPHO, ACLD performs 47% worse, FIDM performs 29% worse, and type 2 FLC performs 18% worse. It is also common practice to use the succession rate to measure how successfully calls are transferred between two BSs. As the number of nodes increases, the accuracy with which calls are transferred declines, primarily because of an increase in data traffic.

Table 4. Handover failure % for varied number of nodes for different techniques

No. of Nodes(Packet/sec)	Handover failure rate(%)				
	WOA-NN[12]	ACLD [15]	FIDM[13]	NT2FLC[14]	MOEPHO[1]
5	4	3	3.2	2.8	2
7.5	4.1	3.5	3.3	2.85	2.2
10	4.2	3.75	3.3	2.9	2.4
12.5	4.3	3.75	3.5	2.95	2.6
15	4.5	3.75	3.75	3	2.7
17.5	4.5	3.85	3.6	3	2.8
20	4.75	4	3.6	3	2.85
22.5	4.85	4.2	3.7	3	2.9
25	5	4.4	3.7	3	2.95

Figure 11. Handover delay in ms for varied number of nodes for different techniques

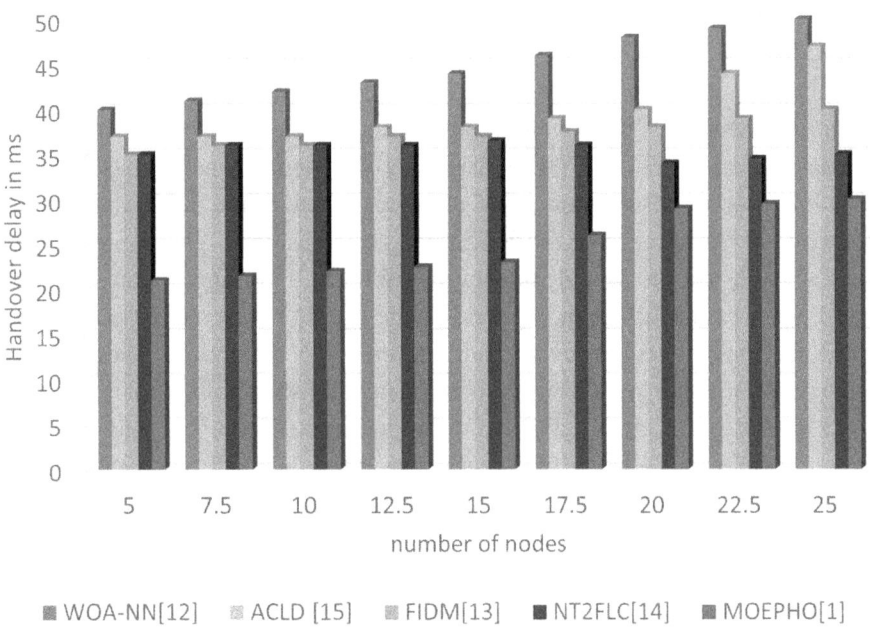

Figure 12 displays the results of a performance analysis of handover accuracy concerning the total number of nodes. WOA-NN, ACLD, FIDM, type 2 FLC, and MOEPHO are all independently shown to achieve 88%, 89%, 94%, 95%, and 98% accurate handover at five nodes, respectively. With 15 nodes, the handover accuracy for WOA-NN, ACLD, FIDM, type 2 FLC, and MOEPHO drops to 83%, 87%, 91%, 93%, and 95%, respectively. However, at 25 nodes, the accuracy rates for WOA-NN, ACLD, FIDM, type 2 FLC, and MOEPHO drop to 80%, 84%, 90%, 90%, and 92%, respectively. It is evident from the results of the performance comparisons that the suggested method (MOEPHO) improves IEEE 802.21's VH operation in heterogeneous networks.

Figure 12. Handover Accuracy in % for varied number of nodes for different techniques

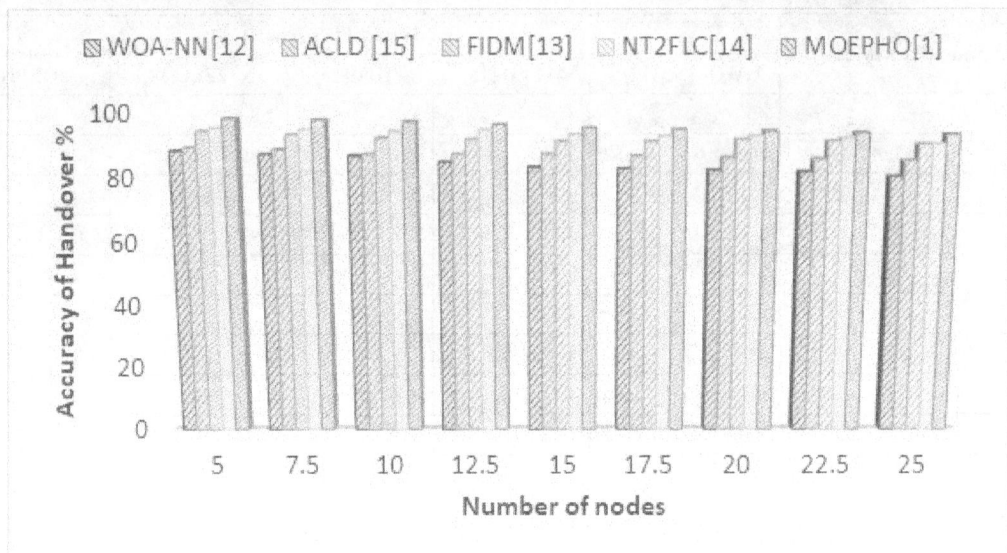

5. CONCLUSION

This chapter discusses the Multi-Objective Emperor Penguin Hanover Optimization (MOEPHO) algorithm for VH management in IEEE 802.21. MOEPHO is designed to function based on several network parameters, including power consumption, connection cost, bandwidth, delay, throughput, the velocity of Mobile Nodes (MN), Bit Error Rate (BER), Signal Noise Ratio (SNR), Received Signal Strength (RSS), and data traffic, in order to provide the best possible result for an accurate handover process. As a result, improved performance has been obtained by the use of NS2 simulation in terms of extremely precise handover, shorter handover latency, and minimal energy consumption. The efficacy of MOEPHO was demonstrated by contrasting it with several previously published works, such as WOA-NN, ACLD, FIDM, and type 2 FLC. Based on the findings of the comparative study, it is evident that the MOEPHO achieves performance that is 13% better than that of the WOA-NN, 9% better than that of the ACLD, and 2% better than that of the FIDM and type 2 FLC. Future research may concentrate on combining two different optimization methods in order to circumvent the challenges presented by individual algorithms when it comes to VH management in heterogeneous networks.

REFERENCES

Ahmed, A., Boulahia, L. M., & Gaiti, D. (2013). Enabling vertical handover decisions in heterogeneous wireless networks: A state-of-the-art and a classification. *IEEE Communications Surveys and Tutorials*, *16*(2), 776–811.

Ahmed, N., & Rikli, N. E. (2018). A QoS Based Algorithm for the Vertical Handover between WLAN IEEE 802.11 e and WiMAX IEEE 802.16 e. *International Journal of Computing and Digital Systems*, *7*(01), 11–22.

Al Emam, F. A., Nasr, M. E., & Kishk, S. E. (2018, December). Adaptive context aware cross-layer vertical handover in heterogeneous networks. In *2018 14th International Computer Engineering Conference (ICENCO)* (pp. 58-63). IEEE.

Aljeri, N., & Boukerche, A. (2018, October). Mobility and handoff management in connected vehicular networks. In *Proceedings of the 16th ACM International Symposium on Mobility Management and Wireless Access* (pp. 82-88).

Anand, R., & Chawla, P. (2022). Bandwidth Optimization of a Novel Slotted Fractal Antenna Using Modified Lightning Attachment Procedure Optimization. In *Smart Antennas* (pp. 379–392). Springer.

Anand, R., Singh, J., Pandey, D., Pandey, B. K., Nassa, V. K., & Pramanik, S. (2022). Modern Technique for Interactive Communication in LEACH-Based Ad Hoc Wireless Sensor Network. In *Software Defined Networking for Ad Hoc Networks* (pp. 55–73). Springer.

Bagubali, A., Verma, T., Anand, A., Prithiviraj, V., & Mallick, P. S. (2018). Performance analysis of handover schemes in heterogeneous networks. *Journal of Circuits, Systems, and Computers*, *27*(11), 1850177.

Bhatt, M. C., Bhardwaj, O., Gupta, P., & Sharma, S. (2018, August). Analysis of handoff prediction algorithms in wireless networks. In *2018 4th International Conference on Computing Sciences (ICCS)* (pp. 40-47). IEEE.

Chattate, I., El Khaili, M., & Bakkoury, J. (2019). Improving modified grey relational method for vertical handover in heterogeneous networks. *International Journal of Advanced Computer Science and Applications*, *10*(2).

Chattate, I., El Khaili, M., & Bakkoury, J. (2019). A New Fuzzy-TOPSIS Based Algorithm for Enhancing Decision Making in a Heterogeneous Network. *Journal of Communication*, *14*(3), 194–201.

Coqueiro, T., Jailton, J., Carvalho, T., & Francês, R. (2019). A fuzzy logic system for vertical handover and maximizing battery lifetime in heterogeneous wireless multimedia networks. *Wireless Communications and Mobile Computing*.

Duong, T. M., & Kwon, S. (2020). Vertical handover analysis for randomly deployed small cells in heterogeneous networks. *IEEE Transactions on Wireless Communications*, *19*(4), 2282–2292.

Khan, B. M., & Bilal, R. (2018). Wireless internet offloading techniques: based on 802.21 medium access control. In Advanced Wireless Sensing Techniques for 5G Networks (pp. 267-283). Chapman and Hall/CRC.

Kunarak, S., & Duangchan, T. (2021, August). Vertical Handover Decision based on Hybrid Artificial Neural Networks in HetNets of 5G. In *2021 IEEE Region 10 Symposium (TENSYMP)* (pp. 1-6). IEEE.

Lahby, M., & Sekkaki, A. (2018). An Efficient Policy for Vertical-Handover-Based Multi-Attribute Utility Theory in Heterogeneous Wireless Networks. In *Advances in Data Communications and Networking for Digital Business Transformation* (pp. 1–20). IGI Global.

Mahajan, P. (2018, November). Review paper on optimization of handover parameter in heterogeneous networks. In 2018 3rd International Innovative Applications of Computational Intelligence on Power, Energy and Controls with their Impact on Humanity (CIPECH) (pp. 1-5). IEEE.

Naresh, M., Reddy, D. V., & Reddy, K. R. (2020, October). A comprehensive study on vertical handover for IEEE 802.21 wireless networks. In *2020 Fourth International Conference on I-SMAC (IoT in Social, Mobile, Analytics and Cloud)(I-SMAC)* (pp. 343-347). IEEE.

Naresh, M., Venkat Reddy, D., & Ramalinga Reddy, K. (2020). Multi-objective emperor penguin handover optimisation for IEEE 802.21 in heterogeneous networks. *IET Communications*, *14*(18), 3239–3246.

Parambanchary, D., & Malleswara Rao, V. (2020). WOA-NN: A decision algorithm for vertical handover in heterogeneous networks. *Wireless Networks*, *26*(1), 165–180.

Parambanchary, D., & Rao, V. M. (2018). WOA-NN: A decision algorithm for vertical handover in heterogeneous networks. *Wireless Networks*, 1–16.

Saeed, M., Kamal, H., & El-Ghoneimy, M. (2018). Novel type-2 fuzzy logic technique for handover problems in a heterogeneous network. *Engineering Optimization*, *50*(9), 1533–1543.

Saeed, R. A. (2019). Handover in a mobile wireless communication network–A Review Phase. *Int. J Comput. Commun. Inf*, *1*(1), 6–13.

Shao, S., Liu, G., Khreishah, A., Ayyash, M., Elgala, H., Little, T. D., & Rahaim, M. (2020). Optimizing handover parameters by Q-Learning for heterogeneous radio-optical networks. *IEEE Photonics Journal*, *12*(1), 1–15.

Shayea, I., Ergen, M., Azizan, A., Ismail, M., & Daradkeh, Y. I. (2020). Individualistic dynamic handover parameter self-optimization algorithm for 5G networks based on automatic weight function. *IEEE Access: Practical Innovations, Open Solutions*, 8, 214392–214412.

Shiwei, G. (2021, July). An Improved KNN Based Decision Algorithm for Vertical Handover in Heterogeneous Wireless Networks. In *2021 40th Chinese Control Conference (CCC)* (pp. 3011-3016). IEEE.

Stamou, A., Dimitriou, N., Kontovasilis, K., & Papavassiliou, S. (2019). Autonomic handover management for heterogeneous networks in a future internet context: A survey. *IEEE Communications Surveys and Tutorials*, *21*(4), 3274–3297.

Yu, H., Ma, Y., & Yu, J. (2019). Network selection algorithm for multiservice multimode terminals in heterogeneous wireless networks. *IEEE Access: Practical Innovations, Open Solutions*, 7, 46240–46260.

Chapter 14
The Role of Two–Dimensional Materials in the Design of Future Wireless Communications Systems

Rafael Vargas-Bernal

ⓘ https://orcid.org/0000-0003-4865-4575

Instituto Tecnológico Superior de Irapuato, Mexico

ABSTRACT

A crucial technology to meet the growing demand for faster wireless communication is terahertz band (0.1–10 THz) communication. New generations of technology, such as 5G and 6G telecommunications, which have faster transmission speeds, larger network capacities, and shorter delays, will find extensive use soon. Due to fabrication and installation restrictions, particularly for smaller sizes, conventional telecommunications devices cannot support the new frequency. Two-dimensional materials have been proposed as the most suitable possibility to design and implement wireless telecommunications devices that meet these requirements. This chapter describes the advances made in the design of antennas, resonators, and electromagnetic interference shielding systems based on graphene and MXenes. Despite the advances achieved so far, future research directions required to commercialize the developed test-stage devices are also described. These materials must be investigated in this century to guarantee the success of 5G and 6G communications.

INTRODUCTION

In this decade, terahertz band communication is expected to become a reality as a crucial wireless technology to support wireless Terabit-per-second (Tbps) links (Akildiz, 2014). Current wireless systems' capacity and spectrum constraints will be reduced by THz band communication, which will also open new opportunities for applications in traditional networking domains and cutting-edge nanoscale communication paradigms. Removing the current wireless systems' spectrum shortages and capacity

DOI: 10.4018/978-1-6684-7000-8.ch014

constraints will also enable a huge spectrum of eagerly anticipated functions in diverse industries. The THz band is the spectral region in the 0.1 to 10 THz range. So far the most studied telecommunications ranges are microwaves and far infrared, however, the range in THz remains one of the least researched bandwidths for wireless communication. THz band communication is expected to be used in 5G and 6G cellular networks, secure terabit wireless communication for military purposes, T-WLAN (terabit wireless LAN), and T-WPAN (terabit wireless personal area networks). In addition, the THz band is intended for the implementation of nanoscale machines, which are defined as nanomachines to wirelessly communicate with one another. Nanomachines are very small functional devices capable of performing basic functions at the nanometer scale including sensing, actuation, data storage, or computation. The size of a nanomachine is in the range of a few hundred cubic nanometers for each component and, at most, a few cubic micrometers for the entire machine. According to cutting-edge research, technologically advanced nanoscale transmitters and antenna elements operate in the THz band. Some specialized applications include nuclear, biological, and chemical defenses as well as the Internet of Nano Things (IoNT), wireless networks-on-chip connectivity, and health monitoring systems. Terahertz band device technologies face numerous challenges, such as the need to develop novel transceiver configurations with low noise capabilities, high sensitivity, and high power, as well as the requirement for multi-band and ultra-broadband antennas to support connections at rates of Gbps (Gigabits-per-second) and Tbps (Terabits-per-second) for THz band with minimal channel path loss. The antennas inside a THz band communication network, like the transceiver, must be capable of working in a transmission bandwidth in the GHz to THz range.

Future wireless communications standards like 6G and beyond have high bandwidth and data rate requirements that can no longer be fulfilled by available moment technologies, which had mostly based on three-dimensional materials found in nature (Abohmra, 2022). The terahertz (THz) band represents the potential frequency range for the implementation of wireless communication networks not feasible in microwave and infrared bands. The recent rediscovery of the potential application in the Terahertz (THz) band of two-dimensional materials has promoted the use of graphene, the transition metal dichalcogenides (TMDs), and perovskites, which have the potential to solve some lengthy issues concerning the development of effective controls for THz wave transmission and detection. Graphene, TMDs (transition metal dichalcogenides), MXenes, and MOFs (Metal-Organic Frameworks) are case studies of 2D materials with significant electrical properties which can be used in THz devices to develop effective systems for future wireless communication systems. Two-dimensional materials, both in their monolayer and multilayer versions, have unique physical properties in the electrical, thermal, and mechanical sectors. These properties are used in the development of transceivers/receivers' radio frequency (RF) front-end components, mixers, modulators, oscillators, switches, as well as amplifiers required for different signal modulators (Zhu, 2020). Since 2D materials have high mobilities, when they are applied as active materials in the channels of RF transistors, high cut-off frequencies, and high gains in analog and RF circuit design can be achieved. 2D materials are the next generation of smart materials for wireless communications, used to design flexible, miniature, wide bandwidth, and reliable patch-type RF monopole antennas with omnidirectional radiation for portable communication devices (Gund, 2019). MXenes, a distinctive family of 2D materials, had already displayed superior qualities across many wireless communications because of their outstanding high flexibility, mechanical stability, electrical conductivity, and ease of processing (He, 2021).

The development of wireless communication technology is facing challenges in the implementation of multipurpose and multifunctional devices (Sa'don, 2020). Essentially, these features are necessary because the contemporary world requires multiple forms of communication and multitasking that require an antenna that can adapt to any environment or frequency. A tunable antenna is therefore appropriate because it performs similarly to many antennas. Due to its ability to operate at many frequencies, the device tends to be smaller, have more functions, and be more beautiful while also reducing the size of the system and the number of components. A special material that has tuning properties is graphene. This has complicated surface conductivity, which is managed by chemical potential, which is the source of this property. It is difficult to implement tunable graphene antennas effectively because most of their properties have only been researched at terahertz frequencies. However, significant developments have been made to increase this potential.

Radar stealth material, also known as microwave absorption material, is capable of absorbing microwave radiation without dispersing or reflecting it (Li, 2021). The basic idea of microwave absorption is to convert microwaves into thermal energy, which is then dissipated into the environment via a variety of absorption methods. Microwave absorption materials for applications must possess a greater sufficiently microwave absorption rate as well as a wide absorption frequency band. They should also possess additional qualities including a low surface density, thinness, environmental friendliness, and mechanical strength.

In recent years, two-dimensional materials have become the most prominent materials in electromagnetic interference (EMI) shielding applications (Kumar, 2019). Layered structures, thin films, and hybrid structures are now being produced using two-dimensional materials, including a focus on the influence of structure, thickness, size, and interfacial interactions on shielding efficiency. Despite notable advancements in this area, there are still several opportunities to investigate useful EMI shielding materials.

Rapid advances in wireless technologies for heat dissipation, electromagnetic shielding, and signal transmission necessitate the designing of novel materials with opposed functions such as mechanical strength, flexibility, and electrical conductivity (Yuan, 2020). Electromagnetic interference (EMI) shielding is a strategic technique that has had one of the most significant influences on the rapid expansion of wireless technology due to the growing physical environmental damage caused by electromagnetic (EM) wave radiation. More efficient EMI shielding materials are therefore critical for resolving this issue. Conducting polymer matrix composites due to their high flexibility, low cost, lightweight, and good corrosion resistance are considered feasible alternatives to traditional metal-based shielding materials for wireless technology devices applied in aircraft, medical equipment, robots, and flexible electronics. However, it is now crucial to address the issue of developing effective, adaptable, and satisfying composite shielding materials. The next generation of reinforcements for these composite materials is the use of two-dimensional materials either of the mono-elemental, heterostructures or hybrid types.

This chapter presents, a comprehensive assessment of current developments and prospects in two-dimensional materials with different designs for EM wave absorption, EMI shielding, and wireless communication. Also, it is discussed how two-dimensional materials function across a range of frequency ranges, as well as how they are used for future wireless communication systems. The chapter has been divided into the following sections: In the section entitled Materials for Traditional Wireless Communication Systems, materials that were commonly valid for radio frequency applications for applications up to GHz but that do not operate satisfactorily at THz are described. Below is the section titled Advantages of 2D Materials Used in Future Wireless Communications Systems, the advantages offered by 2D materials for 5G and 6G applications are discussed. In the section entitled Solutions and Recommendations for

Future Wireless Communications Systems, some of the contributions that have so far been achieved by their Terahertz operation using two-dimensional materials are summarized. In the section entitled Future Research Directions, some of the lines of research that should be developed to optimize the success of two-dimensional materials in wireless telecommunications are discussed. Finally, in the section called Conclusions, the main findings of the chapter are described.

MATERIALS FOR TRADITIONAL WIRELESS COMMUNICATIONS SYSTEMS

The market for portable devices is expanding quickly, and the demand for wireless communication is rising, which is causing a lot of electromagnetic radiation confusion among the many communication channels (Kumar, 2019). This electromagnetic radiation generates unwanted electromagnetic energy, which interacts with the operation condition of surrounding electronic equipment and components, as well as device performance and lifetime. Electronics, communications, aviation, medical technology, and the military now face many difficulties due to electromagnetic radiation. Additionally, this unwanted radiation pollution poses a major risk to living things because it has been linked to various health issues like brain cancer. However, a recent World Health Organization (WHO) research suggests that the effects of radiation on living things are still quite debatable. Current and emerging wireless technologies necessitate the use of appropriate shielding materials that can block harmful radiation while also protecting the functionality and longevity of the devices. Due to the growing demand for portable devices, shielding materials for current-generation electronics must be flexible and lightweight.

Traditional metals such as nickel, iron, copper, silver, and others as well as their alloys are commonly used as EMI shielding materials. However, they suffer from high density and are susceptible to corrosion which limits their potential use in a wider range of compact and intelligent electronic systems (Song, 2021). Conductive polymer matrix composites exhibit EMI shielding properties that are made by integrating highly conductive fillers into a polymer matrix using a variety of specialized processing techniques. Carbon materials, intrinsically conductive materials, and metal materials are the three main types of electrically conductive fillers utilized frequently in CPC.

To combat the problem of unwanted EMI, metals such as aluminum and copper serve as reflecting shields (Kumar, 2019). EM wave absorption, on the other hand, makes use of magnetic materials that have high permeability. Many different materials have been developed for effective EMI shielding, including metallic magnets, ferrites, ceramics, and hybrids. However, their heaviness and lack of flexibility make them unsuitable for the current electronics era. Among the traditional materials used in electromagnetic shielding applications are CNTs (carbon nanotubes), carbon nanofibers (CN), graphite, as well as CB (carbon black). These allow for a reduction of the density and increase the mechanical resistance. Thermoplastics, rubber, and rubber-based composites have applied the CB to operate as a conducting filler. To achieve shielding effectiveness (SE) of 60 dB, it would be necessary to use a material with a thickness of 7 cm for a large filler content making it incompatible and impractical for flexible applications. Subsequently, carbon fibers were proposed for electromagnetic interference (EMI) shielding to exploit their large surface area, low percolation threshold or filler content, high electrical conductivity as well as high mechanical strength. CNTs as well as mixtures with them can be electrical fillers applied to polymer matrices producing highly electrically conductive composites for effective EMI shielding implementations. CNTs outperform traditional carbon fillers for electromagnetic interference (EMI) shielding making use of the high aspect ratio, high electrical conductivity, as well as tunable diameter.

When composited to polymers, a greater EMI SE may be effectively obtained at a low filling. Although other CNTs-based fabrics displayed higher EMI shielding effectiveness, their poor pairings at the CNTs/base interaction, CNT agglomeration, and high cost prevented them from finding practical applications. As a result, more progress in effective EMI shielding materials is required to be used in today's flexible and portable electronics.

Modern wireless communication systems applied to smart devices and the Internet of Things (IoT) require radio frequency (RF) antennas that are thin and bendable (Gund, 2019). Thanks to their notable electrical conductivity, nanostructured materials like conducting polymers, graphene, carbon nanotubes, MXene, and metals have so far been studied. However, most metallic material-based antennas are thick, which restricts their use in portable and tiny electronic equipment. Metal-based implementations for 2.4 GHz Wi-Fi and Bluetooth systems have thicknesses four times the value of skin depth thanks to the fact that 98% of its electrical current circulates through the outer layers. This leads to the manufacture of antenna elements with thin and lightweight characteristics. Thus, the ability to produce tiny and light antennas is constrained by the required thickness (>5 μm). Thus, a thickness much greater than 30 microns is required for a metal-based implementation of a wireless antenna. Conductive 2D nanomaterials are being studied as potential replacements for traditional metals to address the skin depth problem and build light antennas.

Any communication system's integration requires transmitters, detectors, and modulators; the same is true for THz technology (Abohmra, 2022). To implement feasible Terahertz (THz) wireless systems it is necessary to upgrade radio frequency components such as detectors, signal sources, modulators, and amplifiers. Additionally, most common materials used at microwave frequencies perform terribly at frequencies in the Terahertz (THz) range. The most used metal for the implementation of radio frequency and microwave devices is copper. Both electrical wiring and radiation components in the THz band for radiofrequency applications implemented with conventional materials are not feasible due to their high resistance per unit distance and large sizes. A microwave metal antenna is different from a THz metal antenna. THz antennas, unlike microwave antennas, require bias voltage. The antennas in the THz range need using lasers through fiber or air to be excited, whereas CPW (coplanar waveguide) or a microstrip coaxial wire is necessary to excite conventional microwave metal antennas.

ADVANTAGES OF 2D MATERIALS USED FOR WIRELESS COMMUNICATIONS SYSTEMS

Potential developments, particularly in aircraft, aerospace, and vehicles, as well as rapidly expanding upcoming flexible portable electronics, will necessitate flexible and lightweight EMI shielding materials (Kumar, 2019). To ensure that devices operate without interruption from unwanted signals, it is crucial to develop new flexible, lightweight, and highly electrically conductive modern electronics and communication materials. The most cutting-edge EMI shielding will be used in fighter jets and military aircraft with exceptional radiation absorption powers. Effective electromagnetic interference (EMI) shielding based on 2D materials must satisfy this practical demand because of their amazing features, such as enormous aspect ratios, ultrahigh electrical conductivity, extreme thermal conductivity, and outstanding mechanical stiffness. Due to their special qualities, 2D materials favor high-performance electromagnetic interference (EMI) shielding materials.

Since the discovery of graphene in 2004 by Geim and Novoselov, the research on the development and application of 2D nanomaterials has been deeply explored (Kumar, 2019). As a result, graphene was considered an exciting nanomaterial for different fields of science and technology thanks to its unique properties. Graphene has the following properties: a large aspect ratio, a surface area of 2600 $m^2 \cdot g^{-1}$), an electrical conductivity of 6000 $S \cdot cm^{-1}$, thermal conductivity of 5000 $W \cdot m \cdot K^{-1}$, worthy mechanical stiffness, and Young's modulus of 1 TPa. Many research fields, including batteries, supercapacitors, optoelectronic applications, and catalysis, favor graphene due to its special features. Graphene exhibits significant intraband absorption due to the optical transitions presented in the path from the conduction band to the valence band, making it an appealing material for the sensing, modulation, and generation of terahertz waves (Viti, 2021).

Due to the amazing properties of graphene, researchers created further 2D materials that could have similar properties (Kumar, 2019). Since then, scientists have been developing a wide variety of ultrathin 2D nanomaterials among which are hexagonal boron nitride (*h*-BN), hexagonal graphitic carbon nitride (g-C_3N_4), layered metal oxides (LMOs), layered double hydroxides (LDHs), transition metal dichalcogenides (TMDs) such as TaS_2, MoS_2, WS_2, TiS_2, WSe_2, and $MoSe_2$, MXenes, silicene, aluminene, stanene, germanene, bismuthene, covalent organic frameworks (COFs), metal-organic frameworks (MOFs), black phosphorus (BP), and borophene. Applications such as sensors, electronics/optoelectronics, electromagnetic interference (EMI) shielding, energy storage and conversion, biomedicine, and catalysis have benefited from the progress and advances achieved in the synthesis and use of two-dimensional materials. According to current thinking, functional materials are the primary building slabs for such innovation of THz devices (Abohmra, 2022). In this regard, the perovskite family, hexagonal boron nitride (*h*-BN), transition metal dichalcogenides (TMDs), graphene, as well as other 2D materials are being applied to THz applications. Figure 1 depicts the main applications of two-dimensional materials in wireless communications technology that will be developed in the coming decades.

Figure 1. Two-dimensional materials with THz applications

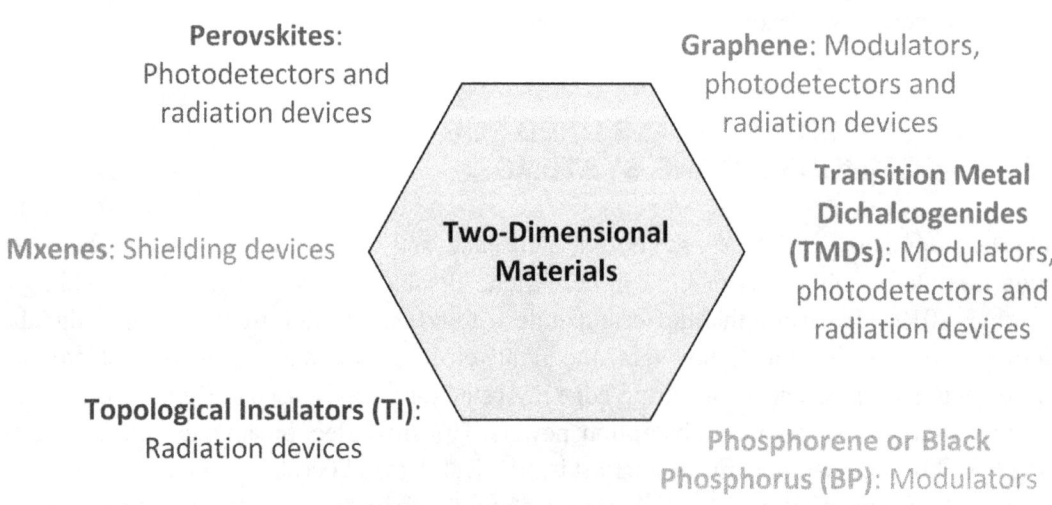

Transition metal dichalcogenides (TMDs) with the chemical formula MX_2 are a type of 2D material, in which M is a group IV transition metal (for instance, Ti, Zr, or Hf), group V transition metal (i.e., Nb, V, or Ta), or group VI transition metal (for instance, Cr, Mo, or W), and X is indeed an element in Group 16 called chalcogen (e.g., S, Se, or Te) (Zhu, 2020). This material with structure X-M-X forms a multilayer composite so that the ends with hexagonal planes chalcogens sandwich metal atoms found in another plane. Transition metal chalcogenides represent a set of nanomaterials with unique electrical, optoelectronic, stoichiometric, and mechanical capabilities. Semiconductor materials with electrical conductivities $<10^{-4}$ S·cm^{-1}, including $MoSe_2$, MoS_2, WSe_2, and WS_2, continue to be studied for applications in electronics, energy storage, and electrocatalysis. 2D metallic TMDCs, such as $NbSe_2$ and NbS_2, have indeed been explored for superconductors and electrical applications under environmental temperature (>1000 S·cm^{-1}).

By taking the A element out of the MAX phase, MXenes can be synthesized as two-dimensional sheet materials (Song, 2021). The valence bond M-X combines metal, ionic, and covalent interactions. More metallic bond elements are present in M-A and A-A bonds. M-A and A-A bonds present smaller strengths than M-X bonds. As a result, the A-layer atoms are more reactive and simpler to peel off. Depending on the n value, MAX phases are separated into 211 (M_2AX), 312 (M_3AX_2), and 413 (M_4AX_3) series, providing different elements and structures. $M_{n+1}X_nT_x$, where T_x stands for the active groups bonded to the surface because of the MAX phase etching (such as = O, -F, and -OH, etc.), can be used to define the generic formula for MXene. Single-layer MXene has diatomic (M) or polyatomic (X) structures, as opposed to reduced graphene oxide's (rGO) monoatomic structure. M-X bond combines metal, ionic, and covalent bonds. Both the M-X bond in MXenes and the C-C bond in the rGO are comparable, but the MXenes's M-X bond has more plentiful and adaptable features than the C-C bond in rGO.

A stable single-element dipole-dipole semiconductor based on phosphorus and called black phosphorus or phosphorene with a honeycomb pattern is also a two-dimensional material of interest. Its bulk version has a direct bandgap of 0.3 eV while the monolayer version has a bandgap of 1.0 eV. These values are intermediate between those achieved by graphene and by transition metal dichalcogenides (Viti, 2021). As a result, it offers intermediate mobility between graphene and TMDs (at room temperature, it achieves a value of 1000 cm^2·V^{-1}·s^{-1}) as well as middle switching features (on-off current ratios of approximately 10^5 are possible in Field-Effect-Transistor's configuration).

SOLUTIONS AND RECOMMENDATIONS FOR FUTURE WIRELESS COMMUNICATIONS SYSTEMS

With conventional solid-state devices, which are heavy and rigid, textile and flexible technologies cannot be realized, but 2D electronics may (Zhu, 2020). 2D electronics can integrate the radiofrequency topologies of WiFi sensing devices and integrated systems using a few basic building blocks. These fundamental building components include active RF components such as signal modulators and demodulators, variable-gain amplifiers, and low-noise amplifiers, and passive RF components such as filters and small antennas. Layered 2D materials, which include TMDs (transition metal dichalcogenides), graphene, and BP (black phosphorus), as well as their associated van der Waals (vdW) heterostructures, are sparking the curiosity to research their applications due to their unique optical and electrical features, due these allow the spreading, manipulation, and sensing of THz waves with previously unheard-of precision (Viti, 2021). Until now, the main wireless applications operating at THz based on 2D materials are shown in Figure 2.

Figure 2. Wireless applications operating in THz based on two-dimensional materials

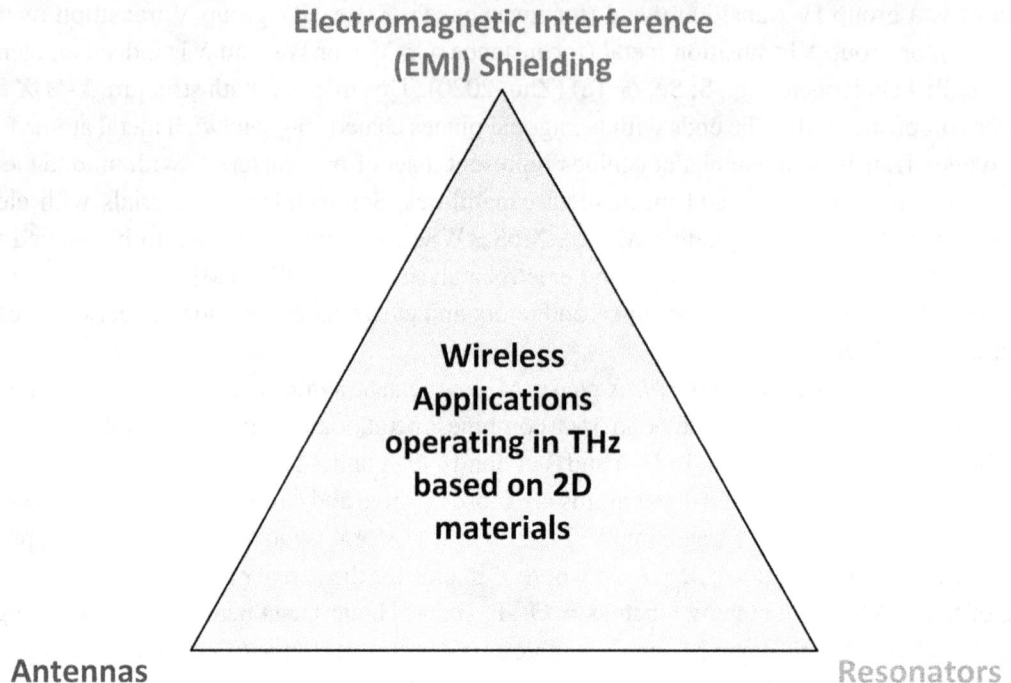

Antennas

Graphene is an exciting 2D material for electrical applications (Luo, 2019). This is driving the development of antennas for applications in the terahertz (THz) range due to their high conductivity, intrinsic tunability, and low skin effect. Until now, several proposals for graphene-based multi-beam tunable antennas operating in the THz range. For these antennas two modulation strategies are possible: 1) the reflector-director model known as the Yagi-Uda antenna, and 2) the reflector-transmission window model. For the case of the proposed antenna, the key attributes are the primary beam direction, front-to-back ratio, peak gain, and frequency response, which are adjusted when the chemical potential of the graphene used is modified. It was also shown that multiple nanoscale wireless communications technology applications soon will be based on devices with graphene-based structures. Therefore, even sophisticated graphene-based multi-beam antennas can be designed through simple implementation methods. A wide frequency range reaching 1.16 to 1.26 THz is produced thanks to increases in chemical potential and resonance frequency of the device. In this way, the graphene-based dipole antenna has a dynamically tunable operating frequency.

For THz applications, graphene has been suggested and investigated (Colmiais, 2022). Graphene has primarily been used for the modeling, analysis, and complex calculation of THz antenna elements. Nevertheless, the modeling procedure and the theoretical hypotheses' proof are extremely complex. Monolayer graphene is used considerably more frequently and its tunability is utilized in the design of THz antennas and waveguides. In the implementation of graphene-based waveguides and antennas, graphene inks/pastes for GHz applications as well as single-layer graphene for THz applications must be used. The great challenges in the manufacture of large-area graphene components and systems are due to quality control and the high difficulty in achieving the perfect monolayer graphene. Kubo's formalism provides the foundation for the modeling of graphene-based antennas for applications in the THz range. In this way, graphene is a worthy competitor for high-frequency applications where radio frequency transistors operating at cut-off frequencies of 427 GHz, passive elements and low-loss interconnects, as well as meta-surfaces operating at 18.9 THz, are required.

Due to the skin depth limit, it seems impossible to obtain excessively thin, flexible, and conformal antennas out of conventional metal materials (He, 2021). To address this weakness, nanomaterials have indeed been investigated for applications in wireless communication. Antennas have been made with thinner patches thanks to the use of graphene and MoS_2. Despite the advances achieved so far, alternative materials tunable in electrical conductivity must be sought to satisfy the requirements of the antennas. The most extensive research has been conducted on patch antennas manufactured of two-dimensional materials. Many polymeric materials, such as polyethylene, rubber, cellulose nano-paper, and many others, should be considered so that flexible patch antennas can increase their efficiency. The mechanical properties of rubber provide patch antennas with the required versatility to be used in wireless communication technologies.

A monopole patch RF antenna made with 2D niobium diselenide ($NbSe_2$) performs well even with sub-micrometer thickness, whose value is less than skin depth when compared to many other materials (Gund, 2019). This prototype antenna achieves omnidirectional RF propagation, a radiation efficiency of 70.6%, a thickness of 855 nm, a sheet resistance of 1.2 Ω sq^{-1}, and its reflection coefficient achieves a value of -46.5 dB. This antenna's resonance frequency is also changed from 2.01 to 2.80 GHz while maintaining the same thickness, length, and less than -10 dB reflection coefficient. This method enables the implementation of miniaturized, flexible, frequency-tunable, omnidirectional antennas based on TMDs for body-centered wearable monopole patch wireless communication systems. To implement antennas made of $NbSe_2$ thin films that offer tunable frequency, omnidirectional propagation as well as maximum performance, it is necessary to operate them at critical temperatures of 3 K reached in intergalactic space. The obtained devices are used in ubiquitous wireless LAN sensor networks for portable and/or wearable electronic parts as well as for space communication systems using superconductor materials.

For applications such as satellites for 6G communications to provide mobile phone services, THz antenna arrays are required (Tabatabaeian, 2021). The critical properties that seek to be resolved through this type of antenna are the increase in harmonic rejection as well as the bandwidth. To increase the antenna bandwidth by 15.2% and to reduce the harmonic return loss between a power divider and an antenna array, a filter in the THz range was introduced which is based on graphene. A total gain of 9.07 dB with a sidelobe level of -5 dB was obtained for a directional pattern. Unwanted harmonics in the array structure can be eliminated using the power dividers' graphene load. The resulting antenna has greater bandwidth and can handle wireless communication because of its 9 dB gain and coverage of 0.45 to 0.5 THz. Furthermore, the proposed antenna was developed using a variety of approaches that reduced the antenna's mutual coupling by even more than 6 dB.

A photo-thermoelectric (PTE) mechanism can be used to detect THz light photons in 2D materials (Viti, 2021). It's significant to note that the main detecting method in 2D material-based designs can be specifically customized. It is possible to preferentially activate different THz detection dynamic behavior in the black phosphorus-field-effect transistor (BP-FET) by trying to adjust the factor that allows relating the axis of polarization of the antenna in the Terahertz (THz) range with the axial direction of the phosphorene armchair. The band structure of black phosphorus is affected by in-plane crystallographic anisotropy, resulting in anisotropy of its thermal (thermal conductance), electrical (effective mass, mobility), and optical (linear dichroism) properties. To perform photothermoelectric detection, it was necessary to test different geometries of detectors to exploit the properties of phosphorene more efficiently. The use of materials with van der Waals heterostructure (vdW) allowed the development of phosphorene-based field effect transistors which reached a signal-to-noise ratio of 43 dB, and noise-equivalent powers (NEPs) with a value of 5 nW·Hz$^{-1/2}$ for frequencies of 3.5 THz or 58 pW·Hz$^{-1/2}$ for frequencies of 0.1 THz, and with response time in microseconds.

To summarize the contributions reported so far by research groups around the world regarding antennas, the main results have been compiled in Table 1. It can be identified that graphene is the main material used to implement antennas. Phosphorene is an emerging material that is being used as a proposal for future development. The frequency range is from 0.45 to 18.9 THz, which depends on the topology used as well as the materials used in the substrate. The frequency range achieved can be a single frequency or tunable over a specific range or different frequency values. The gain ranges from 0.94 dB to 43 dB in the same way depending on the implementation developed. The possible trend in the range of operation in this decade will be feasible in the range of THz in the antennas.

Table 1. Performance parameters of antennas based on two-dimensional materials

Two-dimensional material	Frequency	Gain	Reference
Graphene	1.16 to 1.26 THz	0.94 to 10 dB	Luo, 2019
Graphene	0.45 to 0.5 THz	9 dB	Tabatabaeian, 2021
Graphene	18.9 THz, 13.49 THz, or 5.72 THz	3 dB	Ghosh, 2021
Black phosphorus or Phosphorene	0.1 to 3.5 THz	43 dB	Viti, 2021

Resonators

One of the great challenges to implementing components operating in the terahertz (THz) range is to design materials with high electrical properties as well as tunability across electrical biases. One of these materials is based on a heterostructure between multilayer graphene intercalated with hexagonal boron nitride (*h*-BN-IMLG) (Li, 2017). A capacitor was proposed to serve as a THz resonator using this heterostructure which increases planar conductance by making use of a flat mechanical structure with a reduced number of charge traps. This structure operates as follows: the graphene layers, being conductive, dominate the tunneling presented, and the boron nitride layers, being insulators, are responsible for the decoupling of the generated signals. In this way, the experimental simulation results achieved by the resonator in THz offer a calibration ratio of 4.74% when applying 0.8 V as bias voltage, reaching a quality factor (*Q*) of more than 20, which makes the application feasible of this heterostructure. A significant turning ratio of up to 21% can be achieved by carefully choosing the resonator design and electrical properties of graphene. The tuning ratio and the *Q*-factor conflict.

To detect THz radiation, antenna-coupled mechanical resonators made of nanoscale two-dimensional materials were proposed (Hassel, 2017). Graphene or a 2D material equivalent suspended in a voltage-dependent capacitance is used in the THz detection system to mix electrical signals. Coherent detector operation with a relatively high resonant frequency was discovered as an approach for quantum-restricted THz sensing. The implementation of this detector using two-dimensional materials such as graphene thanks to its outstanding electrical and mechanical properties allows an extraordinary sensitivity that should be highlighted. Among the physical properties that make graphene interesting for its sensitivity are its extremely low weight as well as the high Young's modulus of 1 TPa. Because the presented capacitive behavior must be equated with circuit elements or electromagnetic cavities it is possible to achieve optimal radiation detection. This setup couples a THz regime antenna to a mechanical resonator. The mechanical resonator is pushed by the signal received by the antenna so that the detection is carried out through the mechanical vibrations produced.

To summarize the contributions reported so far by research groups around the world regarding resonators, the main results have been compiled in Table 2. In this table, it can be identified that the most used two-dimensional material to implement resonators is graphene. A frequency range in the interval from 0.2 to 1.4 THz is possible, where the topology and the material used in the substrate have a strong dependence on the value achieved. The range that can be reached can take a single value or a tunable interval. When comparing the value of the quality factor achieved by these 2D materials with those achieved by previously reported superconducting materials, it can be lower or higher than 2 orders of magnitude. The possible trend in the operating range in this decade will be feasible in the THz range in resonators with quality factors even better than other traditional materials.

Table 2. Performance parameters of resonators based on different materials

Material	Frequency	Q-factor	Reference
NbTiN thin film	1 THz	10^4	Brandstetter, 2012
MgB$_2$ thin film	2.5 THz	10^4	Brandstetter, 2012
Graphene	0.2-1.0 THz	10^6	Hassel, 2017
h-BN-Graphene	1.0-1.4 THz	20	Li, 2017

Electromagnetic Interference (EMI) Shielding

To implement electromagnetic interference (EMI) shielding applications it is possible to use two-dimensional materials either in a single form, in heterostructures, or as composite materials using graphene, Mxenes, or MoS$_2$ (Kumar, 2019). Among the practical possibilities is the use of diverse two-dimensional materials with unique electrical (insulators and conductors) and magnetic (diamagnetic and ferrimagnetic) properties, to promote electromagnetic (EMI) shielding materials. For this application, two-dimensional materials represent interesting contenders to materials such as carbon nanotubes, carbon fibers, conducting polymers, magnetic materials, and metals. Such 2D materials have several advantages over conventional in terms of EMI shielding. Their strong electrical conductivity, a crucial component of a successful shield, is their first advantage. Their 2D structure may also provide designers with more creative freedom in developing the required hybrid structural materials. Third, 2D materials are lightweight and flexible, which is important for today's electrical devices. Additionally, the 2D material structure is flexible. The creation of effective 2D materials for EMI shielding applications may present certain difficulties.

As a result of the rapidly growing wireless devices and flexible electronic devices, flexible materials operate well enough as shields for highly demanding applications from an electromagnetic point of view (Yuan, 2020). Future wireless technologies and electronic devices, on the other hand, should be capable of withstanding a broad range of deformations, such as bends, folds, narrows, curvatures, and twists. Because electrical conductivities decrease rapidly with increasing tensile strain, it is incredibly challenging to design materials that can withstand stretching processes without losing good shielding performance. To meet this demand, a fabric with a subtle wrinkle design made of electrospinning polyurethane (PU) nanofibers layered in MXene ($Ti_3C_2T_x$) was developed. The sandwich-shaped, as-prepared MXene/PU fabric has consistent mechanical characteristics and therefore can maintain shielding effectiveness (SE) (>20 dB) while being stretched within a 30% deformation range in the X-band. Due to their remarkable qualities, composite textiles are excellent candidates to develop wireless communications applications including flexible electronic devices as well as robotic joints using electromagnetic shielding. The composite fabrics also have other benefits including reduced weight, simple production, and strong mechanical stability.

Two-dimensional materials show more diverse properties than solid matter structures because at the nanoscale and with a two-dimensional arrangement, electrons move freely in the structural plane or even out of it without any obstruction (Li, 2021). Due to the unusual configurations of these materials, it is possible to take advantage of their active surface, high electrical conductivity, as well as high mechanical resistance to develop radar stealth as well as absorption of electromagnetic waves. Extraordinary components for microwave-absorbing wireless communication systems are based on two-dimensional materials such as MXene, graphene, MoS_2, and their combinations with several other insulating or magnetic materials. The layouts of all composites made from two-dimensional materials have all been defined by the structures of these materials. Such nanocomposites may have enhanced electrical, heat transfer, and mechanical properties in addition to physical properties including large specific surface area, low density, and excellent thermal stability. Most notably, they meet the so-called "light, strong, wide, and thin" characteristics for microwave absorption performance. The higher microwave absorption ability of these materials is primarily due to their functionalized substrate surface, numerous defects, and exhaustive interlayer spacing, which results in dipole polarization. The three primary types of graphene-based nanocomposites used in microwave absorption are nanometal-graphene, conductive polymer-graphene, and conductive polymer-nano metal-graphene. Ferrites, especially spinel based on iron, have been blended with graphene to create composites with high microwave absorption properties. Organic materials could be mixed into the material to create a ternary nanocomposite, which improves the attenuation features and output impedance within the material.

Two-dimensional (2D) materials are ideal for wireless communication and EM attenuation materials due to their unique electrical properties, high aspect ratios, and low weight (He, 2021). MXenes, a unique set of 2D materials, have significant benefits for wireless communications applications in terms of EM wave absorption, EMI shielding, wireless communication and due to their inherent characteristics, among which are high mechanical stability, high electrical conductivity, broad family tree, ease of processing, and high flexibility stand out. Because of MXenes' superior chemical and physical properties, it is now possible to develop pure MXenes and hybrids as electromagnetic absorption materials with precisely controlled structures, such as films, textiles, aerogels, and foams. Avoiding conduction losses is a challenging goal due to the possible decrease in electrical conductivity presented by defects in the two-dimensional structure. These defects modify the transfer of electrons produced by electromagnetic waves in the structure.

Nitrides, carbides, and carbonitrides based on transition metals with two-dimensional structures at the nanoscale are exploited due to their high electrical conductivity, chemical activity, and hydrophilicity for electromagnetic shielding applications (Song, 2021). The binding sites on the surface of these MXenes are fully infiltrated with electrical conductivity at the same time. Inert gas or reducing gas treatment at high temperatures can effectively remove -OH, -F, as well as other functional groups. It will, however, outcome mostly in the oxidation of MXenes.

Due to the rapid development of terahertz technology, it is necessary to provide reliable electromagnetic environments through shielding materials to reduce electromagnetic interference (Zhu, 2021). The biomineralization process inspired the development of hydrogel-like shields to protect material which consists of poly(acrylic acid) and MXene. With its good elastic modulus and reusability, favorable form adaptability and poor adhesion, and quick self-healing potential, the composite hydrogel demonstrated significant application flexibility and dependability. Shielding performance in these materials under humid environments is dominated by absorption which reduces electrical conductivity because the hydrogel introduces a porous structure, and this leads to an internal ecosystem with a large amount of water. However, a very thin nano-gel (0.13 mm) can lead to shielding effectiveness of 45.3 dB across a frequency band of 0.2 to 2.0 THz under a return loss of 23.2 dB.

It was proposed to create a terahertz absorber from an ultra-thin bandwidth MXene/rGO nanocomposite using a sacrificial PMMA circular template (Li, 2022). The 5-layer sandwich-like layout has a licensed bandwidth that covers the entire test band in the range of 0.37 to 2.0 THz with a width of 0.148 mm which is excessively thin even because it achieves an absorption loss value of 57.7 dB. On average, the absorption level is 43.2 dB while the absorption percentage reaches a value of 99.999%, which makes the composite material based on two-dimensional heterostructures spectacular. Furthermore, when compared to other materials investigated thus far, parameters such as effective absorption bandwidth, maximum EMI shielding efficacy, and maximum reflection loss reach the values of 11.1 THz·mm^{-1}, 366.2 dB·mm^{-1}, as well as 389.9 dB·mm^{-1}, respectively. The results achieved to provide the stimulus to continue investigating alternatives that allow the design of high electromagnetic absorption materials for wide bandwidths in the THz range of very thin thicknesses. This work can serve as a blueprint for the construction of high-performance MGF broadband THz absorbers with sandwich-like structures for use in future 6G communications as well as electromagnetic interference shielding, radar absorption, and military detection applications.

To summarize the contributions so far reported by research groups around the world regarding shielding from electromagnetic interference, the main results have been compiled in Table 3. In it, it can be identified that Mxene is the main two-dimensional material used to implement shields against electromagnetic interference. The operating frequency where this shielding is effective is between 0.2 and 2.5 THz, which depends on the topology used as well as the materials used in the substrate. The frequency range achieved can be a single frequency or tunable over a range. The effectiveness of electromagnetic shielding can be low or in the range of 43 dB. The possible trend in the operating range in this decade will be feasible in the THz range in shields with even better shielding effectiveness than other traditional materials.

Table 3. Performance parameters of electromagnetic interference (EMI) shielding based on two-dimensional materials

Material	Frequency	Shielding Effectiveness (SE)	Reference
MXene	0.5-2-5 THz	2.5-3 dB	Li, 2020
MXene	0.2-2.0 THz	45.3 dB	Zhu, 2021
MXene/rGO	0.37-2.0 THz	43.2 dB	Li, 2022

FUTURE RESEARCH DIRECTIONS

Wireless telecommunication and sensor technologies, which serve as the building blocks for IoT systems, show a constant expansion regarding the number of subscribers and the amount of information transmitted (Zhu, 2020). Despite notable advancements in this area, there are still several opportunities to investigate useful EMI shielding applications (Kumar, 2019). The choice of two-dimensional nanoscale material to design components for electromagnetic shielding at a commercial level is based on cost reduction, so alternative materials that offer this possibility will be preferred. However, the quality of interference shielding can definitely be improved even if the cost is high for high-value-added applications. For example, the environment and the shield design will determine the best use of 2D materials. The useful facilitating of two-dimensional nanomaterials and tuning of the different topologies are key components in the design of thin and flexible electromagnetic interference (EMI) shielding materials. Therefore, the complete study of physical properties is crucial for achieving the best shielding capability of 2D nanomaterials. It is here that the development of high-performance barrier material starts and ends. These nanostructures, including MXene, graphene, etc., might be doped with defects that would drastically alter the properties of 2D nanomaterials. Exploring the synthesis of polymer hybrids with regulated doping or defects on all those materials is important to establish a defect/doping synergy impact with the capabilities of the polymeric host. Despite the advancements made so far, technical issues such as low efficiency, sizes in the millimeter range, low environmental stability, low operation stability, poor optical absorption, and low carrier mobility even at room temperature have yet to be resolved (Shi, 2022). Future research directions for two-dimensional materials in THz wireless communications are depicted in Figure 3.

MAX phases and MXenes are more expensive and need more work during preparation (Song, 2021). Large amounts of MXenes are frequently needed to produce the electrical and mechanical properties as well as manufacturing and price issues to produce polymer/MXenes-based nanocomposites with electromagnetic shielding capabilities. It is critical to lower the price of production of dimensional materials based on MXenes and *MAX* phases while also creating novel synthesis methods to achieve industrial-scale manufacturing for wireless communication applications. One of the future research priorities is the synthesis of MXenes that do not include functional groups on their surface for the development of applications in wireless communications. Electromagnetic absorption must be restricted to a certain frequency range to increase the quality of the shielding against interference in the design of devices operated in THz (Shi, 2022). Additionally, most of the existing THz shielding research is focused on reflection techniques, which do not eliminate electromagnetic interference. Hence, research into innovative THz shielding techniques utilizing weak reflection processes is essential, as is research into additional two-dimensional materials than MXene having low density and high shielding coefficients.

Figure 3. Future Research Directions for 2D Materials in Wireless Communications

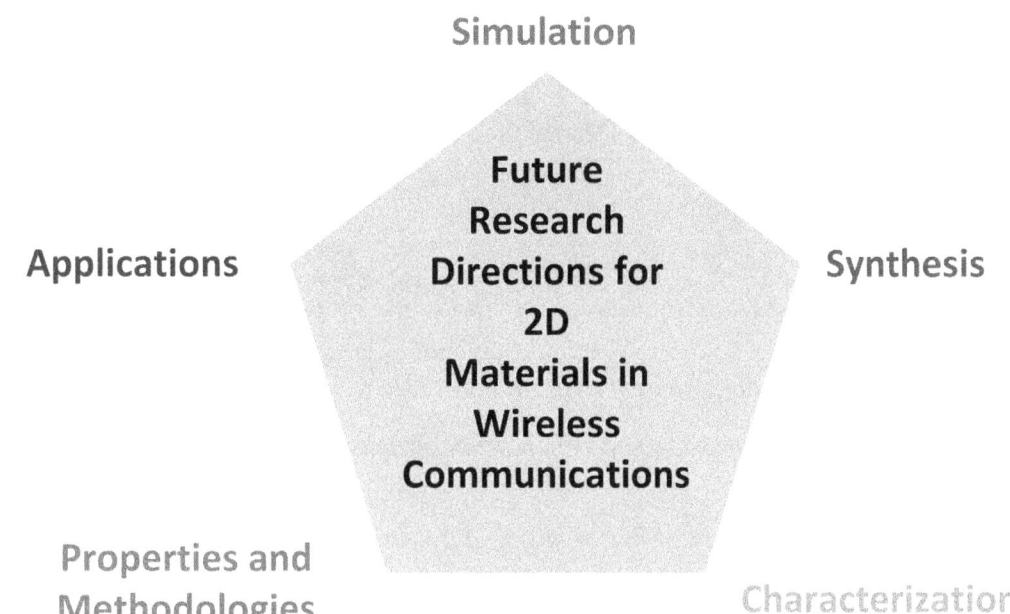

CONCLUSION

This chapter provides an analysis of the possible contributions that two-dimensional materials can offer for the implementation of devices for wireless communications for 5G and 6G technologies as well as for the Internet of Things. 2D materials' newly discovered electrical properties enable a wide variety of RF applications. When compared to traditional RF front-end components, the use of 2D materials for circuits and devices exploits their advantages in terms of price, power efficiency, performance, design, as well as scalability. For the design of soft robotics, sensors, bioimplants, and wearable electronics everywhere on the Internet of Things (IoT) applications, researchers predict that the use of 2D materials will usher in a new paradigm. Lightweight, flexibility, electrical properties, and environmental friendliness are just a few benefits of 2D materials. The use of 2D materials can considerably improve devices for wearable electronics applications operating in the THz range. Also, the main disadvantage of these materials is associated with their high electromagnetic absorption. Two-dimensional materials have special characteristics in THz tunability, modulation, detection, and THz nanosciences, including ultrafast dynamics, extreme field confinement, long-lived collective excitations, and nonlinear light-matter interaction. These characteristics can make two-dimensional materials game-changers for creating novel structures and creating cutting-edge technologies in the future. Finally, it is determined that several issues about the THz communication system need to be resolved as well as the necessity for future development of the current technology in the upcoming years.

ACKNOWLEDGMENT

The author wishes to thank his wife and son for their assistance while he was performing this scientific research and expresses gratitude to the Instituto Tecnológico Superior de Irapuato (ITESI) for their support. The author thanks the researchers who shared their articles to complete this study.

REFERENCES

Abohmra, A., Khan, Z. U., Abbas, H. T., Shoaib, N., Imran, M. A., & Abbasi, Q. H. (2022). Two-Dimensional Materials for Future Terahertz Wireless Communications. *IEEE Open Journal of Antennas and Propagation*, *3*, 217–226. doi:10.1109/OJAP.2022.3143994

Akildiz, I. F., Jornet, J. M., & Han, C. (2014). Terahertz Band: Next Frontier for Wireless Communications. *Physical Communication*, *12*, 16–32. doi:10.1016/j.phycom.2014.01.006

Brandstetter, M., Benz, A., Deutsch, C., Detz, H., Klang, P., Andrews, A. M., Strasser, G., & Unterrainer, K. (2012). Superconducting Microdisks Cavities for THz Quantum Cascade Lasers. *IEEE Transactions on Terahertz Science and Technology*, *2*(5), 550–555. doi:10.1109/TTHZ.2012.2212321

Colmiais, I., Silva, V., Borme, J., Alpuim, P., & Mendes, P. M. (2022). Towards RF Graphene Devices: A Review. *FlatChem*, *35*, 100409. doi:10.1016/j.flatc.2022.100409

Ghosh, S. K., Das, S., & Bhattacharyya, S. (2021). Transmittive-type Triple-band Linear to Circular Polarization Conversion in THz Region using Graphene-based Metasurface. *Optics Communications*, *480*, 126480. doi:10.1016/j.optcom.2020.126480

Gund, G. S., Jung, M. G., Shin, K.-Y., & Park, H. S. (2019). Two-Dimensional Metallic Niobium Diselenide for Sub-Micrometer-Thin Antennas in Wireless Communication Systems. *ACS Nano*, *13*(12), 14114–14121. doi:10.1021/acsnano.9b06732 PMID:31746198

Hassel, J., Oksanen, M., Elo, T., Seppä, H., & Hakonen, P. J. (2017). Terahertz Detection using Mechanical Resonators based on 2D Materials. *AIP Advances*, *7*(6), 065014. doi:10.1063/1.4990405

He, P., Cao, M.-S., Cao, W.-Q., & Yuan, J. (2021). Developing MXenes from Wireless Communication to Electromagnetic Attenuation. *Nano-Micro Letters*, *13*(1), 115. doi:10.100740820-021-00645-z PMID:34138345

Kumar, P. (2019). Ultrathin 2D Nanomaterials for Electromagnetic Interference Shielding. *Advanced Materials Interfaces*, *6*(24), 1901454. doi:10.1002/admi.201901454

Li, G., Amer, N., Hafez, H. A., Huang, S., Turchinovich, D., Mochalin, V. N., Hegmann, F. A., & Titova, L. V. (2020). Dynamical Control over Terahertz Electromagnetic Interference Shielding with 2D $Ti_3C_2T_y$ MXene by Ultrafast Optical Pulses. *Nano Letters*, *20*(1), 636–643. doi:10.1021/acs.nanolett.9b04404 PMID:31825625

Li, J., Zhou, D., Wang, P.-J., Du, C., Liu, W.-F., Su, J.-Z., Pang, L.-X., Cao, M.-S., & Kong, L.-B. (2021). Recent Progress in Two-Dimensional Materials for Microwave Absorption Applications. *Chemical Engineering Journal*, *425*, 131558. doi:10.1016/j.cej.2021.131558

Li, S., Xu, S., Pan, K., Du, J., & Qiu, J. (2022). Ultra-Thin Broadband Terahertz Absorption and Electromagnetic Shielding Properties of MXene/rGO Composite Film. *Carbon, 194*, 127–139. doi:10.1016/j.carbon.2022.03.048

Li, X., Wu, L.-S., & Mao, J.-F. (2017). High-Frequency Analysis of Intercalated Multilayer Graphene (IMLG) and Implication for Tunable Terahertz Resonator Design. *IEEE Access: Practical Innovations, Open Solutions, 5*, 7532–7541. doi:10.1109/ACCESS.2017.2701506

Luo, Y., Zeng, Q., Yan, X., Wu, Y., Lu, Q., Zheng, C., Hu, N., Xie, W., & Zhang, X. (2019). Graphene-based Multi-Beam Reconfigurable THz Antennas. *IEEE Access: Practical Innovations, Open Solutions, 7*, 30802–30808. doi:10.1109/ACCESS.2019.2903135

Sa'don, S. N. H., Jamaluddin, M. H., Kamarudin, M. R., Ahmad, F., Yamada, Y., Karmadin, K., Idris, I. H., & Seman, N. (2020). Characterisation of Tunable Graphene Antenna. [AEÜ]. *International Journal of Electronics and Communications, 118*, 153170. doi:10.1016/j.aeue.2020.153170

Shi, Z., Zhang, H., Khan, K., Cao, R., Zhang, Y., Ma, C., Tareen, A. K., Jiang, Y., Jin, M., & Zhang, H. (2022). Two-Dimensional Materials toward Terahertz Optoelectronic Device Applications. *Journal of Photochemistry and Photobiology C, Photochemistry Reviews, 51*, 100473. doi:10.1016/j.jphotochemrev.2021.100473

Song, P., Liu, B., Shi, X., Cao, D., & Gu, J. (2021). MXenes for Polymer Matrix Electromagnetic Interference Shielding Composites: A Review. *Composites Communications, 24*, 100653. doi:10.1016/j.coco.2021.100653

Tabatabaeian, Z. S. (2021). Graphene Load for Harmonic Rejection and Increasing the Bandwidth in Quasi Yagi-Uda Array THz Antenna for the 6G Wireless Communication. *Optics Communications, 499*, 127272. doi:10.1016/j.optcom.2021.127272

Viti, L., & Vitiello, M. S. (2021). Tailored Nano-Electronics and Photonics with Two-Dimensional Materials at Terahertz Frequencies. *Journal of Applied Physics, 130*(17), 170903. doi:10.1063/5.0065595

Yuan, W., Yang, J., Yin, F., Li, Y., & Yuan, Y. (2020). Flexible and Stretchable MXene/Polyurethane Fabrics with Delicate Wrinkle Structure Design for Effective Electromagnetic Interference Shielding at a Dynamic Stretching Process. *Composite Communications, 19*, 90–98. doi:10.1016/j.coco.2020.03.003

Zhu, L., Farhat, M., Salama, K., & Chen, P.-Y. (2020). 2D Materials-based Radio-Frequency Wireless Communication and Sensing Systems for Internet-of-Thing Applications. In L. Tao & D. Akinwande (Eds.), Emerging 2D Materials and Devices for the Internet of Things (pp. 29–57). Elsevier. 10.1016/B978-0-12-818386-1.00002-3. doi:10.1016/B978-0-12-818386-1.00002-3

Zhu, Y., Liu, J., Guo, T., Wang, J. J., Tang, X., & Nicolosi, V. (2021). Multifunctional $Ti_3C_2T_x$ MXene Composite Hydrogels with Strain Sensitivity toward Absorption-Dominated Electromagnetic-Interference Shielding. *ACS Nano, 15*(1), 1465–1474. doi:10.1021/acsnano.0c08830 PMID:33397098

ADDITIONAL READING

Deng, R., Chen, B., Li, H., Zhang, K., Zhang, T., Yu, Y., & Song, L. (2019). MXene/Co$_3$O$_4$ Composite Material: Stable Synthesis and Its Enhanced Broadband Microwave Absorption. *Applied Surface Science, 488*, 921–930. doi:10.1016/j.apsusc.2019.05.058

Hu, M., Zhang, N., Shan, G., Gao, J., Liu, J., & Li, R. K. Y. (2018). Two-Dimensional Materials: Emerging Toolkit for Construction of Ultrathin High-Efficiency Microwave Shield and Absorber. *Frontiers in Physics, 13*(4), 138113. doi:10.100711467-018-0809-8

Huang, X., Leng, T., Zhu, M., Zhang, X., Chen, J., Chang, K., Aqeeli, M., Geim, A. K., Novoselov, K. S., & Hu, Z. (2015). Highly Flexible and Conductive Printed Graphene for Wireless Wearable Communications Applications. *Scientific Reports, 5*(1), 18298. doi:10.1038rep18298 PMID:26673395

Jia, H., Yang, X., Kong, Q., Xie, L., Guo, Q., Song, G., Liang, L., Chen, J., Li, Y., & Chen, C. (2021). Free-Standing, Anti-Corrosion, Super Flexible Graphene Oxide/Silver Nanowire Thin Film for Ultra-Wideband Electromagnetic Interference Shielding. *Journal of Materials Chemistry. A, Materials for Energy and Sustainability, 9*(2), 1180–1191. doi:10.1039/D0TA09246K

Papadopoulou, K. A., Chroneos, A., Parfitt, D., & Christopoulos, S.-R. G. (2020). A Perspective on MXenes: Their Synthesis, Properties, and Recent Applications. *Journal of Applied Physics, 128*(17), 170902. doi:10.1063/5.0021485

Sa'don, S. N. H., Jamaluddin, M. H., Kamarudin, M. R., Ahmad, F., Yamada, Y., Kamardin, K., & Idris, I. H. (2019). Analysis of Graphene Antenna Properties for 5G Applications. *Sensors (Basel), 19*(22), 4835. doi:10.339019224835 PMID:31698830

Wu, N., Zeng, Z., Kummer, N., Han, D., Zenobi, R., & Nyström, G. (2021). Ultrafine Cellulose Nanofiber-Assisted Physical and Chemical Cross-Linking of MXene Sheets for Electromagnetic Interference Shielding. *Small Methods, 5*(12), 2100889. doi:10.1002mtd.202100889 PMID:34928022

Xu, J., Li, R., Ji, S., Zhao, B., Cui, T., Tan, X., Gou, G., Jian, J., Xu, H., Qiao, Y., Yang, Y., Zhang, S., & Ren, T.-L. (2021). Multifunctional Graphene Microstructures Inspired by Honeycomb for Ultrahigh Performance Electromagnetic Interference Shielding and Wearable Applications. *ACS Nano, 15*(5), 8907–8918. doi:10.1021/acsnano.1c01552 PMID:33881822

KEY TERMS AND DEFINITIONS

Antenna: Link between electric currents traveling through conductive materials and radio waves that are propagating through space when a transmitter or receiver is utilized.

Electromagnetic interference (EMI) shielding: Technique for erecting a wall that stops powerful electromagnetic fields from escaping and interfering with signals and devices that are sensitive.

Graphene: Allotrope or carbon formed by a single layer of carbon atoms located in a hexagonal lattice.

MXene: Two-dimensional inorganic compound with thickness layers of a few atoms based on transition metal carbides, nitrides, or carbonitrides.

Phosphorene: It is the most stable allotrope of phosphorus and a two-dimensional material made of one or more layers of black phosphorus.

Resonator: A system or equipment that exhibits resonance, or resonant activity, or that naturally oscillates at specific frequencies with a larger amplitude.

Transition Metal Dichalcogenide (TMD): Chemical compound consisting of two chalcogen anions (sulfur, selenium, tellurium) and at least one transition metal (molybdenum, tungsten, cobalt, etc.) with two-dimensional structure.

Two-Dimensional (2D) Materials: Crystalline materials made up of single or few layers of atoms that have substantially stronger in-plane interactions than those along the stacking direction

Chapter 15
5G and 6G Wireless Communication

Sobana Sikkanan

https://orcid.org/0000-0001-8237-7140
Karpagam College of Engineering, India

Seerangurayar T.
Hindusthan College of Engineering and Technology, India

Krishna Prabha S.

https://orcid.org/0000-0002-5526-7701
PSNA College of Engineering and Technology, India

Kasthuri M.
PSNA College of Engineering and Technology, India

ABSTRACT

The 6G wireless communication network has shown its tremendous advantages in digital transformation of societies. 6G enables reliable, pervasive, and near instant wireless connectivity by integrating aerial, maritime, and terrestrial communications. The development of cutting-edge technologies like machine learning, blockchain, millimeter wave communication, non-orthogonal multiple access, tera-Hertz communication, quantum communication/quantum machine learning, fog/edge computing, and tactile Internet encourages the demand of beyond 5G and 6G communication. An effective wireless communication system must be capable of satisfying the user requirements such as increase in capacity, efficiency flexibility, coverage, and quality of experience. As the number of users and coverage area increases, the complexity of designing a 6G network is also getting increased. In recent years several research works are applying ML to wireless communication. This chapter discusses about the application of ML in the area of massive MIMO, NOMA,OWC, polar codes, and security of wireless communication.

DOI: 10.4018/978-1-6684-7000-8.ch015

1 INTRODUCTION

The explosive development of Internet increases the demand for bandwidth-intensive wireless communication applications. The analogue communication system was introduced in the 1980s, and every ten years a new generation is introduced. The increased demand for machine-to-machine (M2M) communications and smart devices increases mobile data traffic. 5G networks will lay the groundwork for the introduction of data-driven intelligent systems (Nawaz et al., 2019). It is anticipated that by 2030, 5G will have attained its maximum capacity, and 6G will support all adaptive and intelligent methods. 6G systems secure data and facilitate ease of operation with higher quality of service (QoS) than 5G and beyond communication systems. 6G is 1000 times faster than 5G and has less than 1 ms latency.

Most wireless communication systems today use the Multi-input Multi-output (MIMO) technology to communicate with a large number of users (Gesbert et al., 2003; Rusek et al., 2013). Machine Learning (ML) techniques have been modified for channel estimation (Prasad et al., 2015), signal detection(Liang et al., 2016) and link allocation(Rico-Alvariño & Heath, 2014) tasks in Massive MIMO (mMIMO) systems.

In recent years, information signals have been transmitted across power lines to support long-distance communication, and this technology is known as Power Line Communications (PLC)(Lampe et al., 2016). ML aids in the simplification of the complexity of the bottom-up design approach, MAC layer design, and resource allocation process in PLC networks (Cortés, 2016). ML techniques enable intelligent authentication in time-varying network environments and support new 5G and beyond system applications (Fang, Wang, & Tomasin, 2019).

This chapter covers the fundamentals of several emerging wireless technologies, such as massive MIMO, optical wireless communication, Non-orthogonal Multiple Access (NOMA), power line communication, polar coding, and reconfigurable intelligent surfaces. The chapter also delves into the machine learning frameworks for handling the introduced for dealing with the design and security issues of 6G wireless communication systems.

2 MASSIVE MIMO

Rapid growth in the number of cellular users imposes the need for new emerging technologies to satisfy the user demands. The massive multiple input multiple output (mMIMO) system is a key technology for boosting 5G networks' user capacity. In millimeter wave (mmWave) systems with greater complexity and physical testing, mMIMO performs well (Alkhateeb & Beltagy, 2018; Lamare, 2013; Liu et al., 2016).

Machine learning methods are being applied to both mmwave and mMIMO in an effort to simplify real-time implementation (Liu, 2014; Zappone et al., 2018). ML delivers higher performance with less complexity when compared to the existing techniques, such as game theory, stochastic geometry, and combinatorial optimization. Machine learning is important in MIMO load balancing, beamforming, spectrum optimization, and channel prediction due to its dynamic nature (Jiang et al., 2017; Simeone, 2018).

To integrate ML in a larger mMIMO system, the developers of (Booth, 2019) used software defined radio. They implemented the hardware framework and examined the speed of huge MIMO simulations using the LimeSDR development kit and the Lime Suite signal processing environment. The findings indicate that the bottlenecks between the base station (BS) and user are reducing the throughput and dependability of mmWave (Alkhateeb & Beltagy, 2018). The IRES project uses stationary transmitters and receivers with a moving obstruction to get around this. In order to follow the previous beamforms, a gated recurrent unit (GRU) is implemented, and the likelihood that the following beamform will be blocked is determined. In order to obtain the desired results, the system is first simulated using the DeepMIMO dataset (Alkhateeb, 2019). The data is then utilised to test and train the recurrent neural network (RNN).

For the MIMO detection procedure, Yu-Di Huang et al. (Huang, 2018) presented a clustering approach. They used a modulation-constrained Gaussian mixture model (MC-GMM) and its optimization technique to simplify the clustering algorithm. Another way to reduce label transmission overhead and research label design is label reconstruction. If the channel state information (CSI) of MIMO is ideal, the performance of the label assisted clustering (LAC) receiver approaches the behaviour of the optimal Maximum Likelihood detection (MLD).

In massive multiple-input multiple-output (mMIMO) systems, the overhead of orthogonal pilot-based channel estimation algorithms grows as the number of users rises. Jide Yuan, et al.(Yuan et al., 2019) suggested a machine learning-based time division duplex (ML-TDD) technique to address this issue. The measurement of channel state information in the ML-TDD system is influenced by the temporal channel correlation (CSI). The suggested approach uses an autoregressive (AR) predictor for channel estimation and a conventional neural network (CNN) for pattern extraction. Results showed that ML-based predictors produce improved throughput per user and have extremely high quality predictions for a variety of mobility scenarios.

Prior to the demodulation process, modulation estimation is performed in communication systems. In a MIMO system with poor CSI and correlated channels, authors developed an AdaBoost MLT to estimate the modulation type. Failure to do so results in performance deterioration when predicting the source information (Ha et al., 2019). Measurements of the true and false positive rates, time required, precision, and F-measure are used to assess the effectiveness of modulation detection methods.

To address the preamble collision that occurs in grant-free random access (GFRA) user equipment, Ding, Jie; Qu, et al. suggested a machine learning-based distribute mMIMO system. The task does not necessitate prior CSI and RA user activity knowledge or equipment (UE). By creating a 2D mMIMO system with signal strength and location information, the GFRA preamble problem is resolved. With the aid of deep learning assisted classification models, the relationship between the received preamble signal patterns and multiplicities of distributed mMIMO system is estimated. The authors also suggested using K-means access point (AP) clustering to group nearby APs of colliding RA UEs and sort out each AP cluster separately to decode the data from the colliding RA UEs. This improved the performance of the mMIMO system by decreasing the occurrence of mutual interference (Ding et al., 2021).

Power allocation (PA) and pilot assignment (PS) under the pilot contamination (PC) rule are two significant challenges that massive MIMO encounters. The implementation of the resource allocation policy is proposed in (Santos, 2022) using a machine learning (ML) method with reward Q-Learning. The authors measured the spectrum efficiency, energy efficiency, and bit error rate of various systems and in a range of channel conditions while taking into account diverse situations. According to the simulation results, the PS-PA approach improves multicellular large MIMO system performance.

For better error performance in MIMO detection, Jae-Hyun Ro et al. proposed combining deep neural network (DNN) and machine learning algorithm. In this technique, various training sets are produced by random sampling, and various DNN structures produce K various networks utilising various random data sets. ML makes predictions based on either the majority vote method or the maximum probability approach. The detector producing the highest value is chosen as the final output in the maximum probability technique. The detector with the most votes is chosen as the output in the majority vote technique. In order to boost the error performance of mMIMO without increasing the number of antennas needed, the proposed DNN-based ML method employs single detection to build parallel combinations of many identical detectors (Ro, 2022).

DNN predicts the broadcast symbols based on the received symbols using a supervised learning algorithm. By creating precise weights and biases in DNN, offline learning corrects the discrepancy between input and output symbols. For DNN-based implementation, mMIMO training sequences must be divided into real and imaginary parts because they are typically complex values. As an example N training data sets of the DNN are separated real R [.] and imaginary IM [.] parts and represented in Table. 1.

Table 1. Training set of DNN based MIMO system

Information 1	Information 2	...	Information N
$\text{Re}[y_1]_1$	$\text{Re}[y_1]_2$	...	$\text{Re}[y_1]_N$
$\text{Im}[y_1]_1$	$\text{Im}[y_1]_2$	...	$\text{Im}[y_1]_N$
$\vdots$	$\vdots$	$\vdots$	$\vdots$
$\text{Re}[y_{Nr}]_1$	$\text{Re}[y_{Nr}]_2$	...	$\text{Re}[y_{Nr}]_N$
$\text{Im}[y_{Nr}]_1$	$\text{Im}[y_{Nr}]_2$	...	$\text{Im}[y_{Nr}]_N$
$\text{Re}[h_{11}]_1$	$\text{Re}[h_{11}]_2$	...	$\text{Re}[h_{11}]_N$
$\text{Im}[h_{11}]_1$	$\text{Im}[h_{11}]_2$	...	$\text{Im}[h_{11}]_N$
$\vdots$	$\vdots$	$\vdots$	$\vdots$
$\text{Re}[h_{NrNt}]_1$	$\text{Re}[h_{NrNt}]_2$	...	$\text{Re}[h_{NrNt}]_N$
$\text{Im}[h_{NrNt}]_1$	$\text{Im}[h_{NrNt}]_2$	...	$\text{Im}[h_{NrNt}]_N$

The modulation order and the quantity of antennas utilised on the transmitter side determine how send symbols are encoded. Fig. 1 shows the basic 2×2 MIMO system using maximum likelihood detection (MLD) method. The encoding of a 2×2 MIMO system with Quadrature Phase Shift Keying (QPSK) modulation is shown in Table. 2.

Figure 1. Basic 2×2 MIMO system

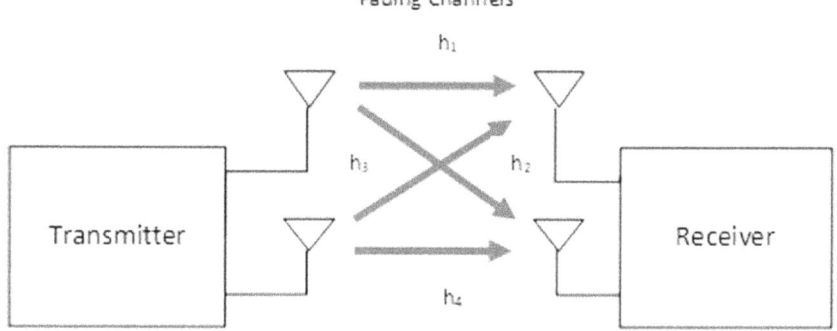

Table 2. Encoding of the transmit symbol in the 2 ×2 MIMO system

Label	Information (T×1)	Information (T×2)
0	00	00
1	00	01
2	00	10
3	00	11
4	01	00
⋮	⋮	⋮
15	11	11

The received signal vector 'y_{ij}' in terms of transmission matrix is given as,

$$\begin{bmatrix} y_1 \\ y_2 \end{bmatrix} = \begin{bmatrix} h_{11} & h_{12} \\ h_{21} & h_{22} \end{bmatrix} + \begin{bmatrix} x_1 \\ x_2 \end{bmatrix} + \begin{bmatrix} n_1 \\ n_2 \end{bmatrix} \tag{1}$$

where x_i represents the signal vector, n_i is the noise vector, and h_{ij} is the transmission channel vector (Sobana & Jeyanthi, 2015). The label with the most votes is chosen as the output of the DNN-based MIMO detection using the majority vote method, and it is displayed as

$$\hat{i} = \arg\max_i \sum_{k=1}^{K} y_{V,i}^{(k)} \tag{2}$$

Similarly the maximum probability method selects the label with highest probability as output represented as

$$\hat{i} = \arg\max_i \beta_i \tag{3}$$

3 NON-ORTHOGONAL MULTIPLE ACCESS (NOMA)

Currently, 5G wireless connection allows for a flexible and quick manner of communication all around the world. High base station capacity, low latency, support for more users, and excellent Quality of Service (QoS) are all features of 5G. By using a technology known as non-orthogonal multiple access (NOMA), which involves an increased receiver complexity, 5G and beyond systems can achieve spectral efficiency and throughput.

NOMA-MIMO Concept

To boost the point-to-point link and the quantity of data streams that are transmitted to the BS, NOMA with MIMO utilises multiple antennas at the UE. For effective functioning, NOMA multiplexes the data stream from a single UE through the spatial domain and divides the UE through the power domain. One strong user (User 1) and one weak user are considered to be present in a 22 MIMO system (User 2). The channel condition of User 1 is therefore better than User 2's (h1>h2), and User 1's power allocation is lower than User 2's (P1P2). Figure 2 illustrates the separation of UEs' power and geographical domains (Kizilirmak, 2016).

Figure 2. NOMA system Model

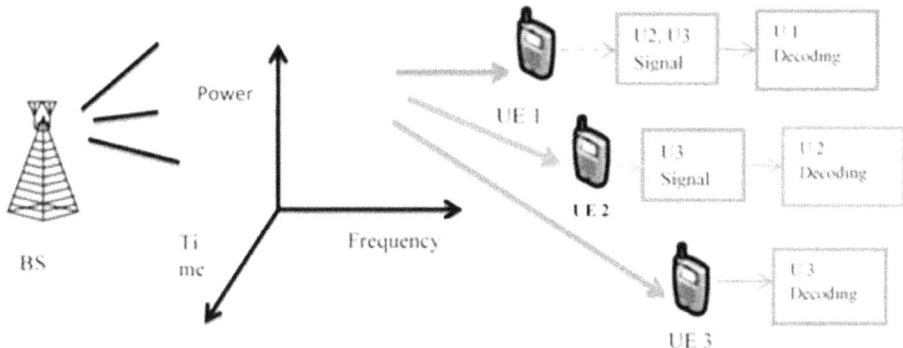

NOMA Downlink System Model

Fig.3. shows the transmitter and receiver blocks of a NOMA system model. The transmitted signal BS can be given as

$$s(t) = \sum_{i=1}^{N} \sqrt{\alpha_i P_T} s_i(t) \tag{4}$$

where $S_i(t)$ is the individual information, α_i is the power allocation coefficient for i^{th} user and P_T is the total power at the BS.

Figure 3. NOMA Block Diagram

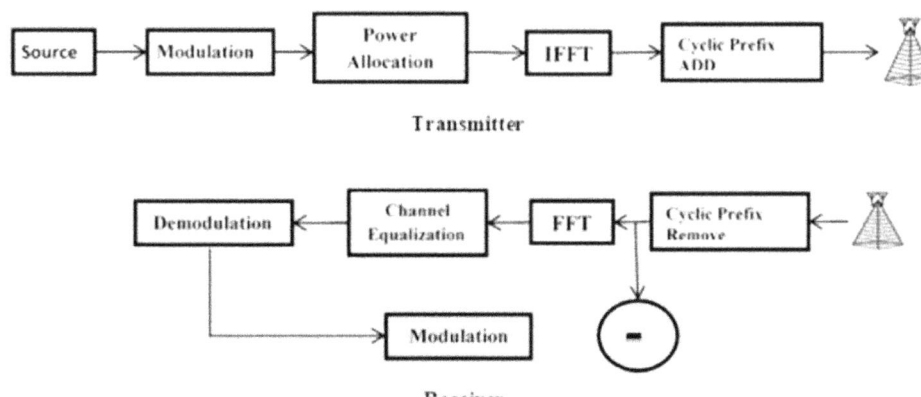

The received signal at the ith user U_i is

$$y_i(t) = s(t)h_i + n_i(t) \tag{5}$$

Where $S(t)$ is the transmitted signal, h_i is the channel attenuation factor and $n_i(t)$ is the additive white Gaussian noise with zero mean and variance of σ^2.

When compared to other users, the farthest user will be assigned the most power. As a result, the signal decoded at the farthest user is its own signal. Other user signals are treated as interference signals. Thus, the signal to noise ratio (SNR) for the Nth user is

$$SNR_{DN} = \frac{P_{DN}h_{DN}^2}{N_0 B + \sum_{i=1}^{N} P_i h_i^2} \tag{6}$$

where N_0 is the noise density and B is the bandwith of the signal.

The nearest user U_1 is assumed to have its own signal, and other users' interference is assumed to be cancelled. As a result, the SNR for the nearest user can be written as,

$$SNR_{D1} = \frac{P_{D1}h_{D1}^2}{N_0 B} \tag{7}$$

The throughput for each User can be given as

$$R_i = B \log_2 \left(1 + \frac{P_i h_i^2}{N + \sum_{i=1}^{N-1} P_i h_i^2} \right) \tag{8}$$

where $N = N_0 B$.

NOMA Uplink System Model

The signal received by the BS for uplink is given as

$$y(t) = \sum_{i=1}^{N} x_i(t) h_i + n(t) \tag{9}$$

where h_i is the channel attenuation factor and $n(t)$ is the additive white Gaussian noise with zero mean and variance of σ^2.

The signal from the nearest user will reach the receiver first. The SNR for User 1 takes into account interference from other users and is given as,

$$SNR_{U1} = \frac{P_{U1}h_{U1}^2}{N_0 B + \sum_{i=1}^{N} P_i h_i^2} \qquad (10)$$

The SNR for the fartest user is written as,

$$SNR_{UN} = \frac{P_{UN}h_{UN}^2}{N_0 B} \qquad (11)$$

Equation (8) gives the throughput of each individual user.

Multiple access techniques are developed as a result of efficient resource allocation among multiple users. The access techniques are classified as Orthogonal Multiple Access (OMA) and Non-Orthogonal Multiple Access (NOMA) by resource allocation methods (Choi, 2008). OMA allows users to access resources in an orthogonal manner within the specified frequency band (FDMA), time slot (TDMA), or code (CDMA). NOMA, on the other hand, allows concurrent resource allocation and supports more interference cancellation schemes, improved spectral efficiency, reduced reception complexity, and increased system capacity.

The combination of massive multiple input multiple output (MIMO)(Huh et al., 2012) and NOMA in a single system results in improved energy and spectral efficiency in 5G and beyond systems. The performance is evaluated by using the same transmission frequency for multiple users and a different transmission frequency for each user (Huang, 2020).

Users in NOMA must decode their own signal as well as other user signals to avoid interference, which increases processing delay. Because the base station must be aware of channel state information, the feedback overhead is also increased (CSI). The efficiency of uplink and downlink NOMA transmissions is determined by successive interference cancellation (SIC) techniques. The error probability of the SIC process will affect transmission efficiency. This can be avoided by limiting the number of users who use the same resources (Luo & Zhang, 2016).

According to recent study, machine learning through clustering provides an improved solution to fast data clustering due to flexibility with more information features of increasing dimensionality (Baştanlar & Ozuysal, 2014). The machine learning approach can figure out the relation between channel information and system performance. A mmWave NOMA transmission system was described in (Ding et al., 2017) to take use of the fact that the channels of mmWave users in one cluster are highly correlated. To implement the mmWave NOMA technique, the authors of (Cui, Ding, & Fan, 2018) proposed a novel K-means enabled machine learning framework for user clustering that makes use of the correlation feature of the users' mmWave channel responses. The beam is then strengthened using a beamforming mechanism based on maximum ratio transmission (MRT). They also proposed a low-complexity K-means-based on-line user clustering algorithm to facilitate newly arriving users in the system, which can substantially reduce computational complexity when compared to the strategy in which the clustering framework is changed completely even if only a single new user arrives in the system (Cui, Ding, Fan et al, 2018).

In future, combining the maximization of the number of user clusters with the mmwave-NOMA system enhances the NOMA system's sum rate. In the future, more advanced on-line and reinforcement learning processes can be employed to update the partition based on dynamic mmWave NOMA scenarios.

4 POWER LINE COMMUNICATION

Since the 1920s, electrical wiring and power lines have been used for network communication, which is now known as Power Line Communications (PLC) (Aalamifar & Lampe, 2017). Smart grid is a growing technology that integrates the power grid with an Internet Protocol (IP)-based communication network to enable broadband wireless applications (Lazaropoulos, 2021). Power grid eliminates the need for optical fibres and phone lines, lowering the cost of establishing a communication network (Bumiller et al., 2010). The use of various devices with different impedances creates a multipath environment in the PLC network, resulting in the generation of impulsive noise (Srinivasa Prasanna, 2009).

Low frequency power signals in PLC must pass through several transformers, causing propagation issues for high frequency signals (Kim, 2011). This issue is mitigated by using intra-level PLC, which maintains the signal properties of the power signal throughout the communication.

Signal repeaters are used to solve the same problem in inter-level PLC systems (Götz et al., 2004). Because of the additional devices used, designing, implementing, and maintaining these PLC systems is expensive. Later, ultra-low-frequency (ULF) signals are used in PLCs (ULF-PLCs), which do not attenuate but support a lower data rate (Zimmermann & Dostert, 2002).

Machine learning (ML) algorithms were used to make the PLC system usable (Ye et al., 2018). Machine learning is used in broadband over power line communication to collect and analyse data. ML algorithms control the signal parameters, allowing the original communication signal to be extracted and reconstructed at the receiver.

The authors of (Seo, 2019) discussed the use of ML in PLC for monitoring and characterization of each PLC layer. They used ML algorithms to assess the performance of grid diagnostics. Power line diagnostics solution based on machine learning (Huo et al., 2019) to identify, measure, and locate power line degradations. Existing machine learning (ML) algorithms, such as support vector machine (SVM) (Hearst et al., 1998), only detect degradation. As a result, it is necessary to identify a suitable ML algorithm capable of detecting degradation and extracting the communication signal from the raw data.

Authors of (Huo, 2019) proposed an automated ML (AutoML) based diagnostics solution to automatically decide the data pre-processing techniques, hyper parameters and ML algorithm that provide better performance. This method achieves accurate results in detecting, establishing and measuring power line degradations.

Lazaropoulos, et.al., (Lazaropoulos, 2021) proposed a neural network based identification method of Overhead Low-Voltage Broadband over Power Lines (OV LV BPL) networks to identify the branch channel attenuation behavior. This is carried out with the help of Deterministic Hybrid Model (DHM). Topology Identification Method (TIM) is proposed to generate the topology database of OV LV BPL network.

5 RECONFIGURABLE INTELLIGENT SURFACES (RIS)

Millimeter wave communication, tiny cells, and massive MIMO are the main technologies influencing the development of 5G and future networks. 6G is introduced to combine wireless communication, sensing, and computation on a single platform. For the launch of 6G technology, several studies are being conducted in terahertz communications, artificial intelligence (AI), ultra-massive multiple-input multiple-output (MIMO), and reconfigurable intelligent surfaces (RISs).

The hardware technique known as reconfigurable electromagnetic (EM) surfaces (RIS) coats things with intelligent surfaces. Wireless networks' energy and spectrum efficiency are enhanced using RIS. To achieve the performance of RIS, massive arrays of antennas placed half a wavelength apart, flat surfaces based on metamaterials, and scattering elements with sizes and inter-wavelength distances less than the wavelength are all employed.

The design of combined transmit and passive beamforming, resource allocation, channel estimate, channel modelling, and wireless network integration all provide technological challenges. The iterative algorithm (Wu & Zhang, 2019), convex optimization (Guo, 2019), alternating optimization method (Shen et al., 2019) and gradient descent approach (Huang et al., 2019) are a few of the current best practises for addressing the aforementioned issues. With certain restrictions, these strategies function effectively in wireless networks that have been upgraded by RIS.

The segregation of user requirements is overlooked by optimization algorithms, which also ignore the dynamic movement of users and presume that data transmission takes place in a familiar context. It is assumed that the BS parameters are random and adhere to a dynamic stochastic environment. Additionally, it is presumed that the BS is aware of the CSI of every channel involved in the communication process.

Due of RISs' submissive character, processing random variables and acquiring CSI might be challenging in practise. Using the current optimization strategies, which are ineffective for the implementation of RIS enhanced wireless network systems in real-time applications, is challenging due to the aforementioned issues.

Numerous studies are being conducted to provide a solution to the problems with RIS-enhanced wireless networks. Reinforcement learning (RL), an AI-based technique, uses its capacity for learning to collect data on the surroundings, past events, and user input (Liu, Liu, Chen et al, 2019) and subsequently raise network quality of service. In RIS, a deep deterministic policy gradient (DDPG)-based approach is developed for tackling the phase shift design problem as well as a DQN technique for increasing wireless network throughput.

A deep reinforcement learning (DRL) based technique improves the communication rate of the RIS augmented network(Taha, 2020). Zhang et al. (Zhang et al., 2022) examined the performance of a DRL-based method using perfect and imperfect channel state information (CSI).

ML approaches are also incredibly useful for addressing problems with wireless networks that are based on RIS. The advantages of applying AI tools in RIS contexts were covered by Gacanin et al. (Gačanin & Di Renzo, 2020). In order to enable automated adaptation to the changing dynamic environment utilising the information gathered from the environment and user input, the authors of (Liu et al., 2021) developed a two-step strategy in ML-enabled RIS-enhanced wireless networks.

To address the present problems with the RIS-enhanced wireless networks, machine learning (ML) employs both supervised and unsupervised learning techniques. In order to handle problems with spectrum sensing (Umebayashi et al., 2018), channel/antenna selection (Thilina et al., 2016), networking association (Abouzar, 2011) and traffic/QoE prediction (Feng et al., 2017) supervised learning algorithms have been implemented in wireless networks. These algorithms include decision tree, regression, random forest, Bayes classification method, K-nearest neighbours (KNN), and support vector machines (SVM).

Additionally, in the RIS-enhanced wireless networks, researchers used unsupervised learning methods such expectation-maximization, principal component analysis (PCA), independent component analysis (ICA), and K-means clustering. These methods can deal with issues with BS deployment, user clustering/association (Liu, Liu, & Chen, 2019), interference cancellation (Li et al., 2017), data aggregation (Morell et al., 2016) and channel state detection (Assra et al., 2016).

To address the present problems with the RIS-enhanced wireless networks, machine learning (ML) employs both supervised and unsupervised learning techniques. Federated learning is an emerging ML approach that finds the statistical training models on remove devices using supervised learning algorithms like decision trees and regression. Large-scale machine learning and distributed optimization are two areas where federated learning is used (Niknam et al., 2020). Each RIS can function as a distributed learner thanks to the decentralised procedure carried out by this algorithm. The produced data are trained, and the aggregating unit receives the local design parameters.

6 OPTICAL WIRELESS COMMUNICATION

Wireless communication will soon be used for video conferencing, mobile video calls, high-speed internet, and other applications. Only a few Mbps can currently be supported via wireless transmission. Nowadays, RF technology that permits connectivity in the ultraviolet range of frequencies is being replaced by optical wireless communication (OWC) technology. OWC is a substitute technology for present wireless RF communication due to its larger bandwidth, ability to operate in unlicensed spectrum, resistance to electromagnetic interference, high level of security, and low cast.

OWC encourages resource reuse to expand network capacity and coverage. Directed line of sight (LOS), Non directed LOS (NLOS), Diffuse LOS, Quasi diffuse LOS, and Multi spot LOS are the various connection configurations in OWC. OWC network is used to construct communication systems for inter- and intra-data centre interconnects, optical access networks, mobile fronthaul, and in-door communications for distances of up to 100Km (Xie et al., 2022).

Fig.4 and Fig, 5 represents the outdoor and indoor system model of optical wireless communication system. At the transmitter the total power emitted from the optical source is written as

$$P_T = BA_s \theta_s \tag{12}$$

where B is the brightness function, A_s is the surface area and θ_s is the emission angle (Musumeci et al., 2019)

Figure 4. Outdoor OWC system model

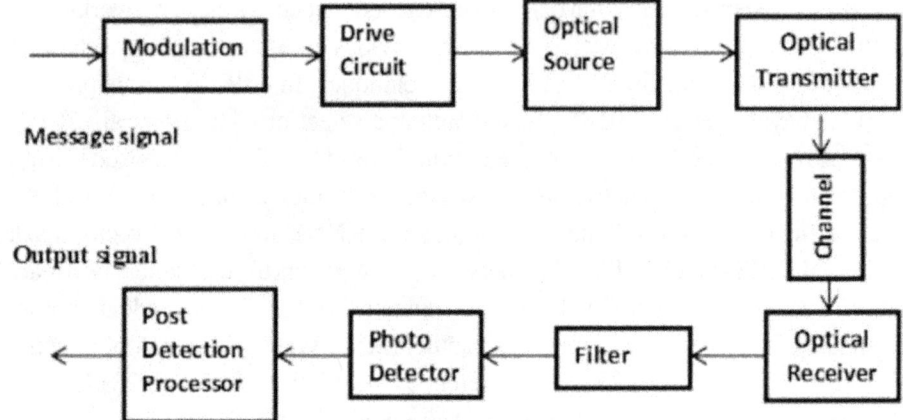

Figure 5. ML in indoor OWC system

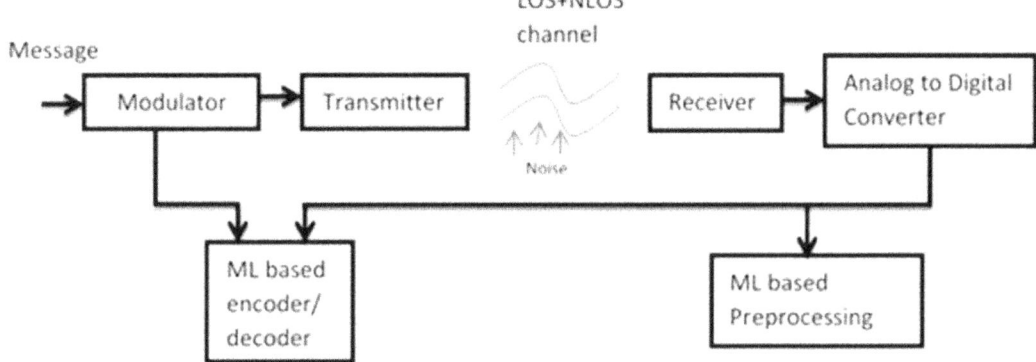

The receiver beam diameter D_R is given as,

$$D_R = D_T \left[1 + \left(\frac{\lambda L}{D_T^2} \right)^2 \right]^{1/2}$$

(13)

Here λ represents the wavelength, L indicates the link range and D_T represents transmitter beam diameter.

The transmitting antenna gain G_T is given by,

$$G_T = \left(\frac{4 D_T}{\lambda} \right)^2$$

(14)

The field intensity of the wave after propagating through the optical link of distance L is written as,

$$I = \frac{G_T P_T}{4 \pi L^2}$$

(15)

Receiving antenna gain is defined in terms of antenna area (A) as,

$$G_R = \left(\frac{4\pi}{\lambda^2} \right) A$$

(16)

The received signal power at the input of photodetector in terms of the above mentioned parameters is written as,

$$P_R = P_T G_T \Delta_T L_P \left(\frac{\lambda}{4\pi L}\right)^2 G_R \Delta_R W \tag{17}$$

where Δ_T and Δ_R represents the transmitting and receiving optics efficiency, L_p is the transmitting loss factor and W indicates the filter transmission factor.

Recently, the performance enhancement of optical wireless communication has benefited from the use of new artificial intelligence (AI) technologies. Deep learning (DL) and machine learning (ML) techniques were both used in the creation of the AI-based OWC system.

Many aspects of OWC, including optical performance monitoring (OPM), modulation format identification (MFI), indoor OWC system, defect detection and management, optical sensing, network automation, and signal processing, find applications for machine learning. BER, OSNR, Dispersion, and phase noise are some of the parameters in the OPM process that are calculated using machine learning. The eye diagram, histogram, scattering diagram, and constellation diagram are all detailed by MRI analysis. Equalization, dispersion correction, pre distortion, auto encoder, and demapping are all applications of machine learning in signal processing.

Visible light communication, infrared communication, and visible light location are all examples of ML uses in indoor OWC. The ML technique is used to construct the physical layer and network layer of the OWC network (Musumeci et al., 2019). In reality, large data modelling feature extraction and analysis cannot be supported by typical ML algorithms (Zheng et al., 2021) . In comparison to other ML algorithms, supervised learning is frequently utilised in the field of OWC because of its capacity to extract output features and determine the link between the input and output training data set (Ye et al., 2018). Critical OWC issues including optical modulation format identification (Skoog et al., 2006), nonlinearity mitigation (Bütler et al., 2021) and performance monitoring (Cheng, 2019) are all addressed by ML.

Support vector machines (SVM) and machine learning (ML) clustering techniques have been used to discover software issues and data virtualization issues (Vela et al., 2018). The wavelength and routing issues that arose in the optical wireless network are resolved by ML when combined with software-defined networking (SDN) (Martín et al., 2019).

The OPM unit is in charge of automatically determining the modulation format used in the intermediate nodes (Khan et al., 2016). Because of the complexity of its implementation and high hardware cost, modulation format identification (MFI) modelling and analysis are challenging in short-range communication systems. ML algorithms make this process simple by utilising previous data history to predict the modulation format (Wang et al., 2019) without the use complex analytical models. Some authors (Huang, 2020) proposed asynchronous amplitude histograms (AAH) and eye-diagram based ML approach for MFI in short reach systems.

In optical signal processing ML helps for reducing the attenuation and footprint occur in CD compensation (Ranzini et al., 2019). ML-based equalisation has been proposed to overcome the problems associated with the nonlinear equalisers and provide better performance with reduced complexity and achieves accuracy in measuring the key parameters of optical communication systems (Kuschnerov, 2020).

The steps involved in ML-based optical parameter measurement in OWC systems were discussed by the authors of (Saif et al., 2020). The optical signal is sampled after being converted to an electrical signal. The data set for training the deep learning algorithm is compiled using the sampled data. To identify the impairments that occurred during modulation, specific features like the power spectrum, histogram, eye diagram, and constellation diagram are extracted in the second step. ML algorithm execute the optical parameter measurement using the extracted features. The signal characteristics may change depending on the time of operation. As a result, ML needs to be regularly updated with the aid of fresh data sets gathered (Klabjan & Zhu, 2020).

The ML OWC model, which is based on an artificial neural network (NN), comprises 8 hidden neurons and a hidden layer. The CD, PMD, and OSNAR of an optical signal are measured using an eye diagram by ML based on ANN method. The training sequence's minimum mean square error (MMSE) is expressed (Wu, 2009) as follows,

$$E_T\left(w\right) = \frac{1}{2}\sum_{m \in T}\sum_{n=1}^{N}\left|y_n\left(x_m, w\right) - d_{nm}\right|^2 \tag{18}$$

For image recognition and sequential data analysis, respectively, convolutional neural networks (CNN) and recurrent neural networks (RNN) are used among the DL algorithms. To improve the effectiveness of the end-to-end learning process, DL introduces the idea of data-driven channel modelling as an alternative to the traditional block-based modelling method. For the purpose of creating a data set from experimental data and enabling self-configuration and adaptive resource allocation, the authors (Wang & Zhang, 2021) also suggested generative adversarial networks (GAN) and deep reinforcement learning (DRL) methods.

The aforementioned talks show that when paired with ML and Big data, OWC method performance improves. Currently, OWC is using the ML algorithm in offline mode. Future ML-based OWC systems will need to be modified to automatically handle real-time network changes.

7 POLAR CODING

The first practical codes for increasing the capacity of a broad class of channels were polar codes. Polar codes and binary memoryless symmetric (BMS) channels were created by Erdal Arkan in 2009 (Arıkan, 2009). The main technique utilised in polar codes is channel polarisation, which has a complexity order of O (N log N). The polar codes have been accepted as an effective channel coding technology by recent 5G technology.

A polar code converts the physical channel into outer channels using a linear block error-correcting code. The first code to accomplish the symmetric binary-input, discrete, memoryless channel (B-DMC) channel capacity is polar code (Pfister, 2017). By applying a linear modification to N independent copies of a B-DMC, this algorithm generates W, N distinct channels, WN(i), $1 \leq I \leq N$. The mutual information is either 0 or 1 as n increases. When I(WN(i)) is close to 1, the percentage of indices I is reported as I(W), but if it is close to 0, it is given as 1-I(W). The encoding and decoding complexity of a polar code is O (N log N). The factor graph of a two-channel polar system is shown in Fig. 6.

For a setup with two equiprobable bits (U$_1$, U2) encoded into $(X_1, X_2) = (U_1 \oplus U_2)$. The encoded bits are then mapped to (Y$_1$, Y2) by two independent BMS channels having the transition probabilities

$$P((Y_1 = y|X_1 = x)) = P(Y_2 = y|X_2 = x) = W(y|x).$$

Figure 6. Factor graph for a two channel polar coder

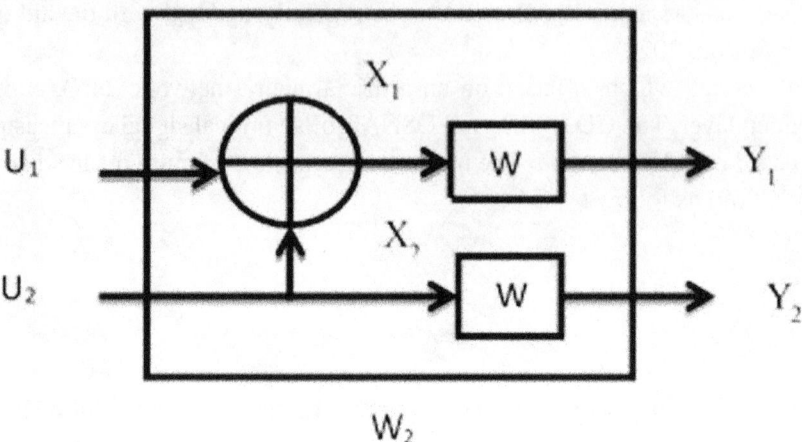

The capacity of the system is given as

$$I(U_1,U_2; Y_1,Y_2) = (X_1,X_2; Y_1,Y_2) = I(X_1; Y_1) + (X_2; Y_2) = 2I(W) \tag{19}$$

The computational cutoff rate and sum cut off rate conditions are shown below

$$R_0(W) = 1 - \log_2(1+Z(W)) \tag{20}$$

$$R_0(W^+) + R_0(W^-) \geq 2R_0(W) \tag{21}$$

At the receiver, polar coding, consecutive cancellation decoding, channel splitting, and polar encoding are utilised to transform a block of bits into polarised bit streams. Polar codes have a simpler decoding process than current turbo codes, and they do error correction more effectively. Due to this, URLLC (ultra-reliable low latency communications), MMTC (massive machine-type communications), and 5G eMBB (enhanced Mobile Broadband) are using polar codes more frequently. Polar codes can deliver up to 5 Gbps, while 5G and beyond 5G systems must support 20 Gbps and several Tbps. This limits the use of Polar codes in cutting-edge wireless communication systems.

With a successive cancellation (SC) decoder, the block sizes must be asymptotically huge for polar code to operate at its maximum capacity. Low-density parity-check code (LDPC) performs better than successive cancellation code in industrial applications. This limits the usability of polar codes in real time industrial applications (Arıkan, 2015). Huawei tested 5G channel coding utilising polar codes in the field in 2016 and was successful in achieving a data throughput of 27 Gbps. For the 5G NR (New Radio) interface, the 3GPP decided to employ polar codes for the eMBB (Enhanced Mobile Broadband) control channels and LDPC for the data channels (Dhuheir & Öztürk, 2018). Figure 7 depicts the block diagram for 5G Polar coding using a CRC coder.

Figure 7. Block Diagram of 5G Polar coding

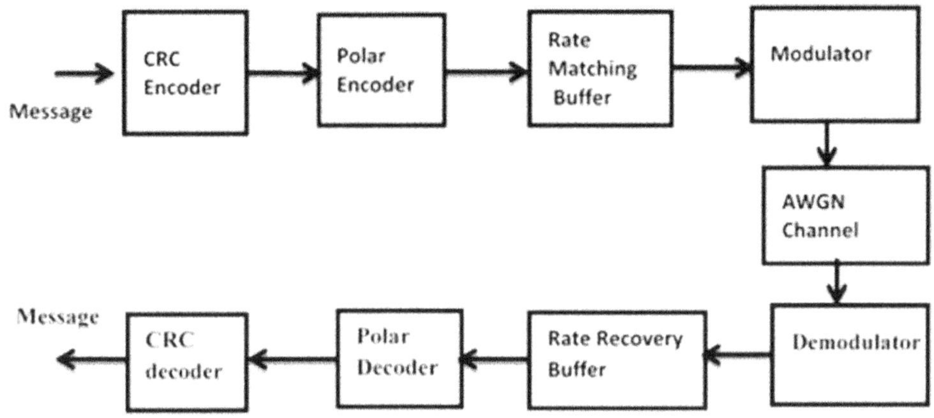

In eMBB scenarios, the decoding of polar codes with short lengths did not produce the desired results. To lower the FER and BER of a decoder, the SC List (SCL) decoding technique was suggested in (Tal & Vardy, 2011).). Cyclic redundancy check SCL (CA-SCL) (Niu & Chen, 2012) and segmented CRC error-correcting aided SCL (SCRC-EC-SCL) (Liu et al., 2020) decoding algorithms were developed to increase the efficiency of Polar coding. The Flip-SC (SCF) decoding algorithm (Afisiadis et al., 2014) helped to minimise the complexity of SCL decoding methods, however it can only flip one incorrect bit at once. In contrast, the Dynamic SCF (DSCF) decoding (Fang, Wang, & Tomasin, 2019) method conducts many flips at once with an increased demand for exponential and logarithmic computations. A deep learning based DSCF (Wang, 2019) was introduced later in which LSTM network and RL networks are used to locate the erroneous bit and avoid the middle state error.

To improve the performance of the Polar decoder, the authors of (He, 2020) proposed a machine learning-based multi-flips SC decoding scheme (ML-MSCF). LSTM and RL use logarithmic and exponential computations to identify and locate the occurrence of first and additional erroneous bits. To deal with multiple flips, a Q Learning-based Q Table was created to determine the best method, and SC decoding approaches were carried out as needed. The process is repeated until the message estimation satisfied the CRC check or the number of repetition reaches the maximum time T.

The block neural network (BlockNN) technique for Polar code was presented in (Jiang, 2019), in which the 2n bit polar code was broken into equal-length subblocks and the decoding of blocks inside the modules was performed in parallel. The decoding of the modules is then done serially. This neural network-based decoder structure is less complicated and does not alter with code length. Arkan recently proposed polarization-adjusted convolutional (PAC) codes as a polar coding approach that outperforms previous polar coding and CRC-based polar decoding methods (Moradi, 2020).

8 COMMUNICATION AND INFORMATION SYSTEMS SECURITY

The connectivity of heterogeneous equipment and devices in 5G and Beyond networks makes them more vulnerable to various spoofing attacks. Because of their complexity and dynamic nature, existing cryptography and authentication systems are unsuitable for today's wireless networks. Low dependability, security overhead, and continuous protection against time-varying aspects are among the issues. ML-based authentication solutions are reliable, low-cost, continuous, simple to implement, and and adaptable to situation.

ML is used in 5G and beyond wireless networks to perform intrusion detection, authentication, and access control. ML recognizes and learns the dynamic changes in the network, resulting in autonomous security management. The use of block chain in machine learning may simplify the process of distributed security management. Adversarial behaviour analysis in ML is carried out in order to forecast malicious attacks by attackers. In decision-making processes such as analytical thinking, distributed computing, client orientation, planning, and resource allocation, ML algorithms may be combined and used. Thus, ML techniques enable 5G and future wireless networks to provide intelligent and autonomous services to humans.

Parametric learning methods require some specific form of training functions. This training approach associates training samples with training functions in order to improve accuracy, simplicity of operation, and use fewer training samples. For example, the authors of (Xiao et al., 2018) analyzed static radio nodes and developed an authentication approach that combines the concepts of logistic regression and transmitter RSSI. The attributes are designed sequentially to eliminate the uncertainty induced by signals' time varying nature. Prediction of static properties and details of previous signal attributes is difficult in 5G and beyond systems, limiting the use of parametric learning methods in the authentication process.

To address this issue, the non-parametric learning method (Jiang et al., 2017) proposes an authentication method that determines security aspects using existing data and does not require prior information. This method gathers data dynamically from a rapidly changing environment and achieves greater flexibility in a real-time authentication environment. The requirement for more training samples renders it unsuitable for 5G applications. The non-parametric learning technique based on kernels provides constant monitoring of system validation and collects data exclusively from those devices (Fang, Wang, & Hanzo, 2019), and reduces the computational complexity.

Non-supervised learning algorithms do not necessitate prior knowledge of the system and do not require labelled outputs. As a result, the authentication procedure significantly reduces latency and energy usage for output labelling. The authors of (Xiao et al., 2016) proposes a Q-learning-based reinforcement learning system that achieves the test threshold and accuracy based on the RSSI without any input, output data, or parameter updates. Figure 8 depicts the classification of machine learning-based security algorithms.

6G communication necessitates the secure transmission of spectrum information among peers waiting to use the same frequency spectrum. This is made possible by ML-based intelligent spectrum sharing, which employs ML algorithms at every network layer. ML secures spectrum sharing information at the physical layer and routing information at the network layer. To enable secure synchronization, the remaining upper layers employ Ml algorithms.

Wireless medium supports Broadcast and distributed environment make them vulnerable to eavesdropping and jamming attacks. Because of the automated nature of machine learning (ML), wireless communication channels may learn and adapt to environmental conditions, providing excellent security against malicious attacks.

Research works in the area of ML based wireless security is needed in various directions. Evasion attacks, Trojan attacks, and causative attacks are examples of malicious attacks. Physical layer privacy, model inversion, and membership inference attacks are all privacy concerns. In wireless security, ML is

Figure 8. Classification of ML algorithm used in security

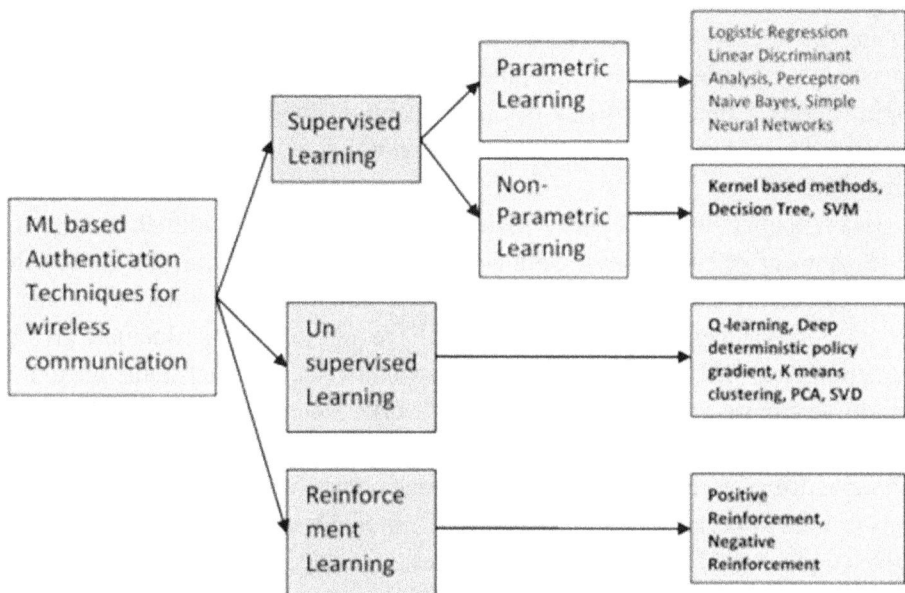

used to identify devices, monitor spectrum, smart jamming, RF finger printing, location identification, authentication, smart eavesdropping, intrusion detection, authentication, IoT security, and anonymity.

In (Adhikari et al., 2019) the authors discussed about Wireless virtualization (WiVi) for high speed and quality of service (QoS) aware wireless communication. WiVi used the Block chain concept to reduce the over loading of wireless channels and ML is used to determine the QoS requirements. The authors of (Pajola et al., 2019) carried out a survey in security attack on ML based wireless communication systems. They concluded that model inversion, model extraction and leak of information to evasion are some of the security attacks affecting the performance of the wireless systems.

Noise- robust classification on compressed spectrogram images was carried out with the help of Siamese convolutional neural networks (CNNs) for distinguishing different wireless signals (Langford et al., 2019). The shift from statistical learning, deep neural frameworks and cognitive systems to artiðcial intelligence was discussed in (Stantchev, 2019).

Li et al.,2021 (Li et al., 2021) and Ogudo et al.,2019 (Ogudo et al., 2019) proposed SVM based Intrusion Detection System (SVM-IDS) for identifying malicious nodes. Zheng et al.,2021 proposed a Network Intrusion Detection Systems (NIDS) which has sensors to detect unusual behavior of packet data and a monitoring system to create a meaningful data (Sengan, 2022).

9 ISSUES AND FUTURE TRENDS

The use of machine learning in wireless communication systems can provide new research areas and solutions, as well as aid in the implementation of 6G wireless communication networks and services. Despite extensive study in the field of machine learning in wireless communication systems, there are still numerous obstacles and unresolved topics to be addressed. An in-depth investigation of the unusually long convergence time of ML approaches, as well as the factors influencing convergence, is required. Long ML time convergence can degrade performance in highly dynamic wireless networks, hence optimising the time convergence is crucial (Elsayed & Erol-Kantarci, 2019). This will affect the time convergence of the wireless network. Similarly, AI-enabled networks influence resource allocation in e-health applications. Extending outside-of-clinic operations with wearable sensors necessitates co-ordinating network resource allocation across several technologies, and ML can help with this (Elsayed & Erol-Kantarci, 2019).

In video stream applications, customers seek fast throughput and minimal latency at the price of security, yet in payment software, consumers need strong security even at the loss of throughput. A network with a big and diverse collection of users will operate in a very dynamic manner, since users' quality of service (QoS) and quality of experience (QoE) requirements may fluctuate greatly. Designing a cross-layer, action-based ML protocol for diverse applications is a crucial challenge in this area, since it must suit multiple requirements while balancing network resources (Tang et al., 2021).

UAVs as an intelligent service (UaaIS) uses sophisticated ML methods to intelligently deliver fundamental services such as wireless communication, edge computing, and edge caching. Due to limited resources, it is critical to perform energy-efficient ML model training and inference for UaaIS, a difficult unresolved topic in the field. When a UAV serves as an edge intelligence trainer, energy-efficient training procedures for all participants, particularly UAVs with low energy, should be developed (Dong et al., 2021).

Acquiring fast and accurate channel state information (CSI) is critical in intelligent reflecting surface (IRS) enhanced wireless systems, particularly MIMO-IRS and multiple input single output (MISO)-IRS networks. Acquiring CSI in IRS enhanced systems necessitates a significant amount of training. In IRS-assisted systems, users in every cluster must share the CSI. Because of IRS's passive nature, acquiring and transferring CSI is a difficult operation. The use of ML and DL techniques to optimise CSI in scenarios other than linear correlations is a significant task (Liu et al., 2021).

10 CONCLUSION

The rapid growth in the area of machine learning techniques exhibit the desire to support 6G wireless communication and enable high speed, accurate, bandwidth efficient long distance data transmission. The integration of ML approach with wireless communication still faces several difficulties, including those related to spectrum allocation, noise detection, and numerous unresolved privacy and security issues. This chapter discusses the demand for machine learning techniques in wireless communication, some applications of ML in communication systems, security issues faced by ML, existing solutions, and security algorithms in overcoming the issues and challenges and some future directions in the emergence of ML based wireless communication technologies. We believe that this chapter will provides enough motivation for future development of research activity on security aspects for 5G and Beyond communication systems using Artificial intelligence.

REFERENCES

Aalamifar, F., & Lampe, L. H.-J. (2017). Optimized WiMAX Profile Configuration for Smart Grid Communications. *IEEE Transactions on Smart Grid*, *8*(6), 2723–2732. doi:10.1109/TSG.2016.2536145

Abouzar, P. (2011). Action-based scheduling technique for 802.15.4/ZigBee wireless body area networks. *2011 IEEE 22nd International Symposium on Personal, Indoor and Mobile Radio Communications*, (pp. 2188-2192). IEEE.

Adhikari, A., Rawat, D. B., & Song, M. (2019). Wireless Network Virtualization by Leveraging Blockchain Technology and Machine Learning. *Proceedings of the ACM Workshop on Wireless Security and Machine Learning*. ACM. 10.1145/3324921.3328790

Afisiadis, O., Balatsoukas-Stimming, A., & Burg, A. P. (2014). A low-complexity improved successive cancellation decoder for polar codes. *2014 48th Asilomar Conference on Signals, Systems and Computers*, (pp. 2116-2120). IEEE. 10.1109/ACSSC.2014.7094848

Alkhateeb, A. (2019). *DeepMIMO: A Generic Deep Learning Dataset for Millimeter Wave and Massive MIMO Applications.* ArXiv.

Alkhateeb, A., & Beltagy, I. (2018). Machine Learning For Reliable Mmwave Systems: Blockage Prediction And Proactive Handoff. *2018 IEEE Global Conference on Signal and Information Processing (GlobalSIP)*, 1055-1059 10.1109/GlobalSIP.2018.8646438

Arıkan, E. (2009). Channel Polarization: A Method for Constructing Capacity-Achieving Codes for Symmetric Binary-Input Memoryless Channels. *IEEE Transactions on Information Theory*, *55*(7), 3051–3073. doi:10.1109/TIT.2009.2021379

Arıkan, E. (2015). Challenges and some new directions in channel coding. *Journal of Communications and Networks (Seoul)*, *17*(4), 328–338. doi:10.1109/JCN.2015.000063

Assra, A., Yang, J., & Champagne, B. (2016). An EM Approach for Cooperative Spectrum Sensing in Multiantenna CR Networks. *IEEE Transactions on Vehicular Technology*, *65*(3), 1229–1243. doi:10.1109/TVT.2015.2408369

Baştanlar, Y., & Ozuysal, M. (2014). Introduction to machine learning. *Methods in Molecular Biology (Clifton, N.J.)*, *1107*, 105–128. doi:10.1007/978-1-62703-748-8_7 PMID:24272434

Booth, J. (2019). *Machine Learning for Reliable MIMO Systems.*

Bumiller, G., Lampe, L. H.-J., & Hrasnica, H. (2010). Power line communication networks for large-scale control and automation systems. *IEEE Communications Magazine*, *48*(4), 106–113. doi:10.1109/MCOM.2010.5439083

Bütler, R. M., Hager, C., Pfister, H. D., Liga, G., & Alvarado, A. (2021). Model-Based Machine Learning for Joint Digital Backpropagation and PMD Compensation. *Journal of Lightwave Technology*, *39*(4), 949–959. doi:10.1109/JLT.2020.3034047

Cheng, S. (2019). Machine Learning for Regenerator Placement Based on the Features of the Optical Network. *2019 21st International Conference on Transparent Optical Networks (ICTON)*, (pp. 1-3). IEEE. 10.1109/ICTON.2019.8840391

Choi, J. (2008). H-ARQ Based Non-Orthogonal Multiple Access with Successive Interference Cancellation. *2008 IEEE Global Telecommunications Conference*, (pp. 1-5). IEEE. 10.1109/GLOCOM.2008. ECP.640

Cortés, J.A. (2016). *Medium Access Control and Layers Above in PLC.*

Cui, J., Ding, Z., & Fan, P. (2018). The Application of Machine Learning in mmWave-NOMA Systems. In *2018 IEEE 87th Vehicular Technology Conference (VTC Spring)*. IEEE. 10.1109/VTCSpring.2018.8417523

Cui, J., Ding, Z., Fan, P., & Al-Dhahir, N. (2018). Unsupervised Machine Learning-Based User Clustering in Millimeter-Wave-NOMA Systems. *IEEE Transactions on Wireless Communications*, *17*(11), 7425–7440. doi:10.1109/TWC.2018.2867180

Dhuheir & Öztürk. (2018). Polar codes analysis of 5G systems. In *2018 6th International Conference on Control Engineering & Information Technology (CEIT)*. IEEE.

Ding, J., Qu, D., Liu, P., & Choi, J. (2021). Machine Learning Enabled Preamble Collision Resolution in Distributed Massive MIMO. *IEEE Transactions on Communications*, *69*(4), 2317–2330. doi:10.1109/ TCOMM.2021.3051202

Ding, Z., Fan, P., & Poor, H. V. (2017). Random Beamforming in Millimeter-Wave NOMA Networks. *IEEE Access: Practical Innovations, Open Solutions*, *5*, 7667–7681. doi:10.1109/ACCESS.2017.2673248

Dong, C., Shen, Y., Qu, Y., Wang, K., Zheng, J., Wu, Q., & Wu, F. (2021). UAVs as an Intelligent Service: Boosting Edge Intelligence for Air-Ground Integrated Networks. *IEEE Network*, *35*(4), 167–175. doi:10.1109/MNET.011.2000651

Elsayed, M., & Erol-Kantarci, M. (2019). AI-Enabled Future Wireless Networks: Challenges, Opportunities, and Open Issues. *IEEE Vehicular Technology Magazine*, *14*(3), 70–77. doi:10.1109/MVT.2019.2919236

Fang, H., Wang, X., & Hanzo, L. H. (2019). Learning-Aided Physical Layer Authentication as an Intelligent Process. *IEEE Transactions on Communications*, *67*(3), 2260–2273. doi:10.1109/TCOMM.2018.2881117

Fang, H., Wang, X., & Tomasin, S. (2019). Machine Learning for Intelligent Authentication in 5G and Beyond Wireless Networks. *IEEE Wireless Communications*, *26*(5), 55–61. doi:10.1109/MWC.001.1900054

Feng, Z., Li, X., Zhang, Q., & Li, W. (2017). Proactive Radio Resource Optimization With Margin Prediction: A Data Mining Approach. *IEEE Transactions on Vehicular Technology*, *66*(10), 9050–9060. doi:10.1109/TVT.2017.2709622

Gačanin, H., & Di Renzo, M. (2020). Wireless 2.0: Toward an Intelligent Radio Environment Empowered by Reconfigurable Meta-Surfaces and Artificial Intelligence. *IEEE Vehicular Technology Magazine*, *15*(4), 74–82. doi:10.1109/MVT.2020.3017927

Gesbert, D., Shafi, M., Da-shan Shiu, Smith, P. J., & Naguib, A. (2003). From theory to practice: An overview of MIMO space-time coded wireless systems. *IEEE Journal on Selected Areas in Communications*, *21*(3), 281–302. doi:10.1109/JSAC.2003.809458

Götz, M., Rapp, M., & Dostert, K. M. (2004). Power line channel characteristics and their effect on communication system design. *IEEE Communications Magazine*, *42*(4), 78–86. doi:10.1109/MCOM.2004.1284933

Guo, H. (2019). Weighted Sum-Rate Maximization for Intelligent Reflecting Surface Enhanced Wireless Networks. *2019 IEEE Global Communications Conference (GLOBECOM)*, (pp. 1-6). IEEE. 10.1109/GLOBECOM38437.2019.9013288

Ha, C.-B., You, Y.-H., & Song, H.-K. (2019). Machine Learning Model for Adaptive Modulation of Multi-Stream in MIMO-OFDM System. *IEEE Access: Practical Innovations, Open Solutions*, *7*, 5141–5152. doi:10.1109/ACCESS.2018.2889076

He, B. (2020). A Machine Learning Based Multi-flips Successive Cancellation Decoding Scheme of Polar Codes. *2020 IEEE 91st Vehicular Technology Conference (VTC2020-Spring)*, (pp. 1-5). IEEE. 10.1109/VTC2020-Spring48590.2020.9128875

Hearst, M. A., Dumais, S. T., Osuna, E., Platt, J., & Scholkopf, B. (1998). Support vector machines. *IEEE Intelligent Systems & their Applications*, *13*(4), 18–28. doi:10.1109/5254.708428

Huang, C., Zappone, A., Alexandropoulos, G. C., Debbah, M., & Yuen, C. (2019). Reconfigurable Intelligent Surfaces for Energy Efficiency in Wireless Communication. *IEEE Transactions on Wireless Communications*, *18*(8), 4157–4170. doi:10.1109/TWC.2019.2922609

Huang, L. (2020). *Modulation format identification under stringent bandwidth limitation based on an artificial neural network*.

Huang, T.-J. (2020). Theoretical Analysis of NOMA Within Massive MIMO Systems. *Wireless Personal Communications*, *112*(2), 777–783. doi:10.100711277-020-07073-z

Huang, Y.-D. (2018). A Machine Learning Approach to MIMO Communications. *2018 IEEE International Conference on Communications (ICC)*, (pp. 1-6). IEEE. 10.1109/ICC.2018.8422211

Huh, H., Caire, G., Papadopoulos, H. C., & Ramprashad, S. A. (2012). Achieving "Massive MIMO" Spectral Efficiency with a Not-so-Large Number of Antennas. *IEEE Transactions on Wireless Communications*, *11*(9), 3226–3239. doi:10.1109/TWC.2012.070912.111383

Huo, Y. (2019). Smart-Grid Monitoring: Enhanced Machine Learning for Cable Diagnostics. *2019 IEEE International Symposium on Power Line Communications and its Applications (ISPLC)*, (pp. 1-6). IEEE. 10.1109/ISPLC.2019.8693287

Huo, Y., Prasad, G., Atanackovic, L., Lampe, L., & Leung, V. C. M. (2019). Cable Diagnostics With Power Line Modems for Smart Grid Monitoring. *IEEE Access: Practical Innovations, Open Solutions*, *7*, 60206–60220. doi:10.1109/ACCESS.2019.2914580

Jiang, C., Zhang, H., Ren, Y., Han, Z., Chen, K.-C., & Hanzo, L. (2017). Machine Learning Paradigms for Next-Generation Wireless Networks. *IEEE Wireless Communications*, 24(2), 98–105. doi:10.1109/MWC.2016.1500356WC

Jiang, Y. (2019). Mind: Model independent neural decoder. In *2019 IEEE 20th International Workshop on Signal Processing Advances in Wireless Communications (SPAWC)*. IEEE. 10.1109/SPAWC.2019.8815537

Khan, F. N., Zhong, K., Al-Arashi, W. H., Yu, C., Lu, C., & Lau, A. P. T. (2016). Modulation Format Identification in Coherent Receivers Using Deep Machine Learning. *IEEE Photonics Technology Letters*, 28(17), 1886–1889. doi:10.1109/LPT.2016.2574800

Kim, Y. (2011). *Iterative Coding for High Speed Power Line Communication Systems*. The Journal of the Institute of Webcasting. *Internet and Telecommunication*, 11, 185–192.

Kizilirmak, R.C. (2016). *Non-Orthogonal Multiple Access (NOMA) for 5G Networks*.

Klabjan, D. & X. Zhu, X. (2020). *Neural Network Retraining for Model Serving*.

Kuschnerov, M. (2020). Advances in Deep Learning for Digital Signal Processing in Coherent Optical Modems. *2020 Optical Fiber Communications Conference and Exhibition (OFC)*, (pp. 1-3). 10.1364/OFC.2020.M3E.2

Lamare, R.C. (2013). *Massive MIMO Systems: Signal Processing Challenges and Future Trends*.

Lampe, L. H.-J., Tonello, A. M., & Swart, T. G. (2016). *Power Line Communications: Principles*. Standards and Applications from Multimedia to Smart Grid. doi:10.1002/9781118676684

Langford, Z. L., Eisenbeiser, L., & Vondal, M. (2019). Robust Signal Classification Using Siamese Networks. *Proceedings of the ACM Workshop on Wireless Security and Machine Learning*. ACM. 10.1145/3324921.3328781

Lazaropoulos, A. (2021). *Information Technology, Artificial Intelligence and Machine Learning in Smart Grid – Performance Comparison between Topology Identification Methodology and Neural Network Identification Methodology for the Branch Number Approximation of Overhead Low-Voltage Broadband over Power Lines Network Topolog*. Trends in Renewable Energy.

Li, G., Liu, F., Sharma, A., Khalaf, O. I., Alotaibi, Y., Alsufyani, A., & Alghamdi, S. (2021). Research on the Natural Language Recognition Method Based on Cluster Analysis Using Neural Network. *Mathematical Problems in Engineering*, 2021, 1–13. doi:10.1155/2021/9982305

Li, J., Zhang, H., & Fan, M. (2017). Digital Self-Interference Cancellation Based on Independent Component Analysis for Co-Time Co-frequency Full-Duplex Communication Systems. *IEEE Access: Practical Innovations, Open Solutions*, 5, 10222–10231. doi:10.1109/ACCESS.2017.2712614

Liang, H.-W., Chung, W.-H., & Kuo, S.-Y. (2016). Coding-Aided K-Means Clustering Blind Transceiver for Space Shift Keying MIMO Systems. *IEEE Transactions on Wireless Communications*, 15(1), 103–115. doi:10.1109/TWC.2015.2467394

Liu, D., Wang, L., Chen, Y., Elkashlan, M., Wong, K.-K., Schober, R., & Hanzo, L. (2016). User Association in 5G Networks: A Survey and an Outlook. *IEEE Communications Surveys and Tutorials*, *18*(2), 1018–1044. doi:10.1109/COMST.2016.2516538

Liu, J. (2014). Seeing the Unobservable: Channel Learning for Wireless Communication Networks. *2015 IEEE Global Communications Conference (GLOBECOM)*, (pp. 1-6). IEEE. 10.1109/GLOBECOM.2014.7417805

Liu, X., Liu, Y., & Chen, Y. (2019). Reinforcement Learning in Multiple-UAV Networks: Deployment and Movement Design. *IEEE Transactions on Vehicular Technology*, *68*(8), 8036–8049. doi:10.1109/TVT.2019.2922849

Liu, X., Liu, Y., Chen, Y., & Hanzo, L. (2019). Trajectory Design and Power Control for Multi-UAV Assisted Wireless Networks: A Machine Learning Approach. *IEEE Transactions on Vehicular Technology*, *68*(8), 7957–7969. doi:10.1109/TVT.2019.2920284

Liu, X., Wu, S., Wang, Y., Zhang, N., Jiao, J., & Zhang, Q. (2020). Exploiting Error-Correction-CRC for Polar SCL Decoding: A Deep Learning-Based Approach. *IEEE Transactions on Cognitive Communications and Networking*, *6*(2), 817–828. doi:10.1109/TCCN.2019.2946358

Liu, Y., Liu, X., Mu, X., Hou, T., Xu, J., Di Renzo, M., & Al-Dhahir, N. (2021). Reconfigurable Intelligent Surfaces: Principles and Opportunities. *IEEE Communications Surveys and Tutorials*, *23*(3), 1546–1577. doi:10.1109/COMST.2021.3077737

Luo, F.-L., & Zhang, C. J. (2016). *Non-Orthogonal Multiple Access (NOMA)*. Concept and Design.

Martín, I., Troia, S., Hernandez, J. A., Rodriguez, A., Musumeci, F., Maier, G., Alvizu, R., & Gonzalez de Dios, O. (2019). Machine Learning-Based Routing and Wavelength Assignment in Software-Defined Optical Networks. *IEEE eTransactions on Network and Service Management*, *16*(3), 871–883. doi:10.1109/TNSM.2019.2927867

Moradi, M. (2020). *Performance and Complexity of Sequential Decoding of PAC Codes*.

Morell, A., Correa, A., Barcelo, M., & Vicario, J. L. (2016). Data Aggregation and Principal Component Analysis in WSNs. *IEEE Transactions on Wireless Communications*, *15*(6), 3908–3919. doi:10.1109/TWC.2016.2531041

Musumeci, F., Rottondi, C., Nag, A., Macaluso, I., Zibar, D., Ruffini, M., & Tornatore, M. (2019). An Overview on Application of Machine Learning Techniques in Optical Networks. *IEEE Communications Surveys and Tutorials*, *21*(2), 1383–1408. doi:10.1109/COMST.2018.2880039

Nawaz, S. J., Sharma, S. K., Wyne, S., Patwary, M. N., & Asaduzzaman, M. (2019). Quantum Machine Learning for 6G Communication Networks: State-of-the-Art and Vision for the Future. *IEEE Access: Practical Innovations, Open Solutions*, *7*, 46317–46350. doi:10.1109/ACCESS.2019.2909490

Niknam, S., Dhillon, H. S., & Reed, J. H. (2020). Federated Learning for Wireless Communications: Motivation, Opportunities, and Challenges. *IEEE Communications Magazine*, *58*(6), 46–51. doi:10.1109/MCOM.001.1900461

Niu, K., & Chen, K. (2012). CRC-Aided Decoding of Polar Codes. *IEEE Communications Letters*, *16*(10), 1668–1671. doi:10.1109/LCOMM.2012.090312.121501

Ogudo, K. A., Muwawa Jean Nestor, D., Ibrahim Khalaf, O., & Daei Kasmaei, H. (2019). A Device Performance and Data Analytics Concept for Smartphones' IoT Services and Machine-Type Communication in Cellular Networks. *Symmetry*, *11*(4), 593. doi:10.3390ym11040593

Pajola, L., Pasa, L., & Conti, M. (2019). Threat is in the Air: Machine Learning for Wireless Network Applications. *Proceedings of the ACM Workshop on Wireless Security and Machine Learning*. ACM. 10.1145/3324921.3328783

Pfister, H.D. (2017). *A Brief Introduction to Polar Codes Notes for Introduction to Error-Correcting Codes*.

Prasad, R., Murthy, C. R., & Rao, B. D. (2015). Joint Channel Estimation and Data Detection in MIMO-OFDM Systems: A Sparse Bayesian Learning Approach. *IEEE Transactions on Signal Processing*, *63*(20), 5369–5382. doi:10.1109/TSP.2015.2451071

Ranzini, S. M., Da Ros, F., Bülow, H., & Zibar, D. (2019). Tunable Optoelectronic Chromatic Dispersion Compensation Based on Machine Learning for Short-Reach Transmission. *Applied Sciences (Basel, Switzerland)*, *9*(20), 4332. doi:10.3390/app9204332

Rico-Alvariño, A., & Heath, R. W. (2014). Learning-Based Adaptive Transmission for Limited Feedback Multiuser MIMO-OFDM. *IEEE Transactions on Wireless Communications*, *13*(7), 3806–3820. doi:10.1109/TWC.2014.2314104

Ro, J.-H. (2022). *Improved MIMO Signal Detection Based on DNN in MIMO-OFDM System*. Computers, Materials & Continua. doi:10.32604/cmc.2022.020596

Rusek, F., Persson, D., Larsson, E. G., Marzetta, T. L., & Tufvesson, F., & Lau, B. K. (2013). Scaling Up MIMO: Opportunities and Challenges with Very Large Arrays. *IEEE Signal Processing Magazine*, *30*(1), 40–60. doi:10.1109/MSP.2011.2178495

Saif, W. S., Esmail, M. A., Ragheb, A. M., Alshawi, T. A., & Alshebeili, S. A. (2020). Machine Learning Techniques for Optical Performance Monitoring and Modulation Format Identification: A Survey. *IEEE Communications Surveys and Tutorials*, *22*(4), 2839–2882. doi:10.1109/COMST.2020.3018494

Santos, H. L. (2022). Machine learning-aided pilot and power allocation in multi-cellular massive MIMO networks. *Physical Communication*, *52*, 101646. doi:10.1016/j.phycom.2022.101646

Sengan, S. (2022). Security-Aware Routing on Wireless Communication for E-Health Records Monitoring Using Machine Learning. *International Journal of Reliable and Quality E-Healthcare*.

Seo, S.-I. (2019). Study on Efficient Impulsive Noise Mitigation for Power Line Communication. *International journal of advanced smart convergence, 8*(2), 199-203.

Shen, H., Xu, W., Gong, S., He, Z., & Zhao, C. (2019). Secrecy Rate Maximization for Intelligent Reflecting Surface Assisted Multi-Antenna Communications. *IEEE Communications Letters*, *23*(9), 1488–1492. doi:10.1109/LCOMM.2019.2924214

Simeone, O. (2018). A Very Brief Introduction to Machine Learning With Applications to Communication Systems. *IEEE Transactions on Cognitive Communications and Networking*, *4*(4), 648–664. doi:10.1109/TCCN.2018.2881442

Skoog, R. A., Banwell, T. C., Gannett, J. W., Habiby, S. F., Pang, M., Rauch, M. E., & Toliver, P. (2006). Automatic Identification of Impairments Using Support Vector Machine Pattern Classification on Eye Diagrams. *IEEE Photonics Technology Letters*, *18*(22), 2398–2400. doi:10.1109/LPT.2006.886146

Sobana, S., & Jeyanthi, K. M. (2015). Novel Multiple-Input Multiple Output Precoding Techniques with Improved Bit Error Rate Performance. *Journal of Computational and Theoretical Nanoscience*, *12*(11), 4794–4802. doi:10.1166/jctn.2015.4441

Srinivasa Prasanna, G. N. (2009). Data communication over the smart grid. *2009 IEEE International Symposium on Power Line Communications and Its Applications*, (pp. 273-279). IEEE. 10.1109/ISPLC.2009.4913442

Stantchev, G. (2019). *Machine Learning for RF Signal Processing: Catching the Third Wave*.

Taha, A. (2020). Deep Reinforcement Learning for Intelligent Reflecting Surfaces: Towards Standalone Operation. *2020 IEEE 21st International Workshop on Signal Processing Advances in Wireless Communications (SPAWC)*, (pp. 1-5). IEEE. 10.1109/SPAWC48557.2020.9154301

Tal, I., & Vardy, A. (2011). List decoding of polar codes. *2011 IEEE International Symposium on Information Theory Proceedings*, (pp. 1-5). IEEE.

Tang, F., Mao, B., Kawamoto, Y., & Kato, N. (2021). Survey on machine learning for intelligent end-to-end communication toward 6G: From network access, routing to traffic control and streaming adaption. *IEEE Communications Surveys and Tutorials*, *23*(3), 1578–1598. doi:10.1109/COMST.2021.3073009

Thilina, K. M., Hossain, E., & Kim, D. I. (2016). DCCC-MAC: A Dynamic Common-Control-Channel-Based MAC Protocol for Cellular Cognitive Radio Networks. *IEEE Transactions on Vehicular Technology*, *65*(5), 3597–3613. doi:10.1109/TVT.2015.2438058

Umebayashi, K., Kobayashi, M., & López-Benítez, M. (2018). Efficient Time Domain Deterministic-Stochastic Model of Spectrum Usage. *IEEE Transactions on Wireless Communications*, *17*(3), 1518–1527. doi:10.1109/TWC.2017.2779511

Vela, A. P., Ruiz, M., & Velasco, L. (2018). *Examples of Machine Learning Algorithms for Optical Network Control and Management*. 2018 20th International Conference on Transparent Optical Networks (ICTON), (pp. 1-4). IEEE. 10.1109/ICTON.2018.8473900

Wang, D., Wang, M., Zhang, M., Zhang, Z., Yang, H., Li, J., Li, J., & Chen, X. (2019). Cost-effective and data size-adaptive OPM at intermediated node using convolutional neural network-based image processor. *Optics Express*, *27*(7), 9403–9419. doi:10.1364/OE.27.009403 PMID:31045092

Wang, D., & Zhang, M. (2021). Artificial Intelligence in Optical Communications: From Machine Learning to Deep Learning. In Frontiers in Communications and Networks. doi:10.3389/frcmn.2021.656786

Wang, X. (2019). Learning to Flip Successive Cancellation Decoding of Polar Codes with LSTM Networks. *2019 IEEE 30th Annual International Symposium on Personal, Indoor and Mobile Radio Communications (PIMRC),* (pp. 1-5). IEEE. 10.1109/PIMRC.2019.8904878

Wu, Q., & Zhang, R. (2019). Intelligent Reflecting Surface Enhanced Wireless Network via Joint Active and Passive Beamforming. *IEEE Transactions on Wireless Communications, 18*(11), 5394–5409. doi:10.1109/TWC.2019.2936025

Wu, X. (2009). Applications of Artificial Neural Networks in Optical Performance Monitoring. *Journal of Lightwave Technology, 27*(16), 3580–3589. doi:10.1109/JLT.2009.2024435

Xiao, L., Li, Y., Han, G., Liu, G., & Zhuang, W. (2016). PHY-Layer Spoofing Detection With Reinforcement Learning in Wireless Networks. *IEEE Transactions on Vehicular Technology, 65*(12), 10037–10047. doi:10.1109/TVT.2016.2524258

Xiao, L., Wan, X., Lu, X., Zhang, Y., & Wu, D. (2018). IoT Security Techniques Based on Machine Learning: How Do IoT Devices Use AI to Enhance Security? *IEEE Signal Processing Magazine, 35*(5), 41–49. doi:10.1109/MSP.2018.2825478

Xie, Y., Wang, Y., Kandeepan, S., & Wang, K. (2022). Machine Learning Applications for Short Reach Optical Communication. *Photonics, 9*(1), 30. doi:10.3390/photonics9010030

Ye, H., Li, G. Y., & Juang, B.-H. (2018). Power of Deep Learning for Channel Estimation and Signal Detection in OFDM Systems. *IEEE Wireless Communications Letters, 7*(1), 114–117. doi:10.1109/LWC.2017.2757490

Yuan, J., Ngo, H. Q., & Matthaiou, M. (2019). Machine Learning-Based Channel Estimation in Massive MIMO with Channel Aging. *2019 IEEE 20th International Workshop on Signal Processing Advances in Wireless Communications (SPAWC),* (pp. 1-5). IEEE. 10.1109/SPAWC.2019.8815557

Zappone, A., Sanguinetti, L., & Debbah, M. (2018). User Association and Load Balancing for Massive MIMO through Deep Learning. *2018 52nd Asilomar Conference on Signals, Systems, and Computers,* (pp. 1262-1266). IEEE. 10.1109/ACSSC.2018.8645483

Zhang, Q., Saad, W., & Bennis, M. (2022). Millimeter Wave Communications With an Intelligent Reflector: Performance Optimization and Distributional Reinforcement Learning. *IEEE Transactions on Wireless Communications, 21*(3), 1836–1850. doi:10.1109/TWC.2021.3107520

Zheng, X., Ping, F., Pu, Y., Wang, Y., Montenegro-Marin, C. E., & Khalaf, O. I. (2021). Recognize and regulate the importance of work-place emotions based on organizational adaptive emotion control. *Aggression and Violent Behavior,* 101557. doi:10.1016/j.avb.2021.101557

Zimmermann, M., & Dostert, K. (2002). A multipath model for the powerline channel. *IEEE Transactions on Communications, 50*(4), 553–559. doi:10.1109/26.996069

Chapter 16
Sustainable Development Method in Healthcare Systems After the COVID–19 Pandemic:
Transformation Through Information Technology

Manpreet Kailay
Lovely Professional University, India

Kamalpreet Kaur Paposa
Lovely Professional University, India

Priyanka Chhibber
Lovely Professional University, India

ABSTRACT

The outbreak of COVID-19 pandemic is an unprecedented shock to the entire economy, majorly the hospital sector. In the existing situation, embedding sustainability in the hospital sector is the most crucial aspect for reducing harm to the environment. Thus, the question is aroused in front of researchers to study more about the health care sector and to find out certain practices that emphasizes on the sustainability of hospitals sector. During research it has been identified that one of the most important factors known as information technology is playing a very vital role in the success of heath care sector and specially in the era of pandemic. Various technological advancements like digital health care, internet of medical things, smart health monitoring, telemedicine, chatbot systems, emotive sensory web and robotics can play a huge role towards strengthening the healthcare sector leading to sustainability of the entire world economy, paving a way towards better well-being of health care professionals.

DOI: 10.4018/978-1-6684-7000-8.ch016

I. INTRODUCTION

The health care sector is considered as one of the largest industry and the availability of good infrastructure at hospitals will ensure the quality of life in each country. The Indian Health Sector consists of hospitals, medical professionals like physicians and doctors, clinic centers, pharmaceutical manufacturer, and medical equipment manufacturer. Health care system of India is divided into two sectors, that is, public and private sector hospitals. These hospitals provide their services at three levels: primary, secondary, and tertiary as described in figure 1 and figure 2 respectively. In the public sector, networks of health care facilities at primary, secondary and tertiary levels provide free and low-cost medical services (Gupta et al., 2021). It follows a tired system of infrastructure of primary health centers, community health centers, district hospitals and medical colleges for delivering services to the patients at large.

The competitiveness of India's health care sector lies in the availability of well-trained medical professionals. In light of this, demand of medical professionals are boosting because of origin of new diseases (Amin, 2020). To attain the demands of the health care sector witnessed an increased spend by government on health and well-being. In nutshell, there is need to integrate different elements of health care in order to ensure equal and affordable access of treatment to all patients on right time and to spend on new innovations and technologies to improve the quality of care delivered to the patients.

During the last two decades the sudden shift towards the techno-driven approaches had accelerated the demand for innovations, and has pushed the limits of technology in almost every field (Khan et al., 2021). Information technology had revolutionized the way of living. Healthcare sector is also not far behind in opting information technology as they have come to realize importance of information required for timely decision making (Itumalla, 2012). Healthcare information technology is widely asserted to be one of the means for improving the quality of healthcare (efficient nursing care, better care co-ordination, and patient safety) and potentially reducing its cost (Palvia, Lowe, Nemati, & Jacks, 2012).

Figure 1. Public Hospital Classification
Source: Authors Owned

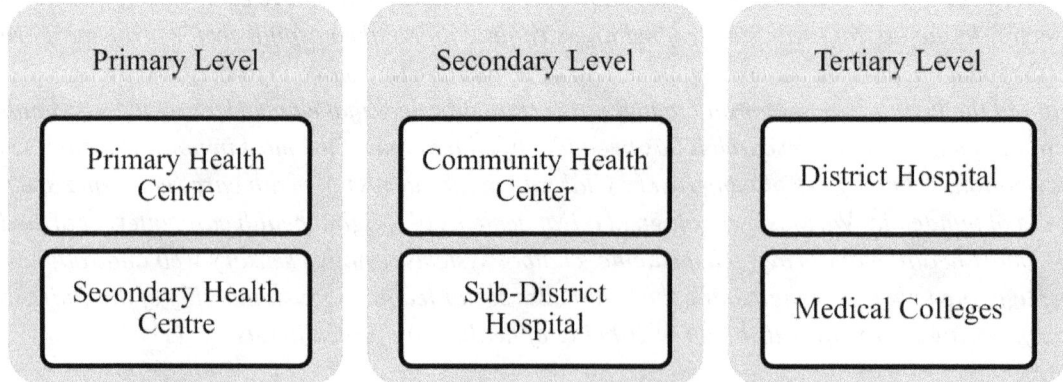

Figure 2. Private Hospital Classification
Source: Authors Owned

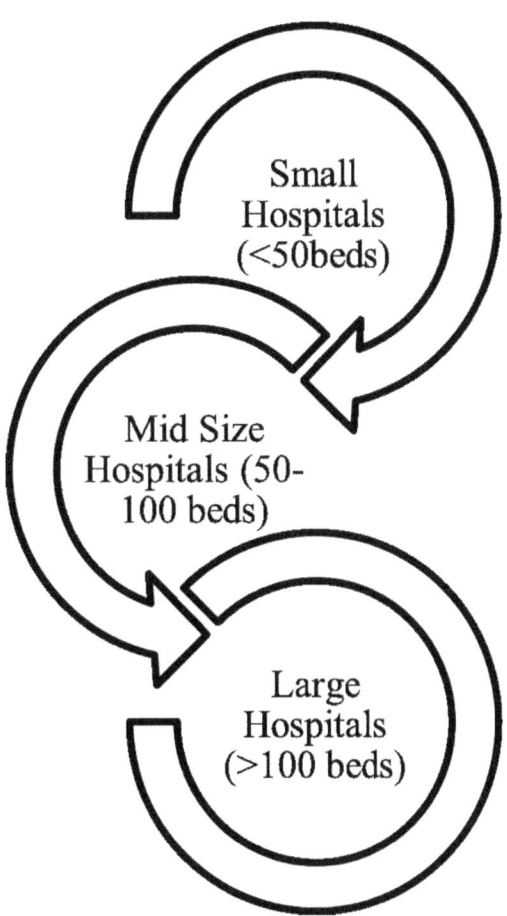

II. HEALTHCARE SECTOR: A GLOBAL CRISIS

Presently, the global population is concerned about Corona Virus Disease-2019 and its long term consequences (A. Kumar & Nayar, 2020). The focus of the World Health Organization is on controlling and mitigating the impact of pandemic on the economy. Corona Virus infectivity measures the potential of one individual to infect others. Initially, it took 67 days to infect one lakh individuals but only next four days to infect next one lakh individuals. This sequence has brought the World figure into 185 million cases today. Given this trend of multiplication, India's health care personnel are abysmally inadequate with single allopathic doctor catering the needs of overall 10, 926 infected individuals (Iyengar, Mabrouk, Kumar, & Venkatesan, 2020). Moreover, the entire healthcare infrastructure consists of 713,986 amounting to 55 beds per 1000 people, situating in India at 155th position among 167 countries. The precarious situation also highlighted the stark imbalance between the demand and availability of hospital beds, ventilators, personal protective equipment kits and trained medical personnel. In light of difficulties faced by hospital sector- decreased revenue, increased cost, increased expenditure on PPE kits, lack of infrastructure and poor quality of services.

The authors have formulated the following research questions for this book chapter after extensive review of literature, the key research questions for this study are: -

RQ1. What are major challenges posed by this scenario for health care sector?

RQ2. Why there is need for Smart wireless technologies?

RQ3. How Smart technologies assists healthcare sector?

RQ4. How information technology will assist the health care professionals in maintaining well-being and hospitals to move towards sustainability?

III. DIGITAL INFRASTRUCTURE AND HEALTHCARE SECTOR: PRE-PANDEMIC CHALLENGES

Health Care has emerged as one of the most progressive and largest service sectors in India. Health care sector of India has been growing at enormous pace. From the 1947, the average growth rate of economy is about 3% per annum. Till the 1980s period, the economic growth rate enhanced to 5.6% per annum (RamaKumar & Kanitkar, 2021). The priority will be to develop effective and sustainable health system that can meet the demands of society with better quality and higher level of health care. In India, public sector plays a dominating role in health care sector. Services delivered at these hospitals are far from quality due to lack of affordability to necessary resources due to the socio-economic conditions of the country (Venkatanarayan Motkuri; Suresh V. Naik, 2010). Even without pandemic, the situation of public sector hospitals is such that there is very limited staff and poor infrastructure (Wilson et al., 2020). The Indian health care infrastructure is not able to keep the pace with the growing population. Access to health care in India is limited due to combination of various insufficiencies- infrastructure, poor health financing, and lack of adequate human resources (Srinivasan & Chandwani, 2014). Additionally, psychological stressors are similar pre-COVID times, the fear of spread of infection to family also prevails, access to particular equipment and machines is also limited for the health care professionals. Bajpai (2014) listed various challenges confronting hospital sector which includes, deficient infrastructure, deficient man-power, patient load, out of pocket expenditures. Moreover, medical doctors and para-medical staff are required to work for long hours as the nature of this profession is such that medical professionals have to work at odd hours throughout day (Shivakumar, 2018).

The first case of Pneumonia was diagnosed in the last week of December, 2019 which was later on named as Novel Corona Virus (Ghosh, Nundy, & Mallick, 2020). But, in the year 2020, January 30th, the World Health Organization declared the outbreak of novel Corona Virus posing higher risk and immediate emergency initiatives to curb the situation. As in the coming next months the intensity of virus is on peak which count for more than eleven thousand positive cases and approx. forty-five hundred deaths in our global countries (Singh et al., 2020). This is unpredictable and unique situation globally, because the virus is getting transmitted just by droplets of cough and sneeze. Due to the vulnerability of this infection and the overall response witnessed globally, both public and private sector hospitals should require to start working together to tackle the situation (Gopalan & Misra, 2020). India's private health care sector has contributed significantly in testing, isolation beds for treatment, medical staff, medical equipment and for about 60% of in- patient care (Dev Mahendra & Sengupta, 2020). During the initial quarter of the arrival of this virus, medical professionals, tele-communication industry, police department have been shown more gratitude. But in the later stages, health care workers are targeted by general public. People did not support if any medical professional goes out to perform moral responsibilities, in fear of that they will take infection to home and society, which is not safe for society (Figure 3). Medical professionals with no option left had to stay in hospitals for their own well-being and well-being of their families. Moreover, the revenue generation and long term sustainability also become a concern for hospitals in this scenario (Sengupta et al., 2021).

COVID-19 hit the people of all ages, affected the employment and more importantly, resulted into loss of huge human lives. Millions of people were infected with this virus and so far, more than a million lost their lives because of pandemic (Hafner et al., 2021). The severity of Covid-19 outbreak response are unimaginable (Kumar et al., 2020). Moreover, health care workers were also not able to cater their own well-being in this tough scenario owing to less time for sleep, disturbed shift work, enhanced working hours (Agarwal et al., 2021). But all these are considered least of the problems, because the vulnerability to infection was considered as the much bigger problem. There was huge lack of availability of health care professionals because most among them were having quarantine time due to exposure to infectious patients. Most recent studies on COVID-19 outbreak demonstrated that health care workers also suffered from anxieties and traumatization due to lack of knowledge about correct drugs for curing patients and insufficient training on various self-care initiatives (Tan et al., 2020).

Figure 3. Psychological threats in front of health care workers
Source: Authors Owned

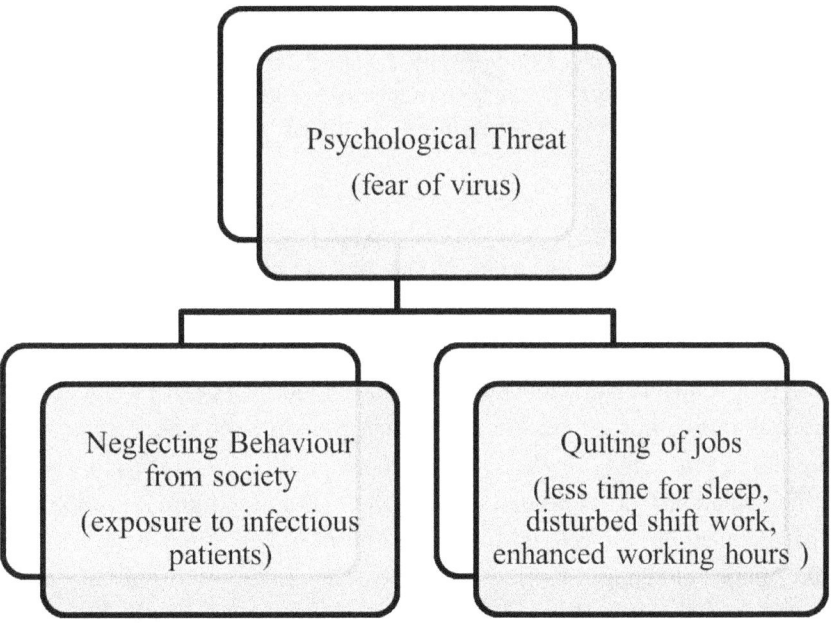

IV. WIRELESS NETWORKING IN HEALTH CARE COMBATING COVID-19

During Novel Corona Virus, it was evident that our health care still lacks in in terms of efficiency and productivity (Rathi et al., 2021). Entire health system is not ready for such a massive disruption, which is going to give such an effect on human lives (Bhambere, Abhishek, & Sumit, 2021). Hospital industry on one side- undergo through providing high number of hospital beds, oxygen tanks, respirators, intensive care units. On the other side supplies of protective suits, masks and testing kits will run short of getting all this arranged (Mohan, 2020). In addition to this, the government spending on health care over last decade has enhanced four folds. The expenditures of hospital industry have also increased in terms of making available all necessary items at workplace.

Additionally, as a protection measure against the global pandemic, many countries have forced to mandate lockdown which eventually resulted in negative impact on the environmental aspects, disrupt livelihood of more than half of the population and overall slowdown the progress of economy. The daily expected amount of generated waste is 3.4kg from each infected individual. The volume of waste arose due to the enhanced use of protective kits including- face masks, hand gloves, rubber boots, white gowns, hand sanitizers, syringes, plastic containers and test kits (Haque, Uddin, Sayem, & Mohib, 2020). The social and economic issues will result in mass unemployment, homelessness, hunger, poverty, chronic stress, anxiety and depression. Due to unemployment at large level, inaccessibility to food is also a major reason of deaths in our country. There are majorly four groups of challenges that every individual is going through- problems related with work, problems related to lockdown, difficulties in family situations, personal well-being and financial problems (Gopalan & Misra, 2020).

As with most of electronic technologies, new devices and transmission mode arise in field of wireless communication every year (Heslop, Howard, Fernando, Rothfield, & Wallace, 2003). Utilizing Information and communication technologies for healthcare and medical purposes has obtained lot of interest and attention during recent years (Hämäläinen, Iinatti, & Kohno, 2011). Awareness of Information Communication Technology is now days widely spread throughout the population. Through the use of ICT, healthcare can receive two benefits: their quality of life is increased and expenses for healthcare facilities are decreased. Though there are lots of wireless applications already in use. These will provide a convenient means in home-care related procedures.

V. TRANSFORMATION TOWARD SUSTAINABLE MODERNIZATION

In today's world, information and communication technologies play an important role in simplifying personnel and professionals lives (Mohammad et al., 2021). Today, the Healthcare sector, more preciously-hospitals are undergoing transformation towards the digital economy. The focus is also shifting from health care to health and well-being. Earlier, healthcare 4.0 involves digitation of records, Internet of Medical Technologies, mobile internet, and 'Health is more important than healthcare' (Viswanadham, 2021). Industry 5.0 is a new visionary concept that seeks to make industry more 'sustainable, human-centric and resilient'. But, the emergence of Novel Corona Virus 2019 has caused major impact on the global economy including major challenges for the medicine. The pandemic eventually led to overcrowded emergency units and intensive care units in the hospitals. In such a scenario face-to-face interactions with the patients is risky and tough to manage. This has raised a question that- how to provide medical care without physically interacting with patients. Healthcare has undergone a paradigm change to reach the new era of smart health management, virtual care, smart monitoring, and smart self-management (Mbunge et al., 2021). In the wake of COVID-19, embracing corporate social responsibilities is right thing for the sustainable growth of society at large. Next, policy making is not just making and implementing on public, it involves mere attention towards the situation people are going through and to involve them in all stages to get most acceptance of policies (Roziqin, Mas'udi, & Sihidi, 2021).

VI. INFORMATION AND COMMUNICATION TECHNOLOGIES INTRODUCING AUTOMATED INNOVATIVE TECHNOLOGIES IN HEALTHCARE SECTOR

Advancements in Information and Communication Technologies have paved way for provision of cost-effective e-services to the people around the globe. These applications include use of cell phones and other communication devices to gather health data, delivery of healthcare information to doctors, researchers and patients. It can help improve clinical outcomes and contribute to better public health monitoring and education (Burney, Mahmood, & Abbas, 2010). Currently, industry pioneers and business leaders are leading towards the Fifth Industrial Revolution or Industry 5.0 on the collaboration human touch, human touch, and efficiency. This revolution is regarded as a return to pre-industrial production, where advanced technologies are utilized by giving preference to customer's demands, especially (Ostergaard, 2019). In this revolution, personalized requirements of customers could fulfill, because customers are more devoted towards customization of products which will provide them satisfaction (Haleem & Javaid, 2019).

It is anticipated that advanced technologies will enable industrialization that will eventually lead to sustainability through human-technology interaction (Madsen & Berg, 2021). It is right time to use automated innovative technologies especially in the hospital sector, as the minimal face-to-face interaction would lead to overall sustainable future. In healthcare, this revolution provides patient-centric healthcare for providing better quality of care. In which, the profession of medicine has also diverting towards personalization of medical device which can used for their personal use such as- diabetes checker, blood pressure testing machines. With regard to the evolution of technologies in the hospitals, patients are held accountable for their self-care at their homes. Homecare is essential for patients receiving remotely to ensure that support was available for their health (Maarup et al., 2021). "Health is more than healthcare" is followed in this revolution (Viswanadham, N. 2021). Healthcare is going through a transformation where automation technology is embedded not to reduce the workload of individuals, but to reach patients at mass level, enhance work efficiency, improve quality of care and sustainability.

Innovative technologies can be used for maintaining well-being of health care professionals which is well explained in a study done on Chinese hospitals (Lin & Lin et al., 2021). It disclosed the automated hand hygiene habits and reminders especially for the pandemic time for the health care workers, such as, Automated training systems, Electronic counting device, and automated hand hygiene monitoring system. With an increase in number of COVID-19 cases, the demand for robots and automated technologies are on peak. Robots can partially take over some of the medical roles, such as, disinfecting or cleaning the hospital wards. Hospitality robots can also be used as a receptionist, clinicians. These robots can deliver their services like providing medicines and gather information related to the patients (Raje et al., 2021).

In the earlier times, these automated technologies as described in figure 4 have been used to perform the tasks efficiently and effectively, to reduce the medical errors, and to handle large amount of data at a time. Automation empowers health providers to offer better services, better patient management and increased work efficiency. Currently, emphasizing on the struggles of the medical professionals embedding human-machine co-working will be of great help. The authors have listed many earlier studies where the information technology, artificial intelligence, internet of medical things, and digital health care has proven as best technological advancements that have enhanced the efficiency and effectiveness levels of hospitals.

Digital Health care: Digital Health care involves the use of innovations in healthcare to support "healing at a distance". In this, audiovisual technologies are involved such as through smart phones and vedio-conferencing to help care for patients allowing physicians and doctors to diagnose the patients virtually. Earlier, these innovative medical care are used to provide Medicare facilities easily accessible to all, and to manage cost and convenience, to reduce the travel time, shorter waiting time (Anthony, B. 2021). In the pandemic, these technologies would enable minimal human interaction in order to lower down the chance of transmission of virus and reduce the chance of infection. Under its umbrella, digital health includes mobile health aps, electronic health records, electronic medical records, telehealth, wearable devices as well as personalized medicines.

Robotics: The current pandemic brings several threats and restrictions to our society. This had created unprecedented demands for hospitals. Machine intelligence is gaining significance in healthcare to combat the virus (Zemmar et al., 2020). Robotics and Artificial intelligence would be incorporated to deal with terrible situation as current pandemic revealed that most of activities in the hospitals involves human-to-human contact, which is dangerous in current scenario (Kaiser et al., 2021). Hospital sector and health care workers have been facing various restrictions such as: lack of workforce, risk of exposure to infection, wearing PPE kits, and maintaining social distancing. Considering all these factors, robotics would provide the best solution to deal with all such hardships (Marin et al., 2021). In hospitals, robots can serve multipurpose in the hospital, they would also be used to disinfect or sanitize the isolated rooms, lifts, tools and equipments which are used on regularly (Chamola et al., 2020). For example: A Autonomous robot, invented by Asimov robots, Kerela that can be used help critical patients in isolation with full support. A humanoid robot can also be applied for delivering food and medicine to COVID19 patients admitted in the hospitals. Secondly, UV disinfection robots have become the first choice for sanitizing healthcare centers.

Real-time hand hygiene reminders and wearable: Real-time hand hygiene reminders and wearable are communication devices worn on the body that are connected to an internet source. It has ability to monitor people's physical health and have sensors on it (Jahnke et al., 2021). The red light on screen or audible beep was used to remind the users (health care professionals) to perform proper hand hygiene before patient contact. For example: iwash is an electronic device which is used for giving reminders to health care professionals for performing hand hygiene.

Radical Technological Innovations (RTIs): (Chatbot systems): Artificial Intelligence is a digital innovation or program that had capability to display human intelligence. Chatbot is a one type of AI-based software used for real conversation between users and devices by means of digital interface. In healthcare sector, Chatbot particularly encompasses a unique platform, that enables an interaction between patients and healthcare providers to constant care service not through face-to-face consultations (Maarup et al., 2021). Through this particular platform one can get advice on which hospital is best for which treatment, which hospital or doctor cures what kind of disease. In the corona virus pandemic, human and machine co-working is essential in order to reduce direct interaction with patients and medical providers. Only patients with intense situations are required to visit doctors personally, rest of the patients with mild symptoms would be cured through these types of platforms.

Internet of Medical Things: It is a network of medical devices that feed vital data in real time to software applications that analyze and communicate with healthcare IT systems (Zounia et al., 2021). It enables the digital transfer and exchange of patient critical care data to enable better care management in real time. Telemedicine is another technology innovation in the hospital sector (Mbunge et al., 2021). IoMT, drones, robots, GPS, Bluetooth are the devices that played primary role in such circumstances to mitigate the impact of COVID-19 outbreak.

Smart Health Monitoring (SHM): SHM are considered as an extension of medical systems in hospitals by which the patients can take under the interpretation without any negligence. SHM is able to monitor real time conditions of Covid-19 patients remotely. Most of the countries prefer tele-consultation and online appointment to avoid hospital visits and public gatherings (Sujith et al., 2020). IoT devices tagged with sensors are used for tracking real time location of medical equipment like wheelchair, oxygen pumps, nebulizers and other monitoring equipments.

Emotive Sensory Web: Sensors are used worldwide as it has tremendously improved the healthcare. The unprecedented shock caused by pandemic, put everyone in a terrible situation. Healthcare professionals could not able to understand how to handle the situation- as they unable to monitor the mental state infected individuals. But, with emotive sensory web- the emotional state of infected individuals can be monitored by medical professionals in order to provide righteous medical services (Mbunge et al., 2021). This study revealed that emerging technologies such as; Artificial Intelligence, Internet of Things, 5G technology, Industry 5.0, big data, Cloud computing are paramount in the development and implementation of sensory emotive web in virtual health care.

Telemedicine: Enables healthcare professionals to evaluate, diagnose and treat patients situated on remote areas. Moreover, the use of mobile technology in organizations has grown rapidly in the last decade. Many healthcare organizations are spending large sums of money in implementing mobile technologies with expectation that nurses will employ the system to enhance individual performance (Connor & Reilly, 2018).

Figure 4. Innovative Technologies in healthcare sector
Source: Author's Own

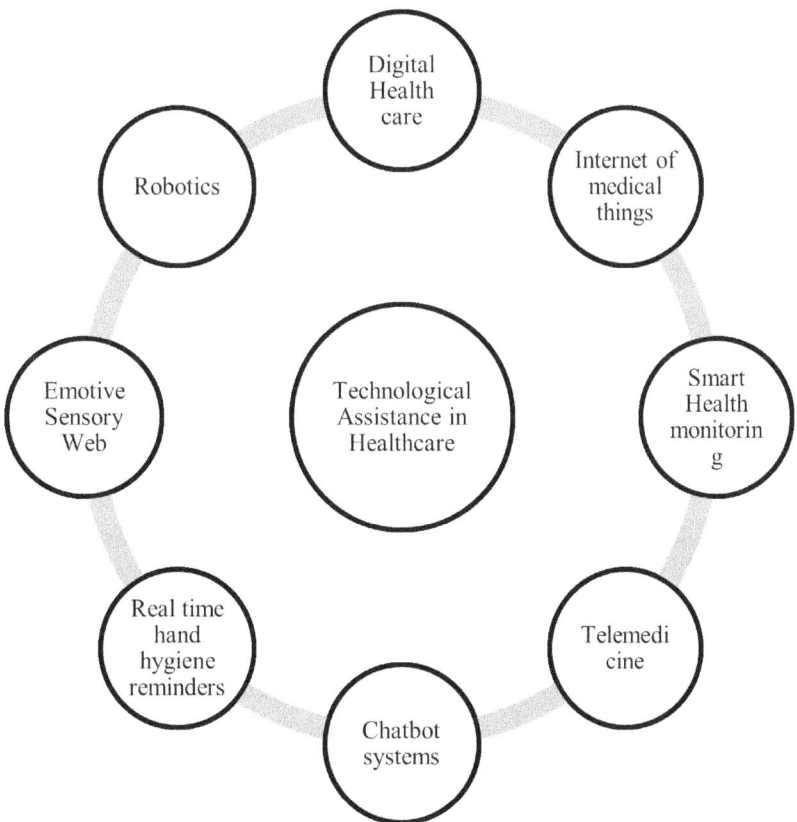

VII. RESEARCH IMPLICATIONS: HOSPITAL MANAGEMENT

With the use of smart technologies, organization's overall efficiency will improve and this will bring a method of innovative care in healthcare. Moreover, implementation of IT will motivate health care workers to perform best by utilizing all their capabilities and innovation because little pressure of them is now transferred to IT's. This enables them to concentrate on other essential services with full empowerment. Moreover, inclusion of smart technologies; hospitals can achieve optimum outcomes on each and every individual patients. It is necessary to achieve greater collaboration within the healthcare system through virtual care and co-ordination, personalized medicines, enhanced efficiency, wearable devices. The research suggests that economy has to come forward for the well-being of whole population in this new normal. This pandemic period made us to think for hygiene, go digitization and sustainability. All the hospital organizations should to appoint RCTF (Rapid Covid Task Force) in their hospitals for keeping an eye on the unpredictable upcoming covid crisis in order to build strategies in advance handle and manage the crisis before handedly. Government should maintain a good stock of life –saving drugs and oxygen concentrators and other necessary medical kits for minimizing the effects of further insecure and threats in future. Healthcare professionals can be trained to handle critical patients on priority basis and provide their services in good faith/ unbiased. Health care professionals required to maintain a balance while delivering services between all the patients. Doctors and nurses should be trained to identify cases that need extra care and that do not; in order to utilize their services more accurately.

Hospitals and healthcare professionals are hit harder through the virus, adopting digital technology will become a good solution for their well-being. Access to Information Technology and Wireless devices to all medical professionals would lead to well-being of population at large. For example: gathering of people can be reduced by providing medical advices through Video-conferencing and telephones. Moreover, Robots can be installed in hospitals in order to serve critical patients by reducing doctor/ nurses and patients physical interaction. Technologies can be used for maintaining hygiene and sanitization at workplace. Government can enhance spending upon the installation of Information Technology, Artificial Intelligence, Internet of Things and Wireless devices for the smooth operations of hospital sector and for preparing them for unfortunate/ unpredictable environment.

VIII. CONCLUSION

The findings establish how the smart technologies connect large masses to tackle the emergency situations. This chapter gives deep insights on new innovations in the smart technologies world that has been globally investigated during crises. The enormous importance of health care sector has made us realize to bring smart technologies in this sector. A broader picture of how technology has driven the industries during the pandemic period, has pushed the use of technologies in every field and in every sector. The health care comprises of hospitals, medical clinics, community health centers, telemedicine, medical tourism and health insurance. The country India is a geographical area is much wider having dense population along with inefficient infrastructure poses a big challenge to handle COVID-19 outbreak (Pai *et al.*, 2020). No doubt, our country is suffering from economic crisis prior to the pandemic period also, but due to this virus the situation went downhill. In was an unexpected announcement for the country when a complete lockdown was announced by our Prime Minister. This announcement is very sudden and from the immediate next day people are required to restrict themselves in their homes. Eventually, the most affected and tremendously changed sector due to this pandemic is the Health care sector.

Figure 5. Innovative Technologies in healthcare sector
Source: Author's Own

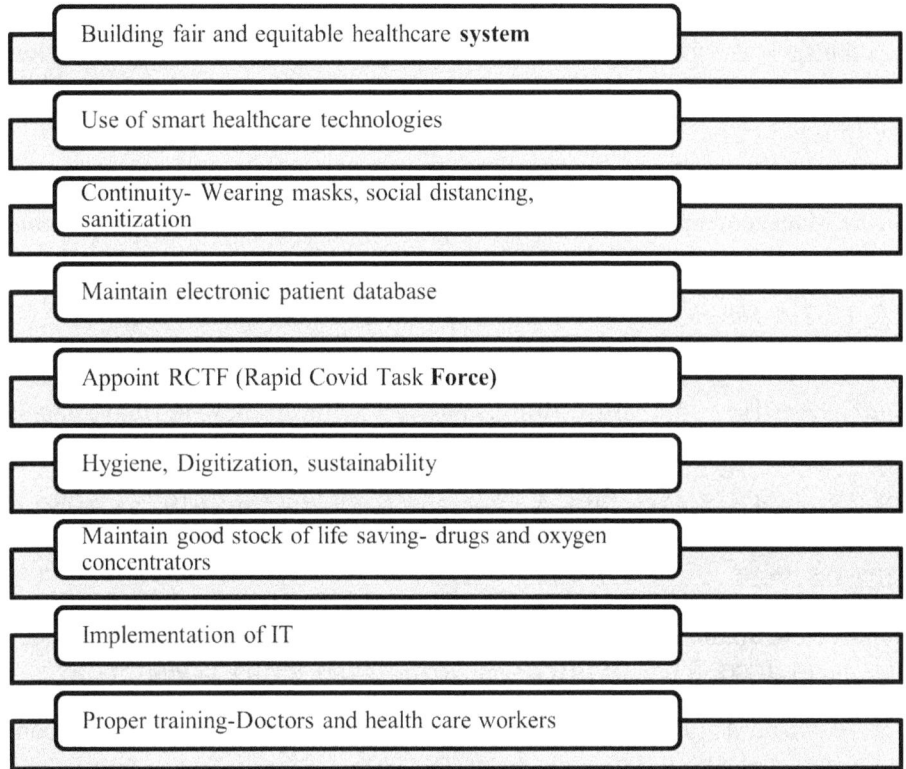

ACKNOWLEDGMENT

- There is no 'competing interest'.
- The research received no specific grant from any funding agency.

REFERENCES

Agarwal, A., Ranjan, P., Saraswat, A., Kasi, K., Bharadiya, V., Vikram, N., Singh, A., Upadhyay, A. D., Baitha, U., Klanidhi, K. B., & Chakrawarty, A. (2021). Are health care workers following preventive practices in the COVID-19 pandemic properly? - A cross-sectional survey from India. *Diabetes & Metabolic Syndrome, 15*(1), 69–75. doi:10.1016/j.dsx.2020.12.016 PMID:33310264

Aguinis, H., Villamor, I., & Gabriel, K. P. (2020). Understanding employee responses to COVID-19: A behavioral corporate social responsibility perspective. *Management Research*, *18*(4), 421–438. doi:10.1108/MRJIAM-06-2020-1053

Amin, S. (2020). The psychology of coronavirus fear: Are healthcare professionals suffering from corona-phobia? *International Journal of Healthcare Management*, *13*(3), 249–256. doi:10.1080/2047 9700.2020.1765119

Bhambere, S., Abhishek, B., & Sumit, H. (2021). Rapid Digitization of Healthcare - A Review of CO-VID-19 Impact on our Health systems. *International Journal of All Research Education and Scientific Methods*, *9*(2), 1457–1459.

Burney, D. S. M. A., Mahmood, N., & Abbas, Z. (2010). Information and Communication Technology in Healthcare Management Systems: Prospects for Developing Countries. *International Journal of Computers and Applications*, *4*(2), 27–32. doi:10.5120/801-1138

Corporation, R. (2021). *The global economic cost of COVID-19 vaccine nationalism.*

Dev Mahendra, S., & Sengupta, R. (2020). COVID-19 Impact on the Indian Economy - Detailed Analysis. *Igidr*, (April), 1–48. https://blog.smallcase.com/the-new-normal-analysis-of-covid-19-on-indian-businesses-sectors-and-the-economy/

Gambhir, R. S., Dhaliwal, J. S., Aggarwal, A., & Anand, S. (2020). Covid-19 : a survey on knowledge. *Awareness and hygiene practices among dental health professionals in an indian scenario*, *71*(2), 223–229. PMID:32519827

Ghosh, A., Nundy, S., & Mallick, T. K. (2020). How India is dealing with COVID-19 pandemic. *Sensors International*, *1*(June), 100021. doi:10.1016/j.sintl.2020.100021 PMID:34766039

Gopalan, H. S., & Misra, A. (2020). COVID-19 pandemic and challenges for socio-economic issues, healthcare and National Health Programs in India. *Diabetes & Metabolic Syndrome*, *14*(5), 757–759. doi:10.1016/j.dsx.2020.05.041 PMID:32504992

Hämäläinen, M., Iinatti, J., & Kohno, R. (2011). Wireless communications in healthcare recent and future topics. *ACM International Conference Proceeding Series*. ACM. 10.1145/2093698.2093796

Haque, S., Uddin, S., Sayem, S., & Mohib, K. M. (2020). ur na l P re. *Biochemical Pharmacology*, *2019*, 104660. doi:10.1016/j.jece.2020.104660 PMID:33194544

Heslop, L., Howard, A., Fernando, J., Rothfield, A., & Wallace, L. (2003). Wireless communications in acute health-care. *Journal of Telemedicine and Telecare*, *9*(4), 187–193. doi:10.1258/135763303322225490 PMID:12952687

Itumalla, R. (2012). Information Technology and Service Quality in Health Care: An Empirical Study of Private Hospital in India. *International Journal of Innovation, Management and Technology*, *3*(4), 1–4. doi:10.7763/IJIMT.2012.V3.269

Iyengar, K., Mabrouk, A., Kumar, V., & Venkatesan, A. (2020). *Since January 2020 Elsevier has created a COVID-19 resource centre with free information in English and Mandarin on the novel coronavirus COVID- 19 .* Elsevier Connect.

Khan, H., Kushwah, K. K., Singh, S., Urkude, H., Maurya, M. R., & Sadasivuni, K. K. (2021). Smart technologies driven approaches to tackle COVID-19 pandemic: a review. *3 Biotech, 11*(2), 1–22. doi:10.1007/s13205-020-02581-y

Kumar, A., & Nayar, K. R. (2020). COVID 19 and its mental health consequences. *Journal of Mental Health (Abingdon, England), 0*(0), 1–2. doi:10.1080/09638237.2020.1757052 PMID:32339041

Kumar, S. U., Kumar, D. T., & Christopher, B. P. (2020). *The Rise and Impact of COVID-19 in India. 7*(May), 1–7. doi:10.3389/fmed.2020.00250

Mohan, D. (2020). *What Will Be the Economic Consequences of COVID-19 for India and the World?* Palvia, P., Lowe, K., Nemati, H., & Jacks, T. (2012). Information Technology Issues in Healthcare: Hospital CEO and CIO Perspectives. *Communications of the Association for Information Systems, 30*. doi:10.17705/1CAIS.03019

Motkuri, V., & Naik, S. V. (2010). Workforce in Indian Health Care Sector. *The Asian Economic Review, 52*(2).

Roziqin, A., Mas'udi, S. Y. F., & Sihidi, I. T. (2021). An analysis of Indonesian government policies against COVID-19. *Public Administration and Policy, 24*(1), 92–107. doi:10.1108/PAP-08-2020-0039

Sandeep Kumar, M., Maheshwari, V., Prabhu, J., Prasanna, M., Jayalakshmi, P., Suganya, P., & Jothikumar, R. (2020). Social economic impact of COVID-19 outbreak in India. *International Journal of Pervasive Computing and Communications, 16*(4), 309–319. doi:10.1108/IJPCC-06-2020-0053

Science, N., Phenomena, C., Pai, C., Bhaskar, A., & Rawoot, V. (2020). Chaos, Solitons and Fractals Investigating the dynamics of COVID-19 pandemic in India under lockdown. *Chaos, Solitons and Fractals: The Interdisciplinary Journal of Nonlinear Science, and Nonequilibrium and Complex Phenomena, 138*, 109988. doi:10.1016/j.chaos.2020.109988

Sengupta, M., Roy, A., Ganguly, A., Baishya, K., Chakrabarti, S., & Mukhopadhyay, I. (2021). Challenges Encountered by Healthcare Providers in COVID-19 Times: An Exploratory Study. *Journal of Health Management, 23*(2), 339–356. doi:10.1177/09720634211011695

Shivakumar, K. N. (2018). Work Life Balance in the Health Care Sector. *Amity Journal of Healthcare Management.* doi:10.13140/RG.2.2.19413.73440

Srinivasan, V., & Chandwani, R. (2014). HRM innovations in rapid growth contexts: The healthcare sector in India. *International Journal of Human Resource Management, 25*(10), 1505–1525. doi:10.10 80/09585192.2013.870308

Tan, B. Y. Q., Chew, N. W. S., Lee, G. K. H., Jing, M., Goh, Y., Yeo, L. L. L., Zhang, K., Chin, H.-K., Ahmad, A., Khan, F. A., Shanmugam, G. N., Chan, B. P. L., Sunny, S., Chandra, B., Ong, J. J. Y., Paliwal, P. R., Wong, L. Y. H., Sagayanathan, R., Chen, J. T., & Sharma, V. K. (2020). Psychological Impact of the COVID-19 Pandemic on Health Care Workers in Singapore. *Annals of Internal Medicine, 173*(April), 19–22. doi:10.7326/M20-1083 PMID:32251513

Viwattanakulvanid, P. (2021). Ten commonly asked questions about Covid-19 and lessons learned from Thailand. *Journal of Health Research, 35*(4), 329–344. doi:10.1108/JHR-08-2020-0363

Weekly, P., & Weekly, P. (2021).. . *Public Health Human Resource.*, *50*(17), 20–23.

Wilson, W., Raj, J. P., Rao, S., Ghiya, M., Nedungalaparambil, N. M., Mundra, H., & Mathew, R. (2020). Prevalence and Predictors of Stress, anxiety, and Depression among Healthcare Workers Managing COVID-19 Pandemic in India: A Nationwide Observational Study. *Indian Journal of Psychological Medicine*, *42*(4), 353–358. doi:10.1177/0253717620933992 PMID:33398224

Compilation of References

1mg. (n.d.). *Online pharmacy India: Buy medicines from India's Trusted Medicine Store.* 1mg. https://www.1mg.com/

3GPP. (2010). Evolved universal terrestrial radio access (e-utra); relay architectures for e- utra (lte-advanced). *3rd Generation Partnership Project (3GPP), Technical Report (TR) 36.806, 04, version 9.0.0.*

Aalamifar, F., & Lampe, L. H.-J. (2017). Optimized WiMAX Profile Configuration for Smart Grid Communications. *IEEE Transactions on Smart Grid, 8*(6), 2723–2732. doi:10.1109/TSG.2016.2536145

Aazhang, B. (2019). Key drivers and research challenges for 6G ubiquitous wireless intelligence (white paper). *Human-Centric Computing and Information Sciences, 9*(1), 1–10. doi:10.1186/s13673-019-0198-1

Abdalla & Ibrahim. (2017). Design and performance evaluation of metamaterial inspired MIMO antennas for wireless applications. *Wireless Personal Communication, 95*(2), 1001–1017.

Abed, A. T., Singh, M. S. J., & Islam, M. T. (2018). Compact fractal antenna circularly polarized radiation for wi-fi and WiMAX communications. *IET Microwaves, Antennas & Propagation, 12*(14), 2218–2224. doi:10.1049/iet-map.2018.5213

Abedian., M. (2017). Compact ultra wideband MIMO dielectric resonator antennas with WLAN band rejection. *IET Microwaves, Antennas & Propagation, 11*(11), 1524–1529.

Abohmra, A., Khan, Z. U., Abbas, H. T., Shoaib, N., Imran, M. A., & Abbasi, Q. H. (2022). Two-Dimensional Materials for Future Terahertz Wireless Communications. *IEEE Open Journal of Antennas and Propagation, 3*, 217–226. doi:10.1109/OJAP.2022.3143994

Abouzar, P. (2011). Action-based scheduling technique for 802.15.4/ZigBee wireless body area networks. *2011 IEEE 22nd International Symposium on Personal, Indoor and Mobile Radio Communications,* (pp. 2188-2192). IEEE.

Adhikari, A., Rawat, D. B., & Song, M. (2019). Wireless Network Virtualization by Leveraging Blockchain Technology and Machine Learning. *Proceedings of the ACM Workshop on Wireless Security and Machine Learning.* ACM. 10.1145/3324921.3328790

Afisiadis, O., Balatsoukas-Stimming, A., & Burg, A. P. (2014). A low-complexity improved successive cancellation decoder for polar codes. *2014 48th Asilomar Conference on Signals, Systems and Computers,* (pp. 2116-2120). IEEE. 10.1109/ACSSC.2014.7094848

Agarwal, A., Ranjan, P., Saraswat, A., Kasi, K., Bharadiya, V., Vikram, N., Singh, A., Upadhyay, A. D., Baitha, U., Klanidhi, K. B., & Chakrawarty, A. (2021). Are health care workers following preventive practices in the COVID-19 pandemic properly? - A cross-sectional survey from India. *Diabetes & Metabolic Syndrome, 15*(1), 69–75. doi:10.1016/j.dsx.2020.12.016 PMID:33310264

Agrawal, S., Gupta, R. D., Parihar, M. S., & Kondekar, P. N. (2017). A wideband high gain dielectric resonator antenna for RF energy harvesting applications. *AEÜ. International Journal of Electronics and Communications*, *78*, 24–31. doi:10.1016/j.aeue.2017.05.018

Aguinis, H., Villamor, I., & Gabriel, K. P. (2020). Understanding employee responses to COVID-19: A behavioral corporate social responsibility perspective. *Management Research*, *18*(4), 421–438. doi:10.1108/MRJIAM-06-2020-1053

Ahamed, M. M., & Faruque, S. (2018). 5G backhaul: requirements, challenges, and emerg-ing technologies. Broadband Communications Networks: Recent Advances and Lessons from Practice, (vol. 43).

Ahmed, A., Bakar, K. A., Channa, M. I., Khan, A. W., & Haseeb, K. (2017). Energy-aware and secure routing with trust for disaster response wireless sensor network. *Peer-to-Peer Networking and Applications*, *10*(1), 216–237. doi:10.100712083-015-0421-4

Ahmed, A., Boulahia, L. M., & Gaiti, D. (2013). Enabling vertical handover decisions in heterogeneous wireless networks: A state-of-the-art and a classification. *IEEE Communications Surveys and Tutorials*, *16*(2), 776–811.

Ahmed, N., & Rikli, N. E. (2018). A QoS Based Algorithm for the Vertical Handover between WLAN IEEE 802.11 e and WiMAX IEEE 802.16 e. *International Journal of Computing and Digital Systems*, *7*(01), 11–22.

Akildiz, I. F., Jornet, J. M., & Han, C. (2014). Terahertz Band: Next Frontier for Wireless Communications. *Physical Communication*, *12*, 16–32. doi:10.1016/j.phycom.2014.01.006

Al Emam, F. A., Nasr, M. E., & Kishk, S. E. (2018, December). Adaptive context aware cross-layer vertical handover in heterogeneous networks. In *2018 14th International Computer Engineering Conference (ICENCO)* (pp. 58-63). IEEE.

Alamouti, S. M. (1998). A simple transmit diversity technique for wireless communications. *IEEE J. Selected Areas Communication*, *16*, 1451-1458.

Al-Bawri, S. S., Islam, M. T., Shabbir, T., Muhammad, G., Islam, M. S., & Wong, H. Y. (2020, October). Hexagonal shaped near zero index (NZI) metamaterial based MIMO antenna for millimeter-wave application. *IEEE Access: Practical Innovations, Open Solutions*, *8*, 181003–181013. doi:10.1109/ACCESS.2020.3028377

Aldmour, I. (2017). Wireless Broadband Tools and Their Evolution Towards 5G Networks. *Wireless Personal Communications*, *95*(4), 4185–4210. doi:10.100711277-017-4058-x

Al-Hourani, A., Kandeepan, S., & Jamalipour, A. (2016). Stochastic Geometry Study on Device-to-Device Communication as a Disaster Relief Solution. *IEEE Transactions on Vehicular Technology*, *65*(5), 3005 3017. doi:10.1109/TVT.2015.2450223

Ali, K., Nguyen, H. X., Vien, Q.-T., Shah, P., & Chu, Z. (2018). Disaster management using D2D communication with power transfer and clustering techniques. *IEEE Access: Practical Innovations, Open Solutions*, *6*, 14643–14654. doi:10.1109/ACCESS.2018.2793532

Aljeri, N., & Boukerche, A. (2018, October). Mobility and handoff management in connected vehicular networks. In *Proceedings of the 16th ACM International Symposium on Mobility Management and Wireless Access* (pp. 82-88).

Alkhateeb, A. (2019). *DeepMIMO: A Generic Deep Learning Dataset for Millimeter Wave and Massive MIMO Applications*. ArXiv.

Alkhateeb, A., & Beltagy, I. (2018). Machine Learning For Reliable Mmwave Systems: Blockage Prediction And Proactive Handoff. *2018 IEEE Global Conference on Signal and Information Processing (GlobalSIP)*, 1055-1059 10.1109/GlobalSIP.2018.8646438

Al-Maitah, M., Semenova, O. O., Semenov, A. O., Kulakov, P. I., & Kucheruk, V. Y. (2018). A hybrid approach to call admission control in 5G networks. *Advances in Fuzzy Systems*, *2018*, 1–7. doi:10.1155/2018/2535127

Altaf, A., Seo, M., & Tilted, A. (2018). D-shaped monopole antenna with wide dual-band dual-sense circular polarization. *IEEE Antennas and Wireless Propagation Letters*, *12*(12), 2464–2468. doi:10.1109/LAWP.2018.2878334

Ameen., M. (2020). Bandwidth and gain enhancement of triple band MIMO antenna incorporating metasurface-based reflector for WLAN/WiMAX applications. *IET Microwaves, Antennas & Propagation*, *14*(13), 1493–1503.

Ameen, M., & Chaudhary, R. K. (2019). Metamaterial-based wideband circularly polarised antenna with rotated V-shaped metasurface for small satellite applications. *Electronics Letters*, *55*(7), 365–366. doi:10.1049/el.2018.7348

Amin, S. (2020). The psychology of coronavirus fear: Are healthcare professionals suffering from corona-phobia? *International Journal of Healthcare Management*, *13*(3), 249–256. doi:10.1080/20479700.2020.1765119

Anand, R., Sindhwani, N., & Saini, A. (2021). Emerging Technologies for COVID-19. *Enabling Healthcare 4.0 for Pandemics: A Roadmap Using AI, Machine Learning, IoT and Cognitive Technologies*, 163-188.

Anand, R., & Chawla, P. (2022). Bandwidth Optimization of a Novel Slotted Fractal Antenna Using Modified Lightning Attachment Procedure Optimization. In *Smart Antennas* (pp. 379–392). Springer. doi:10.1007/978-3-030-76636-8_28

Anand, R., Shrivastava, G., Gupta, S., Peng, S. L., & Sindhwani, N. (2018). Audio watermarking with reduced number of random samples. In *Handbook of Research on Network Forensics and Analysis Techniques* (pp. 372–394). IGI Global. doi:10.4018/978-1-5225-4100-4.ch020

Anand, R., Singh, J., Pandey, D., Pandey, B. K., Nassa, V. K., & Pramanik, S. (2022). Modern Technique for Interactive Communication in LEACH-Based Ad Hoc Wireless Sensor Network. In *Software Defined Networking for Ad Hoc Networks* (pp. 55–73). Springer. doi:10.1007/978-3-030-91149-2_3

Anioke, C., Nnamani, O., & Ani, C. (2015). *Call Drop Minimization Techniques for Handover Calls in Mobile Cellular Networks*, 1–5. UNN. http://www.dspace.unn.edu.ng/handle/123456789/5866

Anitha., R. (2016). A Compact Quad Element Slotted Ground Wideband Antenna for MIMO Applications. *IEEE Transactions on Antennas and Propagation*, *64*(10), 4550–4553.

Anwar, A., Seet, B., Hasan, M. & Li, X. (2019). *A Survey on Application of Non-Orthogonal Multiple Access to Different Wireless Networks*. MDPI.

Arıkan, E. (2009). Channel Polarization: A Method for Constructing Capacity-Achieving Codes for Symmetric Binary-Input Memoryless Channels. *IEEE Transactions on Information Theory*, *55*(7), 3051–3073. doi:10.1109/TIT.2009.2021379

Arıkan, E. (2015). Challenges and some new directions in channel coding. *Journal of Communications and Networks (Seoul)*, *17*(4), 328–338. doi:10.1109/JCN.2015.000063

Arora, S., Sharma, S., & Anand, R. (2023). A Survey on UWB Textile Antenna for Wireless Body Area Network (WBAN) Applications. In *Artificial Intelligence on Medical Data* (pp. 173–183). Springer. doi:10.1007/978-981-19-0151-5_14

Arora, V. K., Sharma, V., & Sachdeva, M. (2019). ACO optimized self-organized tree-based energy balance algorithm for wireless sensor network: AOSTEB. *Journal of Ambient Intelligence and Humanized Computing*, *10*(12), 4963–4975. doi:10.100712652-019-01186-5

Assra, A., Yang, J., & Champagne, B. (2016). An EM Approach for Cooperative Spectrum Sensing in Multiantenna CR Networks. *IEEE Transactions on Vehicular Technology*, *65*(3), 1229–1243. doi:10.1109/TVT.2015.2408369

Babu, S. Z. D. (2022). Analysation of Big Data in Smart Healthcare. In M. Gupta, S. Ghatak, A. Gupta, & A. L. Mukherjee (Eds.), *Artificial Intelligence on Medical Data. Lecture Notes in Computational Vision and Biomechanics* (Vol. 37). Springer. doi:10.1007/978-981-19-0151-5_21

Bagubali, A., Verma, T., Anand, A., Prithiviraj, V., & Mallick, P. S. (2018). Performance analysis of handover schemes in heterogeneous networks. *Journal of Circuits, Systems, and Computers, 27*(11), 1850177.

Balanis, C. A. (2005). *Antenna Theory: Analysis and Design* (3rd ed.). John Wiley & Sons, Inc.

Bamy, C. L., Mbango, F. M., Konditi, D. B. O., & Mpele, P. M. (2021, April). A compact dual-band Dolly-shaped antenna with parasitic elements for automotive radar and 5G applications. *Heliyon, 7*(4), e06793. doi:10.1016/j.heliyon.2021.e06793 PMID:33948514

Banerjee, U., Karmakar, A., & Saha, A. (2020). A review on Circularly Polarized Antennas, trends and advances. *International Journal of Microwave and Wireless Technologies, 12*(9), 922–943. doi:10.1017/S1759078720000331

Banerjee, U., Karmakar, A., Saha, A., & Chakraborty, P. (2019). A CPW-fed compact monopole antenna with defected ground structure and modified parasitic hilbert strip having wideband circular polarization. *AEÜ. International Journal of Electronics and Communications, 110*, 152831. doi:10.1016/j.aeue.2019.152831

Bansal, R., Gupta, A., Singh, R., & Nassa, V. K. (2021). Role and Impact of Digital Technologies in E-Learning amidst COVID-19 Pandemic. *2021 Fourth International Conference on Computational Intelligence and Communication Technologies (CCICT)*, (pp. 194-202). IEEE.10.1109/CCICT53244.2021.00046

BARTLETT. (1941).A dual diversity preselector. *QST*. XXV. 37–39.

Baştanlar, Y., & Ozuysal, M. (2014). Introduction to machine learning. *Methods in Molecular Biology (Clifton, N.J.), 1107*, 105–128. doi:10.1007/978-1-62703-748-8_7 PMID:24272434

Bhambere, S., Abhishek, B., & Sumit, H. (2021). Rapid Digitization of Healthcare - A Review of COVID-19 Impact on our Health systems. *International Journal of All Research Education and Scientific Methods, 9*(2), 1457–1459.

Bhatt, M. C., Bhardwaj, O., Gupta, P., & Sharma, S. (2018, August). Analysis of handoff prediction algorithms in wireless networks. In *2018 4th International Conference on Computing Sciences (ICCS)* (pp. 40-47). IEEE.

Bhatti, R. A., Choi, J., & Park, S. (2009). Quad-Band MIMO Antenna Array for Portable Wireless Communications Terminals. *IEEE Antennas and Wireless Propagation Letters*. 8. 129-132.

Bhowmick, A., Das, G. C., Roy, S. D., Kundu, S., & Maity, S. P. (2020). Allocation of optimal energy in an energy-harvesting cooperative multi-band cognitive radio network. *Wireless Networks, 26*(2), 1033–1043. doi:10.100711276-018-1849-2

Bhowmick, A., Roy, S. D., & Kundu, S. (2015). Performance of secondary user with combined RF and non-RF based energy-harvesting in cognitive radio network. In *2015 IEEE International Conference on Advanced Networks and Telecommuncations Systems (ANTS)*, (pp. 1–3). 10.1109/ANTS.2015.7413665

Bilal., M. (2017). An FSS-Based Non-planar Quad-Element UWB-MIMO Antenna System. *IEEE Antennas and Wireless Propagation Letters, 16*, 987–990.

Bilal, M., Saleem, R., Abbasi, H. H., Shafique, M. F., & Brown, A. K.Bilalet al. (2017). An FSS-Based Nonplanar Quad-Element UWB-MIMO Antenna System. *IEEE Antennas and Wireless Propagation Letters, 16*, 987–990. doi:10.1109/LAWP.2016.2615884

Biswas, A. & Chakraborty, U. (2019).Compact wearable MIMO antenna with improved port isolation for ultra-wideband applications. *IET Microwaves, Antennas & Propagation, 13*(4), 498–504.

Blanch, J. (2003). Exact Representation of Antenna System Diversity Performance from Input Parameter Description. *IET Electronic Letters*, *39*(9), 705-707.

Blanco, D., Rajo-Iglesias, E., Maci, S., & Llombart, N. (2015). Directivity enhancement and spurious radiation suppression in leaky-wave antennas using inductive grid metasurfaces. *IEEE Transactions on Antennas and Propagation*, *63*(3), 891–900. doi:10.1109/TAP.2014.2387422

Boccardi, F., Heath, R. W., Lozano, A., Marzetta, T. L., & Popovski, P. (2014). Five Disruptive Technology Directions for 5G. *IEEE Communications Magazine*, *52*(2), 74–80. doi:10.1109/MCOM.2014.6736746

Bojic, D., Sasaki, E., Cvijetic, N., Wang, T., Kuno, J., Lessmann, J., Schmid, S., Ishii, H., & Nakamura, S. (2013). Advanced wireless and optical technologies for small-cell mobile backhaul with dynamic softwaredefined management. *IEEE Communications Magazine*, *51*(9), 86–93. doi:10.1109/MCOM.2013.6588655

Booth, J. (2019). *Machine Learning for Reliable MIMO Systems*.

Bosshart, P., Daly, D., Gibb, G., Izzard, M., McKeown, N., Rexford, J., Schlesinger, C., Talayco, D., Vahdat, A., Varghese, G., & Walker, D. (2014, July). P4: Programming Protocol-independent Packet Processors. *Computer Communication Review*, *44*(3), 87–95. doi:10.1145/2656877.2656890

Brandstetter, M., Benz, A., Deutsch, C., Detz, H., Klang, P., Andrews, A. M., Strasser, G., & Unterrainer, K. (2012). Superconducting Microdisks Cavities for THz Quantum Cascade Lasers. *IEEE Transactions on Terahertz Science and Technology*, *2*(5), 550–555. doi:10.1109/TTHZ.2012.2212321

Brittain, J. (2008). Electrical Engineering Hall of Fame: Harold H. Beverage. *Proceedings of the IEEE*, *96*, 9.

Bumiller, G., Lampe, L. H.-J., & Hrasnica, H. (2010). Power line communication networks for large-scale control and automation systems. *IEEE Communications Magazine*, *48*(4), 106–113. doi:10.1109/MCOM.2010.5439083

Burney, D. S. M. A., Mahmood, N., & Abbas, Z. (2010). Information and Communication Technology in Healthcare Management Systems: Prospects for Developing Countries. *International Journal of Computers and Applications*, *4*(2), 27–32. doi:10.5120/801-1138

Bütler, R. M., Hager, C., Pfister, H. D., Liga, G., & Alvarado, A. (2021). Model-Based Machine Learning for Joint Digital Backpropagation and PMD Compensation. *Journal of Lightwave Technology*, *39*(4), 949–959. doi:10.1109/JLT.2020.3034047

Caesar, M., Caldwell, D., & Feamster, N. (2005). Design and Implementation of a Routing Control Platform. *2Nd Conference on Symposium on Networked Systems Design & Implementation – (Volume 2*, pp. 15–28). USENIX Association.

CallHealth. (n.d.). *Everything about health*. Call Health. https://www.callhealth.com/

Chae, S. H., Oh, S., & Park, S.-O. (2007). Analysis of Mutual Coupling, Correlations, and TARC in WiBro MIMO Array Antenna. *IEEE Antennas and Wireless Propagation Letters*, *6*, 122–125. doi:10.1109/LAWP.2007.893109

Chair, R., Yang, S. L. S., Kishk, A. A., Lee, K. F., & Luk, K. M. (2006). Aperture fed wideband Circularly Polarized Rectangular Stair-shaped Dielectric Resonator Antenna. *IEEE Transactions on Antennas and Propagation*, *54*(4), 1350–1352. doi:10.1109/TAP.2006.872665

Chandel, & … Gautam, A., & Rambabu, K. (2018). Tapered Fed Compact UWB MIMO-Diversity Antenna with Dual Band-Notched Characteristics. *IEEE Transactions on Antennas and Propagation*, *66*(4), 1677–1684.

Chattate, I., El Khaili, M., & Bakkoury, J. (2019). A New Fuzzy-TOPSIS Based Algorithm for Enhancing Decision Making in a Heterogeneous Network. *Journal of Communication*, *14*(3), 194–201.

Chattate, I., El Khaili, M., & Bakkoury, J. (2019). Improving modified grey relational method for vertical handover in heterogeneous networks. *International Journal of Advanced Computer Science and Applications*, *10*(2).

Chatterjee, S., Maity, S. P., & Acharya, T. (2015). On optimal sensing time and power allocation for energy efficient cooperative cognitive radio networks. In *2015 IEEE International Conference on Advanced Networks and Telecommuncations Systems (ANTS)*, (pp. 1–6). IEEE. 10.1109/ANTS.2015.7413620

Chaudhary, P., Kumar, A., & Yadav, A. (2020). Pattern diversity MIMO 4G and 5G wideband circularly polarized antenna with integrated LTE band for mobile handset. *Progress in Electromagnetics Research M. Pier M*, *89*, 111–120. doi:10.2528/PIERM19111202

Chen, Y. S. & Chang, C. (2016). Design of a four-element multiple-input– multiple-output antenna for compact long-term evolution small-cell base stations. *IET Microwave Antennas Propagation*, *10*(4), 385–392.

Cheng, S. (2019). Machine Learning for Regenerator Placement Based on the Features of the Optical Network. *2019 21st International Conference on Transparent Optical Networks (ICTON)*, (pp. 1-3). IEEE. 10.1109/ICTON.2019.8840391

Cheng, B., & Du, Z. (2021, July). Dual Polarization MIMO Antenna for 5G Mobile Phone Applications. *IEEE Transactions on Antennas and Propagation*, *69*(7), 4160–4165. doi:10.1109/TAP.2020.3044649

Chen, W.-S., Wu, C.-K., & Wong, K.-L. (2001). Novel compact circularly polarized square microstrip antenna. *IEEE Transactions on Antennas and Propagation*, *49*(3), 340–342. doi:10.1109/8.918606

Chen-Xiao, C., & Ya-Bin, X. (2016). Research on load balance method in SDN. *International Journal of Grid and Distributed Computing*, *9*(1), 25–36. doi:10.14257/ijgdc.2016.9.1.03

Chia, S., Gasparroni, M., & Brick, P. (2009). The next challenge for cellular networks: Backhaul. *IEEE Microwave Magazine*, *10*(5), 54–66. doi:10.1109/MMM.2009.932832

Chibber, A., Anand, R., & Arora, S. (2021, September). A Staircase Microstrip Patch Antenna for UWB Applications. In *2021 9th International Conference on Reliability, Infocom Technologies and Optimization (Trends and Future Directions)(ICRITO)* (pp. 1-5). IEEE. 10.1109/ICRITO51393.2021.9596108

Choi, J. (2008). H-ARQ Based Non-Orthogonal Multiple Access with Successive Interference Cancellation. *2008 IEEE Global Telecommunications Conference*, (pp. 1-5). IEEE. 10.1109/GLOCOM.2008.ECP.640

Chu, Z. (2017). D2D cooperative communications for disaster management. In *2017 24th International Conference on Telecommunications (ICT)*, (pp. 1–5). IEEE. 10.1109/ICT.2017.7998227

Coldrey, M., Engstr̈om, U., Helmersson, K. W., Hashemi, M., Manholm, L., & Wallentin, P. (2014). Wireless backhaul in future heterogeneous networks. *Ericsson Review*, *91*, 1–11.

Colmiais, I., Silva, V., Borme, J., Alpuim, P., & Mendes, P. M. (2022). Towards RF Graphene Devices: A Review. *FlatChem*, *35*, 100409. doi:10.1016/j.flatc.2022.100409

Coqueiro, T., Jailton, J., Carvalho, T., & Francês, R. (2019). A fuzzy logic system for vertical handover and maximizing battery lifetime in heterogeneous wireless multimedia networks. *Wireless Communications and Mobile Computing*.

Corporation, R. (2021). *The global economic cost of COVID-19 vaccine nationalism.*

Cortés, J.A. (2016). *Medium Access Control and Layers Above in PLC.*

Cui, J., Ding, Z., & Fan, P. (2018). The Application of Machine Learning in mmWave-NOMA Systems. In *2018 IEEE 87th Vehicular Technology Conference (VTC Spring)*. IEEE. 10.1109/VTCSpring.2018.8417523

Cui., L. (2019). An 8-Element Dual-Band MIMO Antenna with Decoupling Stub for 5G Smartphone Applications. *IEEE Antennas and Wireless Propagation Letters, 18*(10), 2095–2099.

Cui, J., Ding, Z., Fan, P., & Al-Dhahir, N. (2018). Unsupervised Machine Learning-Based User Clustering in Millimeter-Wave-NOMA Systems. *IEEE Transactions on Wireless Communications, 17*(11), 7425–7440. doi:10.1109/TWC.2018.2867180

D.S., C., & Karthikeyan, S. S. (2017). A novel broadband dual circularly polarized microstrip-fed monopole antenna. *IEEE Transactions on Antennas and Propagation, 65*(3), 1410–1415. doi:10.1109/TAP.2016.2647705

Dahrouj, H., Douik, A., Rayal, F., Al-Naffouri, T. Y., & Alouini, M. (2015). Cost-effective hybrid rf/fso backhaul solution for next generation wireless systems. *IEEE Wireless Communications, 22*(5), 98–104. doi:10.1109/MWC.2015.7306543

Dai, L., Wang, B., Yuan, Y., Han, S., Chih-lin, I., & Wang, Z. (2015). Non-orthogonal multiple access for 5G: Solutions, challenges, opportunities, and future research trends. *IEEE Communications Magazine, 53*(9), 74–81. doi:10.1109/MCOM.2015.7263349

Dao, N.-N., Pham, Q.-V., Tu, N. H., Thanh, T. T., Bao, V. N. Q., Lakew, D. S., & Cho, S. (2021). Survey on aerial radio access networks: Toward a comprehensive 6g access infrastructure. *IEEE Communications Surveys and Tutorials, 23*(2), 1193–1225. doi:10.1109/COMST.2021.3059644

Darboe, O., Konditi, D. B. O., & Manene, F. (2019, June). A 28 GHz rectangular microstrip patch antenna for 5G applications. *International Journal of Engineering Research & Technology (Ahmedabad), 12*(6), 854–857.

Das, G., Sharma, A., & Gangwar, R. (2018). Dielectric resonator-based two-element MIMO antenna system with dual band characteristics. *IET Microwaves, Antennas & Propagation, 12*(5), 734-741.

Das., G., Sharma, A., & Gangwar, R. (2018). Wideband self-complementary hybrid ringdielectric resonator antenna for MIMO applications. *IET Microwaves, Antennas & Propagation, 12*(1), 108–114.

Deng, G. (2016). An Ultra-wideband MIMO Antenna with a High Isolation. *IEEE Antennas and Wireless Propagation Letters, 15,* 182-185.

Dev Mahendra, S., & Sengupta, R. (2020). COVID-19 Impact on the Indian Economy - Detailed Analysis. *Igidr,* (April), 1–48. https://blog.smallcase.com/the-new-normal-analysis-of-covid-19-on-indian-businesses-sectors-and-the-economy/

Dhanya, D., & Sankar, P. (2016). Call Drop Improvement in the Cellular Network. International Journal of Advanced Research in Electrical. *Electronics and Instrumentation Engineering, 5*(2), 1117–1123. doi:10.15662/IJAREEIE.2016.0502025

Dhuheir & Öztürk. (2018). Polar codes analysis of 5G systems. In *2018 6th International Conference on Control Engineering & Information Technology (CEIT).* IEEE.

Diawuo, H. A., & Jung, Y. (2018, July). Broadband Proximity-Coupled Microstrip Planar Antenna Array for 5G Cellular Applications. *IEEE Antennas and Wireless Propagation Letters, 17*(7), 1286–1290. doi:10.1109/LAWP.2018.2842242

Dighriri, M., Lee, G., & Baker, T. (2017). *Measurement and Classification of Smart Systems Data Traffic Over 5G Mobile Networks, 1*–363. Springer. doi:10.1007/978-3-319-60137-3

Ding, J., Qu, D., Liu, P., & Choi, J. (2021). Machine Learning Enabled Preamble Collision Resolution in Distributed Massive MIMO. *IEEE Transactions on Communications, 69*(4), 2317–2330. doi:10.1109/TCOMM.2021.3051202

Ding, K., Gao, C., Yu, T., & Qu, D. (2015). Broadband C-shaped circularly polarized monopole antenna. *IEEE Transactions on Antennas and Propagation, 63*(2), 785–790. doi:10.1109/TAP.2014.2380437

Ding, Z., Adachi, F., & Poor, H. V. (2016, January). The Application of MIMO to Non-Orthogonal Multiple Access. *IEEE Transactions on Wireless Communications, 15*(1), 537–552. doi:10.1109/TWC.2015.2475746

Ding, Z., Fan, P., & Poor, H. V. (2017). Random Beamforming in Millimeter-Wave NOMA Networks. *IEEE Access: Practical Innovations, Open Solutions, 5*, 7667–7681. doi:10.1109/ACCESS.2017.2673248

DocsApp. (n.d.). *DocsApp is Now MediBuddy*. MediBuddy. https://m.docsapp.in/

Dong, C., Shen, Y., Qu, Y., Wang, K., Zheng, J., Wu, Q., & Wu, F. (2021). UAVs as an Intelligent Service: Boosting Edge Intelligence for Air-Ground Integrated Networks. *IEEE Network, 35*(4), 167–175. doi:10.1109/MNET.011.2000651

Duan, B. (2020). Evolution and innovation of antenna systems beyond 5G and 6G. *Frontiers of Information Technology and Electronic Engineering, 21*(1), 1–3. doi:10.1631/FITEE.2010000

Duong, T. M., & Kwon, S. (2020). Vertical handover analysis for randomly deployed small cells in heterogeneous networks. *IEEE Transactions on Wireless Communications, 19*(4), 2282–2292.

Dushyant, K., Muskan, G., Gupta, A., & Pramanik, S. (2022). Utilizing Machine Learning and Deep Learning in Cyber security: An Innovative Approach. In M. M. Ghonge, S. Pramanik, R. Mangrulkar, & D. N. Le (Eds.), *Cyber security and Digital Forensics*. Wiley. doi:10.1002/9781119795667.ch12

El Gholb, Y., El Amrani El Idrissi, N., & Ghennioui, H. (2017, March). 5G: An Idea Whose Time Has Come. *International Journal of Scientific and Engineering Research, 8*(3).

Elsayed, M., & Erol-Kantarci, M. (2019). AI-Enabled Future Wireless Networks: Challenges, Opportunities, and Open Issues. *IEEE Vehicular Technology Magazine, 14*(3), 70–77. doi:10.1109/MVT.2019.2919236

Eluwole, O. T., Udoh, N., Ojo, M., Okoro, C., & Akinyoade, A. J. (2018). From 1G to 5G, what next? *IAENG International Journal of Computer Science, 45*(3), 413–434.

Elwi, T. A. (2021, August). Remotely controlled reconfigurable antenna for modern 5G networks applications. *Microwave and Optical Technology Letters, 63*(2), 2018–2023. doi:10.1002/mop.32505

Esper, G., Sweeney, R., Winchell, E, Duffell, J., Kier, S, Lukens, H., & Krupinski, E. (2020). Rapid Systemwide Implementation of Outpatient Telehealth in Response to the COVID-19 Pandemic. *Journal of Healthcare Management, 65*(6), 443-452. doi:10.1097/JHM-D-20-00131

Esselle, K. P. (1996). Circularly polarized higher order rectangular dielectric resonator antenna. *Electronics Letters, 32*(3), 150–151. doi:10.1049/el:19960171

Fang, H., Wang, X., & Hanzo, L. H. (2019). Learning-Aided Physical Layer Authentication as an Intelligent Process. *IEEE Transactions on Communications, 67*(3), 2260–2273. doi:10.1109/TCOMM.2018.2881117

Fang, H., Wang, X., & Tomasin, S. (2019). Machine Learning for Intelligent Authentication in 5G and Beyond Wireless Networks. *IEEE Wireless Communications, 26*(5), 55–61. doi:10.1109/MWC.001.1900054

Farswan, A., Gautam, A. K., Kanaujia, B. K., & Rambabu, K. (2016). Design of Koch Fractal Circularly Polarized Antenna for Handheld UHF RFID reader Applications. *IEEE Transactions on Antennas and Propagation, 64*(2), 771–775. doi:10.1109/TAP.2015.2505001

Feamster, N. Rexford J., & Zegura, E. (2013). The Road to SDN.

Feng, Z., Li, X., Zhang, Q., & Li, W. (2017). Proactive Radio Resource Optimization With Margin Prediction: A Data Mining Approach. *IEEE Transactions on Vehicular Technology, 66*(10), 9050–9060. doi:10.1109/TVT.2017.2709622

Fernandez, S. C., & Sharma, S. K. (2013). Multiband Printed Meandered Loop Antennas with MIMO Implementations for Wireless Routers. *IEEE Antennas and Wireless Propagation Letters, 12*, 96–99. doi:10.1109/LAWP.2013.2243104

Ferreira, R., Joubert, J., & Odendaal, J. W. (2017). A compact dual-circularly polarized cavity-backed ring-slot antenna. *IEEE Transactions on Antennas and Propagation, 65*(1), 364–368. doi:10.1109/TAP.2016.2623654

Firozjae, H. M., Moghaddam, J. Z., & Ardebilipour, M. (2022). Performance Analysis of an UAV-assisted cognitive D2D communication-based Disaster Response Network. *In 2022 30th International Conference on Electrical Engineering (ICEE),* (pp. 665–669). IEEE. 10.1109/ICEE55646.2022.9827018

Flipkart Health. (n.d.). *Health.* Flipkart Health. https://www.sastasundar.com/index.php/user/login/

Fodor, G., Parkvall, S., Sorrentino, S., Wallentin, P., Lu, Q., & Brahmi, N. (2014). Device-to-device communications for national security and public safety. *IEEE Access: Practical Innovations, Open Solutions, 2*, 1510–1520. doi:10.1109/ACCESS.2014.2379938

Ford, E. (2022). Creating Your Personal Board of Directors. *Journal of Healthcare Management, 67*(5), 303-305. doi:10.1097/JHM-D-22-00160

Foschini, G. J. (1996).Layered space-time architecture for wireless communication in a fading environment when using multi-element antennas. *Bell Labs Technical Journal,* 41–59.

Foschini, G. J., & Gans, M. J.Foschini and Gans. (1998). On limits of wireless communications in a fading environment when using multiple antennas. *Wireless Personal Communications, 6*(3), 311–335. doi:10.1023/A:1008889222784

Fu, S., Kong, Q., Fang, S., & Wang, Z. (2014, May). Broadband circularly polarized microstrip antenna with coplanar parasitic ring slot patch for L-band satellite system application. *IEEE Antennas and Wireless Propagation Letters, 13*, 943–946. doi:10.1109/LAWP.2014.2323113

FutureLearn. (n.d.). *Online courses and degrees from top universities.* FutureLearn. https://www.futurelearn.com/

G. Association. (2018). *Mobile backhaul options: spectrum analysis and recommen- dations.* G. Association.

Gačanin, H., & Di Renzo, M. (2020). Wireless 2.0: Toward an Intelligent Radio Environment Empowered by Reconfigurable Meta-Surfaces and Artificial Intelligence. *IEEE Vehicular Technology Magazine, 15*(4), 74–82. doi:10.1109/MVT.2020.3017927

Gambhir, R. S., Dhaliwal, J. S., Aggarwal, A., & Anand, S. (2020). Covid-19 : a survey on knowledge. *Awareness and hygiene practices among dental health professionals in an indian scenario, 71*(2), 223–229. PMID:32519827

Gamboa, J., & Demirkol, I. (2018). Softwarized lte self-backhauling solution and its evaluation. In 2018 IEEE Wireless Communications and Networking Conference. WCNC.

Gao., P. (2014). Compact Printed UWB Diversity Slot Antenna with 5.5-GHz Band-Notched Characteristics. *IEEE Antennas and Wireless Propagation Letters, 13*, 376–379.

Gao, Y., Ma, R., Wang, Y., Zhang, Q., & Parini, C. (2016). Stacked Patch Antenna with Dual-Polarization and Low Mutual Coupling for Massive MIMO. *IEEE Transactions on Antennas and Propagation, 64*(10), 1–1. doi:10.1109/TAP.2016.2593869

Garg, P., & Anand, R. (2011). Energy efficient data collection in wireless sensor network. *Dronacharya Research Journal, 3*(1), 41.

Gaur, P. (2016). A review of Menace of Call Drops in India and possible ways to minimize it. International Journal of Mathematical. *Engineering and Management Sciences, 1*(3), 130–138. doi:10.33889/IJMEMS.2016.1.3-014

Gautam, A. K. (2018). Design of ultra-compact UWB antenna with band-notched characteristics for MIMO applications. *IET Microwave Antennas Propagation,12*(12).1895-1900.

Gesbert, D., Shafi, M., Da-shan Shiu, Smith, P. J., & Naguib, A. (2003). From theory to practice: An overview of MIMO space-time coded wireless systems. *IEEE Journal on Selected Areas in Communications, 21*(3), 281–302. doi:10.1109/JSAC.2003.809458

Ge, X., Tu, S., Mao, G., Wang, C., & Han, T. (2016). 5G ultra-dense cellular networks. *IEEE Wireless Communications, 23*(1), 72–79. doi:10.1109/MWC.2016.7422408

Ghasemi, A., & Sousa, E. S. (2005). Collaborative spectrum sensing for opportunistic access in fading environments. In *First IEEE International Symposium on New Frontiers in Dynamic Spectrum Access Networks*, (pp. 131–136). IEEE. 10.1109/DYSPAN.2005.1542627

Ghosh, A., Nundy, S., & Mallick, T. K. (2020). How India is dealing with COVID-19 pandemic. *Sensors International, 1*(June), 100021. doi:10.1016/j.sintl.2020.100021 PMID:34766039

Ghosh, S. K., Das, S., & Bhattacharyya, S. (2021). Transmittive-type Triple-band Linear to Circular Polarization Conversion in THz Region using Graphene-based Metasurface. *Optics Communications, 480*, 126480. doi:10.1016/j.optcom.2020.126480

Giustiniano, D., Goma, E., Lopez Toledo, A., & Athanasiou, G. (2013). Optimizing TCP performance in multi-AP residential broadband connections via Minislot access. *Journal of computer networks and communications.*

Gopalan, H. S., & Misra, A. (2020). COVID-19 pandemic and challenges for socio-economic issues, healthcare and National Health Programs in India. *Diabetes & Metabolic Syndrome, 14*(5), 757–759. doi:10.1016/j.dsx.2020.05.041 PMID:32504992

Götz, M., Rapp, M., & Dostert, K. M. (2004). Power line channel characteristics and their effect on communication system design. *IEEE Communications Magazine, 42*(4), 78–86. doi:10.1109/MCOM.2004.1284933

Gund, G. S., Jung, M. G., Shin, K.-Y., & Park, H. S. (2019). Two-Dimensional Metallic Niobium Diselenide for Sub-Micrometer-Thin Antennas in Wireless Communication Systems. *ACS Nano, 13*(12), 14114–14121. doi:10.1021/acsnano.9b06732 PMID:31746198

Guo, H. (2019). Weighted Sum-Rate Maximization for Intelligent Reflecting Surface Enhanced Wireless Networks. *2019 IEEE Global Communications Conference (GLOBECOM)*, (pp. 1-6). IEEE. 10.1109/GLOBECOM38437.2019.9013288

Gupta, A., Kaushik, D., Garg, M., & Verma, A. (2020).Machine Learning model for Breast Cancer Prediction. 2020 Fourth International Conference on I-SMAC (IoT in Social, Mobile, Analytics and Cloud) (I-SMAC), (pp. 472-477). IEEE. 10.1109/I-SMAC49090.2020.9243323

Gupta, A. (2019). Script classification at word level for a Multilingual Document. *International Journal of Advanced Science and Technology, 28*(20), 1247–1252. http://sersc.org/journals/index.php/IJAST/article/view/3835

Gupta, A. (2020). An Analysis of Digital Image Compression Technique in Image Processing. *International Journal of Advanced Science and Technology, 28*(20), 1261–1265. http://sersc.org/journals/index.php/IJAST/article/view/3837

Gupta, A., Goyal, B., Dogra, A., & Anand, R. (2023). Proximity Coupled Antenna with Stable Performance and High Body Antenna Isolation for IoT-Based Devices. In *Communication, Software and Networks* (pp. 591–600). Springer. doi:10.1007/978-981-19-4990-6_55

Gupta, A., & Jha, R. K. (2015). A survey of 5G network: Architecture and emerging technologies. *IEEE Access: Practical Innovations, Open Solutions, 3*, 1206–1232. doi:10.1109/ACCESS.2015.2461602

Gupta, A., Singh, R., Nassa, V. K., Bansal, R., Sharma, P., & Koti, K. (2021) Investigating Application and Challenges of Big Data Analytics with Clustering. *2021 International Conference on Advancements in Electrical, Electronics, Communication, Computing and Automation (ICAECA)*, (pp. 1-6). IEEE.10.1109/ICAECA52838.2021.9675483

Gupta, N., Khosravy, M., Patel, N., Dey, N., Gupta, S., Darbari, H., & Crespo, R. G. (2020). Economic data analytic AI technique on IoT edge devices for health monitoring of agriculture machines. *Applied Intelligence*, *50*(11), 3990–4016. doi:10.100710489-020-01744-x

Gupta, S., Patil, S., Dalela, C., & Kanaujia, B. K. (2021). Analysis and Design of Fractal Defected Ground Based Circularly Polarized Antenna for CA band Applications. *International Journal of Microwave and Wireless Technologies*, *13*(4), 397–406. doi:10.1017/S1759078720001142

Ha, C.-B., You, Y.-H., & Song, H.-K. (2019). Machine Learning Model for Adaptive Modulation of Multi-Stream in MIMO-OFDM System. *IEEE Access: Practical Innovations, Open Solutions*, *7*, 5141–5152. doi:10.1109/ACCESS.2018.2889076

Hämäläinen, M., Iinatti, J., & Kohno, R. (2011). Wireless communications in healthcare recent and future topics. *ACM International Conference Proceeding Series*. ACM. 10.1145/2093698.2093796

Haneishi, M., Nambara, T., & Yoshida, S. (1982). Study on ellipticity properties of single-feed-type circularly polarised microstrip antennas. *Electronics Letters*, *18*(5), 191–193. doi:10.1049/el:19820132

Han, R., Zhong, S. S., & Liu, J. (2014). Broadband Circularly Polarized Dielectric Resonator Antenna fed by wideband Switched Line Coupler. *Electronics Letters*, *50*(10), 725–726. doi:10.1049/el.2014.0809

Haque, S., Uddin, S., Sayem, S., & Mohib, K. M. (2020). ur na l P re. *Biochemical Pharmacology*, *2019*, 104660. doi:10.1016/j.jece.2020.104660 PMID:33194544

Hassel, J., Oksanen, M., Elo, T., Seppä, H., & Hakonen, P. J. (2017). Terahertz Detection using Mechanical Resonators based on 2D Materials. *AIP Advances*, *7*(6), 065014. doi:10.1063/1.4990405

Haykin, S. (2014). *Digital communication systems*. Wiley.

He, B. (2020). A Machine Learning Based Multi-flips Successive Cancellation Decoding Scheme of Polar Codes. *2020 IEEE 91st Vehicular Technology Conference (VTC2020-Spring)*, (pp. 1-5). IEEE. 10.1109/VTC2020-Spring48590.2020.9128875

Hearst, M. A., Dumais, S. T., Osuna, E., Platt, J., & Scholkopf, B. (1998). Support vector machines. *IEEE Intelligent Systems & their Applications*, *13*(4), 18–28. doi:10.1109/5254.708428

Heidari, A. A., Heyrani, M., & Nakhkash, M. (2009). A Dual – band Circularly Polarized stub loaded microstrip patch antenna for GPS applications. *Progress in Electromagnetics Research*, *92*, 195–208. doi:10.2528/PIER09032401

He, P., Cao, M.-S., Cao, W.-Q., & Yuan, J. (2021). Developing MXenes from Wireless Communication to Electromagnetic Attenuation. *Nano-Micro Letters*, *13*(1), 115. doi:10.100740820-021-00645-z PMID:34138345

Heslop, L., Howard, A., Fernando, J., Rothfield, A., & Wallace, L. (2003). Wireless communications in acute health-care. *Journal of Telemedicine and Telecare*, *9*(4), 187–193. doi:10.1258/135763303322225490 PMID:12952687

Heydari, S., Jahangiri, P., Arezoomand, A. S., & Zarrabi, F. B. (2016). Circular polarization fractal slot by Jerusalem cross slot for wireless applications. *Progress in Electromagnetics Research Letters*, *63*, 79–84. doi:10.2528/PIERL16070802

Hong, W., Baek, K., & Ko, S. (2017, December). Millimeter-Wave 5G Antennas for Smartphones: Overview and Experimental Demonstration. *IEEE Transactions on Antennas and Propagation*, *65*(12), 6250–6261. doi:10.1109/TAP.2017.2740963

Höyhtyä, M., Apilo, O., & Lasanen, M. (2018). Review of latest advances in 3GPP standardization: D2D communication in 5G systems and its energy consumption models. *Futur. Internet*, *10*(1), 3. doi:10.3390/fi10010003

Huang, L. (2020). *Modulation format identification under stringent bandwidth limitation based on an artificial neural network.*

Huang, X., Niu, K., Si, Z., He, Z., & Dong, C. (2017). *Rate-Splitting Non-orthogonal Multiple Access: Practical Design and Performance Optimization.* Communications and Networking. 11th EAI International Conference, Chongqing, Chin.

Huang, C., Zappone, A., Alexandropoulos, G. C., Debbah, M., & Yuen, C. (2019). Reconfigurable Intelligent Surfaces for Energy Efficiency in Wireless Communication. *IEEE Transactions on Wireless Communications*, *18*(8), 4157–4170. doi:10.1109/TWC.2019.2922609

Huang, T.-J. (2020). Theoretical Analysis of NOMA Within Massive MIMO Systems. *Wireless Personal Communications*, *112*(2), 777–783. doi:10.100711277-020-07073-z

Huang, Y.-D. (2018). A Machine Learning Approach to MIMO Communications. *2018 IEEE International Conference on Communications (ICC)*, (pp. 1-6). IEEE. 10.1109/ICC.2018.8422211

Hua, S., Guo, Y., Liu, Y., Liu, H., & Panwar, S. S. (2011). Scalable video multicast in hybrid 3G/Ad-Hoc networks. *IEEE Transactions on Multimedia*, *13*(2), 402–413. doi:10.1109/TMM.2010.2103929

Huh, H., Caire, G., Papadopoulos, H. C., & Ramprashad, S. A. (2012). Achieving "Massive MIMO" Spectral Efficiency with a Not-so-Large Number of Antennas. *IEEE Transactions on Wireless Communications*, *11*(9), 3226–3239. doi:10.1109/TWC.2012.070912.111383

Huo, Y. (2019). Smart-Grid Monitoring: Enhanced Machine Learning for Cable Diagnostics. *2019 IEEE International Symposium on Power Line Communications and its Applications (ISPLC)*, (pp. 1-6). IEEE. 10.1109/ISPLC.2019.8693287

Huo, Y., Prasad, G., Atanackovic, L., Lampe, L., & Leung, V. C. M. (2019). Cable Diagnostics With Power Line Modems for Smart Grid Monitoring. *IEEE Access: Practical Innovations, Open Solutions*, *7*, 60206–60220. doi:10.1109/ACCESS.2019.2914580

Hussain, R. & Sharawi, M. (2019). An Integrated Slot-Based Frequency-Agile and UWB Multifunction MIMO Antenna System. *IEEE Antennas and Wireless Propagation Letters, 18*(10).2150-2154.

Hussain., R. (2019). Split-ring-resonator-loaded multiband frequency agile slot-based MIMO antenna system. *IET Microwaves, Antennas & Propagation*, *13*(14), 2449–2456.

International Telecommunication Union (2017). *R-REC-P*, pp. 618-13.

International Telecommunication union. (2019). Rec. *ITU-R*, (August), 531–535.

International Telecommunication Union. (2021). *ITU-R recommendation.* p. 1546.

Islam, S. M. R., Avazov, N., Dobre, O. A., & Kwak, K. (2017). Power-Domain Non-Orthogonal Multiple Access (NOMA) in 5G Systems: Potentials and Challenges. In IEEE Communications Surveys & Tutorials, 19(2), pp. 721-742.

Issel, M L. (2020). Value Added of Management to Health Care Organizations. *Health Care Management Review, 45*(2), 95. doi:10.1097/HMR.0000000000000280

Ittipiboon, A., Roscoe, D., Mongia, R. K., & Cuhaci, M. (1994). Circularly polarized dielectric resonator antenna. *Electronics Letters*, *30*(17), 1361–1362. doi:10.1049/el:19940968

Itumalla, R. (2012). Information Technology and Service Quality in Health Care: An Empirical Study of Private Hospital in India. *International Journal of Innovation, Management and Technology*, *3*(4), 1–4. doi:10.7763/IJIMT.2012.V3.269

Iyengar, K., Mabrouk, A., Kumar, V., & Venkatesan, A. (2020). *Since January 2020 Elsevier has created a COVID-19 resource centre with free information in English and Mandarin on the novel coronavirus COVID- 19* . Elsevier Connect.

Jaber, M., Imran, M. A., Tafazolli, R., & Tukmanov, A. (2016). 5G backhaul challenges and emerging research directions: A survey. *IEEE Access: Practical Innovations, Open Solutions*, *4*, 1743–1766. doi:10.1109/ACCESS.2016.2556011

Jaffry, S., Hussain, R., Gui, X., & Hasan, S. F. (2020). A comprehensive survey on moving networks. *IEEE Communications Surveys and Tutorials*.

Jafri., S. (2016). Compact reconfigurable multiple-input multiple-output antenna for ultra-wideband applications. *IET Microwaves, Antennas & Propagation*, *10*(4), 413–419.

Jain, S., Sindhwani, N., Anand, R., & Kannan, R. (2022). COVID Detection Using Chest X-Ray and Transfer Learning. In *International Conference on Intelligent Systems Design and Applications* (pp. 933-943). Springer. 10.1007/978-3-030-96308-8_87

James, J. R., & Hall, P. P. (1989). *Handbook of Microstrip Antenna*. Peter Peregrinus Ltd.

Jandi, Y., Gharnati, F., & Oulad, A. (2017). Said, "Design of a compact dual bands patch antenna for 5G applications." *2017 International Conference on Wireless Technologies, Embedded and Intelligent Systems (WITS)*, (pp. 1-4). Semantic Scholar.

Jehn-Ruey, J. (2011). *Widhi Yahya," Load Balancing and Multicasting Using the Extended Dijkstra's Algorithm in Software Defined Networking*. Springer-Verlag Berlin Heidelberg.

Jensen, M. A. (2016).A history of MIMO wireless communications. *IEEE International Symposium on Antennas and Propagation (APSURSI)* (pp. 681-682). IEEE.

Jhajharia, T., Tiwari, V., Yadav, D., Rawat, S., & Bhatnagar, D. (2018). Wideband Circularly Polarised Antenna with an asymmetric meandered-shaped monopole and defected ground structurefor wireless communuication. *IET Microwaves, Antennas & Propagation*, *12*(9), 1554–1558. doi:10.1049/iet-map.2018.0092

Jiang, Y. (2019). Mind: Model independent neural decoder. In *2019 IEEE 20th International Workshop on Signal Processing Advances in Wireless Communications (SPAWC)*. IEEE. 10.1109/SPAWC.2019.8815537

Jiang, C., Zhang, H., Ren, Y., Han, Z., Chen, K.-C., & Hanzo, L. (2017). Machine Learning Paradigms for Next-Generation Wireless Networks. *IEEE Wireless Communications*, *24*(2), 98–105. doi:10.1109/MWC.2016.1500356WC

Joe. (2020). Does power allocation affect NOMA? *Wireless Communication*. https://ecewireless.blogspot.com/2020/04/how-does-power-allocation-affect-noma.html

Jou, C. F., Wu, J. W., & Wang, C. J. (2009). Novel broadband monopole antennas with dual-band circular polarization. *IEEE Transactions on Antennas and Propagation*, *57*(4), 1027–1034. doi:10.1109/TAP.2009.2015827

Joy, S., Natarajamani, S., & Vaitheeswaran, S. M. (2018). Minkowski fractal circularly polarized planar antenna for GPS application. In *Proceedings of the 8th International Conference on Advances in Computing and Communication* (ICACC), 143, 66–73.

Kamal, S., Mohammed, A. S. B., Bin Ain, M. F., Ullah, U., Hussin, R., Ahmad, Z. A., Othman, M., & Rahman, M. F. A. (2020, September). A novel negative meander line design of microstrip antenna for 28 GHz mmwave wireless communications. *Wuxiandian Gongcheng*, *29*(3), 479–485. doi:10.13164/re.2020.0479

Kamel, M., Hamouda, W., & Youssef, A. (2016). Ultra-dense networks: A survey. *IEEE Communications Surveys and Tutorials*, *18*(4), 2522–2545. doi:10.1109/COMST.2016.2571730

Kang., L. (2015). Compact Offset Microstrip-Fed MIMO Antenna for Band-Notched UWB Applications. *IEEE Antennas and Wireless Propagation Letters*, *14*, 1754–1757.

Karimian, Oraizi, H., Fakhte, S., & Farahani, M. (2013). Novel F-Shaped Quad-Band Printed Slot Antenna for WLAN and WiMAX MIMO Systems. *IEEE Antennas and Wireless Propagation Letters*, *12*, 405–408. doi:10.1109/LAWP.2013.2252140

Karmakar, A., Chakraborty, P., Banerjee, U., & Saha, A. (2019). Combined triple-band circularly polarized and compact UWB monopole antenna. *IET Microwaves, Antennas & Propagation*, *13*(9), 1306–1311. doi:10.1049/iet-map.2018.5459

Kaur, J., Sindhwani, N., Anand, R., & Pandey, D. (2023). Implementation of IoT in Various Domains. In *IoT Based Smart Applications* (pp. 165–178). Springer. doi:10.1007/978-3-031-04524-0_10

Kaur, M., & Sivia, J. S. (2019, February). Minkowski, Giuseppe Peano and Koch Curves based Design of Compact Hybrid Fractal Antenna for Biomedical Applications using ANN and PSO. *International Journal of Electronics and Communications*, *99*, 14–24. doi:10.1016/j.aeue.2018.11.005

Kaur, M., & Sivia, J. S. (2019, March). ANN-based Design of Hybrid Fractal Antenna for Biomedical Applications. *International Journal of Electronics*, *106*(8), 1184–1199. doi:10.1080/00207217.2019.1582712

Kaur, M., & Sivia, J. S. (2019, November). Giuseppe Peano and Cantor set fractals based miniaturized hybrid fractal antenna for biomedical applications using artificial neural network and firefly algorithm. *International Journal of RF and Microwave Computer-Aided Engineering*, *30*(1), 1–11.

Kaur, M., & Sivia, J. S. (2020, January). ANN and FA Based Design of Hybrid Fractal Antenna for ISM Band Applications. *Progress in Electromagnetics Research C*, *98*, 127–140. doi:10.2528/PIERC19110901

Kaushik, K., & Garg, M. Annu, Gupta, A. & Pramanik, S. (2021). Application of Machine Learning and Deep Learning. In M. Ghonge, S. Pramanik, R. Mangrulkar and D. N. Le (eds.) Cyber security: An Innovative Approach, in Cybersecurity and Digital Forensics: Challenges and Future Trends. Wiley.

Kaushik, K., & Garg, M., Gupta, A., & Pramanik, S. (2021). Application of Machine Learning and Deep Learning. In M. Ghonge, S. Pramanik, R. Mangrulkar, & D. N. Le, (eds.), Cyber security: An Innovative Approach, in Cybersecurity and Digital Forensics: Challenges and Future Trends. Wiley.

Kaushik, D., & Gupta, A. (2021). Ultra-secure transmissions for 5G-V2X communications. *Materials Today: Proceedings*. doi:10.1016/j.matpr.2020.12.130

Khan, B. M., & Bilal, R. (2018). Wireless internet offloading techniques: based on 802.21 medium access control. In Advanced Wireless Sensing Techniques for 5G Networks (pp. 267-283). Chapman and Hall/CRC.

Khan, H., Kushwah, K. K., Singh, S., Urkude, H., Maurya, M. R., & Sadasivuni, K. K. (2021). Smart technologies driven approaches to tackle COVID-19 pandemic: a review. *3 Biotech*, *11*(2), 1–22. doi:10.1007/s13205-020-02581-y

Khan, J. A., & Anand, R. (2022). A review on healthcare data privacy and security. *Networking Technologies in Smart Healthcare: Innovations and Analytical Approaches, 79*.

Khan, M., Ginzboorg, P., Järvinen, K., & Niemi, V. (2018). Defeating the downgrade attack on identity privacy in 5G. Lecture Notes in Computer Science (Including Subseries Lecture Notes in Artificial Intelligence and Lecture Notes in Bioinformatics), 11322 LNCS, (pp. 95–119). Springer. doi:10.1007/978-3-030-04762-7_6

Khan., M. (2014). Compact ultra-wideband diversity antenna with a floating parasitic digitated decoupling structure. *IET Microwaves, Antennas & Propagation, 8*(10), 747–753.

Khan, Capobianco, A.-D., Asif, S. M., Anagnostou, D. E., Shubair, R. M., & Braaten, B. D. (2017). A Compact CSRR-Enabled UWB Diversity Antenna. *IEEE Antennas and Wireless Propagation Letters, 16*, 808–812. doi:10.1109/LAWP.2016.2604843

Khan, F. N., Zhong, K., Al-Arashi, W. H., Yu, C., Lu, C., & Lau, A. P. T. (2016). Modulation Format Identification in Coherent Receivers Using Deep Machine Learning. *IEEE Photonics Technology Letters, 28*(17), 1886–1889. doi:10.1109/LPT.2016.2574800

Kim, Y. (2011). *Iterative Coding for High Speed Power Line Communication Systems.* The Journal of the Institute of Webcasting. *Internet and Telecommunication, 11*, 185–192.

Kirubasri, G., Sankar, S., Pandey, D., Pandey, B. K., Singh, H., & Anand, R. (2021, September). A Recent Survey on 6G Vehicular Technology, Applications and Challenges. In *2021 9th International Conference on Reliability, Infocom Technologies and Optimization (Trends and Future Directions)(ICRITO)* (pp. 1-5). IEEE.

Kizilirmak, R. (2016). Non-Orthogonal Multiple Access (NOMA) for 5G Networks. In *Towards 5G Wireless Networks - A Physical Layer Perspective*. IntechOpen.

Kizilirmak, R.C. (2016). *Non-Orthogonal Multiple Access (NOMA) for 5G Networks.*

Klabjan, D. & X. Zhu, X. (2020). *Neural Network Retraining for Model Serving.*

Kliks, A., Musznicki, B., Kowalik, K., & Kryszkiewicz, P. (2018). Perspectives for resource sharing in 5G networks. *Telecommunication Systems, 68*(4), 605–619. doi:10.100711235-017-0411-3

Kordaliv, M. (2014). Common Elements Wideband MIMO Antenna System for WiFi/LTE Access-Point Applications. *IEEE Antennas and Wireless Propagation Letters, 13*, 1601–1604.

Kumar Mongia, R. K., & Ittipiboon, A. (1997). Theoretical and experimental investigations on rectangular dielectric resonator antennas. *IEEE Transactions on Antennas and Propagation, 45*(9), 1348–1356. doi:10.1109/8.623123

Kumar, S. U., Kumar, D. T., & Christopher, B. P. (2020). *The Rise and Impact of COVID-19 in India. 7*(May), 1–7. doi:10.3389/fmed.2020.00250

Kumar, A., & Nayar, K. R. (2020). COVID 19 and its mental health consequences. *Journal of Mental Health (Abingdon, England), 0*(0), 1–2. doi:10.1080/09638237.2020.1757052 PMID:32339041

Kumar, P. (2019). Ultrathin 2D Nanomaterials for Electromagnetic Interference Shielding. *Advanced Materials Interfaces, 6*(24), 1901454. doi:10.1002/admi.201901454

Kumar, P., & Bhowmick, A. (2021). Hybrid Spectrum Access in a CR Enabled Cooperative D2D Network. *IJCS, Wiley, 34*(Issue. 11), 1–14.

Kumar, P., & Bhowmick, A. (2022). Throughput performance of a non-linear energy-harvesting cognitive radio-enabled device-to-device network. *International Journal of Communication Systems, e5124,* doi:10.1002/dac.5124

Kumar, P., Dwari, S., Saini, R. K., Mandal, M. K., & Dual-Sense, D.-B. (2019). Dual-Band Dual-Sense Polarization reconfigurable circularly polarized antenna. *IEEE Antennas and Wireless Propagation Letters, 18*(1), 64–68. doi:10.1109/LAWP.2018.2880799

Kumawat, P., & Joshi, S. (2021). A Novel Dual-Band Orthogonal Polarized Elliptical Patch Antenna Array for 5G Applications. *Journal of Scientific Research, 65*(3), 184–190. doi:10.37398/JSR.2021.650322

Kunarak, S., & Duangchan, T. (2021, August). Vertical Handover Decision based on Hybrid Artificial Neural Networks in HetNets of 5G. In *2021 IEEE Region 10 Symposium (TENSYMP)* (pp. 1-6). IEEE.

Kurt, G. K., Khoshkholgh, M. G., Alfattani, S., Ibrahim, A., Darwish, T. S., Alam, M. S., Yanikomeroglu, H., & Yongacoglu, A. (2021). A vision and framework for the high altitude platform station (haps) networks of the future. *IEEE Communications Surveys and Tutorials*, 23(2), 729–779. doi:10.1109/COMST.2021.3066905

Kuschnerov, M. (2020). Advances in Deep Learning for Digital Signal Processing in Coherent Optical Modems. *2020 Optical Fiber Communications Conference and Exhibition (OFC)*, (pp. 1-3). 10.1364/OFC.2020.M3E.2

Lahby, M., & Sekkaki, A. (2018). An Efficient Policy for Vertical-Handover-Based Multi-Attribute Utility Theory in Heterogeneous Wireless Networks. In *Advances in Data Communications and Networking for Digital Business Transformation* (pp. 1–20). IGI Global.

Laisné, A., Gillard, R., & Piton, G. (2002). Circularly polarized dielectric resonator antenna with metallic strip. *Electronics Letters*, 38(3), 106–107. doi:10.1049/el:20020075

Lamare, R.C. (2013). *Massive MIMO Systems: Signal Processing Challenges and Future Trends*.

Lampe, L. H.-J., Tonello, A. M., & Swart, T. G. (2016). *Power Line Communications: Principles*. Standards and Applications from Multimedia to Smart Grid. doi:10.1002/9781118676684

Langford, Z. L., Eisenbeiser, L., & Vondal, M. (2019). Robust Signal Classification Using Siamese Networks. *Proceedings of the ACM Workshop on Wireless Security and Machine Learning*. ACM. 10.1145/3324921.3328781

Lazaropoulos, A. (2021). *Information Technology, Artificial Intelligence and Machine Learning in Smart Grid – Performance Comparison between Topology Identification Methodology and Neural Network Identification Methodology for the Branch Number Approximation of Overhead Low-Voltage Broadband over Power Lines Network Topolog*. Trends in Renewable Energy.

Lee., J. M. (2012). A Compact Ultra wideband MIMO Antenna With WLAN Band-Rejected Operation for Mobile Devices. *IEEE Antennas and Wireless Propagation Letters*, 11, 990–993.

Lee, J., Han, J.-K., & Zhang, J. (2009). Mimo technologies in 3gpp lte and lte-advanced. *EURASIP Journal on Wireless Communications and Networking*, 2009(1), 1–10. doi:10.1155/2009/302092

Lee, S., & Choi, J. (2018). A 60-GHz Yagi-Uda circular array antenna with omni-direcitional pattern for millimeter-wave WBAN applications. In *2018 IEEE International Symposium on Antennas and Propagation USNC/URSI National Radio Science Meeting*, (pp. 1699–1700). IEEE. 10.1109/APUSNCURSINRSM.2018.8609346

Leggat, S. G., Bartram, T., Casimir, G., Stanton, P. (2010). Nurse perceptions of the quality of patient care: Confirming the importance of empowerment and job satisfaction. *Health Care Management Review, 35*(4), 355-364. doi:10.1097/HMR.0b013e3181e4ec55

Leung, K. W., & Mok, S. K. (2001). Circularly polarized dielectric resonator antenna excited by perturbed annular slot with backing cavity. *Electronics Letters*, 37(15), 934–936. doi:10.1049/el:20010658

Leung, K. W., Wong, W. C., Luk, K. M., & Yung, E. K. N. (2000). Circularly Polarized Dielectric Resonator Antenna excited by dual Conformal Strips. *Electronics Letters*, 36(6), 484–486. doi:10.1049/el:20000453

Lewis, V. A., Schoenherr, K., Fraze, T., Cunningham, A. (2019). Clinical coordination in accountable care organizations: A qualitative study. *Health Care Management Review 44*(2), 127-136. doi:10.1097/HMR.0000000000000141

Li, G. H., Zhai, H. Q., T. L., & Liang, C. H. (2012). A Compact antenna with broad bandwidth and quad- sense circular polarization. *IEEE Transactions on Antennas and Wireless Propagation, 11*(7), 761–794.

Li, J., Chu, Q., & Huang, T. (2012).A Compact Wideband MIMO Antenna with Two Novel Bent Slits. *IEEE Transactions on Antennas and Propagation*, *60*(2), 482-489.

Li., G., Zhai, H., Ma, Z., Liang, C., Yu, R., & Liu, S. (2014). Isolation-Improved Dual-Band MIMO Antenna Array for LTE/WiMAX Mobile Terminals. *IEEE Antennas and Wireless Propagation Letters*, *13*, 1128–1131. doi:10.1109/LAWP.2014.2330065

Li., J. F. (2013). Compact Dual Band-Notched UWB MIMO Antenna with High Isolation. *IEEE Transactions on Antennas and Propagation*, *61*(9), 4759–4766.

Liang, H., Yang, S., Li, L., &Gao, J. (2019). Research on routing optimization of WSNs based on improved LEACH protocol. *Eurasip Journal on Wireless Communications and Networking, 2019*(1). doi:10.1186/s13638-019-1509-y

Liang, Y.-C., Peh, E., Hoang, A., & Zeng, Y. (2007). Sensing-throughput tradeoff for cognitive radio networks. In *IEEE International Conference on Communications*. (pp. 5330-5335). IEEE.

Liang, H.-W., Chung, W.-H., & Kuo, S.-Y. (2016). Coding-Aided K-Means Clustering Blind Transceiver for Space Shift Keying MIMO Systems. *IEEE Transactions on Wireless Communications*, *15*(1), 103–115. doi:10.1109/TWC.2015.2467394

Liang, W., Jiao, Y. C., Luan, Y., Tian, C., & Dual-Band, A. (2015). A Dual-Band Circularly polarized complementary antenna. *IEEE Antennas and Wireless Propagation Letters*, *14*, 1153–1156. doi:10.1109/LAWP.2015.2392787

Liang, Z., Li, Y., & Long, Y. (2014). Multiband monopole mobile phone antenna with circular polarization for GNSS application. *IEEE Transactions on Antennas and Propagation*, *62*(4), 1910–1917. doi:10.1109/TAP.2014.2299821

Li, B., Hao, C.-X., & Sheng, X.-Q. (2009). A dual-mode quadrature-fed wideband circularly polarized dielectric resonator antenna. *IEEE Antennas and Wireless Propagation Letters*, *8*, 1036–1038. doi:10.1109/LAWP.2009.2030700

Lien, S.-Y., Chien, C.-C., Tseng, F.-M., & Ho, T.-C. (2016). 3GPP device-to-device communications for beyond 4G cellular networks. *IEEE Communications Magazine*, *54*(3), 29–35. doi:10.1109/MCOM.2016.7432168

Li, G., Amer, N., Hafez, H. A., Huang, S., Turchinovich, D., Mochalin, V. N., Hegmann, F. A., & Titova, L. V. (2020). Dynamical Control over Terahertz Electromagnetic Interference Shielding with 2D $Ti_3C_2T_y$ MXene by Ultrafast Optical Pulses. *Nano Letters*, *20*(1), 636–643. doi:10.1021/acs.nanolett.9b04404 PMID:31825625

Li, G., Liu, F., Sharma, A., Khalaf, O. I., Alotaibi, Y., Alsufyani, A., & Alghamdi, S. (2021). Research on the Natural Language Recognition Method Based on Cluster Analysis Using Neural Network. *Mathematical Problems in Engineering*, *2021*, 1–13. doi:10.1155/2021/9982305

Li, J., Zhang, H., & Fan, M. (2017). Digital Self-Interference Cancellation Based on Independent Component Analysis for Co-Time Co-frequency Full-Duplex Communication Systems. *IEEE Access: Practical Innovations, Open Solutions*, *5*, 10222–10231. doi:10.1109/ACCESS.2017.2712614

Li, J., Zhou, D., Wang, P.-J., Du, C., Liu, W.-F., Su, J.-Z., Pang, L.-X., Cao, M.-S., & Kong, L.-B. (2021). Recent Progress in Two-Dimensional Materials for Microwave Absorption Applications. *Chemical Engineering Journal*, *425*, 131558. doi:10.1016/j.cej.2021.131558

Lin., G. S. (2017). Isolation Improvement in UWB MIMO Antenna System Using Carbon Black Film. *IEEE Antennas and Wireless Propagation Letters*, *16*, 222–225.

Ling, X. M., & Li, R. L. (2011). A Novel Dual-Band MIMO Antenna Array with Low Mutual Coupling for Portable Wireless Devices. *IEEE Antennas and Wireless Propagation Letters*, *10*, 1039–1042. doi:10.1109/LAWP.2011.2169035

Link Budget Analysis in Mobile Communication System (2006). Hoation DAI *International conference on communications Technology*. doi:10.1109/ICCT.2006.341977

Lin, Y.-D., & Hsu, Y.-C. (2000). Multihop cellular: A new architecture for wireless communications. In *Proceedings IEEE INFOCOM 2000. Conference on Computer Communications. Nineteenth Annual Joint Conference of the IEEE Computer and Communications Societies (Cat. No. 00CH37064)*, (vol. 3, pp. 1273–1282).

Li, S., Xu, S., Pan, K., Du, J., & Qiu, J. (2022). Ultra-Thin Broadband Terahertz Absorption and Electromagnetic Shielding Properties of MXene/rGO Composite Film. *Carbon*, *194*, 127–139. doi:10.1016/j.carbon.2022.03.048

Li, T. W., Lai, C. L., & Sun, J. S. (2005). Study of Dual-Band Circularly Polarized Microstrip Antenna. In *Proceedings of the European Conference on Wireless Technology*, (pp. 79 – 80). Paris, France: IEEE.

Li, T., & Chen, Z. N. (2018, December). Metasurface-based shared-aperture 5G S-/K-band antenna using characteristic mode analysis. *IEEE Transactions on Antennas and Propagation*, *66*(12), 6742–6750. doi:10.1109/TAP.2018.2869220

Liu, L. Cheung, A. & Yuk, T. (2013). Compact MIMO Antenna for Portable Devices in UWB Applications.*IEEE Transactions on Antennas and Propagation*, *61*(8), 4257-4264.

Liu, L., Cheung, A., & Yuk, T. (2014). Compact multiple-input–multiple-output antenna using quasi-self-complementary antenna structures for ultra-wideband applications. *IET Microwave Antennas Propagation*, *8*(13), 1021-1029.

Liu, L., Cheung, A., & Yuk, T. (2015). Compact MIMO Antenna for Portable UWB Applications with Band-Notched Characteristic. *IEEE Transactions on Antennas and Propagation*, *63*(5), 1917-1924.

Liu., D. Q. (2018). Dual-Band Platform-Free PIFA for 5G MIMO Application of Mobile Devices. *IEEE Transactions on Antennas and Propagation*, *66*(11), 6328–6333.

Liu, D., Wang, L., Chen, Y., Elkashlan, M., Wong, K.-K., Schober, R., & Hanzo, L. (2016). User Association in 5G Networks: A Survey and an Outlook. *IEEE Communications Surveys and Tutorials*, *18*(2), 1018–1044. doi:10.1109/COMST.2016.2516538

Liu, J. (2014). Seeing the Unobservable: Channel Learning for Wireless Communication Networks. *2015 IEEE Global Communications Conference (GLOBECOM)*, (pp. 1-6). IEEE. 10.1109/GLOCOM.2014.7417805

Liu, X., Li, F., & Na, Z. (2017). Optimal resource allocation in simultaneous cooperative spectrum sensing and energy harvesting for multichannel cognitive radio. *IEEE Access: Practical Innovations, Open Solutions*, *5*, 3801–3812. doi:10.1109/ACCESS.2017.2677976

Liu, X., Liu, Y., & Chen, Y. (2019). Reinforcement Learning in Multiple-UAV Networks: Deployment and Movement Design. *IEEE Transactions on Vehicular Technology*, *68*(8), 8036–8049. doi:10.1109/TVT.2019.2922849

Liu, X., Liu, Y., Chen, Y., & Hanzo, L. (2019). Trajectory Design and Power Control for Multi-UAV Assisted Wireless Networks: A Machine Learning Approach. *IEEE Transactions on Vehicular Technology*, *68*(8), 7957–7969. doi:10.1109/TVT.2019.2920284

Liu, X., Wu, S., Wang, Y., Zhang, N., Jiao, J., & Zhang, Q. (2020). Exploiting Error-Correction-CRC for Polar SCL Decoding: A Deep Learning-Based Approach. *IEEE Transactions on Cognitive Communications and Networking*, *6*(2), 817–828. doi:10.1109/TCCN.2019.2946358

Liu, Y., Liu, X., Mu, X., Hou, T., Xu, J., Di Renzo, M., & Al-Dhahir, N. (2021). Reconfigurable Intelligent Surfaces: Principles and Opportunities. *IEEE Communications Surveys and Tutorials*, *23*(3), 1546–1577. doi:10.1109/COMST.2021.3077737

Liu, Y., Pan, G., Zhang, H., & Song, M. (2016). On the Capacity Comparison Between MIMO-NOMA and MIMO-OMA. *IEEE Access: Practical Innovations, Open Solutions*, *4*, 2123–2129. doi:10.1109/ACCESS.2016.2563462

Li, X., Wu, L.-S., & Mao, J.-F. (2017). High-Frequency Analysis of Intercalated Multilayer Graphene (IMLG) and Implication for Tunable Terahertz Resonator Design. *IEEE Access: Practical Innovations, Open Solutions*, *5*, 7532–7541. doi:10.1109/ACCESS.2017.2701506

Long, Q., Chen, Y., Zhang, H., & Lei, X. (2019). Software Defined 5G and 6G Networks: A Survey. *Mobile Networks and Applications*. doi:10.100711036-019-01397-2

Long, S., McAllister, M., & Liang, S. (1983). The resonant cylindrical dielectric cavity antenna. *IEEE Transactions on Antennas and Propagation*, *31*(3), 406–412. doi:10.1109/TAP.1983.1143080

Lota, J., Sun, S., Rappaport, T. S., & Demosthenous, A. (2017, November). 5G Uniform Linear Arrays With Beamforming and Spatial Multiplexing at 28, 37, 64, and 71 GHz for Outdoor Urban Communication: A Two-Level Approach. *IEEE Transactions on Vehicular Technology*, *66*(11), 9972–9985. doi:10.1109/TVT.2017.2741260

Ludwig, A. C. (1975). The definition of Cross – polarization. *IEEE Transactions on Antennas and Propagation*, *21*(1), 116–119. doi:10.1109/TAP.1973.1140406

Lu, J. H., & Chang, B. S. (2017). Planar compact square-ring tag antenna with circular polarization for UHF RFID applications. *IEEE Transactions on Antennas and Propagation*, *65*(2), 432–441. doi:10.1109/TAP.2016.2633162

Luo, H. & Zhong, L. (2015). Isolation Enhancement of a Very Compact UWB-MIMO Slot Antenna with Two Defected Ground Structures. *IEEE Antennas and Wireless Propagation Letters*, 14.

Luo, F.-L., & Zhang, C. J. (2016). *Non-Orthogonal Multiple Access (NOMA)*. Concept and Design.

Luo, Y., Zeng, Q., Yan, X., Wu, Y., Lu, Q., Zheng, C., Hu, N., Xie, W., & Zhang, X. (2019). Graphene-based Multi-Beam Reconfigurable THz Antennas. *IEEE Access: Practical Innovations, Open Solutions*, *7*, 30802–30808. doi:10.1109/ACCESS.2019.2903135

Mahajan, P. (2018, November). Review paper on optimization of handover parameter in heterogeneous networks. In 2018 3rd International Innovative Applications of Computational Intelligence on Power, Energy and Controls with their Impact on Humanity (CIPECH) (pp. 1-5). IEEE.

Mak, K. M., & Luk, K. M. (2009, October). A circularly polarized antenna with wide axial ratio beamwidth. *IEEE Transactions on Antennas and Propagation*, *57*(10), 3309–3312. doi:10.1109/TAP.2009.2029370

Mandelbrot, B. (1982). The fractal geometry of nature, 1186–1189. W. H. Freeman and Company

Manteghi, M. & Rahamat, Y.. (2005). Multiport Characteristics of a Wideband Cavity Backed Annular Patch Antenna for Multi polarization Operations. *IEEE Transactions on Antennas and Propagation*, *1*, 466–474.

Manteghi, M. & Tahmat-Samii, Y. (2006). Novel Compact Tri-Band Two-Element and Four-Element MIMO Antenna Designs. *IEEE Antennas and Propagation Society International Symposium*, (pp. 4443-4446). IEEE.

Mao, Y., Clerckx, B., & Li, V. O. K. (2018). Energy Efficiency of Rate-Splitting Multiple Access, and Performance Benefits over SDMA and NOMA. *2018 15th International Symposium on Wireless Communication Systems (ISWCS)*, (pp. 1-5). IEEE. 10.1109/ISWCS.2018.8491100

Mao, Y., Clerckx, B., & Li, V. O. (2018). Rate-splitting multiple access for downlink communication systems: Bridging, generalizing, and outperforming SDMA and NOMA. *J Wireless Com Network*, *2018*(1), 133. doi:10.118613638-018-1104-7 PMID:30996723

Martín, I., Troia, S., Hernandez, J. A., Rodriguez, A., Musumeci, F., Maier, G., Alvizu, R., & Gonzalez de Dios, O. (2019). Machine Learning-Based Routing and Wavelength Assignment in Software-Defined Optical Networks. *IEEE eTransactions on Network and Service Management*, *16*(3), 871–883. doi:10.1109/TNSM.2019.2927867

Marzetta, T. L. (2010). Noncooperative cellular wireless with unlimited numbers of base station antennas. *IEEE Transactions on Wireless Communications*, *9*(11), 3590–3600. doi:10.1109/TWC.2010.092810.091092

Mathew, S. (2019, March 19). Mfine app: Connects patients to doctors instantly. *The Hindu*. https://www.thehindu.com/sci-tech/health/mfine-app-60-seconds-to-a-doctor/article26568179.ece

Mazen, K., Emran, A., Shalaby, A. S., & Yahya, A. (2021). Design of Multi-band Microstrip patch Antenna for Mid-band 5G Wireless Communication. *International Journal of Advanced Computer Science and Applications*, *12*(5), 459–469. doi:10.14569/IJACSA.2021.0120557

Meivel, S., Sindhwani, N., Valarmathi, S., Dhivya, G., Atchaya, M., Anand, R., & Maurya, S. (2023). Design and Method of 16.24 GHz Microstrip Network Antenna Using Underwater Wireless Communication Algorithm. In *Cyber Technologies and Emerging Sciences* (pp. 363–371). Springer. doi:10.1007/978-981-19-2538-2_36

Mesodiakaki, A., Kassler, A., Zola, E., Ferndahl, M., & Cai, T. (2016). Energy efficient line-of-sight millimeter wave small cell backhaul: 60, 70, 80 or 140 ghz?" In *2016 IEEE 17th International Symposium on A World of Wireless, Mobile and Multime- dia Networks (WoWMoM),* (pp. 1–9). IEEE. 10.1109/WoWMoM.2016.7523521

Moghaddam, J. Z., Usman, M., & Granelli, F. (2018). A Device-to-Device Communication-Based Disaster Response Network. *IEEE Trans. Cogn. Commun. Netw.*, *4*(2), 288–298. doi:10.1109/TCCN.2018.2801339

Mohan, D. (2020). *What Will Be the Economic Consequences of COVID-19 for India and the World?* Palvia, P., Lowe, K., Nemati, H., & Jacks, T. (2012). Information Technology Issues in Healthcare: Hospital CEO and CIO Perspectives. *Communications of the Association for Information Systems*, *30*. doi:10.17705/1CAIS.03019

Moradi, M. (2020). *Performance and Complexity of Sequential Decoding of PAC Codes.*

Morell, A., Correa, A., Barcelo, M., & Vicario, J. L. (2016). Data Aggregation and Principal Component Analysis in WSNs. *IEEE Transactions on Wireless Communications*, *15*(6), 3908–3919. doi:10.1109/TWC.2016.2531041

Morgado, A., Huq, K. M. S., Mumtaz, S., & Rodriguez, J. (2018). A survey of 5G technologies: Regulatory, standardization and industrial perspectives. *Digital Communications and Networks*, *4*(2), 87–97. doi:10.1016/j.dcan.2017.09.010

Motkuri, V., & Naik, S. V. (2010). Workforce in Indian Health Care Sector. *The Asian Economic Review*, *52*(2).

Mujawar, M. (2021a). *Antenna Array Design for Massive MIMO System in 5G Application.* Taylor & Francis Group. . doi:10.1201/9781003175155-18

Mujawar, M. (2022). *Arrow Shaped Dual-Band Wearable Antenna for ISM Applications.* Wiley. . doi:10.1002/9781119792581.ch8

Mujawar, M., & Thangavel, G. (2022). *Multiband Slot Microstrip Antenna for Wireless Applications.* Springer. . doi:10.1007/978-3-030-76636-8_3

Mujawar, M. (2021b). *Compact Microstrip Patch Antenna Design with Three I-, Two L.* One E- and One F-Shaped Patch for Wireless Applications. doi:10.1201/9781003093558-7

Mujawar, M., Dommeti, V., Naz, M., & Muduli, A. (2022). Design and performance comparison of arrays of circular, square and hexagonal meta-material structures for wearable applications. *Journal of Magnetism and Magnetic Materials*, *553*, 169235. doi:10.1016/j.jmmm.2022.169235

Müller, C. F., Galaviz, G., Andrade, Á. G., Kaiser, I., & Fengler, W. (2018). Evaluation of Scheduling Algorithms for 5G Mobile Systems. *Studies in Systems, Decision, and Control*, *143*, 213–233. doi:10.1007/978-3-319-74060-7_12

Musumeci, F., Rottondi, C., Nag, A., Macaluso, I., Zibar, D., Ruffini, M., & Tornatore, M. (2019). An Overview on Application of Machine Learning Techniques in Optical Networks. *IEEE Communications Surveys and Tutorials*, *21*(2), 1383–1408. doi:10.1109/COMST.2018.2880039

Nakano, H., Nogami, K., Arai, S., Mimaki, H., & Yamauchi, J. (1986, June). A spiral antenna backed by reflector a conducting plane. *IEEE Transactions on Antennas and Propagation*, *34*(6), 791–796. doi:10.1109/TAP.1986.1143893

Nallagonda, S., Roy, S. D., Kundu, S., Ferrari, G., & Raheli, R. (2013). Cooperative spectrum sensing with censoring of cognitive radios in Rayleigh fading under majority logic fusion. In *Proc. of IEEE Nineteenth National conference on Communications (NCC)*, (pp. 1-5). IEEE.

Nandi, S. & Mohan, A. (2017). CRLH Unit Cell Loaded Triband Compact MIMO Antenna for WLAN/WiMAX Applications. *IEEE Antennas and Wireless Propagation Letters*, *16*, 1816–1819.

Nandi., S. & Mohan, A. (2017). A Compact Dual-Band MIMO Slot Antenna for WLAN Applications. *IEEE Antennas and Wireless Propagation Letters*, *16*, 2457–2460.

Naresh, M., Reddy, D. V., & Reddy, K. R. (2020, October). A comprehensive study on vertical handover for IEEE 802.21 wireless networks. In *2020 Fourth International Conference on I-SMAC (IoT in Social, Mobile, Analytics and Cloud) (I-SMAC)* (pp. 343-347). IEEE.

Naresh, M., Venkat Reddy, D., & Ramalinga Reddy, K. (2020). Multi-objective emperor penguin handover optimisation for IEEE 802.21 in heterogeneous networks. *IET Communications*, *14*(18), 3239–3246.

Nasimuddin, C. Z. N., Chen, Z. N., & Qing, X. (2011, October). Symmetric aperture Antenna for Broadband Circular Polarization. *IEEE Transactions on Antennas and Propagation*, *59*(10), 3932–3936. doi:10.1109/TAP.2011.2163757

Nawaz, F., Ibrahim, J., Junaid, M., Kousar, S., Parveen, T., & Ali, M. A. (2020). A review of vision and challenges of 6G technology. *International Journal of Advanced Computer Science and Applications*, *11*(2), 643–649. doi:10.14569/IJACSA.2020.0110281

Nawaz, S. J., Sharma, S. K., Wyne, S., Patwary, M. N., & Asaduzzaman, M. (2019). Quantum Machine Learning for 6G Communication Networks: State-of-the-Art and Vision for the Future. *IEEE Access: Practical Innovations, Open Solutions*, *7*, 46317–46350. doi:10.1109/ACCESS.2019.2909490

Nezhad, MHassani, H. R. (2010). A Novel Tri-band E-Shaped Printed Monopole Antenna for MIMO Application. *IEEE Antennas and Wireless Propagation Letters*, *9*, 576–579. doi:10.1109/LAWP.2010.2051131

Nguyen, T., Tran, N., Loven, L., Partala, J., Kechadi, M. T., & Pirttikangas, S. (2020). Privacy-aware blockchain innovation for 6G: Challenges and opportunities. *2nd 6G Wireless Summit 2020: Gain Edge for the 6G Era, 6G SUMMIT 2020*, (pp. 1–5). IEEE. doi:10.1109/6GSUMMIT49458.2020.9083832

Nie, G., Tian, H., Sengul, C., & Zhang, P. (2017). Forward and backhaul link optimiza- tion for energy efficient ofdma small cell networks. *IEEE Transactions on Wireless Communications*, *16*(2), 1080–1093. doi:10.1109/TWC.2016.2636821

Niknam, S., Dhillon, H. S., & Reed, J. H. (2020). Federated Learning for Wireless Communications: Motivation, Opportunities, and Challenges. *IEEE Communications Magazine*, *58*(6), 46–51. doi:10.1109/MCOM.001.1900461

Nishiyama, H., Ito, M., & Kato, N. (2014). Relay-by-smartphone: Realizing multihop device-to-device communications. *IEEE Communications Magazine*, *52*(4), 56–65. doi:10.1109/MCOM.2014.6807947

Niu, K., & Chen, K. (2012). CRC-Aided Decoding of Polar Codes. *IEEE Communications Letters*, *16*(10), 1668–1671. doi:10.1109/LCOMM.2012.090312.121501

Ogudo, K. A., Muwawa Jean Nestor, D., Ibrahim Khalaf, O., & Daei Kasmaei, H. (2019). A Device Performance and Data Analytics Concept for Smartphones' IoT Services and Machine-Type Communication in Cellular Networks. *Symmetry*, *11*(4), 593. doi:10.3390ym11040593

Open Networking Foundation. (n.d.). *Open Networking*. https://www.opennetworking.org/.

OpenFlow Switch Specification. (n.d.). Version 1.4.0. https://www. opennetworking.org /images/stories/downloads/sdn-resources/onf-specifications/openflow/openflow-spec-v1.4.0.pdf.

Oraizi, H., & Hedayati, S. (2012). Circularly polarized multiband microstrip antenna using the square and Giuseppe Peano fractals. *IEEE Transactions on Antennas and Propagation*, *60*(7), 3466–3470. doi:10.1109/TAP.2012.2196912

Oraizi, H., & Pazoki, R. (2013, March). Wideband circularly polarized aperture-fed rotated stacked patch antenna. *IEEE Transactions on Antennas and Propagation*, *61*(3), 1048–1054. doi:10.1109/TAP.2012.2229378

Ou, C., Cai, X., & Qian, K. (2017). Two-Element Compact Antennas Decoupled with a Simple Neutralization Line. *Progress In Electromagnetics Research*, *65*, 63–68. doi:10.2528/PIERL16111801

Ozpinar, H., Aksimsek, S., & Tokan, N. T. (2020, March). A Novel Compact, Broadband, High Gain Millimeter-Wave Antenna for 5G Beam Steering Applications. *IEEE Transactions on Vehicular Technology*, *69*(3), 2389–2397. doi:10.1109/TVT.2020.2966009

Pajola, L., Pasa, L., & Conti, M. (2019). Threat is in the Air: Machine Learning for Wireless Network Applications. *Proceedings of the ACM Workshop on Wireless Security and Machine Learning*. ACM. 10.1145/3324921.3328783

Pandey, B. K. (2022). Effective and Secure Transmission of Health Information Using Advanced Morphological Component Analysis and Image Hiding. In M. Gupta, S. Ghatak, A. Gupta, & A. L. Mukherjee (Eds.), *Artificial Intelligence on Medical Data. Lecture Notes in Computational Vision and Biomechanics* (Vol. 37). Springer. doi:10.1007/978-981-19-0151-5_19

Paracha, K. N., Abdul Rahim, S. K. A., Soh, P. J., Kamarudin, M. R., Tan, K. G., Lo, Y. C., Islam, M. T., & Low Profile, A. (2019). A Low Profile, Dual-band, dual polarized antenna for indoor/outdoor wearable application. *IEEE Access: Practical Innovations, Open Solutions*, *7*, 33277–33288. doi:10.1109/ACCESS.2019.2894330

Parambanchary, D., & Malleswara Rao, V. (2020). WOA-NN: A decision algorithm for vertical handover in heterogeneous networks. *Wireless Networks*, *26*(1), 165–180.

Park, B. C., & Lee, J. H. (2011). Omnidirectional circularly polarized antenna utilizing zeroth-order resonance of epsilon negative transmission line. *IEEE Transactions on Antennas and Propagation*, *59*(7), 2717–2721. doi:10.1109/TAP.2011.2152337

Pathania, V. (2022). A Database Application of Monitoring COVID-19 in India. In M. Gupta, S. Ghatak, A. Gupta, & A. L. Mukherjee (Eds.), *Artificial Intelligence on Medical Data. Lecture Notes in Computational Vision and Biomechanics* (Vol. 37). Springer. doi:10.1007/978-981-19-0151-5_23

Pellegrini, A., Brizzi, A., Zhang, L., Ali, K., Hao, Y., Wu, X., Constantinou, C. C., Nechayev, Y., Hall, P. S., Chahat, N., Zhadobov, M., & Sauleau, R. (2013, August). Antennas and Propagation for Body-Centric Wireless Communications at Millimeter-Wave Frequencies: A Review [Wireless Corner]. *IEEE Antennas & Propagation Magazine, 55*(4), 262–287. doi:10.1109/MAP.2013.6645205

Perez, J. (2021). Leadership in Healthcare: Transitioning From Clinical Professional to Healthcare Leader. *Journal of Healthcare Management 66*(4), 280-302. doi:10.1097/JHM-D-20-00057

Peristerianos, Theopoulos, A., Koutinos, A. G., Kaifas, T., & Siakavara, K. (2016). Dual-Band Fractal Semi-Printed Element Antenna Arrays for MIMO Applications. *IEEE Antennas and Wireless Propagation Letters, 15,* 730–733. doi:10.1109/LAWP.2015.2470681

Pfister, H.D. (2017). *A Brief Introduction to Polar Codes Notes for Introduction to Error-Correcting Codes.*

Pourahmadazar, J., Ghobadi, C., Nourinia, J., Felegari, N., & Shirzad, H. (2011). Broadband CPW-Fed circularly polarized square slot antenna with inverted-L strips for UWB applications. *IEEE Antennas and Wireless Propagation Letters, 10*(May), 369–372. doi:10.1109/LAWP.2011.2147271

Pouyanfar, N. (2013). Broadband square-slot circularly polarized antenna for WiMAX and WLAN applications. *Microwave and Optical Technology Letters, 55*(9), 2191–2195. doi:10.1002/mop.27805

Pozar, D. (1998). *Microwave Engineering* (2nd ed.). John Wiley.

Practo. (n.d.). *Video consultation with doctors, Book doctor appointments, order medicine, diagnostic tests.* Practo. https://www.practo.com/

Prajapati, P. R., Murthy, G. G. K., Patnaik, A., & Kartikeyan, M. V. (2015). Design and testing of a compact circularly polarized microstrip antenna with fractal defected ground structure for L-band applications. *IET Microwaves, Antennas & Propagation, 9*(11), 1179–1185. doi:10.1049/iet-map.2014.0596

Prasad, A., Kunz, A., Velev, G., Samdanis, K., & Song, J. (2014). Energy-efficient D2D discovery for proximity services in 3GPP LTE-advanced networks: ProSe discovery mechanisms. *IEEE Vehicular Technology Magazine, 9*(4), 40–50. doi:10.1109/MVT.2014.2360652

Prasad, B., Bhowmick, A., Roy, S. D., & Kundu, S. (2016). Performance of cognitive relay network with novel hybrid spectrum access schemes with imperfect CSI. *International Journal of Communication Systems, 25*(11), 1761–1776. doi:10.1002/dac.3142

Prasad, B., Roy, S. D., & Kundu, S. (2014). Outage and SEP of secondary user with imperfect channel estimation and primary user interference. In *Proc. IEEE CONECCT,* (pp. 1-6). IEEE. 10.1109/CONECCT.2014.6740333

Prasad, R., Murthy, C. R., & Rao, B. D. (2015). Joint Channel Estimation and Data Detection in MIMO-OFDM Systems: A Sparse Bayesian Learning Approach. *IEEE Transactions on Signal Processing, 63*(20), 5369–5382. doi:10.1109/TSP.2015.2451071

Prasad, S., Meenakshi, M., Adhithiya, N., Rao, P. H., Ganti, R. K., & Bhaumik, S. (2021, April). mmWave multibeam phased array antenna for 5G applications. *Journal of Electromagnetic Waves and Applications, 35*(13), 1802–1814. doi:10.1080/09205071.2021.1917004

Raghavan, R., Verma, D. C., Pandey, D., Anand, R., Pandey, B. K., & Singh, H. (2022). Optimized building extraction from high-resolution satellite imagery using deep learning. *Multimedia Tools and Applications, 81*(29), 1–15. doi:10.100711042-022-13493-9

Rajagopalan, A. & Gupta, G. (2007). Increasing Channel Capacity of an Ultra-wideband MIMO System Using Vector Antennas. *IEEE Transactions on Antennas and Propagation, 55*(10), 2880-2887.

Raleigh, G. G., & Cioffi, J. M. (1998). Spatio-temporal coding for wireless communication. *IEEE Transactions on Communications, 46*(3), 357–366. doi:10.1109/26.662641

Ranzini, S. M., Da Ros, F., Bülow, H., & Zibar, D. (2019). Tunable Optoelectronic Chromatic Dispersion Compensation Based on Machine Learning for Short-Reach Transmission. *Applied Sciences (Basel, Switzerland), 9*(20), 4332. doi:10.3390/app9204332

Rao, A., Auti, S., Koul, A., & Sabnis, G. (2016). High availability and load balancing in SDN controllers. *Int J Trend Res Dev, 3*(2), 310–314.

Rao, P. N., & Sarma, N. V. S. N. (2008). Fractal boundary circularly polarized single feed microstrip antenna. *Electronics Letters, 44*(12), 1710–1711.

Ravipati, C. B., & Shafai, L. (1999). A wide Bandwidth Circularly Polarized Microstrip Antenna Using a Single Feed, In *Proceeding of the IEEE Antennas and Propagation Society International Symposium*, 1 (pp. 244–247). 10.1109/APS.1999.789126

Razafimahatratra, S., (2015). On-body propagation characterization with an H-plane Substrate Integrated Waveguide (SIW) horn antenna at 60 GHz. In *2015 European Microwave Conference (EuMC)*, (pp. 211–214). IEEE. 10.1109/EuMC.2015.7345737

Raza, M., Awais, M., Ali, K., Aslam, N., Paranthaman, V. V., Imran, M., & Ali, F. (2020). Establishing effective communications in disaster affected areas and artificial intelligence based detection using social media platform. *Future Generation Computer Systems, 112*, 1057–1069. doi:10.1016/j.future.2020.06.040

RCA Review. (1931).Diversity telephone receiving system of R.C.A. Communications, Inc. *Proc. IRE, 19*, 562–584.

Reddy, V. V., & Sarma, N. V. S. N. (2014). Single feed circularly polarized poly fractal antenna for wireless applications. *International Journal of Computer and Information Technology, 8*(11), 1710–1713.

Reddy, V. V., & Sarma, N. V. S. N. (2014). Triband circularly polarized Koch fractal boundary microstrip antenna. *IEEE Antennas and Wireless Propagation Letters, 13*, 1057–1060. doi:10.1109/LAWP.2014.2327566

Richtmyer, R. D. (1939). Dielectric resonators. *Journal of Applied Physics, 10*(6), 391–398. doi:10.1063/1.1707320

Rico-Alvariño, A., & Heath, R. W. (2014). Learning-Based Adaptive Transmission for Limited Feedback Multiuser MIMO-OFDM. *IEEE Transactions on Wireless Communications, 13*(7), 3806–3820. doi:10.1109/TWC.2014.2314104

Roddy, D. (2017). *Satellite Communications* (4th ed.). Mc GrawHill.

Ro, J.-H. (2022). *Improved MIMO Signal Detection Based on DNN in MIMO-OFDM System*. Computers, Materials & Continua. doi:10.32604/cmc.2022.020596

Row, J. S., & Ai, C. Y. (2004). Compact Design of Single-Feed Circularly Polarised Microstrip Antenna. *IEEE Electronics Letters, 40*(18), 1093–1094. doi:10.1049/el:20045602

Roziqin, A., Mas'udi, S. Y. F., & Sihidi, I. T. (2021). An analysis of Indonesian government policies against COVID-19. *Public Administration and Policy, 24*(1), 92–107. doi:10.1108/PAP-08-2020-0039

Rui, X., Li, J., & Wei, K. (2016). Dual-band dual-sense circularly polarized square slot antenna with simple structure. *Electronics Letters, 52*(8), 578–580. doi:10.1049/el.2015.4499

Rusek, F., Persson, D., Larsson, E. G., Marzetta, T. L., & Tufvesson, F., & Lau, B. K. (2013). Scaling Up MIMO: Opportunities and Challenges with Very Large Arrays. *IEEE Signal Processing Magazine*, *30*(1), 40–60. doi:10.1109/MSP.2011.2178495

Rusmono, E. S., & Marani, T. (2020, July). Design of multiband MIMO antenna for 5G millimeter-wave application. *IOP Conference Series. Materials Science and Engineering*, *852*(1), 012154. doi:10.1088/1757-899X/852/1/012154

Ryu. (n.d.). Component-based Software Defined Networking Framework. https: // osrg.github.io/ryu/.

Sa'don, S. N. H., Jamaluddin, M. H., Kamarudin, M. R., Ahmad, F., Yamada, Y., Karmadin, K., Idris, I. H., & Seman, N. (2020). Characterisation of Tunable Graphene Antenna. [AEÜ]. *International Journal of Electronics and Communications*, *118*, 153170. doi:10.1016/j.aeue.2020.153170

Sabella, D., Rapone, D., Fodrini, M., Cavdar, C., Olsson, M., Frenger, P., & Tombaz, S. (2016). Energy management in mobile networks towards 5G. In *Studies in Systems* (Vol. 50). Decision and Control. doi:10.1007/978-3-319-27568-0_17

Saeed, M., Kamal, H., & El-Ghoneimy, M. (2018). Novel type-2 fuzzy logic technique for handover problems in a heterogeneous network. *Engineering Optimization*, *50*(9), 1533–1543.

Saeed, R. A. (2019). Handover in a mobile wireless communication network–A Review Phase. *Int. J Comput. Commun. Inf*, *1*(1), 6–13.

Saif, W. S., Esmail, M. A., Ragheb, A. M., Alshawi, T. A., & Alshebeili, S. A. (2020). Machine Learning Techniques for Optical Performance Monitoring and Modulation Format Identification: A Survey. *IEEE Communications Surveys and Tutorials*, *22*(4), 2839–2882. doi:10.1109/COMST.2020.3018494

Sandeep Kumar, M., Maheshwari, V., Prabhu, J., Prasanna, M., Jayalakshmi, P., Suganya, P., & Jothikumar, R. (2020). Social economic impact of COVID-19 outbreak in India. *International Journal of Pervasive Computing and Communications*, *16*(4), 309–319. doi:10.1108/IJPCC-06-2020-0053

Santos, H. L. (2022). Machine learning-aided pilot and power allocation in multi-cellular massive MIMO networks. *Physical Communication*, *52*, 101646. doi:10.1016/j.phycom.2022.101646

Sarkar, Singh, A., Saurav, K., & Srivastava, K. V. (2015). Four-element quad-band multiple-input–multiple-output antenna employing split-ring resonator and inter-digital capacitor. *IET Microwaves, Antennas & Propagation*, *9*(13), 1453–1460. doi:10.1049/iet-map.2015.0189

Science, N., Phenomena, C., Pai, C., Bhaskar, A., & Rawoot, V. (2020). Chaos, Solitons and Fractals Investigating the dynamics of COVID-19 pandemic in India under lockdown. *Chaos, Solitons and Fractals: The Interdisciplinary Journal of Nonlinear Science, and Nonequilibrium and Complex Phenomena*, *138*, 109988. doi:10.1016/j.chaos.2020.109988

Sengan, S. (2022). Security-Aware Routing on Wireless Communication for E-Health Records Monitoring Using Machine Learning. *International Journal of Reliable and Quality E-Healthcare*.

Sengupta, M., Roy, A., Ganguly, A., Baishya, K., Chakrabarti, S., & Mukhopadhyay, I. (2021). Challenges Encountered by Healthcare Providers in COVID-19 Times: An Exploratory Study. *Journal of Health Management*, *23*(2), 339–356. doi:10.1177/09720634211011695

Seo, S.-I. (2019). Study on Efficient Impulsive Noise Mitigation for Power Line Communication. *International journal of advanced smart convergence, 8*(2), 199-203.

Shao, S., Liu, G., Khreishah, A., Ayyash, M., Elgala, H., Little, T. D., & Rahaim, M. (2020). Optimizing handover parameters by Q-Learning for heterogeneous radio-optical networks. *IEEE Photonics Journal*, *12*(1), 1–15.

Sharawi, M. S. (2013). Printed Multi-Band MIMO Antenna Systems and Their Performance Metrics. *IEEE Antennas and Propagation Magazine, 55*(5), 218-232.

Sharawi, M. S., Numan, A. B., Khan, M. U., & Aloi, D. N. (2012). A Dual-Element Dual-Band MIMO Antenna System with Enhanced Isolation for Mobile Terminals. *IEEE Antennas and Wireless Propagation Letters, 11*, 1006–1009. doi:10.1109/LAWP.2012.2214433

Sharma, A. & Biswas, A. (2017). Wideband multiple-input–multiple-output dielectric resonator antenna. *IET Microwaves, Antennas & Propagation, 11*(4), 496–502.

Sharma, S., Puthal, D., Jena, S. K., Zomaya, A. Y., & Ranjan, R. (2017). Rendezvous-based routing protocol for wireless sensor networks with mobile sink. *The Journal of Supercomputing, 73*(3), 168–1188. doi:10.100711227-016-1801-0

Shayea, I., Ergen, M., Azizan, A., Ismail, M., & Daradkeh, Y. I. (2020). Individualistic dynamic handover parameter self-optimization algorithm for 5G networks based on automatic weight function. *IEEE Access: Practical Innovations, Open Solutions, 8*, 214392–214412.

Shen, H., Xu, W., Gong, S., He, Z., & Zhao, C. (2019). Secrecy Rate Maximization for Intelligent Reflecting Surface Assisted Multi-Antenna Communications. *IEEE Communications Letters, 23*(9), 1488–1492. doi:10.1109/LCOMM.2019.2924214

Sherwood, R., Gibb, G., Yap, K.-K., Appenzeller, G., Casado, M., McKeown, N., & Parulkar, G. (2009). *Flowvisor: A network virtualization layer. OpenFlow Switch Consortium, Tech. Rep.*

Shivakumar, K. N. (2018). Work Life Balance in the Health Care Sector. *Amity Journal of Healthcare Management.* doi:10.13140/RG.2.2.19413.73440

Shiwei, G. (2021, July). An Improved KNN Based Decision Algorithm for Vertical Handover in Heterogeneous Wireless Networks. In *2021 40th Chinese Control Conference (CCC)* (pp. 3011-3016). IEEE.

Shi, Z., Zhang, H., Khan, K., Cao, R., Zhang, Y., Ma, C., Tareen, A. K., Jiang, Y., Jin, M., & Zhang, H. (2022). Two-Dimensional Materials toward Terahertz Optoelectronic Device Applications. *Journal of Photochemistry and Photobiology C, Photochemistry Reviews, 51*, 100473. doi:10.1016/j.jphotochemrev.2021.100473

Shoaib, Shoaib, I., Shoaib, N., Xiaodong Chen, & Parini, C. G. (2014). Design and Performance Study of a Dual-Element Multiband Printed Monopole Antenna Array for MIMO Terminals. *IEEE Antennas and Wireless Propagation Letters, 13*, 329–332. doi:10.1109/LAWP.2014.2305798

Shukla, A., Ahamad, S., Rao, G. N., Al-Asadi, A. J., Gupta, A., & Kumbhkar, M. (2021). Artificial Intelligence Assisted IoT Data Intrusion Detection. *2021 4th International Conference on Computing and Communications Technologies (ICCCT),* (pp. 330-335). IEEE. 10.1109/ICCCT53315.2021.9711795

Siddique, U., Tabassum, H., Hossain, E., & Kim, D. I. (2015, October). Wireless backhauling of 5G small cells: Challenges and solution approaches. *IEEE Wireless Communications, 22*(5), 22–31. doi:10.1109/MWC.2015.7306534

Simeone, O. (2018). A Very Brief Introduction to Machine Learning With Applications to Communication Systems. *IEEE Transactions on Cognitive Communications and Networking, 4*(4), 648–664. doi:10.1109/TCCN.2018.2881442

Singh, A., Singh, S. P., Tripathi, U. N., & Mishra, M. (2017). Optimizing Call Drops in Cellular Networks using Artificial Intelligence based Handover Schema. *Ijarcce, 6*(1), 286–290. doi:10.17148/IJARCCE.2017.6155

Singh, K., Nirmal, A. V., & Sharma, S. V. (2017). A. V. Nirmal, and S. V. Sharma. "Link margin for wireless radio communication link. *ICTACT Journal on Communication Technology, 8*(3), 3. doi:10.21917/ijct.2017.0232

Sitompul, P. P., Sri Sumantyo, J. T., Kurniawan, F., Santosa, C. E., Manik, T., Hattori, K., Gao, S., & Liu, J. Y. (2019). Circularly Polarized Circularly-Slotted-Patch Antenna with Two Asymmetrical Rectangular Truncations for Nanosatellite Antenna. *Progress in Electromagnetics Research*, *C*, 90, 225–236. doi:10.2528/PIERC18120503

Sklar, B. and Harris, F. (2022). Digital Communications, Prentice Hall

Skoog, R. A., Banwell, T. C., Gannett, J. W., Habiby, S. F., Pang, M., Rauch, M. E., & Toliver, P. (2006). Automatic Identification of Impairments Using Support Vector Machine Pattern Classification on Eye Diagrams. *IEEE Photonics Technology Letters*, *18*(22), 2398–2400. doi:10.1109/LPT.2006.886146

Sobana, S., & Jeyanthi, K. M. (2015). Novel Multiple-Input Multiple Output Precoding Techniques with Improved Bit Error Rate Performance. *Journal of Computational and Theoretical Nanoscience*, *12*(11), 4794–4802. doi:10.1166/jctn.2015.4441

Song, P., Liu, B., Shi, X., Cao, D., & Gu, J. (2021). MXenes for Polymer Matrix Electromagnetic Interference Shielding Composites: A Review. *Composites Communications*, *24*, 100653. doi:10.1016/j.coco.2021.100653

Sonia, D. & Gupta, D. (2021). Performance Enhancement of NOMA: A 5G Candidate. *Thirty Sixth National Convention of Electronics and Telecommunication Engineers on Antenna Design for Efficient Communication and Networking*, (pp. 91-103). The Institute of Engineers, Bhatinda Local Centre, Under the Aegis off Electronics & Telecommunication Divisional Board.

Sonia, S. (2022). Non-Orthogonal Multiple Access Technique in Wireless Communication. In *Emerging Trends in Wireless Communication,* 1-22. Central West Publication.

Sree, G. N. J., & Nelaturi, S. (2021, May). Design and experimental verification of fractal based MIMO antenna for low sub 6-GHz 5G applications. *International Journal of Electronics and Communications*, *137*(10), 153797. doi:10.1016/j.aeue.2021.153797

Sreekanth, N., Rama Devi, J., Shukla, A., Mohanty, D. K., Srinivas, A., Rao, G. N., Alam, A., & Gupta, A. (2022). (2022). Evaluation of estimation in software development using deep learning-modified neural network. *Applied Nanoscience*. doi:10.100713204-021-02204-9

Srinivasa Prasanna, G. N. (2009). Data communication over the smart grid. *2009 IEEE International Symposium on Power Line Communications and Its Applications*, (pp. 273-279). IEEE. 10.1109/ISPLC.2009.4913442

Srinivasan, V., & Chandwani, R. (2014). HRM innovations in rapid growth contexts: The healthcare sector in India. *International Journal of Human Resource Management*, *25*(10), 1505–1525. doi:10.1080/09585192.2013.870308

Srivastava, GMohan, A. (2016). Compact MIMO Slot Antenna for UWB Applications. *IEEE Antennas and Wireless Propagation Letters*, *15*, 1057–1060.

Stamou, A., Dimitriou, N., Kontovasilis, K., & Papavassiliou, S. (2019). Autonomic handover management for heterogeneous networks in a future internet context: A survey. *IEEE Communications Surveys and Tutorials*, *21*(4), 3274–3297.

Stantchev, G. (2019). *Machine Learning for RF Signal Processing: Catching the Third Wave.*

Study on channel model for frequencies 0.5 to 100 GHz, (2017). *3GPP TR 38.901 version 14.3.0 Release 14,*

Su, S., Lee, C., & Chang, F.-S. (2012). Printed MIMO-Antenna System Using Neutralization-Line Technique for Wireless USB-Dongle Applications. *IEEE Transactions on Antennas and Propagation, 60*(2),456-463.

Su, S., Lee, S., & Chang, F. (2012). Printed MIMO-Antenna System Using Neutralization-Line Technique for Wireless USB-Dongle Applications. *IEEE Transactions on Antennas and Propagation*, 60(2), 456-463.

Sudha, T., Vedavathy, T. S., & Bhat, N. (2004). Wideband single-fed circularly polarized patch antenna. *Electronics Letters*, *40*(11), 648–649. doi:10.1049/el:20040407

Sullivan, J. L., Adjognon, O. L., Engle, R. L., Shin, M. H., Afable, M. K., Rudin, W., White, B., Shay, K., & Lukas, C. V. (2018). Identifying and overcoming implementation challenges: Experience of 59 noninstitutional long-term services and support pilot programs in the Veterans Health Administration. *Health Care Management Review, 43*(3), 193-205. doi:10.1097/HMR.0000000000000152

Sun, Fang, H.-S., Lin, P.-Y., & Chuang, C.-S. (2016). Triple-Band MIMO Antenna for Mobile Wireless Applications. *IEEE Antennas and Wireless Propagation Letters*, *15*, 500–503. doi:10.1109/LAWP.2015.2454536

Sun, W.-J., Yang, W.-W., Chu, P., & Chen, J.-X. (2019, October). A wideband stacked dielectric resonator antenna for 5G applications. *International Journal of RF and Microwave Computer-Aided Engineering*, *29*(7), e21897. doi:10.1002/mmce.21897

Swiggy. (2023). Swiggy App for design. [Application].

Swindlehurst, A. L., Ayanoglu, E., Heydari, P., & Capolino, F. (2014). Millimeter-wave massive mimo: The next wireless revolution? *IEEE Communications Magazine*, *52*(9), 56–62. doi:10.1109/MCOM.2014.6894453

Tabatabaeian, Z. S. (2021). Graphene Load for Harmonic Rejection and Increasing the Bandwidth in Quasi Yagi-Uda Array THz Antenna for the 6G Wireless Communication. *Optics Communications*, *499*, 127272. doi:10.1016/j.optcom.2021.127272

Taha, A. (2020). Deep Reinforcement Learning for Intelligent Reflecting Surfaces: Towards Standalone Operation. *2020 IEEE 21st International Workshop on Signal Processing Advances in Wireless Communications (SPAWC)*, (pp. 1-5). IEEE. 10.1109/SPAWC48557.2020.9154301

Taheribakhsh, M. (2020). 5G Implementation: Major Issues and Challenges. *25th International Computer Conference, Computer Society of Iran (CSICC)*.

Tal, I., & Vardy, A. (2011). List decoding of polar codes. *2011 IEEE International Symposium on Information Theory Proceedings*, (pp. 1-5). IEEE.

Tan, B. Y. Q., Chew, N. W. S., Lee, G. K. H., Jing, M., Goh, Y., Yeo, L. L. L., Zhang, K., Chin, H.-K., Ahmad, A., Khan, F. A., Shanmugam, G. N., Chan, B. P. L., Sunny, S., Chandra, B., Ong, J. J. Y., Paliwal, P. R., Wong, L. Y. H., Sagayanathan, R., Chen, J. T., & Sharma, V. K. (2020). Psychological Impact of the COVID-19 Pandemic on Health Care Workers in Singapore. *Annals of Internal Medicine*, *173*(April), 19–22. doi:10.7326/M20-1083 PMID:32251513

Tang, TLin, K. (2014). An Ultra wideband MIMO Antenna with Dual Band-Notched Function. *IEEE Antennas and Wireless Propagation Letters*, *13*, 1076–1079.

Tang, F., Mao, B., Kawamoto, Y., & Kato, N. (2021). Survey on machine learning for intelligent end-to-end communication toward 6G: From network access, routing to traffic control and streaming adaption. *IEEE Communications Surveys and Tutorials*, *23*(3), 1578–1598. doi:10.1109/COMST.2021.3073009

Tao, J. & Feng, Q. (2017). Compact Ultra wideband MIMO Antenna with Half-Slot Structure. *IEEE Antennas and Wireless Propagation Letters*, *16*, 792-795.

Taori, R., & Sridharan, A. (2015). Point-to-multipoint in-band mmwave backhaul for 5G networks. *IEEE Communications Magazine*, *53*(1), 195–201. doi:10.1109/MCOM.2015.7010534

Tarkaa, N., & Mom, J. (2018). Comparative Analysis of Drop-Call Probability Due to Handover and Other Factors. *International Journal of Innovative Research in Science, Engineering and Technology, 7*(7), 8029–8040. doi:10.15680/IJIRSET.2018.70707069

Tarpara, N. M., Rathwa, R. R., & Kotak, D. N. A. (2018, April). Design of slotted microstrip patch antenna for 5G Application. *International Research Journal of Engineering and Technology, 5*(4), 2827–2832.

Tennenhouse, L., & Wetherall, D. J. (1996). Towards an Active Network Architecture. *Computer Communication Review, 26*(2), 5–18. doi:10.1145/231699.231701

Teresa, P. M., & Umamaheswari, G. (2022, September 03). Compact Slotted Microstrip Antenna for 5G Applications Operating at 28 GHz. *Journal of the Institution of Electronics and Telecommunication Engineers, 68*(5), 3778–3785. doi:10.1080/03772063.2020.1779620

Tezergil, B., & Onur, E. (2022). Wireless Backhaul in 5G and Beyond: Issues, Chal- lenges and Opportunities. *IEEE Communications Surveys and Tutorials, 24*(4), 2579–2632. doi:10.1109/COMST.2022.3203578

Thakur, J. P., & Park, J.-S. (2006). An advance design approach for circular polarization of the microstrip antenna with unbalance DGS Feedlines. *IEEE Antennas and Wireless Propagation Letters, 5*, 101–103. doi:10.1109/LAWP.2006.872425

Thi, C. H. L., Ta, S. X., Nguyen, X. Q., Nguyen,, K. K., & & C. D-N. (2021). Design of compact broadband dual-polarized antenna for 5G applications. *International Journal of RF and Microwave Computer-Aided Engineering, 31*(5), e22615.

Thilina, K. M., Hossain, E., & Kim, D. I. (2016). DCCC-MAC: A Dynamic Common-Control-Channel-Based MAC Protocol for Cellular Cognitive Radio Networks. *IEEE Transactions on Vehicular Technology, 65*(5), 3597–3613. doi:10.1109/TVT.2015.2438058

Tighezza, M., Rahim, S. K. A., & Islam, M. T. (2018, January). Flexible Wideband Antenna for 5G Applications. *Microwave and Optical Technology Letters, 60*(1), 38–44. doi:10.1002/mop.30906

Toktas, A. (2017). G-shaped band-notched ultra-wideband MIMO antenna system for mobile terminals. *IET Microwave Antennas Propagation, 11*(5), 718-725.

Torres, A. E., Marante, F., Tazón, A., & Vassal'lo, J. (2015). New microstrip radiator feeding by electromagnetic coupling for circular polarization. *AEÜ. International Journal of Electronics and Communications, 69*(12), 1880–1884. doi:10.1016/j.aeue.2015.09.016

Tracxn. (n.d.). *Technology + human-in-the-loop for deal discovery*. Tracxn. https://tracxn.com/

Tripathi, A., Sindhwani, N., Anand, R., & Dahiya, A. (2023). Role of IoT in Smart Homes and Smart Cities: Challenges, Benefits, and Applications. In *IoT Based Smart Applications* (pp. 199–217). Springer. doi:10.1007/978-3-031-04524-0_12

Tripathi, S., Mohan, A., & Yadav, A. (2015). A Compact Koch Fractal UWB MIMO Antenna with WLAN Band-Rejection. *IEEE Antennas and Wireless Propagation Letters*, 14.

Ullah, U., Ain, M. F., & Ahmad, Z. A. (2017). A review of wideband circularly polarized dielectric resonator antennas. *China Communications, 14*(6), 65–79. doi:10.1109/CC.2017.7961364

Umebayashi, K., Kobayashi, M., & López-Benítez, M. (2018). Efficient Time Domain Deterministic-Stochastic Model of Spectrum Usage. *IEEE Transactions on Wireless Communications, 17*(3), 1518–1527. doi:10.1109/TWC.2017.2779511

Vainieri, M, Ferrè, F, Giacomelli, G., & Nuti, S. (2019). Explaining performance in health care: How and when top management competencies make the difference. *Health Care Management Review, 44*(4), 306-317. doi:10.1097/HMR.0000000000000164

Varshney, G., Singh, R., Pandey, V. S., & Yaduvanshi, R. S. (2020). Circularly polarized Two-Port MIMO dielectric resonator antenna. *Progress in Electromagnetics Research M. Pier M*, *91*, 19–28. doi:10.2528/PIERM20011003

Vaughan, R. & Andersen, J. (2003). Channels Propagation and Antennas for Mobile Communications. IET.

Veeraiah, V., Ahamad, G. P. S., Talukdar, S. B., Gupta, A., & Talukdar, V. (2022) Enhancement of Meta Verse Capabilities by IoT Integration. *2022 2nd International Conference on Advance Computing and Innovative Technologies in Engineering (ICACITE)*, (pp. 1493-1498). IEEE. 10.1109/ICACITE53722.2022.9823766

Veeraiah, V., Khan, H., Kumar, A., Ahamad, S., Mahajan, A., & Gupta, A. (2022). Integration of PSO and Deep Learning for Trend Analysis of Meta-Verse. *2022 2nd International Conference on Advance Computing and Innovative Technologies in Engineering (ICACITE)*, (pp. 713-718). IEEE. 10.1109/ICACITE53722.2022.9823883

Veeraiah, V., Kumar, K. R., Lalitha, K. P., Ahamad, S., Bansal, R., & Gupta, A. (2022). Application of Biometric System to Enhance the Security in Virtual World. *2022 2nd International Conference on Advance Computing and Innovative Technologies in Engineering (ICACITE)*, (pp. 719-723). IEEE. 10.1109/ICACITE53722.2022.9823850

Veeraiah, V., Rajaboina, N. B., Rao, G. N., Ahamad, S., Gupta, A., & Suri, C. S. (2022).Securing Online Web Application for IoT Management. *2022 2nd International Conference on Advance Computing and Innovative Technologies in Engineering (ICACITE)*, (pp. 1499-1504). IEEE. 10.1109/ICACITE53722.2022.9823733

Vela, A. P., Ruiz, M., & Velasco, L. (2018). *Examples of Machine Learning Algorithms for Optical Network Control and Management*. 2018 20th International Conference on Transparent Optical Networks (ICTON), (pp. 1-4). IEEE. 10.1109/ICTON.2018.8473900

Verma, A., Gupta, A., Kaushik, D., & Garg, M. (2021). Performance enhancement of IOT based accident detection system by integration of edge detection. *Materials Today: Proceedings*. doi:10.1016/j.matpr.2021.01.468

Viti, L., & Vitiello, M. S. (2021). Tailored Nano-Electronics and Photonics with Two-Dimensional Materials at Terahertz Frequencies. *Journal of Applied Physics*, *130*(17), 170903. doi:10.1063/5.0065595

Viwattanakulvanid, P. (2021). Ten commonly asked questions about Covid-19 and lessons learned from Thailand. *Journal of Health Research*, *35*(4), 329–344. doi:10.1108/JHR-08-2020-0363

Waldman, J. D, Kelly, F., Arora, S., Smith, H. L. (2010). The shocking cost of turnover in health care. *Health Care Management Review, 35*(3), 206-211. doi:. doi:10.1097/HMR.0b013e3181e3940

Wang, D., & Zhang, M. (2021). Artificial Intelligence in Optical Communications: From Machine Learning to Deep Learning. In Frontiers in Communications and Networks. doi:10.3389/frcmn.2021.656786

Wang, J. A. D. O. (2017). Blockchain-enabled wireless communications: a new paradigm towards 6G. *American Journal of Roentgenology, 186*(2), 227–236. https://pubmed.ncbi.nlm.nih.gov/28459981

Wang, X. (2019). Learning to Flip Successive Cancellation Decoding of Polar Codes with LSTM Networks. *2019 IEEE 30th Annual International Symposium on Personal, Indoor and Mobile Radio Communications (PIMRC)*, (pp. 1-5). IEEE. 10.1109/PIMRC.2019.8904878

Wang., H. (2015). A Wideband Compact WLAN/WiMAX MIMO Antenna Based on Dipole With V-shaped Ground Branch. *IEEE Transactions on Antennas and Propagation*, *63*(5), 2290–2295.

Wang, D., Wang, M., Zhang, M., Zhang, Z., Yang, H., Li, J., Li, J., & Chen, X. (2019). Cost-effective and data size-adaptive OPM at intermediated node using convolutional neural network-based image processor. *Optics Express*, *27*(7), 9403–9419. doi:10.1364/OE.27.009403 PMID:31045092

Wang, L., Guo, Y. X., & Sheng, W. (2012). Tri-band circularly polarized annular slot antenna for GPS and CNSS applications. [Early access]. *IEEE Antennas and Wireless Propagation Letters*, 1–1. doi:10.1109/LAWP.2012.2200869

Wang, W., Wu, Y., Wang, W., & Yang, Y. (2021, June). Isolation Enhancement in Dual-band Monopole Antenna for 5G Applications. *IEEE Transactions on Circuits and Systems II*, 68(6), 1867–1871. doi:10.1109/TCSII.2020.3040164

Wang, Z., Dong, Y., & Itoh, T. (2021). Metamaterial-based, miniaturised circularly polarised antennas for RFID application. *IET Microwaves, Antennas & Propagation*, 15(6), 547–559. doi:10.1049/mia2.12064

Wang, Z., & Yuandan, D. (2020). A Dual Band Circularly Polarized Ring Antenna based on Composite Right and Left Handed Metamaterials. *IET Microwaves, Antennas & Propagation*, 10(8), 363–375. doi:10.2528/mia2.08292

Wani, Z., Abegaonkar, M. P., & Koul, S. K. (2018). A 28-GHz Antenna for 5G MIMO Applications. *Progress in Electromagnetics Research*, 78, 73–79. doi:10.2528/PIERL18070303

Weekly, P., & Weekly, P. (2021).. . *Public Health Human Resource.*, 50(17), 20–23.

Wilson, W., Raj, J. P., Rao, S., Ghiya, M., Nedungalaparambil, N. M., Mundra, H., & Mathew, R. (2020). Prevalence and Predictors of Stress, anxiety, and Depression among Healthcare Workers Managing COVID-19 Pandemic in India: A Nationwide Observational Study. *Indian Journal of Psychological Medicine*, 42(4), 353–358. doi:10.1177/0253717620933992 PMID:33398224

Wu., Y., Zhang, B., & Ding, K. (2018). SIW-tapered slot antenna for broadband MIMO applications. *IET Microwaves, Antennas & Propagation*, 12(4), 612–616.

Wu, F., Xu, L., Kumari, S., & Li, X. (2017). A privacy-preserving and provable user authentication scheme for wireless sensor networks based on Internet of Things security. *Journal of Ambient Intelligence and Humanized Computing*, 8(1), 101–116. doi:10.100712652-016-0345-8

Wu, J., Li, J., & Ji, X. (2018). Security for cyberspace: Challenges and opportunities. *Frontiers of Information Technology and Electronic Engineering*, 19(12), 1459–1461. doi:10.1631/FITEE.1840000

Wu, Q., & Zhang, R. (2019). Intelligent Reflecting Surface Enhanced Wireless Network via Joint Active and Passive Beamforming. *IEEE Transactions on Wireless Communications*, 18(11), 5394–5409. doi:10.1109/TWC.2019.2936025

Wu, X. (2009). Applications of Artificial Neural Networks in Optical Performance Monitoring. *Journal of Lightwave Technology*, 27(16), 3580–3589. doi:10.1109/JLT.2009.2024435

Wu, X. Y., Akhoondzadeh-Asl, L., & Hall, P. S. (2011). Printed Yagi–Uda array for on-body communication channels at 60 GHz. *Microwave and Optical Technology Letters*, 53(12), 2728–2730. doi:10.1002/mop.26443

Wu, Y.-T., & Chu, Q.-X.Wu and Chu. (2014). Dual-band multiple input multiple output antenna with slitted ground. *IET Microwaves, Antennas & Propagation*, 8(13), 1007–1013. doi:10.1049/iet-map.2013.0340

Wu, Z., Lu, K., Jiang, C., & Shao, X. (2018). Comprehensive Study and Comparison on 5G NOMA Schemes. *IEEE Access: Practical Innovations, Open Solutions*, 6, 18511–18519. doi:10.1109/ACCESS.2018.2817221

Xiao, L., Li, Y., Han, G., Liu, G., & Zhuang, W. (2016). PHY-Layer Spoofing Detection With Reinforcement Learning in Wireless Networks. *IEEE Transactions on Vehicular Technology*, 65(12), 10037–10047. doi:10.1109/TVT.2016.2524258

Xiao, L., Wan, X., Lu, X., Zhang, Y., & Wu, D. (2018). IoT Security Techniques Based on Machine Learning: How Do IoT Devices Use AI to Enhance Security? *IEEE Signal Processing Magazine*, 35(5), 41–49. doi:10.1109/MSP.2018.2825478

Xie, Y., Wang, Y., Kandeepan, S., & Wang, K. (2022). Machine Learning Applications for Short Reach Optical Communication. *Photonics*, 9(1), 30. doi:10.3390/photonics9010030

Xiong, X., Li, X., Zhang, W., & Zhang, H. (2018). Enhance dual-band circularly polarized broadband antenna by using parasitic patch. *IET Microwaves, Antennas & Propagation, 12*(13), 2085–2088. doi:10.1049/iet-map.2018.5186

Xu, R., Li, J., Qi, Y. X., Guangwei, Y., & Yang, J. J. (2017). A design of triple-wideband triple-sense circularly polarized square slot antenna. *IEEE Antennas and Wireless Propagation Letters, 16*, 1763–1766. doi:10.1109/LAWP.2017.2674677

Xu, R., Li, J.-Y., Liu, J., Zhou, S.-G., Xing, Z.-J., & Wei, K. (2018). A design of dual-wideband planar printed antenna for circular polarization diversity by combining slot and monopole modes. *IEEE Transactions on Antennas and Propagation, 66*(8), 4326–4331. doi:10.1109/TAP.2018.2836670

Yang, Z., Chen, M., Saad, W., & Shaikh-Bahaei, M. (2020). Downlink Sum-Rate Maximization for Rate Splitting Multiple Access (RSMA). *ICC 2020 - 2020 IEEE International Conference on Communications (ICC),* (pp. 1-6). IEEE. 10.1109/ICC40277.2020.9149417

Yang, J., & Ulukus, S. (2011). Optimal packet scheduling in an energy harvesting communication system. *IEEE Transactions on Communications, 60*(1), 220–230. doi:10.1109/TCOMM.2011.112811.100349

Yang, S. S., Kishk, A. A., & Lee, K. F. (2008, June). Wideband circularly polarized antenna with L-shaped slot. *IEEE Transactions on Antennas and Propagation, 56*(6), 1780–1783. doi:10.1109/TAP.2008.923340

Yang, W., Che, W., Jin, H., Feng, W., & Xue, Q. (2015). A polarization-reconfigurable dipole antenna using polarization rotation AMC structure. *IEEE Transactions on Antennas and Propagation, 63*(12), 5305–5315. doi:10.1109/TAP.2015.2490250

Ye, H., Li, G. Y., & Juang, B.-H. (2018). Power of Deep Learning for Channel Estimation and Signal Detection in OFDM Systems. *IEEE Wireless Communications Letters, 7*(1), 114–117. doi:10.1109/LWC.2017.2757490

Yong-Xin, G., Lei, B., & Xiang Quan, S. (2009, August). Broadband circularly polarized annular-ring microstrip antenna. *IEEE Transactions on Antennas and Propagation, 57*(8), 2474–2477. doi:10.1109/TAP.2009.2024584

Young, L., Robinson, L., & Hacking, C. (1973). Meander-Line Polarizer. Antennas and Propagation. *IEEE Transactions on., 21*(3), 376–378. doi:10.1109/TAP.1973.1140503

Yuan, J., Ngo, H. Q., & Matthaiou, M. (2019). Machine Learning-Based Channel Estimation in Massive MIMO with Channel Aging. *2019 IEEE 20th International Workshop on Signal Processing Advances in Wireless Communications (SPAWC),* (pp. 1-5). IEEE. 10.1109/SPAWC.2019.8815557

Yuan, W., Yang, J., Yin, F., Li, Y., & Yuan, Y. (2020). Flexible and Stretchable MXene/Polyurethane Fabrics with Delicate Wrinkle Structure Design for Effective Electromagnetic Interference Shielding at a Dynamic Stretching Process. *Composite Communications, 19*, 90–98. doi:10.1016/j.coco.2020.03.003

Yu, B., Yang, K., Sim, C., & Yang, G. (2018, January). A Novel 28 GHz Beam Steering Array for 5G Mobile Device With Metallic Casing Application. *IEEE Transactions on Antennas and Propagation, 66*(1), 462–466. doi:10.1109/TAP.2017.2772084

Yu, H., Ma, Y., & Yu, J. (2019). Network selection algorithm for multiservice multimode terminals in heterogeneous wireless networks. *IEEE Access: Practical Innovations, Open Solutions, 7*, 46240–46260.

Zander, J., & Forchheimer, R. (1988). The SOFTNET project: a retrospect. In *Electrotechnics, 1988. Conference Proceedings on Area Communication, EUROCON 88., 8th European Conference on,* (pp. 343–345). 10.1109/EURCON.1988.11172

Zanotto, B., Etges, S., da Silva, A., Marcolino, Zago, M., & Polanczyk, C. (2021). Value-Based Healthcare Initiatives in Practice: A Systematic Review. *Journal of Healthcare Management, 66*(5), 340-365. doi:10.1097/JHM-D-20-00283

Zappone, A., Sanguinetti, L., & Debbah, M. (2018). User Association and Load Balancing for Massive MIMO through Deep Learning. *2018 52nd Asilomar Conference on Signals, Systems, and Computers,* (pp. 1262-1266). IEEE. 10.1109/ACSSC.2018.8645483

Zeain, M. Y., Abu, M., Zakaria, Z., Al-Gburi, A. J. A., Syahputri, R., & Toding, A. (2020, October). Design of a wideband strip helical antenna for 5G applications. *Bulletin of Electrical Engineering and Informatics*, *9*(No. 5), 1958–1963. doi:10.11591/eei.v9i5.2055

Zeng, J., Lv, T., Liu, R. P., Su, X., Peng, M., Wang, C., & Mei, J. (2018). Investigation on Evolving Single-Carrier NOMA into Multi-Carrier NOMA in 5G. *IEEE Access: Practical Innovations, Open Solutions*, *6*, 48268–48288. doi:10.1109/ACCESS.2018.2868093

Zeng, M., Yadav, A., Dobre, O. A., Tsiropoulos, G. I., & Poor, H. V. (2017, October). Capacity Comparison Between MIMO-NOMA and MIMO-OMA With Multiple Users in a Cluster. *IEEE Journal on Selected Areas in Communications*, *35*(10), 2413–2424. doi:10.1109/JSAC.2017.2725879

Zhang, S. (2009). Ultra-wideband MIMO/Diversity Antennas with a Tree-Like Structure to Enhance Wideband Isolation. *IEEE Antennas and Wireless Propagation Letters*, *8*, 1279–1282.

Zhang, SPedersen, G. (2016). Mutual Coupling Reduction for UWB MIMO Antennas with a Wideband Neutralization Line. *IEEE Antennas and Wireless Propagation Letters*, *15*, 166–169.

Zhang, H., Liao, Y., & Song, L. (2017). D2D-U: Device-to-device communications in unlicensed bands for 5G system. *IEEE Transactions on Wireless Communications*, *16*(6), 3507–3519. doi:10.1109/TWC.2017.2683479

Zhang, Q., Saad, W., & Bennis, M. (2022). Millimeter Wave Communications With an Intelligent Reflector: Performance Optimization and Distributional Reinforcement Learning. *IEEE Transactions on Wireless Communications*, *21*(3), 1836–1850. doi:10.1109/TWC.2021.3107520

Zheng, L. & Tse, D. (2003). Diversity and multiplexing: A fundamental trade-off in multiple-antenna channels. *IEEE Transaction Information Theory*. *49*(5), 1073–1096.

Zheng, X., Ping, F., Pu, Y., Wang, Y., Montenegro-Marin, C. E., & Khalaf, O. I. (2021). Recognize and regulate the importance of work-place emotions based on organizational adaptive emotion control. *Aggression and Violent Behavior*, 101557. doi:10.1016/j.avb.2021.101557

Zhou, X., Quan, X., & Li, R. (2012). A Dual-Broadband MIMO Antenna System for GSM/UMTS/LTE and WLAN Handsets. *IEEE Antennas and Wireless Propagation Letters*, *11*, 551-554.

Zhu, L., Farhat, M., Salama, K., & Chen, P.-Y. (2020). 2D Materials-based Radio-Frequency Wireless Communication and Sensing Systems for Internet-of-Thing Applications. In L. Tao & D. Akinwande (Eds.), Emerging 2D Materials and Devices for the Internet of Things (pp. 29–57). Elsevier. 10.1016/B978-0-12-818386-1.00002-3. doi:10.1016/B978-0-12-818386-1.00002-3

Zhu, Y., Wang, X., Zhang, Z., Chen, X., & Chen, Y. (2017). A rate-splitting non-orthogonal multiple access scheme for uplink transmission. *2017 9th International Conference on Wireless Communications and Signal Processing (WCSP)*, (pp. 1-6). 10.1109/WCSP.2017.8171078

Zhu, Y., Zhang, Z., Wang, X., & Liang, X. (2017). A low-complexity non-orthogonal multiple access systems based on rate splitting. *2017 9th International Conference on Wireless Communications and Signal Processing (WCSP)*, (pp. 1-6). 10.1109/WCSP.2017.8171135

Zhu., J. (2016). Compact Dual-Polarized UWB Quasi-Self-Complementary MIMO/Diversity Antenna with Band-Rejection Capability. *IEEE Antennas and Wireless Propagation Letters*, *15*, 905–908.

Zhu, Y., Liu, J., Guo, T., Wang, J. J., Tang, X., & Nicolosi, V. (2021). Multifunctional $Ti_3C_2T_x$ MXene Composite Hydrogels with Strain Sensitivity toward Absorption-Dominated Electromagnetic-Interference Shielding. *ACS Nano*, *15*(1), 1465–1474. doi:10.1021/acsnano.0c08830 PMID:33397098

Zhu, Yang, X., Song, Q., & Lui, B. (2017). Compact UWB-MIMO antenna with metamaterial FSS decoupling structure. *J Wireless Communication Network*, *2017*(1), 1. doi:10.118613638-017-0894-3

Zimmermann, M., & Dostert, K. (2002). A multipath model for the powerline channel. *IEEE Transactions on Communications*, *50*(4), 553–559. doi:10.1109/26.996069

About the Contributors

Ram Krishan, presently working as Assistant Professor and Head, Department of Computer Science, Mata Sundri University Girls College, Mansa, Punjab, India (A Constituent College of Punjabi University, Patiala). Dr. Ram obtained his Ph.D in Computer Science and Engineering from Guru Kashi University, Talwandi Sabo in 2017 and M.Tech. in Computer Engineering from Punjabi University, Patiala in 2009. He has rich teaching experience of more than 15 years. He has authored two academic books and also published more than 35 research papers in various International/National Journals, Conference Proceedings and Book Chapters. Dr. Ram has also edited five research books in the field of wireless communication and computing. His research areas include wireless communication, cloud computing and antenna design.

Manpreet Kaur is working as an Assistant Professor in the Department of Electronics & Communication Engineering at Yadavindra Department of Engineering, Punjabi University Guru Kashi Campus, Talwandi Sabo, Punjab, India.

Shilpa Mehta has recently completed a PhD and is currently a Teaching Assistant at Auckland University of Technology, New Zealand. Mehta has 5 years of teaching experience at different reputed colleges and universities in India. Mehta's research interests are RFIC and microwave circuits. Mehta has hands-on experience with different cadence technologies like 28nm, 65nm, 130nm, and the SiGe 8HP process, and also has a strong interest in solving circuits using optimization algorithms.

* * *

Nischal Adil is currently an Assistant Professor at Rungta College of Engineering and Technology in the Department of Computer Science and Engineering. Obtained master's of technology in computer science and engineering from RSR Rungta College of Engineering and Technology, Bhilai. Has a good morale standing and public record; highly interested in multidisciplinary programs of study, common areas of interest are artificial intelligence, cryptography, cloud computing, cyber security, computer networks.

Shahanawaj Ahamad is an active educator and researcher in the field of Software Engineering, Computer Science, and Information Technology with 17 years of experience in diverse educational setups. He completed 3 master's qualifications followed by a Ph.D. degree in Computer Science specializing in Software Engineering from Department of Computer Science, Jamia Millia Islamia Central University, New Delhi, India in the year 2008. He has contributed to publish 70 research articles and 3 books. He is designated as Asst. Professor and Program Coordinator of Software Engineering in College of Computer Science and Engineering, University of Hail, Saudi Arabia. He has been contributing significantly to various academic and administrative responsibilities, and a member of several scientific and research organizations including fellowship of British Computer Society, UK, and a Senior Life membership of Computer Society of India. His research interest includes software engineering, software aging and program analysis, application of machine learning, IoT and cloud computing. |

Rohit Anand is currently working as an Assistant Professor in the Department of Electronics and Communication Engineering at G.B.Pant Engineering College (Government of NCT of Delhi), New Delhi, India. He has teaching experience of more than 18 years including UG and PG Courses. He is a Life Member of Indian Society for Technical Education (ISTE). He has published 6 book chapters in reputed books and 8 papers in Scopus/SCI Indexed Journals. He has chaired a Session in AICTE-Sponsored International Conference. His research areas include Electromagnetic Field Theory, Antenna Theory and Design, Wireless Communication, Image Processing, Optical Fiber Communication, Machine Learning etc

Silki Baghla is an Associate Professor at JCDM College of Engineering, Sirsa. She is Ph.D. in Electronics and Communication Engineering from I.K.G Punjab Technical University, Kapurthala (Punjab). Her research areas include wireless communication, heterogeneous networks, vertical handover, and multiple attributes decision-making algorithms.

Utsab Banerjee comes from Kolkata, West Bengal, India. He received his B.Tech degree in Electronics and Communication Engineering and M.Tech degree in Communication Engineering, both from Netaji Subhash Engineering College, under West Bengal University of Technology, Kolkata, West Bengal, India in 2010 and 2013, respectively. He completed his PhD in 2021, from the Department of Electronics and Communication Engineering, Tripura University (A Central University), Suryamaninagar, Tripura, India. He is currently working as an Assistant Professor in the Department of Electronics and Communication Engineering, MVJ College of Engineering, Bangalore, Karnataka, India. His research interests include analysis and design of compact antennas for wideband applications, ultra-wideband antennas, fractal antennas, various Circularly Polarized antennas, such as wide band, multiband, as well as antennas having special applications such as GPS, RFID etc, circularly polarized dielectric resonator antennas. He is also active in the study of electromagnetic wave theory and the antenna theory. He has published a number of peer-reviewed journal papers and conference articles. He has been awarded with the "Research Excellence Award" by the Institute of Scholars (INSc).

Anindita Bhattacharjee received M.Tech degree in Electronics and Communication Engineering from Tripura University, India in 2018. She is currently a Research Scholar in the same department. She received her Bachelor in Engineering degree in Electronics and Communication Engineering from Tripura Institute of Technology, under Tripura University in 2014. Her research interests include analysis and design of compact antennas for wideband applications, ultra-wideband (UWB) antennas, fractal antennas, Vivaldi antenna, MIMO antenna. She has published a number of peer-reviewed journal papers and conference articles.|

Abhijit Bhowmick received his B.E (Hons) degree in Electronics and Telecommunication Engineering in 2002 from Burdwan University, West Bengal, India, M.Tech. degree in Telecommunication Engineering in 2009 and PhD in Dec., 2016 from NIT Durgapur. He worked for Cubix Control System Pvt. Ltd. from 2004 to 2006. He joined in the Department of Electronics and Comm. Engineering, Bengal College of Engg. and Tech., Durgapur as a Lecturer in 2006. Thereafter, he joined VIT University, Vellore, TN, India in the School of Electronics Engineering (SENSE) in June, 2016 and is currently serving as an Associate Professor there. His research interests include Cognitive Radio Networks focusing on Spectrum Sensing and Spectrum Sharing issues, Cooperative Communications in Cognitive Radio Networks, Energy harvesting in wireless network, D2D communication, and Physical layer security issues in Wireless Networks, and UAV assisted communication. He has published more than forty (40) research papers in various journals and conferences. He is a reviewer of several IEEE, Springer, Wiley and Elsevier conferences and journals.

Priyanka Chhibber is an Associate Professor of Mittal School of Business, (ACBSP USA, accredited) at the Lovely Professional University- Phagwara, Punjab. In addition to her teaching experience she worked with textile manufacturing concern and served as an HRD officer at SEL Manufacturing Co. Ltd, Chandigarh Road, Ludhiana, India. Other than teaching and Industry experience, also acted as a resource person in the subject of creativity, innovation, mentoring and leadership. Also won best paper award for the paper presented titled gender diversity: An approach towards agile women employees, in international conference.

Dharmesh Dhabliya is currently working as a Profesor at Vishwakarma Institute of Information Technology. Dhabliya is also pursuing a Doctorate In Business Administration from SBSS Swizerland.

N. Gireesh is currently working as Professor and Head, Department of ECE, Mohan Babu University (Erstwhile Sree Vidyanikethan Engineering College), Tirupati. He obtained Doctoral degree from Sri Venkateswara University Tirupati in 2016. He published 15 articles in reputed Journals. His areas of interest are signal processing, biomedical, process control and soft computing techniques.

Geentanjali Goyal is interested in developing beautiful designs and unique ideas.

Vikas Goyal has done B.Tech & M.Tech in Computer Science & Engineering. Currently working as Associate Prof. in CSE deptt. of Govt. Engg. College (M.I.M.I.T) at Malout. He has 18 Years of Teaching Experience.

Ankur Gupta has received the B.Tech and M.Tech in Computer Science and Engineering from Ganga Institute of Technology and Management, Kablana affiliated with Maharshi Dayanand University, Rohtak in 2015 and 2017. He is an Assistant Professor in the Department of Computer Science and Engineering at Vaish College of Engineering, Rohtak, and has been working there since January 2019. He has many publications in various reputed national/ international conferences, journals, and online book chapter contributions (Indexed by Scopus, ESCI, ACM, DBLP, etc). He is doing research in the field of cloud computing, data security & machine learning. His research work in M.Tech was based on biometric security in cloud computing.

Dinesh Kumar Gupta is a Professor at JCDM College of Engineering, Sirsa (Haryana). He is Ph.D. in Electronics and Communication Engineering. His research areas include antenna, optical communications, wireless communication, artificial intelligence, and machine learning techniques.

| **Sushil Janardan** is an experienced Assistant Professor with a demonstrated history of working in the Faculty of Engineering. Skilled in Microsoft Operating Systems, Theory of Computation, Compiler Design, C++, Java Enterprise Edition, Research, and Java. Strong education professional with a Master's of Engineering focused in Computer Technology And Application.

Purva Joshi is a PhD scholar at the University of Pisa, Italy. She was a research scholar at Wroclaw University of Science and Technology, Wroclaw, Poland, and her research title was "non-parametric system identification methods for adaptive control". She completed her bachelor's study in instrumentation and control engineering, at Government Engineering College, Gandhinagar, India. After graduation, she joined the Institute of Technology, Nirma University, Ahmedabad, India, for her post-graduation in instrumentation and control engineering. She has also more than 3 years of teaching experience in various reputed universities. She worked on a project of ISRO based on a satellite network simulator using features like uplink power control (ULPC) and Adaptive Coding and Modulation (ACM). Her research interests are haptic and tactile network, system identification, 5G and beyond 5G network, automation, and robotics."

Manpreet Kailay is a currently a Doctoral fellow at the institute Lovely Professional University, Punjab, India. Her interest area is the Sustainable Human Resource Management. Prior to this, she holds a Bachelor and Master degree in Commerce. Manpreet has also qualified UGC-NET, 2020 in the field of Commerce. She also presented papers in the International Conferences hosted by the institutes such as- Indian Institute of Management, Ranchi- 2020 and National Institute of Technology, Sri Nagar- 2021, National Institute of Technology, Warangal- 2021. She also received 'Best Paper Award' titled 'Sustainable Human Resource Management: A Bibliometric Analysis'. Personal Email: kailaymanpreet.95@gmail.com

Anirban Karmakarhas completed his PhD in Engineering from Jadavpur University, Kolkata, India, in 2015. He has more than 13 years of teaching experience and is cur-rently holding the post of Assistant Professor in the department of Elec-tronics & Communication Engineering at Tripura University (A Central University), India. He has almost 40 research articles in refereed journals and international conference proceedings. He has served as a reviewer in different international journals. He was awarded best paper award from different interna-tional conferences. He is a Senior Member of IEEE and has organized different workshops in the capacity of a convener and completed various UGC-funded research projects. Currently four research scholars are pursuing PhD under his guidance. His areas of inter-est include planar and fractal wideband antennas, arrays, circular polarized antennas, DRA, etc.

Ashok Koujalagi is an Author, Professor, Principal, IT Director, Engineer, and Researcher. Koujalagi obtained an M.Sc by Research degree from Bangalore Central University & Ph.D. from the Central University of Allahabad, and has authored 6 books among 3 are International and published 1 Indian patent on IoT, having more than 35 research publications & Conference Proceedings indexed in Scopus & UGC-Care listed Journals. And presently working as an Assistant Professor in the Department of

Computer Science & Engineering at K L University, Vijayawada, Andhra Pradesh, India. Koujalagi is an active member of 21 International Professional Bodies, Editorial Board Member of 18 International Journals, Peer-Reviewal Board Member of 20 International Journals, and Advisory Board Member of 5 International Journals. Koujalagi has also been invited by 4 international conferences as a "Conference Organizing Committee Member", and invited as a speaker of 16 international Conferences. The main research areas are Computer Networking, Cloud Computing, and IoT.

Pradeep Kumar received his B.E degree in Electronics and Telecommunication Engineering in 2007 from Globus Engineering College, Madhya Pradesh, India, MS degree in Antenna and Wave Propagation in 2015 from VIT Vellore. He worked for Koneru Lakshmaiah Educational Foundation (KLEF), Vijayawada from 2017 to 2019, and currently he is perusing PhD in VIT Vellore. His research interests include Cognitive Radio Networks focusing on Spectrum Sensing and Spectrum Sharing issues, Co-operative Communications in Cognitive Radio Networks, Energy harvesting in wireless network, D2D communication, and Non-orthogonal multiple access (NOMA). He has published research papers in several journals and conferences.

Kasthuri M. received her B.E degree in Electronics and Instrumentation Engineering in 2004 and M.E degree in VLSI System from P.S.N.A College of Engineering and Technology, Anna University Chennai, Tamil Nadu, India, in 2004 and 2009 respectively.Working as Assistant Professor in Electronics and Communication Engineering Department, P.S.N.A College of Engineering and Technology, Tamil Nadu, India from 2010 to till date. Her research interests include: VLSI system Designs and testing of VLSI.

Venkatanaresh M. is currently working as Assistant Professor in the Department of ECE, Mohan Babu University (Erstwhile Sree Vidyanikethan Engieering College), Tirupati. He completed his M.Tech in Communication Systems at S. V. U. College of Engineering, S. V. University, Tirupati. He is presently pursuing Ph.D at S.V. U. College of Engineering, S. V. University, Tirupati. His research interests include Communication Systems, Remote Sensing.

Mehaboob Mujawar completed his B.E degree in Electronics and Telecommunication Engineering in 2015 from Don Bosco College of Engineering, Fatorda- Goa and had secured 2nd Rank in Electronics Telecommunication Department. He has completed M.E in Industrial Automation and Radio Frequency Engineering from Goa College of Engineering in 2017 and had secured 3rd Rank to Goa University. He has published four books. He has Published/Presented 28 papers in the National/International journals and Conferences. He has published 6 book chapters with reputed publishers like CRC Press, Springer, Wiley and Nova Publishers USA. He is reviewer of reputed journals and has worked as reviewer for international conferences. He is a member of IETE, ISTE, IAENG and INSC. He has published one Indian Patent. He has received Seven Awards for Excellent Performance in Academics and Research. He has 1.8 years industrial experience and 4 years teaching experience, currently working as Assistant Professor at Goa College of Engineering –Farmagudi since July 2018. His area of interest is Antenna Design, Printed Circuit Board Design, Microwave and RF Engineering.

Hritik Ranjan Nanda is a computer interested guy who is also interested in designing and drawing.

M. Naresh is working as Assistant Professor in the Department of ECE, Matrusri Engineering College, saidabad, Hyderabad. He obtained his Ph. D in ECE from JNTU H in 2021. His research interests include Wireless Communications, Signal Processing.

Jay Kumar Pandey is an active Researcher and Academician having 12 years of Experience (as an Academician and Researcher). Exploring various real-time-based applications of Instrumentation, Control, & Renewable energy. Developed various Models for efficient application of PV using MatLab & SciLab. Published several research articles on Solar, Communication & Biomedical in Scopus & Web of Science (SCIE) journals and Book Chapters like CRC, Springer & IIP-INSC International Publishers. His domain of expertise lies in efficient Solar PV modules, biomedical applications, and advanced Communication systems.

Kamalpreet Kaur Paposa is a Professor (Assistant) in Human Resource Management at Mittal School of Business, Lovely Professional University, Punjab, India. Her area of research is HRM Practices, Human Resource Development, Skill Gap, Mentoring, Contemporary Leadership and Sustainable HRM. Additionally she has a deep rooted interest in research works related to service marketing, TQM and Industry4.0. She has published and presented research papers in various national and international journals like Management and Labor Studies (Sage), International Journal of Public Sector Performance Management (Inderscience) to a name a few. Email: kamalpreet.paposa@gmail.com

Subuh Pramono was born in Indonesia. He received the B. Eng degree in Electrical Engineering from Telkom University, Indonesia in 2003, and the M. Eng degree in Electrical Engineering from Institut Teknologi Bandung (ITB), Indonesia, in 2009. He has published three books and three Indonesia patents.Now, he is currently pursuing his Ph.D at Graduate School of Science and Engineering, Chiba University, Japan. He has published 84 papers in the national/international journals and conferences. He is currently a lecturer at Department of Electrical Engineering, Faculty of Engineering, Universitas Sebelas Maret, Surakarta-Indonesia. He is a member of IEEE, IEEE Comsoc, and IEEE Antennas & Propagation Society. His main areas of research interests are wireless and mobile communications, antenna and propagation, wireless sensor networks, and synthetic aperture radar.

Anuradha Saha received the gold medal for securing 1st rank in M.Tech.degree in Mechatronics from West Bengal University of Technology, Kolkata in 2009 and Ph.D. degree in Engineering from Jadavpur University in 2017. She has been working as an assistant professor in the department of Applied Electronics and Instrumentation Engineering since 2008, wherein between she has served as full-time research scholar from 2012 to 2015 at the department of Electronics and Telecommunication Engineering of Jadavpur University, funded by UGC. She is the author of around 35 publications in top international journals and conference proceedings. Currently four research scholars are pursuing Ph.D. under her guidance and two have already been awarded their degrees. Her research interests include Signal Processing, Artificial Intelligence, Pattern Recognition, Cognitive Robotics and Human-Computer-Interaction. She is the reviewer of some renowned journals including IEEE Transactions on Fuzzy Systems, IEEE Transactions on Emerging Topics in Computational Intelligence, IETE Journal of Research, IEEE Access, International Journal of Electronics and Communications, International Journal of RF and Microwave Computer-Aided Engineering, Microwave and Optical Technology Letters and so on.

Anuradha Saha received the Ph.D. (Engg.) degree from Jadavpur University, Kolkata, India in 2017 and B. Tech degree in Electronics and Communication Engineering from West Bengal University of Technology, India in 2006. She was awarded gold medal in M. Tech. degree in Mechatronics from National Institute of Technical Teachers' Training and Research, Kolkata, India in 2009. She is currently an Assistant Professor in the Department of Applied Electronics and Instrumentation Engineering, Netaji Subhash Engineering College, affiliated under MAKAUT (formerly known as West Bengal University of Technology). She was awarded Junior Research Fellowship by University Grants Commission, India under the scheme – University with Potential for Excellence – Phase-II, 2012, India from August, 2012 to July, 2015. She has over 17 publications in international journal and conference proceedings. She served as the reviewer of IEEE Transactions on Fuzzy Systems and IEEE Transactions on Emerging Topics in Computational Intelligence. Her principal research interests include Signal Processing, Human-Computer-Interaction, Artificial Intelligence and Pattern Recognition.

Jyoti Saharan received her B-Tech in Electronics and Communication Engineering from the Guru Jambheshwar University of Science & Technology, Hisar in 2018. Currently, she is a Research Scholar (ECE) at JCDM College of Engineering under the affiliation of Chaudhary Devilal University in Sirsa, Haryana and her area of interest include Wireless Communication, Next Generation Mobile Communication Networks like 5G, and 6G.

Sobana Sikkanan received her B.E in Electronics and Communication Engineering and M.E in Applied Electronics from R.V.S College of Engineering and Technology, Anna University Chennai, Tamil Nadu, India, in 2005 and 2007 respectively. She worked as an assistant professor in Electronics and Communication Engineering Department, P.S.N.A College of Engineering and Technology, Tamil Nadu, India from 1/6/2007 to 10/8/2018. She worked as an assistant professor senior grade in Electronics and Communication Engineering Department, Bannari Amman Institute of Technology, Sathyamangalam from 11/8/2018 to 10/5/2019.She is now working as an associate professor in Electronics and Communication Engineering Department, Adithya Institute of Technology,Coimbatore. She has published 28 technical papers in major international journals and conferences and also guided more than 16 B.E and M.E projects. Her current research is focused on Signal Processing and Wireless Communication. Her research interests include: MIMO communication systems, precoding techniques, and interference cancellation.

Krishna Sikkannan completed the M.Sc, M.Phil, Ph.D, ME, SET. Has nearly twenty years of teaching experience and published nearly 36 research articles.

Nidhi Sindhwani is working as an Assistant Professor (ECE) in Amity School of Engineering and Technology, New Delhi, India.

Devendra Somwanshi completed his Bachelor's degree in Electronics Instrumentation & Control in 2007 from Govt. Eng. College, Bikaner. He did his Master's Degree in Electronics Instrumentation & Control from Thapar University, Patiala in 2009 and is currently pursuing his PhD in Artificial Intelligence. He started his teaching career as Assistant Professor, in JNIT, Jaipur (RTU) in 2009, during this time he authored 3 technical books. He joined Poornima Group, as M. Tech. Coordinator and worked in the Advanced Studies and Research Center for more than 8 years. During this time he coordinated and supported to design and execute the curriculum for M. Tech and Ph. D. course work. He guided more than 35 M. Tech. research scholars till now. He has expertise in Artificial Intelligence, Data Analysis, Image Processing, Instrumentation and Process Control. He has published more than 70 research papers in National and International conferences and journals including IEEE, ACM, Springer, Elsevier etc. Currently He is the Registrar of Poornima College of Engineering & Poornima Group of Institutions, Jaipur.

K R Sudhindra received Bachelor of Engineering degree in Electronics and communication Engineering from Mysore University, India in 1999 and M.Sc (Engg). by research in faculty of Electrical Engineering sciences from Visvesvaraya Technological University, Belgaum India in 2007. He has done is Ph.D from Visvesvaraya Technological University, Belgaum, India in 2014. He has around 20 years of experience in the field of electronics and telecommunications with expertise in GSM BSS, GPRS, EDGE, CDMA, WCDMA, LTE, 5G and RF Planning and Optimization, Testing and measurement. He had worked in ZTE Telecom India private Ltd. during 2007 to 2010 and Idea Cellular Ltd. during 2011 to 2014. Currently he is working as an Associate Professor in the Department of Electronics and communication Engineering, B.M.S. College of Engineering, Bangalore, India. His research interests include signal processing, wireless communication and network optimization. Dr. Sudhindra K R is an associate member IETE since 2002 and working as IEEE branch counselor, B.M.S.C.E, Bangalore. He has published 20 research articles in leading International Journals and Conferences. Associate Professor in BMS College of Engineering

M. N. Suma has received her B.E. in Electronics and Communications from Bangalore University, M.Tech. and Ph.D. degree in Electronics Engineering from Visveswaraya Technological University, Belagavi, India. She is currently working as Professor in the Department of Electronics and Communications Engineering, BMS College of Engineering, Bengaluru, India. She has over 22 years of teaching experience in teaching UG and PG course and 2 years of industry experience as application engineer. Her areas of interest include Wireless Communications and Digital Signal processing, Software defined radio. Dr. Suma M N is an active IEEE senior member, Member IETE. She has published 24 papers in International Journals and conferences.

Seerangurayar. T. received his B.E in Agriculture Engineering from Tamil Nadu Agricultural University, Coimbatore, Tamil Nadu, India. Received his M.S in Agricultural from Sultan Quaboos University, Muscat, Oman. Now he is working as an Assistant Professor, Department of Agriculture Engineering, Bannari Amman Institute of Technology, Sathyamangalam for 3 years. He has published 6 technical papers in major international journals and conferences. His current research is focused on food Engineering.

Satyanarayana T. V. V. is currently working as Professor in the department of ECE, Mohan Babu University(Erstwhile Sree Vidyanikethan Engineering College), Tirupati. He obtained his Ph.D in ECE from JNTU Hyderabad and M.Tech in Communication and Radar Systems from Acharya Nagarjuna University in 2015 and 2007 respectively. His research interest include Communication Systems, Signal Processing, Machine Learning and IoT.

Rafael Vargas-Bernal received a Bachelor's degree in Communications and Electronics Engineering from the University of Guanajuato in 1995, and the degrees of Master of Science and Doctorate in Science with a Specialty in Electronics from the National Institute of Astrophysics, Optics and Electronics (INAOE) in 1997 and 2000, respectively. Since January 2002, he has been a professor-researcher at the Higher Technological Institute of Irapuato (ITESI), Mexico, and, psssarticularly since 2006, he has worked in the Department of Materials Engineering where he has established himself as a researcher. He has authored 17 articles in journals, 36 chapters in books, and about 100 conference articles. He is a member of the National System of Researchers (SNI-Mexico). He regularly serves as a reviewer of scientific articles in RSC Advances and Royal Society Open Science as well as Standards in Semiconductor Equipment and Materials International (SEMI). His research interests include two-dimensional materials, nanomaterials, space materials, composite materials, gas sensors, and biosensors.

Vivek Veeraiah is the Founder and CEO of EdVista. A veteran professional in the sector with a long-standing record for innovation and creativity within the sphere, Vivek's leadership of EdVista represents a rare combination of industry expertise and cutting edge contemporary offerings to all businesses and students in the sector.

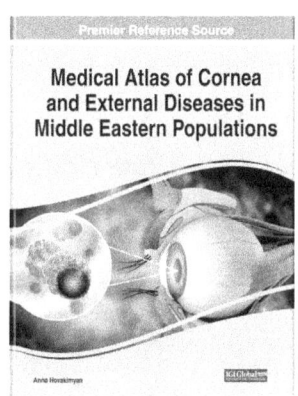

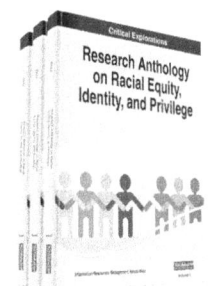

CPSIA information can be obtained
at www.ICGtesting.com
Printed in the USA
BVHW022323080323
660070BV00003B/5

9 781668 470008